COMPUTER GRAPHICS
AND
CHEMICAL STRUCTURES

COMPUTER GRAPHICS AND CHEMICAL STRUCTURES

DATABASE MANAGEMENT SYSTEMS
CAS REGISTRY   CHEMBASE
REACCS   MACCS-II
CHEMTALK

STANLEY V. KASPAREK

CHEMINFO
Cedar Grove, New Jersey

A Wiley-Interscience Publication
JOHN WILEY & SONS
New York · Chichester · Brisbane · Toronto · Singapore

Data retrieved from STN(R) International-CAS ONLINE(R) are copyrighted by the American Chemical Society and are reproduced in this book by permission of the American Chemical Society. Data retrieved from the Molecular(R) Design Ltd. (MDL) databases are proprietary and are reproduced in this book by permission of MDL. Molecular Design Ltd. has not approved or disapproved the presentation of their softwares in this book.

If any registered names or trademarks used in this book are not specifically indicated by TM or R marks, they are not to be regarded as unprotected by law.

No warranties and/or liabilities whatsoever apply concerning the material presented in this book.

Copyright © 1990 by John Wiley & Sons, Inc.

All rights reserved. Published simultaneously in Canada.

Reproduction or translation of any part of this work
beyond that permitted by Section 107 or 108 of the
1976 United States Copyright Act without the permission
of the copyright owner is unlawful. Requests for
permission or further information should be addressed to
the Permission Department, John Wiley & Sons, Inc.

Library of Congress Cataloging in Publication Data:

Kasparek, Stanley V.
    Computer graphics and chemical structures: database management
  systems / Stanley V. Kasparek.
      p.   cm.
    Includes bibliographical references.
    ISBN 0-471-62822-0
    1. Chemical structure — Data processing. 2. Information storage
  and retrieval systems — Chemical structure. I. Title.
  QD471.K37   1989
  541.2'2'028566 — dc20   89-36570
                                                   CIP

Printed in the United States of America

10 9 8 7 6 5 4 3 2 1

To all whose tireless efforts in developing hardware and software for computer graphics made it possible to build and search databases of chemical structures, to all who enthusiastically use these systems, and to all who hope for more to come.

PREFACE

Data and information management has been an integral part of the work of scientists of every discipline. In the last two decades which have been, appropriately, termed the "information explosion" decades, the computer technology has become indispensable in dealing with the everincreasing amount of information from various sources.

The information publishing industry which provides services to the scientists has been the catalyst for developing computer systems for storage and retrieval of textual data. These systems were, at first, employed in-house by the industry for collecting, abstracting, editing, and publishing information for the scientists. Subsequently, in a fruitful cooperation with system development companies, sophisticated systems were developed that allowed for searching the information by specific fields of the text data. The systems have become available to the end-user, the scientist, and information specialist, online through companies specializing in providing online access to the computer databases.

The science of chemistry has a need for not only text and numeric data but also for structural data which are unique for this scientific discipline. The chemist thinks, talks, and writes in structural formulas. The chemist needs to search for not only the numeric and text data but also for exact structures, substructures, and generic structures. The chemist, and increasingly, the biologist, pharmacologist, and research physician have to analyze and correlate structures to perform structure-activity relationship studies.

Previously, various coding systems have been used to store and retrieve chemical structures from computerized databases. However, these systems fell short of the actual requirements for search and retrieval. Especially, they were not easily accessible and usable by the end-user. Therefore, computer graphics systems have been developed to convert chemical structures into a digital form, and store the digital form in the computer and recreate the structures in their usual structural image on retrieving and displaying them. It has become possible to draw the structures on the terminal screen and use these graphic queries to search for the structures, retrieve the structures, and manipulate them on the screen. These systems have become so advanced that three-dimensional images of the structures can be handled in the process of molecular modeling.

The introduction of personal computers and softwares that can emulate various graphics terminals further revolutionized this process of chemical structure database management. Very powerful and inexpensive microcomputers can be employed to run these systems either as standalone microcomputer systems, or to access mini/mainframe hosts where the systems are installed.

This book describes in detail and gives many examples of the use of both the microcomputer and the mini/mainframe systems.

The book deals with CAS ONLINE-Registry File developed by Chemical Abstracts Service. The CAS ONLINE system has become the standard for the online systems of data and information search and retrieval.

The book treats ChemBase[R], MACCS-II, REACCS, and ChemTalk[R], the systems developed by Molecular Design Ltd.

The book will be of use to a wide circle of readers.

The practising chemist will find a how-to reference for using the systems since computers have become equivalent to the common instruments like IR, UV or NMR used by bench chemists. The systems are being used more and more in chemical and pharmaceutical industries. Although the systems were at first at the disposal of computer specialists and information specialists, the bench chemist has to personally use the systems these days to keep pace with the information explosion and with the progress in computer technology and information processing.

The biologist, the pharmacologist, and the research physician are using these systems more and more to acquire data and information for their research projects.

The information and library specialists will find the book to be an indispensable reference source.

The student will find the book helpful in studying for classes as well as in preparing for a carreer in an industrial or academic laboratory.

The computer specialist will find in the book a reference to a subject which is increasingly fascinating, i.e., computer graphics systems.

The personal computer enthusiast will find in the book a subject which is interesting to read about as these systems are becoming available for installing on a variety of microcomputers.

The Molecular Design Limited's (MDL) programs have become the standard for chemical information systems in the chemical and pharmaceutical industries and in academia.

The systems are of significant importance to scientists working with chemical substances not only because they allow for developing personal or corporate databases but also for the possibility of acquiring databases developed for these systems by MDL. The databases, e.g., **CHIRAS, CLF, METALYSIS**, contain reactions for asymmetric syntheses, reactions from the current literature, and/or organometallic synthetic reactions. Biologically active substances are covered by **DRUG DATA REPORT** and **CIPSLINE** databases developed jointly with Prouse Science Publishers.

Although this book does not describe ChemText[R], the wordprocessing program by MDL, it is an example of what can be accomplished with ChemText. The entire manuscript has been created in ChemText including the structures and figures within the text. Structures and text have been captured through ChemTalk[R] from CAS ONLINE and converted into files that can be used in ChemText. Part One contains numerous examples of the captured structures and text. The final manuscript has been printed on the Apple LaserWriter[R] to produce camera-ready copy for printing. If it had not been for ChemText, this book would have been much more difficult to produce.

The book covers all the latest improvements in the systems, i.e., CPSS Version 1.3, MACCS-II Version 1.5 including the Power Search Module, and REACCS Version 7.1.

STANLEY V. KASPAREK

Cedar Grove, New Jersey
October 1989

ACKNOWLEDGMENTS

In the first place, I thank the members of Molecular Design Limited, particularly:
Dr. Guenter Grethe, Director, Scientific Applications, for his support during the preparation of this book; his cooperation made the author's task much more easier; Guenter also critically reviewed the Parts describing the MDL systems, especially REACCS; many of his suggestions have made the presentation better and more complete;
Donna del Rey for answering all the author's cries for help and for many suggestions on how to improve the Parts discussing ChemBase and ChemTalk; Donna also is the creator of the screen shots within the text of ChemBase;
David Dorsett for reviewing Part Two, ChemBase;
David Hughes for critically reviewing Part Three, MACCS-II;
many of their ideas have been included in the text;
Rich Asvitt for reviewing the sections on the VAX operating system;
Steven Goldby, President and CEO, who was receptive to the idea of this project and gave his approval and support of it from its start.
I thank MDL for giving me access to the systems and for providing me with the up to date manuals.

I thank Chemical Abstracts Service for donating the online time for generating the examples for building and searching structures in Part One, and for critically reviewing it.

Thanks are also due to John Wiley & Sons, Inc.: Dr. Theodore P. Hoffman, Senior Editor; Ed Cantillon of the SCI/TECH Production team; and to Penny Lerner, Copy Editor.

CONTENTS

INTRODUCTION	**xxix**
PART ONE — CAS REGISTRY	**1**
1 SYSTEM BACKGROUND	**3**
CAS ONLINE	3
STN INTERNATIONAL	4
2 SYSTEM DESIGN AND DEVELOPMENT	**7**
REGISTRY FILE	7
3 HARDWARE AND SOFTWARE	**12**
HARDWARE FOR ACCESSING CAS ONLINE	12
TERMINALS	13
MICROCOMPUTERS	13
GRAPHICS INPUT DEVICES	13
MODEMS	13
PRINTERS	14
EMULATION SOFTWARE	14
EMULATION SOFTWARE SET UP PARAMETERS FOR STN	15
4 SYSTEM ACCESS	**16**
ACCESSING STN INTERNATIONAL	16
LOGIN PROCEDURE	17
LOGOFF PROCEDURE	19
THE HOME FILE	20
THE SYSTEM PROMPTS	20
CORRECTING TYPING ERRORS	20
5 SYSTEM COMMANDS	**21**
NOVICE AND EXPERT VERSION OF SYSTEM COMMANDS	21
HELP! HELP!	23
THE PANIC BUTTONS	23
INTERRUPTING COMMAND EXECUTION	24
EXITING FROM SUBCOMMAND MODE	24
COMMAND SYNTAX	24
6 GRAPHIC STRUCTURES	**26**
WHAT IS A STRUCTURE GRAPH	26
WHAT IS A NODE	27
DEFAULT NODE VALUES IN GRAPHIC STRUCTURES	28
HYDROGEN AS A VALUE FOR NODES	29

BONDS IN GRAPHIC STRUCTURES	29
DEFAULT BOND VALUE	29
BOND TYPE	29
STEREOCHEMISTRY	29
WHAT IS A CONNECTION TABLE	29
DISPLAY OF GRAPHIC STRUCTURES	31

7 STRUCTURE BUILDING METHODS — 33

HOW TO INITIATE STRUCTURE BUILDING	34
THE MENU	35
HOW TO INITIATE THE MENU	35
DESCRIPTION OF THE MENU	36
CREATING STRUCTURE GRAPHS ON THE MENU	38
THE BASICS OF THE MENU TECHNIQUE	39
KEYBOARD COMMANDS	39
FREE-HAND DRAW	39
HOW TO SWITCH MODES	39
HOW TO END STRUCTURE BUILDING	40

8 CREATING STRUCTURES — 43

STRUCTURE SKELETONS	43
CHAIN STRUCTURES	44
HOW TO CREATE SEPARATE STRUCTURAL FRAGMENTS	44
HOW TO BOND SEPARATE FRAGMENTS	45
RING STRUCTURES	46
SINGLE RINGS	46
RINGS WITH >8 NODES	47
FUSED RING SYSTEMS	47
FUSED RINGS WITH >8 NODES	48
FUSED SYSTEMS WITH TWO RINGS	50
FUSED SYSTEMS WITH TWO 3-8 MEMBERED RINGS	50
FUSED SYSTEMS WITH TWO >8-MEMBERED RINGS	60
FUSED SYSTEMS WITH THREE OR MORE RINGS	64
FUSED SYSTEMS WITH 3-8 MEMBERED RINGS	64
THREE OR MORE >8-MEMBERED FUSED RINGS	69
MORE TECHNIQUES ON FUSING RINGS	70
FUSION ONTO ADJACENT NODES OF A RING	70
FUSION ONTO NONADJACENT NODES OF A RING	72
BRIDGED RINGS	73
SPIRO RING SYSTEMS	75
HOW TO USE THE MENU GRAPH COMMAND	75
RING TEMPLATES	76
HOW TO USE RING TEMPLATES	76
FREE-HAND DRAWING OF STRUCTURES	80
SPECIFICATION OF VALUES AT NODES	83
VALUES AT SINGLE AND MULTIPLE NODES	83

CONTENTS

THE MENU NODE TEMPLATE	84
COMMON SUBSTITUENTS AND FUNCTIONAL GROUPS	84
ASSIGNMENT OF SYSTEM SHORTCUTS	89
ASSIGNING BONDS IN STRUCTURE GRAPHS	90
WHAT IS BOND VALUE	90
NORMALIZED BONDS IN REGISTRY FILE	90
NORMALIZED BONDS IN RINGS AND RING SYSTEMS	95
NORMALIZED BONDS IN TAUTOMERIC SYSTEMS	97
NORMALIZED BONDS IN RING TAUTOMERIC SYSTEMS	98
NORMALIZED BONDS IN DELOCALIZED CHARGE SYSTEMS	100
HOW TO SPECIFY BOND VALUES	100
BOND EXACT VALUES	100
NORMALIZED BOND VALUE	101
CREATING STRUCTURES FROM MODELS	108
REGISTRY SYSTEM MODELS	109
HOW TO RECALL REGISTRY FILE MODELS	109
LIMITS ON USING REGISTRY NUMBERS AS MODELS	109
USER MODELS	112
HOW TO SAVE USER MODEL	112
HOW TO RECALL USER MODELS	114
FILE OF USER MODELS	116
SUMMARY OF STRUCTURE BUILDING AT THIS POINT	116

9 CREATING GENERIC STRUCTURES — 118

HOW TO DEFINE VARIABLES IN GENERIC STRUCTURES	118
SYSTEM DEFINED VARIABLE ATOMS	118
SYSTEM DEFINED GENERIC RING FRAGMENTS	118
ASSIGNMENT OF VARIABLE VALUES AT NODES	118
USER DEFINED VARIABLES	119
THE COMMAND VARIABLE	122
VARIABLE ATOMS IN Gk GROUPS	122
DISPLAY OF Gk GROUPS	122
VARIABLE SHORTCUTS IN Gk GROUPS	123
GENERIC RINGS IN Gk GROUPS	124
NODE NEGATION	124
STRUCTURAL FRAGMENTS IN Gk GROUPS	124
USING Gk GROUPS WITHIN Gk GROUPS	130
THE COMMAND REPEATING	131
REPEATING SINGLE ATOMS/SHORTCUTS-SINGLE ATTACHMENT	132
REPEATING MULTIATOM FRAGMENTS-SINGLE ATTACHMENT	136
REPEATING MULTIATOM FRAGMENTS-MULTIPLE ATTACHMENTS	137
VARIABLE BOND VALUES IN GENERIC STRUCTURES	138
VARIABLE BOND EXACT OR NORMALIZED	139
BOND TYPES IN GENERIC STRUCTURES	142
NODE SPECIFICATION IN GENERIC STRUCTURES	144

STRUCTURE ATTRIBUTE COMMANDS	146
NODE SPECIFICATION	147
RING SPECIFICATION	147
HYDROGEN COUNT	147
THE CONNECT COMMAND	152
THE DEGREE OF SUBSTITUTION AT NODES	152
THE TYPE OF SUBSTITUTION	155
SELECTIVE FUSION OF RINGS BY THE CONNECT COMMAND	157
THE CHARGE COMMAND	158
THE DELOCALIZED CHARGE COMMAND	159
THE HYDROGEN MASS COMMAND	160
THE MASS COMMAND	161
THE VALENCE COMMAND	162
STRUCTURE ATTRIBUTES IN Gk GROUPS	163
HOW TO DISPLAY STRUCTURE GRAPHS	165
DISPLAYING STRUCTURE ATTRIBUTES	165
DISPLAYING STRUCTURE IMAGE AND ATTRIBUTES	166
DISPLAYING STRUCTURE CONNECTION TABLE	167
DISPLAYING STRUCTURE DATA	167
10 CORRECTING/MODIFYING STRUCTURES	**168**
HOW TO DELETE NODES	168
HOW TO DELETE BONDS	170
HOW TO DELETE CHAINS	171
HOW TO DELETE RINGS	172
HOW TO DELETE STRUCTURE GRAPH	173
HOW TO DELETE STRUCTURE ATTRIBUTES	174
RESPECIFICATION OF NODE AND BOND VALUES	175
CHANGE IN CHAIN STRUCTURES	175
CHANGE IN RING STRUCTURES	175
HOW TO MOVE STRUCTURE GRAPHS ON SCREEN	175
HOW TO RESHAPE DISTORTED STRUCTURE GRAPHS	181
11 STRUCTURES BY STN EXPRESS	**184**
WHAT IS STN EXPRESS	184
LIMITS ON USING STN EXPRESS GRAPHICS	184
OPERATION OF STN EXPRESS	184
CREATING STRUCTURES	185
FREE-HAND DRAWING	185
SPECIFYING ATOMS	187
SPECIFYING SHORTCUTS	187
SPECIFYING VARIABLES	187
USING TEMPLATES	187
DRAWING BONDS	190
QUICK DRAW	191
EDITING STRUCTURES	192

CONTENTS

UTILITIES	194
DEFINING GENERIC GROUPS	195
MODIFYING Gk GROUPS	199
DEFINING REPEATING GROUPS	200
UPLOADING QUERY STRUCTURE TO REGISTRY FILE	201
UPLOADING REACTION QUERY FOR CASREACT	204
12 STRUCTURE SEARCHING	**207**
INITIATING SEARCH	207
SCOPE OF SEARCH IN THE REGISTRY FILE	207
SAMPLE SEARCH	208
SEARCH STATISTICS AND SEARCH PROJECTION	209
FULL SEARCH	210
RANGE SEARCH	212
BATCH SEARCH	215
DISPLAYING ANSWERS TO BATCH SEARCH	217
TYPE OF SEARCH IN THE REGISTRY FILE	217
EXACT SEARCH	218
FAMILY SEARCH	218
SUBSTRUCTURE SEARCH (SSS)	218
LIMITS OF SEARCH	226
DISPLAY OF ANSWERS TO A STRUCTURE SEARCH	226
DISPLAY OF SUBSTANCE INFORMATION	228
DISPLAY OF DOCUMENT INFORMATION	230
SCAN FORMAT	231
DEFAULT SUB FORMAT	232
REG FORMAT	232
RANGE OF ANSWERS IN SUB FORMAT	232
SEPARATE ANSWERS IN SAMPLE FORMAT	234
BIBLIOGRAPHIC FORMAT	235
INDEX TERMS FORMAT	236
ABSTRACT FORMAT	237
ALL FORMAT	237
DISPLAY BROWSE	239
HOW TO SAVE ANSWER SET	240
HOW TO RECALL SAVED ANSWER SET	240
HOW TO DISPLAY SAVED ANSWER FILE NAMES	241
HOW TO DELETE SAVED ANSWER FILE	241
PRINTING SEARCH ANSWERS	242
HOW TO DELETE PRINT REQUEST	243
HOW TO DISPLAY PRINT REQUESTS	244
13 SEARCH STRATEGY	**245**
SEARCHING FOR ONE-COMPONENT STRUCTURES	246
SEARCHING WITH A SINGLE STRUCTURE QUERY	246
SEARCHING WITH MULTIPLE-STRUCTURE QUERY	246

BOOLEAN OPERATORS IN STRUCTURE SEARCH	247
SEARCHING FOR MULTICOMPONENT STRUCTURES	248
SEARCHING WITH BOOLEAN OPERATORS	248
SEARCHING STRUCTURE QUERIES COMBINED WITH SCREENS	254
GRAPH MODIFIER SCREENS	255
HOW TO SPECIFY SCREENS	256
CROSSOVER INTO OTHER STN FILES	257
CA AND CAOLD FILES	257
CASREACT	258
CURRENT-AWARENESS SEARCHES	259
ONLINE CURRENT-AWARENESS SEARCH	260
AUTOMATIC CURRENT-AWARENESS SEARCH (SDI)	260
ADDENDUM	**262-A1**

PART TWO   CHEMBASE   263

1 SYSTEM BACKGROUND   265

CPSS - CHEMIST'S PERSONAL SOFTWARE SERIES	265
MOLECULAR DESIGN LTD.	265

2 SYSTEM DESIGN   268

CHEMBASE	268
STRUCTURES	268
STEREOISOMERIC STRUCTURES	269
REACTIONS	269
NUMERIC AND TEXTUAL DATA	269
DATABASE STRUCTURE	270
ID - THE RECORD IDENTIFIER	271
SPECIAL DATA FIELDS	271
DATA FIELD NAMES	271
DATA FIELD TYPES	271
CHEMBASE FILES	272
MOLECULE FILES AND REACTION FILES	272
SDFILES AND RDFILES	273
DATA FILES	273
LIST FILES	273
FORM FILES	273
TABLE FILES	273
TEMPLATE FILES	274
POSTSCRIPT FILES	274
DATABASE FILES	274
INPUT/OUTPUT SYSTEM	274
DATABASE OUTPUT	274
DATABASE INPUT	276

CONTENTS

3 SYSTEM OPERATION — 277
- MAIN MENU — 277
- MOLECULE EDITOR — 278
- REACTION EDITOR — 278
- FORM EDITOR — 278
- TABLE EDITOR — 278
- MENU OPERATION — 278
- SELECTING AND ACTIVATING COMMANDS — 278
- PROMPT BOXES — 280
- FUNCTION KEYS — 280
- CURSOR CONTROL KEYS — 283
- ESCAPE — 283
- TEXT EDITING KEYS — 283
- ERROR MESSAGES — 283
- SYSTEM SETTINGS — 284
- CONFIGURATION FILE — 286
- THE CLIPBOARD — 287
- CREATING A CLIPBOARD — 288
- STARTING CHEMBASE — 290
- THE USE COMMAND — 291
- THE VIEW COMMAND — 291
- THE EDIT COMMAND — 291

4 CREATING STRUCTURES — 292
- MOLECULE EDITOR — 292
- BUILDING STRUCTURES — 292
- DRAWING STRUCTURE SKELETONS — 292
- SPECIFYING ATOMS — 293
- SPECIFYING BONDS — 295
- STEREOCHEMISTRY IN CHEMBASE VERSION 1.3 — 296
- HOW TO SPECIFY HYDROGENS — 298
- USING FUNCTION KEY F9 — 298
- USING MODELS IN BUILDING STRUCTURES — 300
- SYSTEM MODELS — 301
- HOW TO USE SYSTEM MODELS — 301
- USING FUNCTION KEY F7 — 304
- USER MODELS — 305
- TEMPLATES — 305
- SUBSTITUENTS — 306
- COMPLETE STRUCTURES — 308
- CHIRAL DESCRIPTOR — 308
- CHARGE, ISOTOPE, AND VALENCE SPECIFICATION — 309
- CHARGED ATOMS AND FREE RADICALS — 309
- ISOTOPES — 310
- VALENCE — 310

TEXT DESCRIPTOR	311
BUILDING GENERIC STRUCTURES	314
5 CREATING REACTIONS	**319**
BUILDING REACTIONS	319
REACTION EDITOR	321
EDITING REACTIONS	321
SAVING REACTIONS IN RXNFILE	325
USING MODELS	325
COMPLETE STRUCTURES	325
STRUCTURAL FRAGMENTS	326
SYSTEM TEMPLATES	328
USER TEMPLATES	331
COMPLETE STRUCTURAL REACTION	331
TEXT DESCRIPTOR	331
6 CORRECTING/MODIFYING STRUCTURES	**332**
DELETING ATOMS AND BONDS	332
DELETING FRAGMENTS	332
GROUP ABBREVIATION	332
RESIZING STRUCTURES	334
RESHAPING AND MOVING STRUCTURES ON THE SCREEN	335
ROTATING STRUCTURES ON THE SCREEN	336
7 CORRECTING/MODIFYING REACTIONS	**338**
REARRANGING DEFAULT REACTION LAYOUT	338
RESIZING STRUCTURES IN REACTIONS	339
8 SETTING UP DATABASES	**340**
DEFINING A DATABASE	340
MOLECULES DATABASE	342
REACTIONS DATABASE	344
NUMERIC AND TEXTUAL DATABASE	344
ESCAPE	344
MODIFYING DATABASES	345
PASSWORD CHANGE	345
9 CREATING FORMS AND TABLES	**346**
WHAT IS A FORM	346
WHAT IS A BOX	346
CREATING FORMS	346
MORE ON CREATING FORMS	347
WHAT IS A TABLE	349
WHAT IS A COLUMN	350
CREATING TABLES	351
MORE ON CREATING TABLES	352
SAVING/RECALLING FORMS AND TABLES	353
USING FORM/TABLE FOR CURRENT OPERATIONS	354
SAVING FORMS	354

RECALLING FORMS	355
SAVING TABLES	355
RECALLING TABLES	355
LINKING FORMS AND TABLES TO DATA FIELDS	356
FORM BOX OPTIONS	357
TABLE BOX OPTIONS	358
SET FIELD	360 |</p>

10 BUILDING DATABASES   361

DATA REGISTRATION	361
REGISTRATION OF NEW RECORDS	361
REGISTRATION OF UPDATED/EDITED RECORDS	363
DELETING REGISTERED RECORDS	365
REGISTRATION USING FILES	366
STRUCTURES	366
TEXT AND NUMERIC DATA	367
TRANSFERRING DATA	368
EXTERNAL FIELD NAME	368
USING METAFILES IN CHEMBASE 1.3	368
DEFINING FIELD NAME FOR METAFILES IN DATABASE	369
REGISTRATION OF METAFILE STRUCTURES	369
DISPLAYING METAFILE STRUCTURES	370

11 MOLECULE SEARCHING   372

SEARCH DOMAIN	372
LISTS	373
EXACT STRUCTURE SEARCHING	374
INITIATING EXACT STRUCTURE SEARCH	374
PRINTING RETRIEVED DATA	375
SAVING RETRIEVED DATA	377
RECALLING SAVED DATA	378
MOLFILE	378
FORM FILE	378
LIST FILE	379
DATA FILE	379
SDFILES AND RDFILES	379
SUBSTRUCTURE SEARCHING	380
INITIATING SUBSTRUCTURE SEARCH	381
ISOMERISM IN SSS	386
STEREOISOMERS	386
GEOMETRIC ISOMERS	386
TAUTOMERISM	388
VARIABLES IN SSS	389
USING LISTS IN SSS	391
MORE ON USING LISTS	392

12 REACTION SEARCHING — 394
EXACT REACTION SEARCHING — 394
INITIATING EXACT REACTION SEARCH — 396
SUBSTRUCTURE REACTION SEARCHING — 396
INITIATING REACTION SUBSTRUCTURE SEARCH — 397
REACTION TRANSFORM SEARCHING — 398
WHEN TO DO RSS — 400
WHEN TO DO TSS — 402

13 DATA SEARCHING — 403
DATA RETRIEVAL — 403
RETRIEVING ID — 403
RETRIEVING FORMULA — 404
DATA SEARCH — 405
SEARCHING FORMULA — 406
SEARCHING DATA — 407
SEARCHING COMPLEX QUERIES — 407
SEARCHING NUMERIC DATA — 408
SEARCHING TEXT DATA — 408
SEARCHING REACTION SYMBOLS — 410
SEARCHING REACTION KEYWORDS — 411

PART THREE   MACCS-II — 413

1 SYSTEM BACKGROUND — 415
HARDWARE — 416

2 SYSTEM DESIGN — 417
STRUCTURES — 417
SEMA NAME — 417
STRUCTURE SKELETONS — 418
STEREOCHEMISTRY — 418
GEOMETRIC ISOMERISM — 419
TAUTOMERIC STRUCTURES — 420
NUMERIC AND TEXTUAL DATA — 421
DATABASE STRUCTURE — 422
SPECIAL DATA FIELDS — 422
INTERNAL REGISTRY NUMBER — 423
NAME — 424
KEYS — 424
MOLECULAR FORMULA — 425
MOLECULAR WEIGHT — 425
DATA FIELDS — 425
FIXED DATA FIELDS — 425
EXTERNAL REGISTRY NUMBERS — 426
FLEXIBLE DATA FIELDS — 426

CONTENTS

DATA FIELD DESCRIPTORS	426
DATA TYPE AND FORMAT	426
FORMAT SYMBOLS	427
FILES IN MACCS-II	428
SYSTEM FILES	429
DATABASE FILES	429
USER FILES	430
LISTFILES	430
BINARY FILES	430
NUMERIC FILES	431
EXTREG FILES	431
MOLFILES	431
DATFILES	431
3 SYSTEM OPERATION	**432**
MAIN MENU	432
DRAW MODE	432
STRUCTURE MENU	433
QUERY MENU	433
ATTACH MODE	433
REGISTER MODE	434
SEARCH MODE	434
DATA MODE	435
PLOT MODE	435
SELECTING OPTIONS	436
GRAPHIC SELECTION OF OPTIONS	437
KEYBOARD SWITCH	438
SWITCHING FROM GRAPHIC TO KEYBOARD SELECTION	440
KEYBOARD SELECTION	441
STARTING MACCS-II	442
ABORT AND EXIT	442
SETTINGS	442
4 CREATING STRUCTURES	**444**
STRUCTURE MENU	444
HYDROGENS IN STRUCTURES	447
DISPLAYING ATOM LOCANTS	447
DRAWING STRUCTURE SKELETONS	447
CREATING CHAINS	448
CREATING RINGS	448
ASSIGNING ATOM VALUES	449
KEYBOARD ASSIGNMENT OF ATOM VALUES	450
MODELS IN STRUCTURE BUILDING	451
STRUCTURES AS MODELS	451
TEMPLATES	452
CREATING STEREOISOMERIC STRUCTURES	455

CREATING R AND S ISOMERS	455
CREATING STRUCTURES OF RACEMATES	455
CREATING STRUCTURES OF UNDEFINED CONFIGURATION	455
HYDROGENS IN STEREOISOMERS	456
E AND Z ISOMERS	456
MORE ON DRAWING STRUCTURES	457
CONTINUOUS BUILDING OF STRUCTURES	457
ATTACH MODE	458
MORE ON ATTACH MODE	458
SAVING STRUCTURES IN MOLFILES	463

5 CREATING GENERIC STRUCTURES — 465

SYSTEM VARIABLES	465
USER VARIABLES	465
ATOM NEGATION	465
HYDROGEN COUNT	466
SUBSTITUTION AND NO SUBSTITUTION	467
RELATIVE CONFIGURATION	469
VARIABLE BONDS	470
UNSATURATION AT ATOMS	471
RING ATOMS	472
ISOLATED RINGS	473
REPEATING ATOMS	474
MORE ON QUERY MENU OPTIONS	475
GENERIC STRUCTURES WITH DEFINED R GROUPS	476
ATTACHING R GROUPS TO ROOT	477
DEFINING R GROUPS	478
OCCURRENCE OF R GROUPS	481
IF NOT R GROUP THEN HYDROGEN	483
MORE ON RGROUP MENU	484

6 CORRECTING/MODIFYING STRUCTURES — 495

DELETING STRUCTURES, ATOMS, AND BONDS	495
CLEANING STRUCTURES	495
RESIZING STRUCTURES	495
REORIENTING STRUCTURES ON THE SCREEN	495
MOVING STRUCTURES ON THE SCREEN	496
MOVING ATOMS IN STRUCTURES	497
MANIPULATING GROUPS IN STRUCTURES	497
GROUP MODE	498

7 STRUCTURE SEARCHING — 503

EXACT STRUCTURE SEARCHING	504
SEARCH IN REGISTER OR SEARCH MODE?	505
DISPLAYING SINGLE HIT STRUCTURE	506
DISPLAYING FIXED DATA FIELDS	508
DISPLAYING FLEXIBLE DATA FIELDS	509

DISPLAY OF MULTIPLE HITS	511
SEARCHING FOR PARENT OF SALTS	511
SEARCHING CONFIGURATIONAL ISOMERS	511
DISPLAYING MULTIPLE RETRIEVED STRUCTURES	513
SEARCHING TAUTOMERS	516
SUBSTRUCTURE (SSS) SEARCHING	517
SSS WITH GRAPHIC STRUCTURE QUERIES	517
INITIATING SUBSTRUCTURE SEARCH	517
SSS WITH GENERIC RGROUP QUERY	519
SIMILARITY SEARCH	522
SUBSIMILARITY SEARCHES	524
SUPERSIMILARITY SEARCH	525
STEREOCHEMISTRY IN SSS	525
INTERRUPTING AND RESUMING SEARCH	525
USING LISTS IN SEARCH MODE	526
WORKING WITH LISTS	526
WRITING LISTFILE	527
READING LISTFILES	530
LISTFILE FORMAT	530

8 PLOTTING STRUCTURES — 532

SELECTING PLOTTING DEVICE	532
FORMATTING THE PAGE	533
FORMATTING STRUCTURES	534
SPECIFYING STRUCTURES TO PLOT	534

9 DATA SEARCHING — 535

RETRIEVAL WITH REGNO QUERY	535
SEARCHING WITH NAME QUERY	537
SEARCHING WITH KEYS QUERIES	539
SEARCHING WITH FORMULA QUERIES	540
RETRIEVING DATA IN DATA FIELDS	540
SEARCH DATA IN SEARCH OR DATA MODE?	542
DATA FIELDS INDEX	543
DEFINING REFERENCE LIST IN DATA SEARCH	544
DATA SEARCHING TECHNIQUES	544
SEARCHING TEXT DATA	545
COLUMN RANGE SEARCHING	545
ALPHANUMERIC RANGE SEARCHING	546
SEARCHING NUMERIC DATA	547

10 SETTING UP DATABASES — 550

PRIVATE DATABASES	550
PUBLIC DATABASES	550
SETTING UP A DATABASE	550
INITIATING DATABASE SETUP	551
DEFINING FIXED DATA FIELDS	551

DEFINING FLEXIBLE DATA FIELDS	553
11 UPDATING DATABASES	**555**
REGISTERING STRUCTURES	555
REGISTERING ALPHANUMERIC DATA	558
RE-REGISTERING RECORDS IN DATABASE	562
DELETING STRUCTURES	562
DELETING DATA	567
12 TRANSFERRING DATA	**568**

PART FOUR   REACCS   573

1 SYSTEM BACKGROUND	**575**
2 SYSTEM DESIGN	**578**
MOLECULES	578
REACTIONS	578
NUMERIC AND TEXTUAL DATA	579
DATABASES	582
REACCS FILES	584
PROGRAM FILES	584
TEMPORARY FILES	584
TEMPORARY LIST FILE	585
TEMPORARY QUERY FILE	585
TEMPORARY OMIT LIST FILE	585
PERMANENT FILES	585
LISTFILE	586
MOLFILE	586
RXNFILE	586
RDFILE	587
FORM FILE	587
3 SYSTEM OPERATION	**588**
GLOBAL OPTIONS	589
KEYBOARD GLOBAL SWITCHES	590
KEYBOARD GLOBAL OPTIONS	591
SETTINGS	592
USER PROFILE	595
STARTING REACCS	596
CHANGING DATABASE	598
EXITING REACCS	598
4 CREATING MOLECULES	**599**
BUILDING EXACT MOLECULES	599
ATOM VALUES	599
CHARGES ISOTOPES AND ABNORMAL VALENCE	600
STEREOBONDS	600
UNSPECIFIED STEREOCENTERS	600

TEMPLATES	600
MODELS	600
CORRECT/MODIFY STRUCTURES	601
CLEAN STRUCTURES	601
MOVE ATOMS IN STRUCTURES	602
MOVE ENTIRE STRUCTURES	602
RESIZE STRUCTURES	602
REORIENT STRUCTURES	602
MANIPULATE GROUPS IN STRUCTURES	602
HYDROGENS AND NUMBERS DISPLAY IN STRUCTURES	602
TRUNCATE COMMON SUBSTITUENTS	603
CREATING GENERIC MOLECULES	604
VARIABLE ATOM VALUES	604
ATOM NEGATION	604
DEGREE OF SUBSTITUTION	604
RELATIVE CONFIGURATION	604
VARIABLE BONDS	604
SAVING QUERY TEMPORARILY	605
SAVING QUERY PERMANENTLY	605
BUILDING MOLECULES - THE QUICK WAY	606
5 CREATING REACTIONS	**608**
BUILDING REACTIONS	608
EDITING REACTIONS	608
SPECIFYING REACTION CENTERS	610
ATOM/ATOM MAPPING	616
MORE ON ATOM/ATOM MAPPING	619
BUILDING REACTIONS - THE QUICK WAY	622
6 CREATING FORMS AND TABLES	**625**
ABOUT DISPLAY FORMATS	625
CREATING FORMS	626
STRUCTURE BOX	627
DRAWING BOXES	627
ASSIGNING STRUCTURE FIELD TO BOX	629
ASSIGNING DATATYPES TO BOX	629
MULTIPLE DATA DISPLAY	632
EDITING BOX DATA FIELDS	632
MORE ON FORM MENU OPTIONS	632
SAVING FORMS/TABLES IN FILES	632
RECALLING FORMS/TABLES	633
CREATING TABLES	633
ASSIGNING DATATYPE AND LABEL	634
ASSIGNING LABEL	635
MULTIPLE DATA DISPLAY	635
MORE ON TABLE MENU OPTIONS	636

7 SEARCH TECHNIQUES — 637
- GRAPHIC QUERIES — 637
- SEARCHING WITH GRAPHIC QUERIES — 637
- NUMERIC AND TEXT QUERIES — 638
- REGISTRY NUMBERS AS LIST DESCRIPTORS — 639
- REGISTRY NUMBERS — 639
- LIST FILE — 639
- LISTFILE — 639
- CURRENT LIST — 639
- REFERENCE LIST — 639
- STRUCTURE FILES — 640
- DATA DESCRIPTOR — 640
- DATA-SEARCH EXPRESSION — 641
- SPECIAL DATA-SEARCH EXPRESSION — 642
- SEARCHING WITH ALPHANUMERIC QUERIES — 642
- SEARCH TYPES — 643
- SIMPLE SEARCH — 643
- CONVERSION SEARCH — 644
- COMBINATION SEARCH — 645
- DESIGNATING TEMPORARY LIST FILES — 646
- SEARCH IN MULTIPLE DATABASES — 646

8 MOLECULE SEARCHING — 648
- EXACT STRUCTURE SEARCHING — 648
- DISPLAY OF SEARCH RESULTS IN MAIN MODE — 649
- SWITCHING DISPLAY FORMS — 650
- STEREOISOMER SEARCHING — 651
- TAUTOMER SEARCHING — 651
- DISPLAY OF SEARCH RESULTS IN SEARCH MODE — 652
- MORE ON DISPLAY OPTIONS IN VIEWLIST MODE — 653
- VIEWING HITS IN GLOBAL SEARCH — 655
- WORKING WITH LISTS AND LISTFILES — 658
- DEFINING THE REFERENCE LIST — 659
- SAVING HIT LIST IN LISTFILE — 661
- RECALLING SAVED LISTFILES — 661
- CHECKING HIT LISTS IN CURRENT SESSION — 663
- DELETING LISTS — 664
- DEFINING CURRENT LIST — 665
- SUBSTRUCTURE SEARCHING (SSS) — 665
- STEREOISOMERS IN SSS — 665
- SEARCHING MOLECULES IN DATABASE SUBSET — 667
- COMBINING LISTS — 669
- COMBINATION SEARCHES — 670
- SIMILARITY SEARCH — 673

9 REACTION SEARCHING — 674

EXACT REACTION SEARCH	674
SIMPLE SEARCH	674
SUBSET SEARCH	675
CONVERSION SEARCH	677
STEREOISOMERS IN EXACT REACTION SEARCHING	678
SUBSTRUCTURE REACTION SEARCH	678
SIMPLE SEARCH	680
CONVERSION SEARCH	681
STEREOISOMERS IN SUBSTRUCTURE REACTION SEARCHING	681
REACTION CENTER SEARCHING	682
ATOM/ATOM MAPPING IN RSS	685
COMBINATION SEARCHES	687
SIMILARITY SEARCH	692
10 DATA SEARCHING	**697**
EXACT MOLECULE AND REACTION RETRIEVAL	697
REGISTRY NUMBER QUERY	697
DEFAULT REGISTRY NUMBER	699
LIST FILE AND LISTFILE QUERY	701
FORMULA QUERY	701
DATA QUERY	701
SUBSTRUCTURE MOLECULE AND REACTION SEARCH	701
FORMULA SEARCH	701
KEYS SEARCH	702
NAME SEARCH	704
DISPLAYING DATATYPE TREENAMES	705
DATA SEARCHING	706
DISPLAYING DATA IN VIEWLIST MODE	716
11 PLOTTING STRUCTURES	**718**
SELECTING STRUCTURES FOR PLOTTING	720
SPECIFYING APPEARANCE OF PLOTTED STRUCTURES	720
SELECTING PLOTTING DEVICE	720
FORMATTING PAGE FOR PLOTTING	721
12 UPDATING DATABASES	**722**
REGISTERING MOLECULES	722
REGISTERING REACTIONS	723
REGISTERING DATA	724
DELETING MOLECULES AND REACTIONS	726
TRANSFERRING STRUCTURES AND DATA	727
SINGLE MOLECULE TRANSFER	727
SINGLE REACTION TRANSFER	728
MULTIPLE STRUCTURES/DATA TRANSFER	728
RDFILES	728
WRITING RDFILE	729
WRITING PC-RDFILE	730

DATATYPE TRANSLATION TABLE	730
PREPARING REACCS TRANSLATION TABLE	731
READING RDFILES	732

PART FIVE CHEMTALK 735

1 SYSTEM BACKGROUND — 737
TERMINAL EMULATION — 737
DATA RETRIEVAL FROM MACCS-II — 737
FILE TRANSFER — 738

2 SYSTEM DESIGN — 739
CHEMTALK FILES — 739
DATA FILES — 739
METAFILES — 739
LOGIN FILES — 739

3 SYSTEM OPERATION — 740
MAIN MODE — 740
SETTINGS — 740
MORE ON MAIN MODE — 743
TERMINAL MODE — 743

4 TERMINAL OPERATIONS — 745
SWITCHING TO TERMINAL MODE — 745
TERMINAL SETTINGS — 745
MORE ON TERMINAL FUNCTIONS — 747

5 MACCS-II DATA RETRIEVAL — 749
FORM DESIGN — 749
EXACT STRUCTURE RETRIEVAL — 751
CURRENT LIST IN CHEMTALK — 755

6 TRANSFERRING FILES — 761
TRANSFER FROM MACCS-II TO CHEMBASE — 761
SDFILE FORMAT — 765
TRANSFER FROM CHEMBASE TO MACCS-II — 766
TRANSFER FROM REACCS TO CHEMBASE — 767
TRANSFER FROM CHEMBASE TO REACCS — 769

7 CHEMTALK AND CAS REGISTRY — 771
TRANSFER FROM CAS ONLINE TO CPSS — 771
STRUCTURE QUERIES FOR CAS REGISTRY SEARCHES — 772
CREATING EXACT STRUCTURE QUERIES — 772
CREATING GENERIC STRUCTURE QUERIES — 773
DEFINING Gk GROUPS — 774
UPLOADING STRUCTURE QUERY TO CAS REGISTRY — 777
CREATING METAFILES — 780

INDEX 781

INTRODUCTION

This treatise of the computer systems for storage and retrieval of chemical structures and numeric and textual data associated with the chemical structures is organized in five Parts.

Part One describes the CAS ONLINE Registry File. The System's historical background; its design and development from an in-house to a public online system; its operation by means of menu selected or keyboard commands; the techniques for building graphic exact structures and generic (Markush) structures; searching techniques; search strategy; and the options in displaying the retrieved structures and data are discussed.
Numerous examples are given for creating exact structures and generic structures and many real computer sessions to search and retrieve structures and data are shown with detailed description and explanation. The unique concept of normalized bonds is discussed in detail with many examples of structures to which it has to be applied to retrieve structures which can be drawn in more than one way, i.e. resonance isomers, tautomers, and systems with overlapping conjugated and tautomeric bonds. The batch search technique to override the system's search limits is explained. Current awareness and selective dissemination of information (SDI) search technique is described.
The CAS ONLINE System is continuously improved and expanded so that today it is possible to search for chemical reactions reported in the literature and for the structures and data in the Beilstein compendium. The new features that have been introduced in the Registry File after the manuscript was completed are covered in the Addendum, p. 262-A1 to 262-A5.

Parts Two, Three, and Four describe the Molecular Design Limited (MDL) systems, i.e. ChemBase$^{(R)}$, MACCS-II, and REACCS. These are Database Management Systems (DBMS) for storage and retrieval of chemical structures, chemical reactions, and numeric and textual data associated with the structures or reactions. The Systems perceive stereochemical structures so that it is possible to search selectively for configurational isomers. They have the capability for searching simple generic structures or complex (Markush) generic structures with defined R groups.
Searches can be performed for reaction centers undergoing bond change and, by employing atom/atom mapping, for any structural fragment with atom and bond change. Similarity searches of molecular structures and reactions can be also carried out.

Part Five describes ChemTalk$^{(R)}$, a package of terminal emulation and telecommunication softwares which can be used to establish a link between the Systems which run on mini/mainframe computers (MACCS-II and REACCS) and that which runs on microcomputers (ChemBase). Structures and data can be exchanged between these systems through ChemTalk. ChemTalk can be also used to transfer structures and data from CAS ONLINE and incorporate them in documents prepared in ChemText, a wordprocessing system developed by MDL for microcomputers, or for building graphic structure queries offline and uploading them into CAS Registry File search.

Numerous examples of the following are given and discussed in detail in the Parts: building graphic exact structures and generic structures; reactions; searching exact structures and substructures or exact reactions or substructures in reactions; searching reaction centers and/or atom/atom mapped reactions; and similarity searching.

The concept of substructure searching is discussed in detail in Part Two with many substructure search examples. This discussion is applicable to all Parts.

PART ONE

CAS REGISTRY

1   SYSTEM BACKGROUND

CAS ONLINE

CAS ONLINE is a database system consisting of computer-readable files containing scientific and technical data and information from worldwide documents abstracted by Chemical Abstracts Service (CAS). The files can be searched for the data and information via remote computer terminals using a variety of search techniques.

There are many files in the CAS ONLINE system. Those most relevant to searching chemical substance information are shown in Figure 1/1.

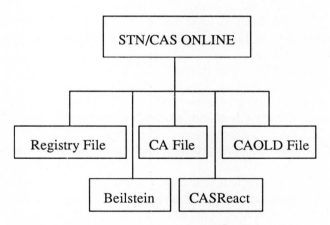

Figure 1/1. CAS ONLINE and Beilstein Files

Registry File, the file that is the major subject of Part One, contains molecular structures, identifying numbers (CAS Registry Numbers), and names of chemical substances reported in the chemical literature since 1962. Information on substances reported from 1957 through 1961 are currently being added to the Registry File. This file is searchable with graphics structure queries as well as with alphanumeric queries.

CA File contains data of documents abstracted by Chemical Abstracts Service since 1967. This file is a text file searchable by subject terms and bibliographic terms to retrieve references, abstracts, and index terms of the documents.

CAOLD File contains references to substances cited in CA prior to 1967. Completion of this file is in progress. Only CA accession numbers, Registry Numbers for substances, and document (patent) type are searchable in this file.

CASREACT File contains information on organic synthetic reactions reported in the literature since 1985. This file is searchable using Registry Numbers of the structures located in the Registry File, or alphanumeric queries.

BEILSTEIN File has been recently added to the STN International system. It is compatible with the Registry File. It can be searched by the same techniques as the Registry File using structural queries built either in the Registry File or the Beilstein File by the methods we will discuss in the later sections of this Part One.

BEILSTEIN File contains information on more than 350,000 heterocyclic substances from the *Beilstein Handbook of Organic Chemistry*, Main Volume and Supplements 1-4.

STN INTERNATIONAL

CAS ONLINE and BEILSTEIN are accessible through STN International, The Scientific and Technical Information Network.

STN International is a cooperative effort of three organizations, participants in STN, to share databases online with the scientific and technical community. STN International makes it possible for users from all over the world to access online the databases that originated in the United States, Europe, and Japan, thus giving an easy and direct access to a variety of data and information.

STN International is a nonprofit organization founded in 1983.

The current participants are:

American Chemical Society-Chemical Abstracts Service (CAS), Columbus, Ohio, USA

CAS was founded in 1907 and has grown into a major world center for scientific and technical information.

CAS abstracts scientific and technical literature from all over the world and makes the information available to the scientific and technical community. In addition to the computer system CAS ONLINE, CAS publishes *Chemical Abstracts* and other printed sources of information.

Fachinformationszentrum Karlsruhe (FIZ Karlsruhe), Karlsruhe West Germany

FIZ Karlsruhe was established in 1977. It produces and disseminates information in energy, physics, mathematics, and related fields and offers over 50 databases for online searching.

Japan Information Center of Science and Technology (JICST) Tokyo, Japan

JICST, Tokyo was founded in 1957 and is now the central information organization in Japan for advancing science and technology. JICST's activities in collecting, processing, and disseminating scientific and technical information culminated in the development of an online retrieval system widely used in Japan.

The following databases/files are available in STN (as of May, 1989):

SYSTEM BACKGROUND

AGPAT-CAS Online Agrochemical Patents File, 1987-Present
APILIT-API Literature File 1964-present (Subscribers)
APILIT2-API LITERATURE FILE 1964-PRESENT (NONSUBSCRIBERS)
APIPAT-API Patent File 1964-present (Subscribers)
APIPAT2-API PATENT FILE 1964-PRESENT (Nonsubscribers)
BEILSTEIN-File of heterocyclic compounds from 1830 thru 1959
BIBLIODATA-German National Bibliography from 1972-present
BIOCAS-BIOSIS/CAS Registry Number Concordance File
BIOQUIP-DECHEMA Bio Equipment suppliers Data Bank
BIOSIS-The BIOSIS Previews (R) File 1969-present
CA-The Chemical Abstracts File for 1967-present
CAOLD-The pre-1967 Chemical Abstracts File
CAPREVIEWS-Preview File for the Chemical Abstracts File
CASREACT-The Chemical Abstracts Reaction Search Service
CHEMLIST-Regulated Chemicals Listing
CJACS-Chemical Journals of the American Chemical Society
CJAOAC-Chemical Journals of the AOAC
CJRSC-Chemical Journals of the Royal Society of Chemistry
CJVCH-ChemicalJournals of VCH Verlagsgesellschaft
CJWILEY-Chemical Journals of John Wiley & Sons, Inc.
COMPENDEX-Computerized Engineering Index from 1969-present
CONF-Conferences in Energy, Physics, Mathematics etc.
CSCHEM-ChemSources-USA and International (Chemicals)
CSCORP-ChemSources-USA and International (Company Directory)
CSNB-Chemical Safety News Base from 1981-present
DECHEMA-Chem. Engineering, Biotech. file from 1976-present
DEQUIP-Dechema Equipment Suppliers Data Bank
DERES-DECHEMA Research Data Bank
DETEQ-Dechema Environmental Technology Equipment Data Bank
DIPPR-AIChE Design Inst. Physical Property Data File
ENERGIE-FIZ Karlsruhe Energy file from 1976-present
ENERGY-DOE ENERGY file from 1974-present
FBR-German R&D Ministry reports from 1985-present
FHGPUBLICA-Publications of the Fraunhofer-Society 1982
FORIS-Research in social sciences from 1973 - present
FORKAT-BMFT Foerderkatalog
GEOREF-GeoRef - Geological Reference File from 1785-present
GFI-GFI Gmelin Formula Index and Complete Catalog
HOME-The default login file. Contains no data.
ICONDA-International Construction Database from 1976-present
IFICDB-The IFI Comprehensive Database from 1950-present
IFIPAT-The IFI Patent Database from 1950-present
IFIREF-The IFI Uniterm and U.S. Class Reference File
IFIRXA-The IFI Reassignment and Reexamination Database

IFIUDB-The IFI Uniterm Database from 1950-present
INFORBW-Baden-Wuerttemberg Institutes Research in Progress
INIS-International Nuclear Information System 1970-present
INPADOC-Bibliographic, Family, Legal Status from 1968-
INPAMONITOR-INPADOC file for SDI
INSPEC-INSPEC File from 1969-present
JANAF-JANAF Thermochemical Tables, 3rd Edition
JICST-E-JICST English File on Sci. & Tech. in Japan
KKF-The Plastics Rubber Fibres File from 1973-present
LBIBLIO-Bibliodata learning File
LCA-The CA Learning File
LCJO-Learning File for the CHEMICAL JOURNALS ONLINE
LPHYS-The PHYS Learning File
LREGISTRY-The REGISTRY Learning File
MATBUS-Materials Business File from 1983-present
MATH-Zentralblatt fuer Mathematik from 1972-present
MEET-EI Engineering Meetings File from 1982-present
METADEX-METADEX File from 1966-present
MONUDOC-Factual documentation on the care of monuments
NBSTHERMO-NBS Tables of Chem. Thermodyn. Props., 1982
NTIS-Government Reports Announcements 1974-present
PATDPA-The German Patent Database from 1968-present
PHARMPAT-CAS Online Pharmaceutical Patents File, 1987-
PHYS-Physics Briefs File from 1979-present
REGISTRY-The CAS Registry File of substances
RSWB-Regional planning and building construction
SDIM-SDIM database in metallurgy from 1979 - present
SIGLE-Grey Literature in Europe from 1981 - present
SILICA-Ceramics and Glass Database from 1975 - present
SITRAFO-University R&D-projects for technology transfer
SOLIS-German literature in social sciences 1945-present
TA-The Technology Assessment Database
TITUS-The textile database in English from 1968 - present
TOXLIST-File name change - see "CHEMLIST"
UFORDAT-Environment Research in Progress from 1974 -present
ULIDAT-Environmental Literature from 1976-present
VADEMECUM-Educational and Research Institutions, F.R. Germany
VTB-Verfahrenstechnische Berichte from 1966-present

2 SYSTEM DESIGN AND DEVELOPMENT

REGISTRY FILE

The Registry File is an integral part of the Registry System which was developed by Chemical Abstracts Service in several stages.

The Registry I was the first system which was put in operation in 1965 in-house to track, detect, and eliminate duplicate substances to be indexed in the *Chemical Abstracts Indexes* in relation to their bibliographic references and other information given in the printed *Chemical Abstracts* of scientific and technical publications.

The registration and storage of chemical structures of conventional chemical substances in the System were accomplished by conversion of the structures to a digital code in form of a connection table *(vide infra)*. Each structure was also assigned a unique computer-generated Registry Number with which each structure was identified in the System.

The Registry I was enhanced in 1968 by inclusion of conventions for representing polymers and coordination compounds. The System was known as Registry II.

The Registry II began to be utilized in-house in the process of producing *Chemical Abstracts Indexes* to reduce cost and to cut the time for their publication (the time was cut from years to months). The Registry II made it possible to determine whether a structure abstracted from a publication had been already entered into the System. It was also possible to retrieve the chemical name associated with the registered structure. If a structure was identical to that one to be indexed, the chemical name associated with the structure in the System was used to index it. Thus, an identical substance, so often encountered in the chemical literature, need not be named again and again for the *Chemical Abstracts Indexes*.

Further enhancements were made in 1974 to improve the representation of the connection table in the System to support generation of chemical systematic names for each registered structure and to allow for generation of structure graphs from the connection tables. Changes were also made in the identification process for ring systems in a structure and in generation of textual stereochemical descriptors for stereoisomers (structure graphs in the Registry System are two-dimensional). This enhanced System was labeled Registry III and all structures in the System were converted to this version.

Hand-in-hand with the explosion of information and computer systems to handle the increasing amount of scientific and technical information, efforts were undertaken to give the scientific community an access to this wealth of information. DIALOG, ORBIT, and BRS Systems offered online searching of *Chemical Abstracts* by using alphanumeric (text) queries to retrieve bibliographic references and subject and keyword terms of chemical substances. However, it became obvious that the chemical community would benefit most from having a system allowing searches of chemical structures by using graphic structure queries.

In 1980, the Registry III was transformed from in-house system to one accessible by outside users. In the initial version of the Registry File, it was not possible to create graphic structure queries; however, the answers could be displayed as graphic chemical structures. The queries were formulated using fragment screens. The screens are numeric codes of common structural fragments that were defined and collected in a screen dictionary. The user had to dissect the structure to be searched into fragments and assign screen codes to the fragments matching those in the dictionary. The screen codes were combined to formulate a query which was submitted to the SEARCH Subsystem of the Registry File.

Although the screens can still be used to formulate a query, the current version of the Registry File accepts graphic structures as queries and most of the searches are performed with structure graphs as queries. However, in some cases the structure graphs are combined with screens to create a highly specific query.

The Registry File contains chemical structures as well as text information associated with each structure in the File. All structures that have been abstracted from the world's literature and indexed in the printed *Chemical Abstracts Indexes* since 1965 are registered in the File. The File is updated weekly with up to 15,000 new structures from the current literature. A project is underway to add the structures indexed prior to 1965 into the File. The Registry File includes all classes of substances, the major ones being:

1. organic compounds
2. inorganic compounds
3. coordination compounds
4. salts and addition compounds
5. alloys
6. minerals
7. mixtures of substances
8. polymers

It is to be noted that the textual and numeric data related to the chemical structures/substances contained in the Registry File can also be searched by using text search techniques. However, these techniques will not be the subject of this book.

References to the 10 most recent documents can be displayed in the Registry File. The bibliographic data are not searchable in the Registry File; they can be searched only in the CA File. It is also possible to perform a structure search in the Registry File and crossover with the Registry Numbers retrieved in the File into the bibliographic files, CA File, CAOLD File, and/or CASREACT. The CAS Registry Numbers, however, cannot be used for searching in the Beilstein File.

The Registry File database is composed of basic units of information, i.e., chemical substance records. Each record is composed of data fields. A record represents the total of data and information for one chemical substance. A data field represent one unit of unique data.

The following data fields are comprised in each record of the Registry File:
structure (STR), Registry Number (RN), molecular formula (MF), element count (EC), periodic group (PG), number of components (NC), formula weight (FW), substance class

SYSTEM DESIGN AND DEVELOPMENT

identifier (CI), complete chemical name and fragments (IN), name synonyms (SY), stereochemistry indicator (ST). Figure 1/2 shows the composition of the Registry File.

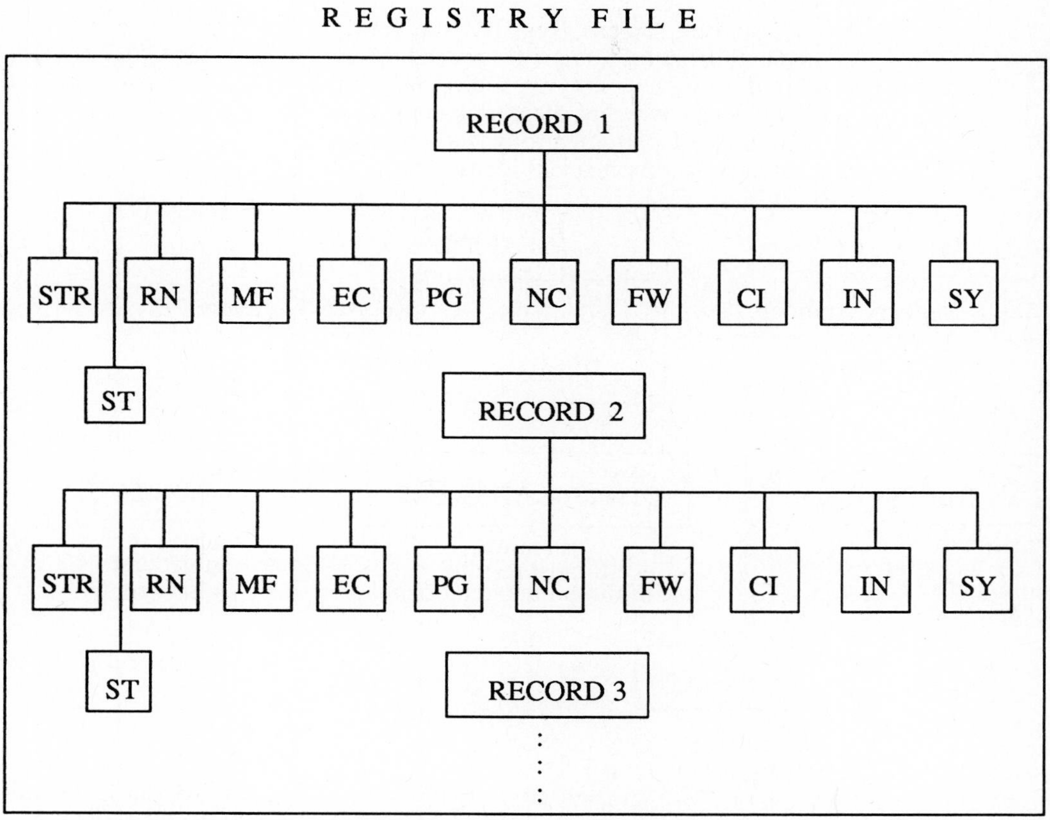

REGISTRY FILE

Figure 1/2. Composition of Registry File

The search and retrieve capabilities of the CAS ONLINE System are illustrated in Figure 1/3.
Because the Registry File contains a huge number of structures and the number will be growing with progressing time, special search techniques have been built into the System to speed up the searching in an efficient way.
The search mechanism is based on dividing the Registry File into segments (this division is not apparent to the user, it is only an internal division of the File). Each segment of the File is allocated to a minicomputer and searching of the segments proceeds in parallel by sequential searching. In sequential search, each structure is evaluated if it meets the search requirements. If it does, it is transferred to the Answer Set and the next structure is evaluated. If it does not meet the search requirements, it is set aside and is not examined

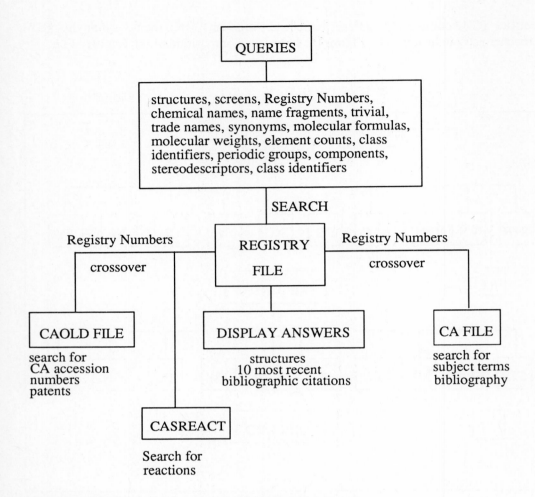

Figure 1/3. Search and crossover with Registry Numbers

further. When all the segments of the File are searched, all answers are pooled and transferred to the main computer for further manipulation by the user. Figure 1/4 illustrates the search process.

The search of the structures in the segments of the File in the minicomputers proceeds in two stages as is illustrated in Figure 1/5.

The first stage is a screen search based on dissecting the structure query into fragments (similarly to the User fragmentation in the initial version of the System) and comparing these fragments with those of the structures in the File. When similar fragments are recognized, the structure is selected for the second stage of the search. If no similarities are found, the structure is eliminated (screened-out) from further consideration for the second stage of the search.

SYSTEM DESIGN AND DEVELOPMENT

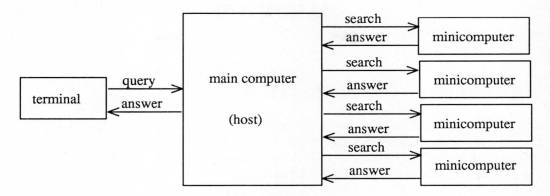

Figure 1/4. CAS Registry File system for search and retrieval

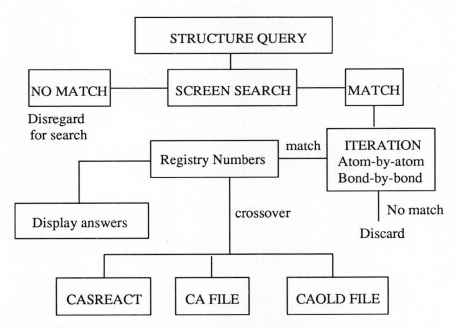

Figure 1/5. Search steps and crossover in Registry File

The second stage of the search is iteration, i.e., comparison atom-by-atom and bond-by-bond of the candidate structures from the first stage of the search with the structure query. If a structure is found which is identical with the structure query and all its requirements, it is pooled into an Answer Set labeled Ln which can be displayed upon completion of the search.

3 HARDWARE AND SOFTWARE

HARDWARE FOR ACCESSING CAS ONLINE

Virtually any terminal equipped with a modem or any microcomputer equipped with a modem and a communication software can be used to access STN International - CAS ONLINE. However, we will discuss only equipment with graphics capability.
The Registry File contains chemical structures and we want to be able to display the structure diagrams on our equipment as real graphics. Chemical Abstracts Service has developed a system that renders the appearance of chemical structures in the form and shape according to the conventional representation of chemical structures by chemists.
Text terminals using the ASCII protocol can also be used, however, the displayed structures are not graphics but are composed of text characters. The folllowing example shows the difference in the representation of chemical structures (Figure 1/6).

structure display on text terminal	structure display on graphics terminal

Figure 1/6. Structure representation

With graphics equipment, we also will be able to create graphic structures on the screen and use them as queries to search for structures in the Registry File.
The equipment with graphics capability can be either a terminal or a microcomputer. Both of these must be compatible with TEKTRONIX PLOT10 graphics, i.e., the equipment must be compatible with TEKTRONIX 40XX or 41XX series of terminals and must be capable to emulate at least the major features of these terminals.
PLOT10 is a trademark for the product developed by TEKTRONIX, a research, development, and production company of terminals and software packages. PLOT10 is a collection of subroutines that provides a high-level interface for utilizing the advanced

features of graphics terminals manufactured by TEKTRONIX. PLOT10 became the standard for graphics software industry because it frees programmers from developing individual subroutines to mimic the features of TEKTRONIX software.

TERMINALS

The terminals available on the market have either PLOT10 capability as a standard feature or they can be adapted, if they do not have this standard, to PLOT10 compatibility.

MICROCOMPUTERS

Microcomputers are more and more used both as stand-alone computers and as terminals to connect with a host computer. Some of the micros have standard PLOT10 compatibility. Those that do not have it can be adapted to emulate both a terminal and PLOT10 features.

There is a great variety of microcomputers on the market of types and makes which grow almost daily. Any micro that has a standard graphics board or can be retrofitted with one of at least 512x240 pixels resolution may be ready to use. It is to be noted, however, that the Emulation Software *(vide infra)* must support the micro or be compatible with it.

GRAPHICS INPUT DEVICES

To create graphic structures on the screen of a terminal or microcomputer by the Menu Driven Technique (see p. 38) or by the DRAW Command (see p. 80), we need a graphics input device which can move a crosshair cursor about the screen. Although the crosshairs can be manipulated by the arrow keys on the keyboard, the movement is too slow. Peripheral graphics input devices, i.e., a mouse, light pen, track ball, or a tablet with a stylus are the common GIN devices. Not all of them are, however, supported by the Emulation Softwares *(vide infra)*. The supplier of the software will be able to provide the necessary guidance in selecting a GIN.

MODEMS

A terminal or a microcomputer must be equipped with a device, a modem, that connects it with a host computer via ordinary telephone lines.

Any asynchronous modem of 300, 1200, or 2400 baud can be used. A common choice is the Hayes Modem or modems that recognize the Hayes Modem commands. A modem connected to a terminal does not need any additional software to manage the communication. However, the microcomputers need a special software to interface the modem and manage the communication. The Emulation Software packages *(vide infra)* contain the communication software. However, some of them support only the Hayes type of modem, other are not specific. The supplier of the Emulation Software should be contacted for the requirements.

PRINTERS

The CAS ONLINE System displays structure graphs on the screen of a terminal or microcomputer. The screen can be copied through a dot-matrix printer, thus, a hard copy of the structures can be made. The printer must be able to do a bit map screen dump. The Emulation Software (*vide infra*) includes screen dump routines that interface the printer. Therefore, the Emulation Software must specifically support the printer that we wish to use in our work with CAS ONLINE.
It is recommended that the supplier of the Emulation Software be consulted to select a printer.
CAS ONLINE has a capability of printing structures offline on a high-speed printer at the host-computer site. The prints are of very high quality and are delivered by mail to the User.

EMULATION SOFTWARE

The microcomputers that do not have the PLOT10 compatibility as standard feature have to be run with a special software that emulates PLOT10, i.e., the TEKTRONIX 40XX and 41XX series of terminals.
The Emulation Software package includes a communication software which links a micro with a host and manages the interaction.
The following Emulation Software packages are available for IBM PCs and IBM compatible microcomputers:

ChemTalk by: Molecular Design Ltd.
       2132 Farallon Drive
       San Leandro, CA 94577, USA
       Telephone: (415) 895-1313
STN EXPRESS by: STN International
       P. O. Box 02228
       Columbus, Ohio 43202, USA
       Telephone: (800) 848-6538
PC-PLOT III/IV by: MicroPlot Systems Co.
       659-H Park Meadow Rd.
       P. O. Box 615
       Westerville, OH 43081, USA
       Telephone: (614) 882-4786
TGRAF05/07 by: Grafpoint
       4300 Stevens Creek Blvd.
       San Jose, CA 95129, USA
       Telephone: (408) 249-7951
EMU-TEK by: Graphics Innovations, Inc.
       P. O. Box 615
       Stanton, CA 90680, USA

HARDWARE AND SOFTWARE

Telephone: (714) 995-3900

The following Emulation Software is available for the Macintosh microcomputer:

Tekalike by: Mesa Graphics
    P. O. Box 506
    Los Alamos, NM 87544, USA
    Telephone: (505) 672-1998
GriffinTerminal 100 by: Metaresearch Inc.
    1100 SE Woodward
    Portland, OR 97202, USA
    Telephone: (503) 232-1712
VersaTerm by: Peripherals Computers and Supplies Inc.
    2457 Perkiomen Avenue
    Mt. Penn, PA 19606, USA
    Telephone: (215) 779-0522

The Apple II series of microcomputers can also be used to work with graphics. However, since the screen resolution is low the graphics are not as good as graphics generated on the other micros. The following Emulation Software package is available for the Apple II series:

Tekterm by: Fountain Computer
    1901 Kipling
    Lakewood, CO 80215, USA
    Telephone: (303) 232-8346

EMULATION SOFTWARE SET UP PARAMETERS FOR STN

The Emulation Software requires configuration parameters to be set up that are specific for the hardware we wish to use. For example, the type of microcomputer, type of graphics board, type of modem and its baud rate have to be entered in the Set mode of the software according to the instructions by the software. In addition, the STN System requires communication parameters that have to be set up in the Set mode of the Emulation Software to obtain optimum performance with the System. The parameters are as follows:

Parity: even
Duplex: full
Data bits: 7
Stop bits: 1
GIN mode terminate: cr

4   SYSTEM ACCESS

ACCESSING STN INTERNATIONAL

Online information search and retrieval is a method of accessing computer readable files by connecting with a host computer where the files reside through remote terminals and telephone lines. The usual equipment configuration for accessing, searching, retrieving, and displaying or printing data and information is illustrated in Figure 1/7.

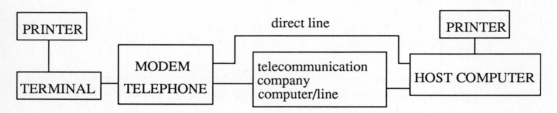

Figure 1/7. Setup of online system

The user can employ either a direct, dedicated telephone line or one of the telecommunication companies, e.g., Telenet, Tymnet, or Datex-P, to call up the host. Commands and queries are typed at the terminal and transmitted to the host computer. There are many terminals/users linked with the host at one time. The telecommunication company computers and the host direct the communication so that each user uses a slice of real computer time at one time while the other user is performing other tasks not requiring the real computer time, e.g., typing commands, answering system prompts, printing, etc. The final result is that to one particular user, the use of the host seems to be exclusively his/her while, in fact, many other users access the same computer and system files at the same time. The search and retrieval process is interactive. The user responds to the system prompts and questions and the system responds to the user directives. Answers can be reviewed selectively one by one and queries can be modified if the answers do not meet the search requirements. Answers can be displayed on the screen of the terminal, printed at the user's site, or a request can be made for offline printing by high-speed laser printers at the host computer site. The prints are then mailed to the requester.
Each user of the system is assigned an identification code (ID) and a password to access the system and is charged for the use of the System. The following organizations can be contacted to obtain details on the contract conditions and the telecommunication companies for accessing STN International:

STN International
2540 Olentangy River Road
P.O.Box 02228
Columbus, Ohio 43202
USA

STN International
Postfach 2465
D-7500 Karlsruhe 1
Federal Republic of Germany

STN International
c/o The Japan Information Center
of Science and Technology
5-2,Nagatacho 2-chome
Chiyoda, Tokyo 100
Japan

CSIRO Information Resources Unit
314 Albert Street
East Melbourne
Victoria 3002
Australia

LOGIN PROCEDURE

The login procedure involves turning the terminal or microcomputer on, dialing the number of the telecommunication network, and, upon receiving the prompt from the network (the nature of the prompt varies for each network), connecting to the host computer and entering the ID and password. STN International responds by a request for a terminal identifier.
There are three terminal identifiers in the System prompt:
1, 2, and 3.

Use number 2 for any terminal or microcomputer with PLOT10 capability, including the HP 2647 and 2648.

Number 1 is used only for the HP 2647 or 2648 loaded with the CAS software. Number 3 is for text terminals only which cannot handle graphics although they can display structures composed of alphanumerics.

This book presumes the use of the terminal identified by number 2, the most commonly used terminal or microcomputer with PLOT10 capability.

Example: *Login in STN International*

This example uses Telenet as the communication network. The login procedure is different for each of the telecommunication networks or for direct dial of the host. Your contract agreement with STN International will supply instructions for login procedures.

1. Dial Telenet number.
2. When dial tone indicates connection has been made or CONNECT appears on the screen, enter:
<cr><cr> (cr stands for carriage return)(enter).
Telenet responds:
TELENET
614 17k (this node identifier code may change from time to time and for different locations).
TERMINAL = (type your terminal identifier, see Telenet codes for terminals) followed by <cr>.
@ (this is Telenet prompt), type C (for "connect") followed by STN accession code followed by cr): C 614 21X <cr>.
Telenet responds:
@ 614 21X CONNECTED
Enter: <cr>
System:
LOGINID: (type your ID followed by <cr>)
System:
PASSWORD: (type your password followed by <cr>)
System:
TERMINAL (ENTER 1,2,3 OR ?):
Type:2 <cr>
System:
COPY AND CLEAR PAGE, PLEASE (your response will depend on the Emulation Software you are using; see software instructions)
System:

* * * * * * * Welcome to STN International   * * * * *

Free Connect Time in JICST-E -- See NEWS 8

JICST has Expanded the Document Delivery Service
See NEWS 9 for Details

CSNB, MONUDOC, SDIM and SIGLE Files are Now Available
See NEWS 13 - 16 for Details

Free Connect Hours in the GeoRef File in May -- See NEWS 21

Changes in Contact, Prices for JPS, JTS -- See NEWS 20

Free SDI Searches in the CJO Files -- See NEWS 12

A New C13NMR/IR Workshop is Available -- See NEWS 10

Changes to the CAS ONLINE Academic Program -- See NEWS 11

New Structure Search interaction now available in the Registry File and LRegistry File -- See NEWS 17 for Details

Enhancements to the BEILSTEIN File -- See NEWS 22 for Details

CASREACT now being updated biweekly. SDI's now available in this file -- See NEWS 18 for Details

Current STN and File Workshop Schedule -- See NEWS 1 Direct Dial Login (614) 447-3781

12:09:56 COPY AND CLEAR PAGE, PLEASE

* * * * * * * * * * * STN Columbus * * * * * * * * *

FILE 'HOME' ENTERED AT 12:09:52 ON 12 MAY 89

LOGOFF PROCEDURE

To terminate connection with STN, we use the LOGOFF Command at the System prompt =>:

Example: *Logoff STN*

System:=>
User:logoff <cr>

(You will be reminded by the System that all data will be lost unless saved):
System:
ALL L-NUMBERED QUERIES AND ANSWER SETS ARE DELETED AT LOGOFF.
LOGOFF? Y/N
User:y <cr>

System:
STN INTERNATIONAL LOGOFF AT 12:18:00 ON 12 MAY 89

THE HOME FILE

The last System message in LOGIN informs us that we have entered the HOME File. The HOME File does not contain any searchable data. It is the file into which we are always connected first when we login into STN International. In the HOME File, we can display the latest news on the System and its files and we can conduct certain operations related to the usage of the System at a lower charge than in the other files of the System. In the HOME File, we issue the command to be transferred to the Registry File (*vide infra*).

THE SYSTEM PROMPTS

To communicate with the System, we enter commands and other instructions at the System prompts.
We will encounter two kinds of prompts in STN and in the Registry File:
=> is the prompt which allows the major commands of the System to be entered. This prompt always stands alone.
: is the prompt at which subcommands within some of the major commands can be entered. The : prompt is preceded by a line of options or instructions for the User when we use the NOVICE version of commands. In the EXPERT version of commands, the : prompt stands alone and we have to know how to respond for the next step of our interaction with the System.

CORRECTING TYPING ERRORS

Typing errors can be corrected, before the carriage return (enter) is pressed, by the following methods:
1. Use the backspace key to erase characters and retype.
2. Enter a $ sign after the incorrect entry and continue typing of the correct version on the same line (everything before the $ sign is ignored by the System).
3. Enter a $ sign and carriage return (enter) to start anew at the System prompt.

5   SYSTEM COMMANDS

The operating system for STN International is the Messenger Software which provides access to the System and to the databases in STN by executing a set of commands.

NOVICE AND EXPERT VERSION OF SYSTEM COMMANDS

The System offers two versions of command names, the NOVICE and EXPERT version.
The NOVICE version consists of typing on the keyboard the command name in full or at least the first four letters of the command name. The System responds with detail prompts and messages related to the operation being conducted. The System offers a line of command options and, upon the User's selection of one of the options, continues in the dialog with the User to complete the operation. The most commonly used (and "safe") options are enclosed in parentheses. Their selection requires only to enter a period (.) or a space by depressing the space bar followed by <cr> to be activated. These options are the defaults of the System. A question mark (?) can also be entered to obtain additional messages and information related to the operation.
The EXPERT version of commands consists of typing only the first three letters of the command name, or in three cases, only the first letter of the command name. The System does not offer options and messages to the User; it expects the User to know how to proceed further. The EXPERT version should be used only by those well versed in the System techniques. The versions can be switched from one to the other at any point of an operation just by typing the appropriate number of letters of the command name.
This book will use only the NOVICE version of the commands. The use of the EXPERT version follows naturally after one feels comfortable with all of the System techniques.
The System accepts the commands only at the System prompts.
Commands entered at the => prompt are executed and return either an immediate result according to the function of the command or invoke a subsystem of commands that are entered at the : prompt.
Some commands are executable in all files of the CAS ONLINE database; some are accepted only in specific files.

COMMANDS ACCEPTED IN THE HOME FILE AT => PROMPT

DELETE	Delete saved or current session items
*DISPLAY	Display saved or current session items
FILE	Specify the search and display file
HELP	For help on how to use the system
LOGOFF	End the online session
NEWS	Display current news about the system
*ORDER	Order an original document or a copy

SAVE	Save an L-numbered query or answer set
*SEND	Send a message to STN staff
SET	Set terminal and interaction options
?	The same as HELP

COMMANDS ACCEPTED IN THE REGISTRY FILE AT => PROMPT

ACTIVATE	Assign L#s to saved query or answer set
BATCH	Request a batch search
DELETE	Delete saved or current session items
*DISPLAY	Display saved or current session items
FILE	Specify the search and display files
HELP	For help on how to use the system
LOGOFF	End the online session
NEWS	Display current news about the system
*ORDER	Order an original document or a copy
*PRINT	Print answers offline
QUERY	Define a search question (query)
SAVE	Save an L-numbered query or answer set
SCREEN	Define a set of screens for searching
*SDI	Request searches be run on file updates
*SEARCH	Perform a search
*SEND	Send a message to STN staff
SET	Set terminal and interaction options
*STRUCTURE	Enter system of subcommands to create structures
?	The same as HELP

*Note: The asterisked commands in the above list (the * is not part of the command) invoke subcommand options or a question/answer prompts by the System. Some of the subcommand options are displayed in a line menu. The following is a list of the STRUCTURE Subcommands for building structure graphs by Keyboard Commands:

STRUCTURE SUBCOMMANDS ACCEPTED AT : PROMPT

GRAPH	Create structure graphs
DRAW	Draw new nodes and bonds
NODE	Specify node values (atoms)
BOND	Assign bond types and values
VARIABLE	Define Gk groups
REPEATING	Define repeating Gk groups
CHARGE	Specify charge
MASS	Specify mass
VALENCE	Specify valence
HCOUNT	Specify number of hydrogens attached

SYSTEM COMMANDS

HMASS	Specify hydrogen of abnormal mass
DLOC	Specify delocalized charge
NSPEC	Specify node as ring, chain, or either
RSPEC	Specify ring(s) in structure as isolated
CONNECT	Specify the number of connections to a node and the bond type of connections
DELETE	Delete node, bond, attribute, ring, chain, or entire structure
DISPLAY	Display image, attributes, or connection table
RECALL	Use a copy of a Registry Number or another structure (by name) as a model
SET	Change the default bond value or specify how Gk groups should be displayed
MOVE	Move a ring, chain, node, or entire structure
HELP	List the STRUCTURE Commands
END	Exit the STRUCTURE Command and return to the => prompt
MENU	Display menu for structure building

Note that DELETE, DISPLAY, and SET Commands have a different function at the : prompt than at the => prompt.

HELP! HELP!

At any point of the session, if we feel uncertain what our action should be, we can request help by typing HELP or ?. The System will respond with instructions how to proceed. When we enter a command which is incorrect either for the particular kind of operation or because of typing error, the System issues an error message. Most of them are self-explanatory or we can request more information by entering HELP.

Advice and help can also be obtained from the CAS Search Assistance Desk by telephone: (800) 848-6533.

THE PANIC BUTTONS

Endless loops, the nightmare in which a computer begins to smolder because of a program or a command that sends the computer into repeated operations, *ad infinitum,* does not happen.

However, we still can be concerned with how to quickly get out of an operation because, after all, we have to pay for our interaction with the system.

Example: *Disconnecting from the System*

At the System Prompt =>:
System:=>
User:logoff <cr>

In Menu Mode:
User:select END
System:Ln STRUCTURE CREATED (or NO STRUCTURE CREATED)
System:=>
User:logoff <cr>

In Keyboard Command Mode:
ENTER (DIS), GRA, NOD, BON, OR ?:
User:end <cr>
System:Ln STRUCTURE CREATED (or NO STRUCTURE CREATED)
System:=>
User:logoff <cr>

At LOGOFF, the System will ask whether we wish to save the data generated in the session or to abandon them.

Interrupting command execution

To interrupt a command execution while the command or the operation invoked by the command is active use the BREAK KEY on the keyboard. You will be returned to the System prompt => or : depending on the operation being conducted.
The break key should be used with caution since it may result in disconnecting from the System and loosing all data.

Exiting from subcommand mode

The LOGOFF *does not* work at the : prompt.
To exit from subcommand mode use the END Command at the System prompt :. You will be returned to the => prompt.
System: :
User:end <cr>
System:=>
User:logoff <cr>

COMMAND SYNTAX

The following applies to the usage of commands and User input as shown in the Examples of Part One.

Any command or response to System messages is typed (commands and responses can be typed in lower or uppercase) by the User after the prompts : and => followed by carriage return <cr> except where indicated.

SYSTEM COMMANDS

Spaces and commas shown in the commands and specifications should be entered as shown. After we gain more experience, some spaces, commas, or dashes are optional.

Stacking commands can be done by separating commands on one line by a semicolon (;). Stacking specifications by a comma (,) is allowed within the commands that have subcommands. A subcommand affects all the stacked specifications until another subcommand is typed.

When an incorrect command or specification is entered, the System issues an Error Message. However, everything prior to the error has been executed.

Defaults are shown in parentheses. They are activated by entering a period (.) followed by <cr>.

The Key in the Examples shows abbreviations of commands that can be used in place of the full command.

The Examples show System output in uppercase, the User input in lowercase, with some exceptions, e.g., to differentiate element symbols, they are shown in uppercase.

History of the commands used by the User in the current session can be requested at the prompt =>:
=>display history <cr>

Help with commands can be requested at the System prompt by:
? commands <cr>

6   GRAPHIC STRUCTURES

Chemical structures are contained in the computer-readable file of the Registry System, i.e., the Registry File, in a digital form since, of course, it is only numbers which a computer can deal with.

It is not within the scope of this volume to discuss the systems for converting chemical structures into the digital forms in which they are entered into the System, stored, searched for, and reassembled back into graphics to be displayed on the screen.

We will only briefly describe the connection table that is the basis of storing and retrieving chemical structures, and some of the terms which we shall frequently use in the following discussion.

Graphic structures in the Registry File can be described as graphs, or structure graphs.

WHAT IS A STRUCTURE GRAPH

A structure graph is an arrangement of nodes and lines representing, respectively, the atoms or groups of atoms and bonds (or connectivities) forming a molecule.

Chemical substances are represented in the Registry File by structure graphs. Graphics capability of our hardware makes it possible to search for the substance structures by creating graphic structure queries and to retrieve and display answers to the queries in form of graphic chemical structures.

A substance may be represented in the Registry File by a single structure or by multiple structures, i.e., a substance may be composed of one component or more than one component.

An example of an one-component substance is the structure graph, as displayed by the System, shown in Figure 1/8.

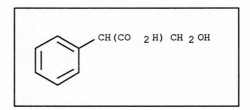

Figure 1/8. Displayed structure

Addition compounds, salts, solvates, mixtures, and copolymers are examples of substances with more than one component (in the *Chemical Abstracts Formula Indexes* they are known as the dot-disconnected formulas). Each component is represented by a structure

CAS REGISTRY

and each of the structures is assigned a Registry Number in addition to a Registry Number for the whole substance. The entire structure is searchable and displayable either as one unit or as the separate components. The example (Figure 1/9) is a two-component structure displayed as one unit:

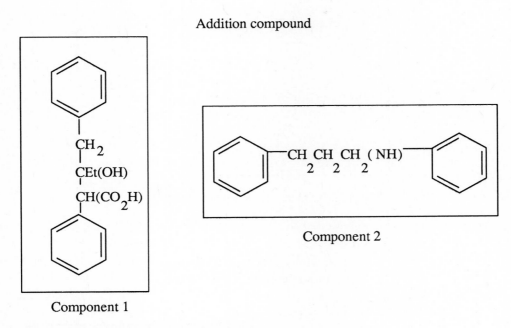

Figure 1/9. Two-component substance

Salts are represented as one-component structure if the cation is a single atom fragment or if it is a single atom with hydrogen connections. Thus, sodium, lithium, and other metal salts, or ammonium salts are displayed as one structure (Figure 1/10). However, alkylammonium salts are multicomponent substances (Figure 1/11).

Nodes in structure graphs may be represented by letters or, in case of implied carbon atom, by the junction of bonds.

WHAT IS A NODE

The term node has been introduced in chemistry in developing theory of graphs and nodal nomenclature system.

The term node is derived from the Latin "nodus" and has the meaning of a knot, bulge, or intersection of branches. It is applied to single skeletal units, i.e., atoms of a structure. It may also be applied to groups of atoms forming important structural units. It is the simplest unit of a graph representing either a single atom or a group of atoms. Values, i.e., elements can be assigned to the nodes.

Salt with single atom cation

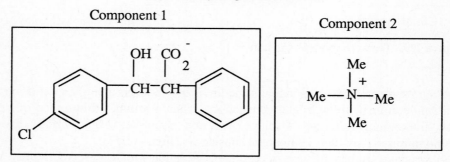

Figure 1/10. One-component structure

Salt with single atom fragment
with non-hydrogen attachments

Figure 1/11. Two-component structure

DEFAULT NODE VALUES IN GRAPHIC STRUCTURES

All nodes in created structure queries are automatically assigned by the System the value of the carbon atom unless otherwise specified by the User.

GRAPHIC STRUCTURES

HYDROGEN AS A VALUE FOR NODES

Although in the Registry System hydrogens are not represented as nodes, the Registry File allows for specifying hydrogens and any of the element symbols of the periodic table as node values in structures. To specify hydrogens in structures, we also use special methods which will be described later (see HCOUNT, p. 147, Shortcuts, p. 85, and Gk groups, p. 119).

BONDS IN GRAPHIC STRUCTURES

Bonds are the connectivities between single nodes or structural fragments, represented by solid, broken, or wavy lines in graphic structures. We will discuss the meaning of these lines and the bonds they represent in detail (see p. 90,142). Bond values, i.e., unspecified, single, double, triple, normalized, and bond types, i.e., ring or chain can be assigned to bonds.

DEFAULT BOND VALUE

When we create a structure graph, the System assigns automatically the value "unspecified" to all bonds in the structure unless otherwise directed by the User. The unspecified bonds are represented by wavy lines.

BOND TYPE

Bond type is determined by the nature of the structure having the bond.
Bonds in ring structures are ring bonds by implication, all other bonds are chain bonds unless otherwise specified.

STEREOCHEMISTRY

In the Registry System, all structures are two-dimensional graphs with nostereo bonds or stereocenters. Stereoisomers in the Registry File are differentiated by text descriptors only. It is possible to search for stereoisomers by using these alphanumeric stereodescriptors which are part of the records of the substances in the Registry File.

WHAT IS A CONNECTION TABLE

The connection table is a form of topological coding of chemical structures. It is a list of the nodes and bonds in structures and a description how the nodes and bonds are connected to each other and with those in their neighborhood.
In converting structures to connection tables, all nodes except hydrogens in a structure are assigned numbers and each bond in the structure is assigned a value (1,2,3) for single, double, and triple bond.
A simple example of a structure and its connection table is shown in Figure 1/12.

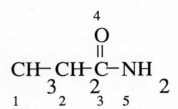

Atom Number	Atom Symbol	Bond Connection	Attached Atom Number	Bond Connection	Attached Atom	Bond Connection	Attached Atom
1	C	1	2				
2	C	1	1	1	3		
3	C	1	2	2	4	1	5
4	O	2	3				
5	N	1	3				

Figure 1/12. A simple connection table

Coding of structures and generation of connection tables have developed from this simple form to sophisticated systems, especially at Chemical Abstracts Service, to register chemical structures. Those readers interested in detailed reading on this subject will find the following references as the basic ones:

Leiter, D. P., Morgan, H. L., and Stobaugh, R. E.
Installation and operation of a registry for chemical compounds
J. Chem. Doc. 5, 238-242(1965)
Gluck, D. J.
A chemical structure storage and search system developed at Du Pont
J. Chem. Doc. 5, 43-51(1965)
Morgan, L. H.
The generation of a unique machine description for chemical structures-A technique developed at Chemical Abstracts Service
J. Chem. Doc. 5, 107-113(1965)

Many relevant references are found in:
Stobaugh, R. E., *et al.*
The Chemical Abstracts Service Chemical Registry System
J. Chem. Inf. Comput. Sci. 28, 180-7(1988)

The connection table for each structure registered in the Registry File can be displayed as shown in the example (Figure 1/13).

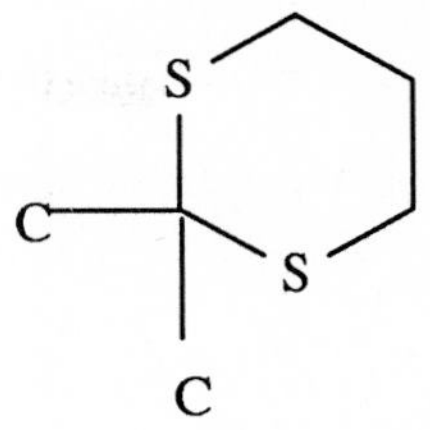

GRAPHIC STRUCTURES

```
                    ******CONNECTIONS******

     NOD    SYM      NOD/BON    NOD/BON    NOD/BON
      1      S        2  RSE     6  RSE
      2      C        3  RSE     1  RSE
      3      C        4  RSE     2  RSE
      4      C        5  RSE     3  RSE
      5      S        6  RSE     4  RSE
      6      C        8  CSE     7  CSE    1  RSE
                      5  RSE
      7      C        6  CSE
      8      C        6  CSE
```

Figure 1/13. Connection table of registered structure

DISPLAY OF GRAPHIC STRUCTURES

In interacting with the System, the dialog between the User and the System, textual data, graphics structure queries, and retrieved graphics structures are displayed on the screen of our terminal.

The display operates on the basis of a page. The screen of the terminal or microcomputer is the page. The default values for the screen/page are:
line length 74 (characters)
page length 33 (lines).
The default page values can be reset by the following commands at the => prompt:

Example: *Set page*

=>set page x (x=number of lines 20-250)
=>set line x (x=number of characters 8-99)

Each page displays a heading as follows:
24 MAY 89 15:08:33   STN INTERNATIONAL   P0001

The heading can be toggled on/off by the following command at the => prompt:

Example: *Turn page heading off/on*

=>set heading off
=>set heading on

When the page is filled or the System determines that more space is needed to continue the display, a prompt appears on the screen asking the User to provide a blank new page:

COPY AND CLEAR PAGE, PLEASE

At this point, we can print on our printer all that is displayed on the screen (dump the screen). Each of the Emulation Softwares has a dump screen routine. We can also save all or parts of our session on disk and print after we disconnect from the System. The manual of the software should be consulted to determine how to print the screen or save the session on disk. Of course, we do not have to print the screen; we can continue without printing or saving.

The response to COPY AND CLEAR PAGE also depends on the Emulation Software (consult the manual).

7 STRUCTURE BUILDING METHODS

Graphic chemical structures in the Registry File are built by following the usual pattern of drawing structures by hand. The chemist has an idea of the structure to be drawn. In composing structures, the chemist draws lines/bonds and indicates the nature of atoms in the structure by element symbols. The chemist creates chain and ring structural fragments, joins them to create larger systems, substitutes a fragment with another one to create substituted structures. To facilitate drawing, the chemist may use stencils and templates. In hand drawing, the chemist uses paper or a blackboard and a drawing instrument.

In building structures in the Registry File, we use the terminal or microcomputer screen as our paper or blackboard and a mouse or light pen as the drawing instruments. We create chains, rings, and by joining these, we create larger structure systems. We specify the nature of atoms, and the value of bonds in structures. We use templates to facilitate structure building.

The same techniques are applicable in the Beilstein File.
We can employ the following methods:
Method 1: Menu driven structure building
Method 2: Free-hand drawing of structures
Method 3: Keyboard commands for structure building
Method 4: Combination of 1, 2, and 3
Method 5: Build structure query offline in STN Express or ChemTalk Plus and upload the query in STN

The methods of structure building are shown in Figure 1/14.

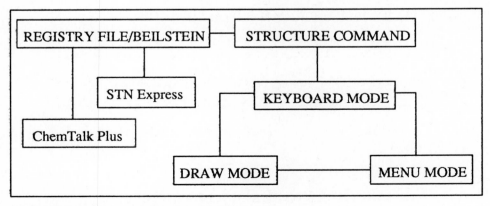

Figure 1/14. Structure building methods

We can build a structure graph entirely in the Menu mode, or in the Keyboard Command mode, we can partially build a graph in the Menu mode and switch to the Keyboard Command mode, and *vice versa*.

However, it is to be noted that structures created in STN Express and uploaded into Registry File (or Beilstein) cannot be modified in these files and structures created in Registry File (or Beilstein) cannot be modified in STN Express. The same applies for ChemTalk Plus.

STN Express is described in Section 11 (p. 184).
ChemTalk Plus is described in Part Five (p. 737).

HOW TO INITIATE STRUCTURE BUILDING

The structure building online is possible only in Registry File (or LREG, or Beilstein), therefore, we have to transfer to one of these files either from the HOME file or from the file we may presently be in.

We enter the Registry File by executing the FILE Command.

(Note that there is a special file for learning the Registry File at a reduced cost, i.e., LREG.)

Example: *Entering Registry File*

Command: FILE REGISTRY
Key:file reg (or lreg, or beilstein))

=>file reg
COST IN U. S. DOLLARS   SINCE FILE ENTRY   TOTAL SESSION
FULL ESTIMATED COST   0.50   0.50
FILE 'REGISTRY' ENTERED AT 17:50:45 ON 12 DEC 87
COPYRIGHT 1985 BY THE AMERICAN CHEMICAL SOCIETY
=>

Once we are in the Registry File, we may execute the STRUCTURE Command.

Example: *Initiating structure building*

Command: STRUCTURE

=>structure
ENTER NAME OF STRUCTURE TO BE RECALLED (NONE):none
(or . for the default)(see p. 109 for structure recall)
ENTER (DIS), GRA, NOD, BON, OR ?:

STRUCTURE BUILDING METHODS

The System responds with the ":" prompt for entering subcommands. Since we have used the NOVICE version of Commands, the ":" prompt is preceded by the line of subcommand options for structure building. The ":" prompt would stand alone if we used the EXPERT version of commands.

The ENTER (DIS), GRA, NOD, BON, OR ?: prompt will drive our structure building. We can proceed in entering the Keyboard Commands or we enter the Menu.

THE MENU

Menu driven building of graphic structures is based on a menu which is displayed on the screen and from which we select commands and specifications to instruct the System to create and display structure graphs.

Some of the commands and specifications needed to complete a structure graph are not available on the Menu and have to be entered on the keyboard.

HOW TO INITIATE THE MENU

The Menu can be invoked only within the STRUCTURE Command.

Example: *Initiating Menu*

If we have just connected to STN or, presently, are in another STN database, we transfer to the Registry File:
=>file reg

COST IN US DOLLARS	SINCE FILE ENTRY	TOTAL SESSION
FULL ESTIMATED COST	0.50	0.50

FILE REGISTRY ENTERED AT 14:25:33 ON 6 JAN 88
COPYRIGHT 1985 BY THE AMERICAN CHEMICAL SOCIETY
=>structure
ENTER NAME OF STRUCTURE TO BE RECALLED (NONE):none
ENTER (DIS), GRA, NOD, BON OR ?:menu (vt) (hp)

The option "vt" is entered if we use the VT640 terminal or the hardware that can emulate VT640 (see p. 12). The "hp" option is valid for any HP terminal with GIN mode and HP erase protocol. By specifying vt or hp, we activate certain features, i.e., highlighting of the selected items on the Menu, and automatic replace/erase of the items in structure graphs, which are not available on the other types of terminals.

The Menu is displayed on the screen (Figure 1/15). If a structure has been recalled (p. 109), it is displayed with the Menu.

```
┌─────────────────────────────────────────────────────┐
│   Area A                                            │
│─────────────────────────────────────────────────────│
│                              HELP    KEY            │
│                              SHIFT   END            │
│                        D.1   RECALL                 │
│                              REFRESH                │
│                              SHORTCUTS              │
│                             ─────────────────       │
│                              C1  C2  C3             │
│                              C4  C5  C6             │
│        Area B          D.2   R4  R5  R6             │
│                              R7  R8  GRAPH          │
│                             ─────────────────       │
│                              C   N   O              │
│                              Si  P   S              │
│                        D.3   F   Cl  Br             │
│                              X   M   Gk             │
│                              A   Q   NODE           │
│                             ─────────────────       │
│                              S   SE    N            │
│                        D.4   D   DE    T            │
│                              UNSPEC BOND            │
│                              R   C    RC            │
│                             ─────────────────       │
│                              CHARGE   MASS          │
│                        D.5   HCOUNT   NSPEC         │
│                              VALENCE  RSPEC         │
│─────────────────────────────────────────────────────│
│ R=Refresh S=Select K=Key F=From T=To M=Move D=Delete│
│                    Area C                           │
└─────────────────────────────────────────────────────┘
```

Figure 1/15. The Menu

DESCRIPTION OF THE MENU

The Menu is divided into areas. Each of the areas has a distinct function in the creation of structure graphs. The areas and their functions are described below.

Area A - Text Display and Entry

This section of the Menu displays System prompts, error, and other messages from the System. It also serves as the input line for the User when text entries are required.

Area B - Structure Graph Display

STRUCTURE BUILDING METHODS

The structure graphs are displayed on this section of the Menu and manipulated by the Menu templates and Keyboard Commands.

The Areas C-D contain commands and items called templates.

Area C - Single Key Commands

Depressing the single letter key *without carriage return* invokes the following functions:

R (Refresh)....redisplays the Menu and redraws (cleans) displayed structure graphs
S (Select).....activates the Menu options in Area D
K (Key)........activates Text Area A for input
F (From).......start drawing a bond at cursor position 1
T (To).........end the bond at cursor position 2

Area D.1

HELP....invokes the Help System
KEY.....activates Area A for input
SHIFT...moves structure graphs about Area B
END.....ends STRUCTURE Command, assigns Ln to graphs, exits the Menu to => prompt
RECALL......recalls structure graphs from System or User File
REFRESH.....redisplays the Menu and redraws (cleans) graphs
SHORTCUTS...displays the System Shortcuts

Area D.2

C1......creates one separate node fragment or one-node chain substituent
C2-C6...creates 2- to 6-membered chain
R4-R8...creates 4- to 8-membered ring
GRAPH...invokes the Command Graph in Area A

Area D.3

C,N,O...specification for carbon, nitrogen, oxygen value at nodes
Si,P,S..specification for silicon, phosphorus, sulfur value at nodes
F,Cl,Br.specification for fluorine, chlorine, bromine values at nodes

X,M,Gk..specification for any halogen value (X), any metal value (M), and user variable group (G) at nodes
A,Q.....specification for any value at nodes, except hydrogen (A), or hydrogen and carbon (Q)
NODE....invokes the command NODE in Area A.

Area D.4

S,SE,N..specification for single (S), single exact (SE), and normalized (N) bonds
D,DE,T..specification for double (D), double exact (DE), and triple (T) bond values
UNSPEC..specification for unspecified bond value
BOND....invokes the Command Bond in Area A
R,C,RC..specification for ring (R), chain (C), ring or chain bond type

Area D.5

CHARGE,MASS.....invokes the Attribute Commands in Area A
HCOUNT,NSPEC....invokes the Attribute Commands in Area A
VALENCE,RSPEC...invokes the Attribute Commands in Area A

CREATING STRUCTURE GRAPHS ON THE MENU

Items on the Menu are selected by moving the mouse cursor or the crosshair cursor that is displayed with the Menu (these alternatives depend on the kind of the Emulation Software in use).
The following terms will be used to describe the technique of using the Menu:

Point will mean to move the cursor and place it over the Menu item that we wish to use.

Select will mean to depress the key S on the keyboard without carriage return.
(Note that some of the Emulation Softwares allow for programming one of the mouse buttons to function as the letter S.)

Place will mean to move the cursor to a position in Area B.

Depress will mean to depress the key of one of the letters in Area C *without carriage return*.

Type will mean to type entries in Area A *followed by carriage return*.

STRUCTURE BUILDING METHODS

THE BASICS OF THE MENU TECHNIQUE

1. *Point* to a Menu item.
2. *Select*. The item will be highlighted on the HP and VT type of terminals but not on other type of terminals.
3. *Place* the selected item in Area B. When we place/select on a blank position in Area B, a structure graph fragment is created. When there is already a graph, a separate fragment is created when we place outside of the existing graph. When we place on a node or bond of the existing structure graph, we build on that single graph
4. *Depress* R if needed to redraw the structure graph (on other than HP or VT terminals).

KEYBOARD COMMANDS

Building of structure graphs by Keyboard Commands is accomplished by typing subcommands and specifications at the prompt line:
ENTER (DIS), GRA, NOD, BON, OR ?:
followed by carriage return.

FREE-HAND DRAW

Drawing of structure skeletons can be carried out both in Menu mode and Keyboard Commands mode of structure building. The graph is completed in the Menu or Keyboard Command mode or both.

HOW TO SWITCH MODES

To switch from Keyboard Commands to Menu:

Example: Switching from Keyboard Commands to Menu

ENTER (DIS), GRA, NOD, BON, OR ?:menu (vt)(hp)
The Menu will be displayed.

If we are in the Menu and wish to execute a Keyboard Command without leaving the Menu: (Figure 1/16).
To switch from Menu to Keyboard Commands mode: (Figure 1/17).
The system will prompt:
COPY AND CLEAR PAGE, PLEASE
The response depends on the Emulation Software. We set the screen to text, i.e., alpha

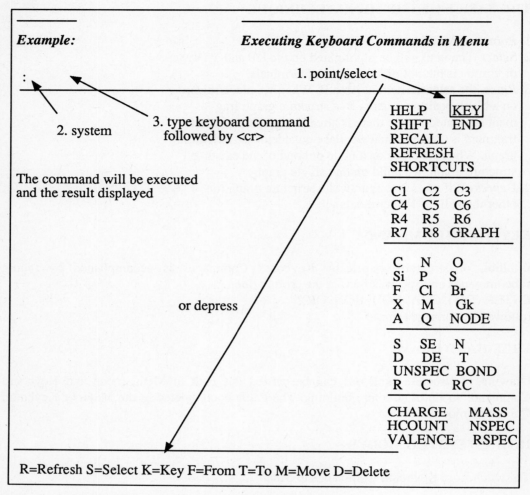

Figure 1/16. Switching to Keyboard mode in the Menu

mode.
ENTER (DIS), GRA, NOD, BON, OR ?:
At this prompt, we can proceed in entering Keyboard Commands.
To switch to DRAW mode, see p. 80.

HOW TO END STRUCTURE BUILDING

The structure building process is terminated by exit from the STRUCTURE Command. This can take place under the following circumstances:

1. We are satisfied that the structure graph meets our goals and wish to transfer into the

STRUCTURE BUILDING METHODS

Example:	Switching from Menu to Keyboard Mode
	HELP KEY SHIFT END RECALL REFRESH SHORTCUTS
	C1 C2 C3 C4 C5 C6 R4 R5 R6 R7 R8 GRAPH
	C N O Si P S F Cl Br X M Gk A Q NODE
1. depress E	S SE N D DE T UNSPEC BOND R C RC
	CHARGE MASS HCOUNT NSPEC VALENCE RSPEC
R=Refresh S=Select K=Key F=From T=To M=Move D=Delete	

Figure 1/17. Switching from Menu to Keyboard Mode

SEARCH Subsystem.
2. We want to terminate the STRUCTURE Command session and save the graph for future use.
3. We wish to abandon the structure graph and start building a new one, or we want to leave structure building process.
The STRUCTURE Command is terminated by the subcommand END (Figure 1/18). The system issues the => prompt at which we are ready to enter the commands that are valid at this prompt. We can, of course, re-enter the STRUCTURE Command at this prompt since we are still in the Registry File.
The Menu disappears from the screen and we receive the message:
L1 STRUCTURE CREATED
=>

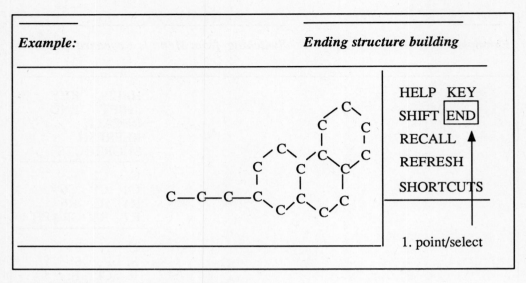

Figure 1/18. Ending structure building

Alternatively, depress K and type "end" in Area A after the System prompt ":".

In Keyboard Mode:
ENTER (DIS), GRA, NOD, BON, OR ?:end
L1 STRUCTURE CREATED
=>

The System assigns the created structure an Ln which is used to refer to the structure in our interaction with the System.
All structure graphs and the Ln are automatically erased when we disconnect from STN, unless we save them.
When no structure has been created the System responds to the END Command:
NO STRUCTURE CREATED
=>

8   CREATING STRUCTURES

In the following discussion of creating structure graphs, we will emphasize the Menu mode. This method offers the User an easy and convenient technique of selecting commands and specifications from the Menu without memorizing all the details of the Keyboard Commands. However, since the Keyboard Commands offer a very fast method of structure building, we will include also those in the examples of creating structures. Those readers who will gain detail knowledge of the Keyboard Commands may find that it is faster and easier to build structures in this mode than in the Menu mode.

The Keyboard Commands are shown under the banner **KEYBOARD**.

A description is given of the command syntax followed by the "Key" description how the commands can be entered, in an abbreviated version, in Keyboard mode at the System prompt:
ENTER (DIS), GRA, NOD, BON, OR?:
or, in the Menu mode, after the System prompt ":" in Area A.
Examples of creating structure graphs are given in the Menu mode followed by the appropriate Keyboard Command and specification.
Structure building in STN Express will be described in Section 11.

STRUCTURE SKELETONS

Structure skeletons, i.e., chains, single rings, fused rings, bridged rings, and spiro rings are created by selecting from the Area D.2 of the Menu (see p. 36).
All nodes will have the default value of carbon and all bond values will be unspecified unless otherwise designated by the User.
For the purpose of creating structures in the Examples with an appearance of being drawn by hand on paper, we have used the command SET BOND SE which automatically assigns to all graphs single bonds.
The structure skeletons are always automatically displayed in the Menu mode. However, the node numbers that we frequently need to manipulate the graph are not displayed in the Menu mode unless we request the display by SET NUM ON (Figure 1/19).
If we use in Figure 1/19 SET NUM OFF, the node numbers will be turned off.
In the Keyboard mode, the structure graph is displayed only on request by the command DISPLAY (Figure 1/20). The node numbers are always on, however, they can be turned off by DISPLAY NUM OFF for one display. The next display will be again with node numbers.

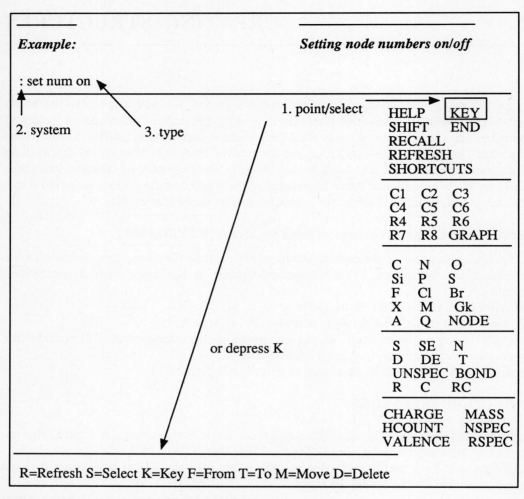

Figure 1/19. Set up of node numbers on/off

CHAIN STRUCTURES

To create straight and/or branched chains, we point/select the C1-C6 templates in Area D.2 and place/select them on the Area B (Figures 1/21 - 1/25).
The substitution in Figure 1/23 can be on either terminal node, the node 1 or 6, to achieve the same extension.

HOW TO CREATE SEPARATE STRUCTURAL FRAGMENTS

Separate structural fragments can be chains, rings, or ring fused systems, a single node (e.g., a metal in acid salts). The consequence of having two separate fragments created in one STRUCTURE Command is discussed in the section "Structure Searching" (p. 207).

CREATING STRUCTURES

```
KEYBOARD  DISPLAY (NUM OFF)
          Key:dis (num off)

ENTER (DIS), GRA, NOD, BON, OR ?: dis
```

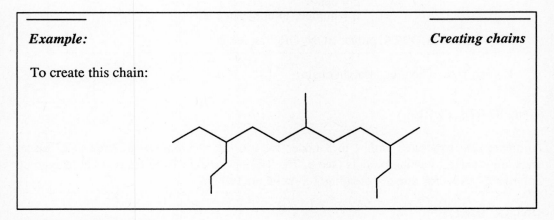

Figure 1/20. Displaying graph with node numbers

Example: *Creating chains*

To create this chain:

Figure 1/21. Branched chain

Here, we learn how to create them. By this technique, we can create a variety of graphs, e.g., branched chains, substituted rings, etc., by joining separate fragments at the appropriate locants (Figures 1/26, 1/27).

HOW TO BOND SEPARATE FRAGMENTS

To join two structural fragments, we form a bond between the desired nodes (Figure 1/27).

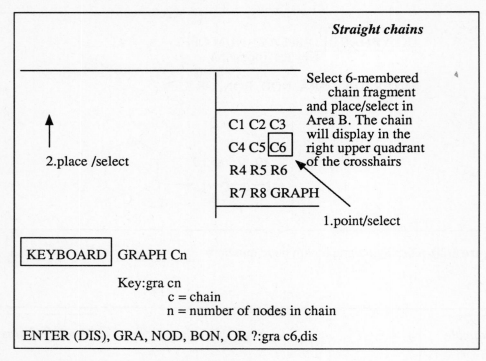

Figure 1/22. Creating straight chain

RING STRUCTURES

To create ring structures with 4 to 8 nodes we use the R templates in Area D.2. Larger rings are created by using models (see p. 76, 108) or are built from chains. The Keyboard Commands allow for directly creating 3-8 membered rings.

SINGLE RINGS

Single rings are created by selecting the R4-R8 templates from the menu and placing them in the Area B (Figure 1/28). Substitution of rings is simply carried out by selecting the chain templates C1-C6 and placing them on the node in the ring where we wish the substitution (Figure 1/29). Rings can be also substituted by another ring by creating two separate ring fragments and forming a bond between them (Figures 1/30, 1/31).
Note: on searching single ring query in Substructure Search (see p. 218) fusion at the ring is allowed, thus, we retrieve both single and fused ring systems unless we use the Structure Attribute Ring Specific (see p. 147) to prevent fusion.
We can also join two separate rings with a chain (Figure 1/32).

CREATING STRUCTURES

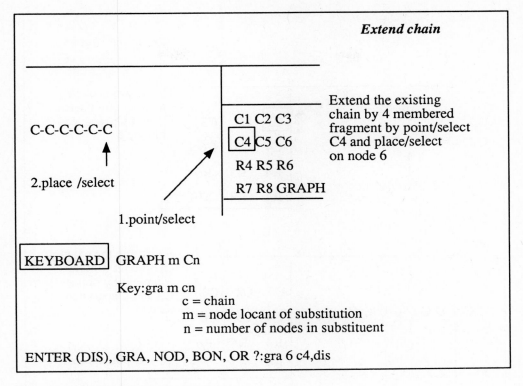

Figure 1/23. Extending straight chain

RINGS WITH >8 NODES

Rings with more than 8 nodes are created from chains of the appropriate length by forming a bond between chain nodes. A bond between the terminal nodes produces an unsubstituted ring having the number of nodes in the chain (Figures 1/33, 1/34). A bond formed at other than terminal nodes of the chain produces a substituted ring. We can also used the Fragment File and the System models (see p. 76, 108) for larger rings.

If we did not move the nodes (Figure 1/33) and form the 1-10 bond, the bond would be underneath the chain and the shape of the graph would not change to ring. The graph could still be submitted to SEARCH since the shape of structure graphs does not affect recognition of the created bonds by the System.

To shape the rings use MOVE Command (see p. 180).

FUSED RING SYSTEMS

Fused ring systems with 4- to 8-membered rings are created by selecting the R templates

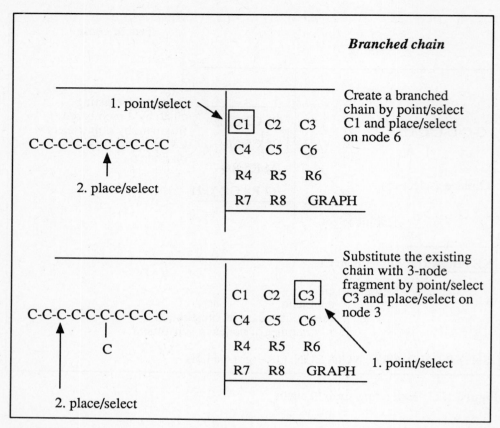

Figure 1/24. Creating branched chain

from the menu and placing them on the bond of the existing ring where we wish the fusion to take place (Figures 1/35, 1/36).

Some bonds in fused rings cannot be used for fusion of another rings. If the nodes at either end of a bond have three or more ring bonds to other ring nodes, the bond cannot be used for place/select for another ring to be fused. For example, in the structure shown in Figure 1/37 the bonds marked with the arrows connect nodes that have three ring bonds. These fused system have to be created by using the From and the To technique (Figures 1/38, 1/39).

FUSED RINGS WITH >8 NODES

These systems are created by substituting a ring with a chain and forming a bond between the ring and the chain. If all the rings in the system are larger than 8 nodes, we create chains of the appropriate length and form bonds to create rings (Figure 1/40). We can also free-hand draw the graph (see p. 80) or use the System models (see p. 76, 108).

CREATING STRUCTURES

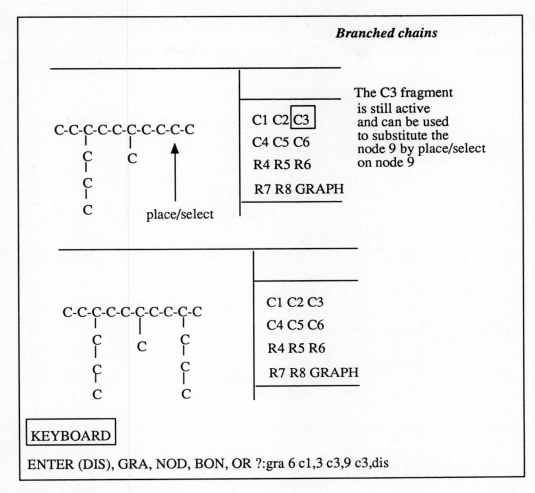

Figure 1/25. Branching chains

KEYBOARD

The Keyboard Commands technique for creating fused ring systems allows for very convenient and fast creation of these systems once we master the details of these commands.
Fused rings are categorized in the Registry File into two classes because the technique of creating them differs. One class of fused rings is systems composed of two fused rings and the other class is systems with three or more fused rings.

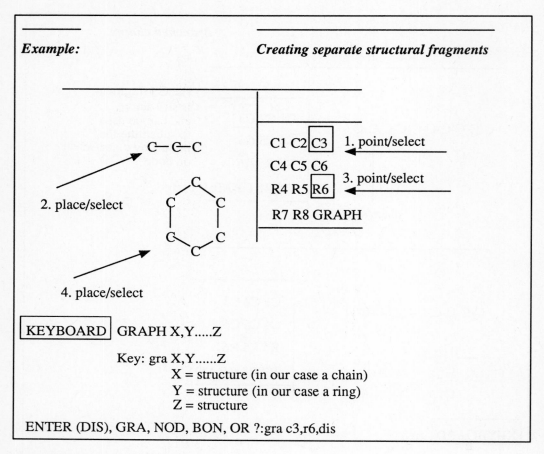

Figure 1/26. Creating separate fragments

FUSED SYSTEMS WITH TWO RINGS

These systems are further divided into systems with 3-8 membered fused rings and systems with larger than 8-node rings since, again, the technique of creating them is different.

FUSED SYSTEMS WITH TWO 3-8 MEMBERED RINGS

When two rings are fused, the fusion involves the sides of the rings. For fused systems with two rings, the side of fusion does not affect the outcome of the fusion, i.e., we always arrive at the same structure. For example, the fusion of two benzene rings can take place at any side of the rings to obtain equivalent structures (Figure 1/41).
To create these fused systems, we use the Keyboard Graph Command and indicate the size of rings (3-8) to be fused.

CREATING STRUCTURES

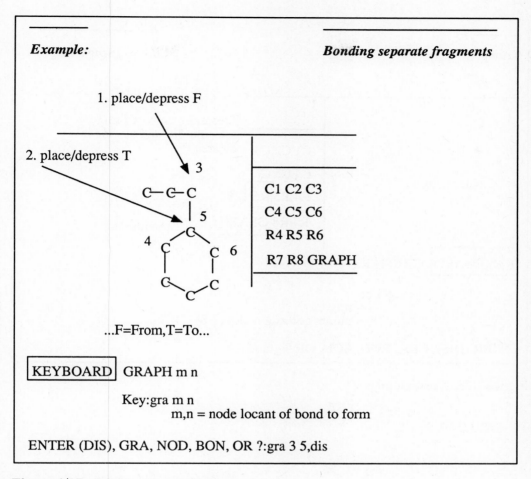

Figure 1/27. Joining separate fragments

Example: **Fusing two 3-8 membered rings**

KEYBOARD   GRAPH Rmn

    Key:gra rmn
        r=ring
        m,n=3-8
ENTER (DIS),GRA,NOD,BON,OR ?:

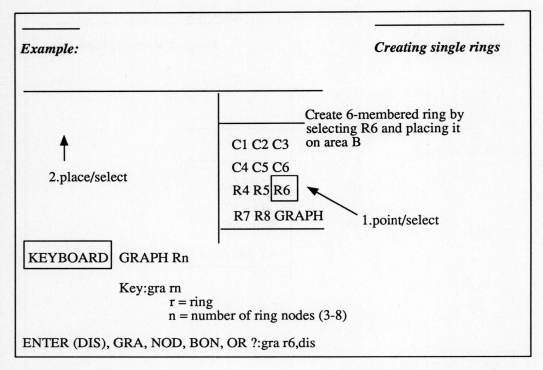

Figure 1/28. Creating rings

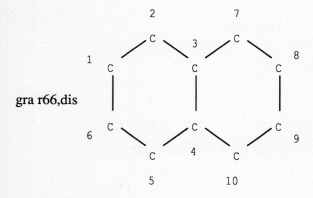

When the rings are of different sizes, we can specify on which side of the system we want to have the rings. The ring cited first will be on the left side in the structure.

ENTER (DIS), GRA, NOD, BON, OR ?

CREATING STRUCTURES

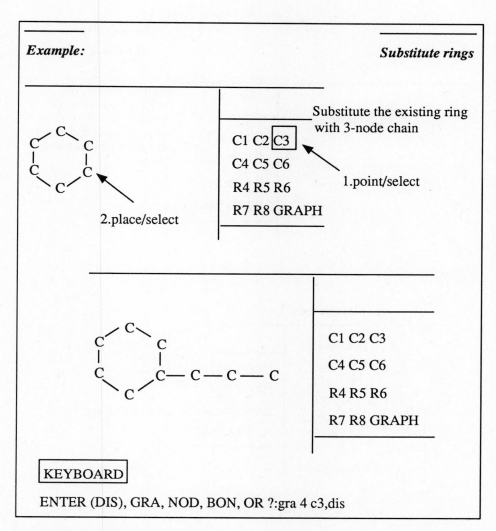

Figure 1/29. Substituting rings

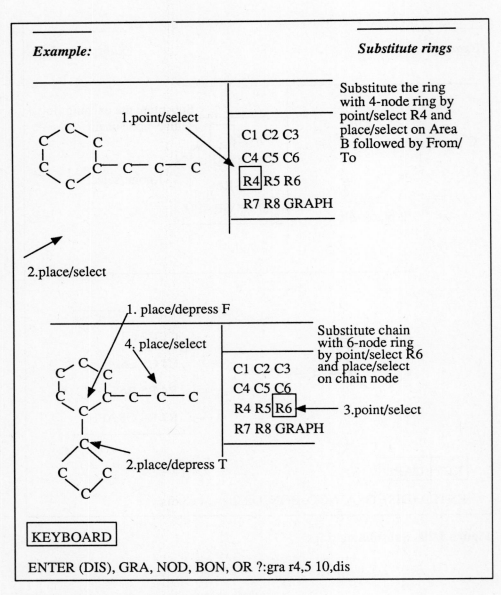

Figure 1/30. Substituting rings with rings

CREATING STRUCTURES

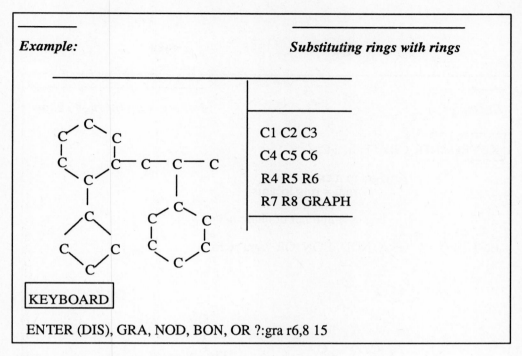

Figure 1/31. Substituting rings with rings

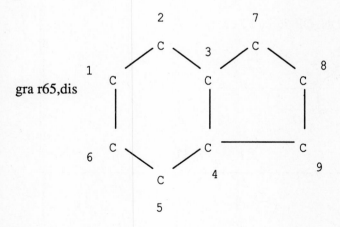

ENTER (DIS), GRA, NOD, BON, OR ?

Example: *Joining rings through chains*

KEYBOARD   GRAPH m n Cx

  Key:gra m n cx
   m,n = ring locants
   c = chain
   x = number of chain nodes

ENTER (DIS), GRA, NOD, BON, OR ?:gra r6,r6,dis

```
       2                         8
    1 C  3                    7 C  9
    C   C                     C    C
    |   |                     |    |
    C   C                     C    C
    6 C  4                   12 C  10
       C                         C
       5                        11
```

ENTER (DIS), GRA, NOD, BON, OR ?:gra 3 7 c2,dis

```
       2    13   14          8
    1 C  3            7 C  9
    C   C — C — C — C    C
    |   |               |    |
    C   C               C    C
    6 C  4             12 C  10
       C                   C
       5                  11
```

Figure 1/32. Joining rings

CREATING STRUCTURES

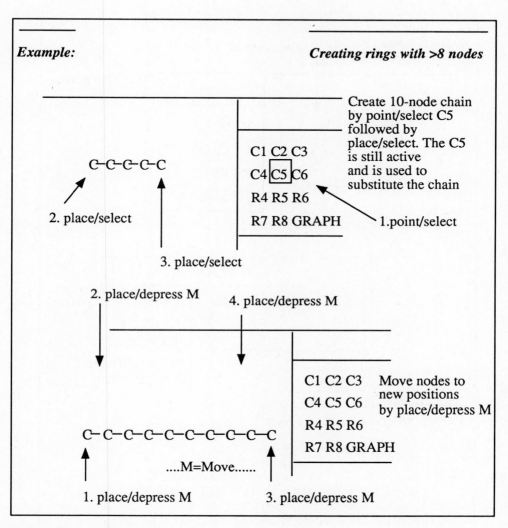

Figure 1/33. Creating large rings

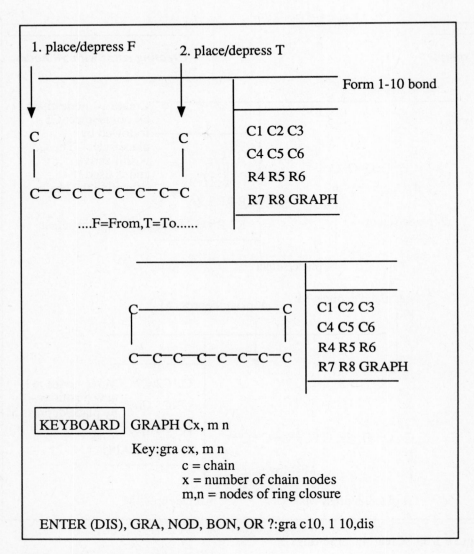

Figure 1/34. Creating large rings

CREATING STRUCTURES

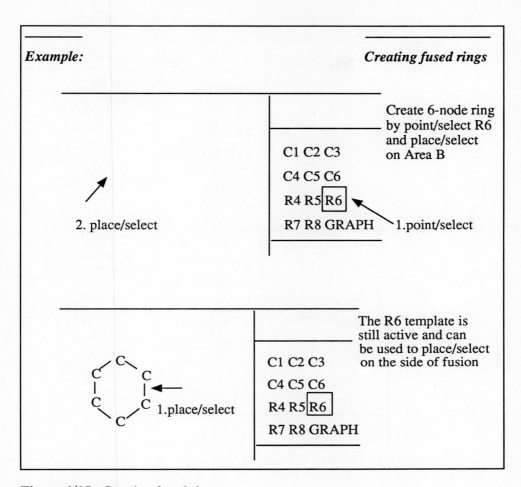

Figure 1/35. Creating fused rings

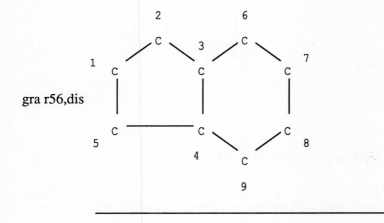

gra r56,dis

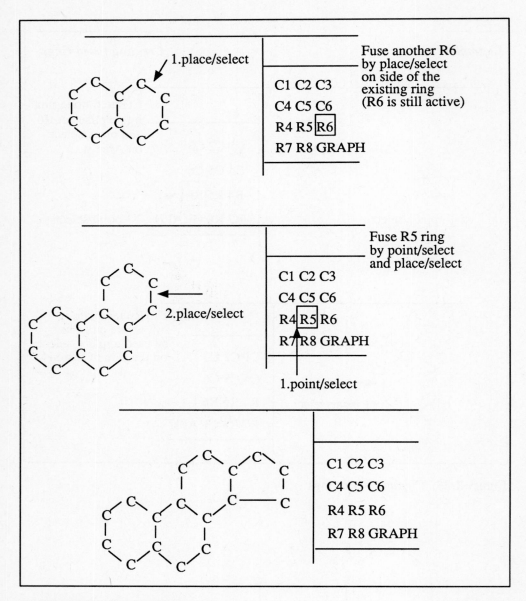

Figure 1/36. Creating fused rings

FUSED SYSTEMS WITH TWO >8-MEMBERED RINGS

Fused ring systems that include rings with more than 8 nodes may be created in various ways, e.g., by substituting the base ring with a chain and forming a bond to arrive at the

CREATING STRUCTURES

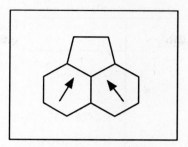

Figure 1/37

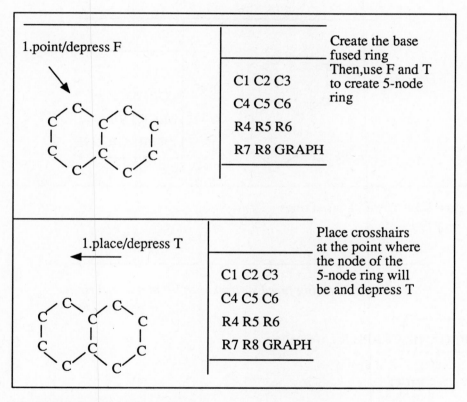

Figure 1/38. Creating fused rings

fused ring. The technique is similar to creating single large rings as discussed on p. 47.

1. Create the base ring of the system.
2. Attach a chain of the appropriate length.
3. Form a bond between the terminal chain node and the ring.

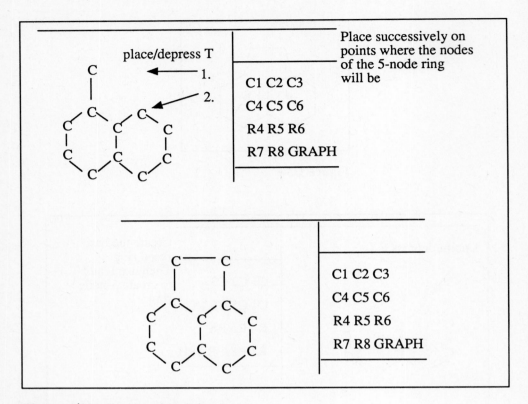

Figure 1/39. Creating fused rings

Example:Creating fused systems with >8-membered rings

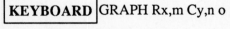 GRAPH Rx,m Cy,n o

    Key:gra rx,m cy,n o
        r=ring
        c=chain
        x,y=number of nodes
        m=node locant of substitution
        n,o=node locant of bond formation

ENTER (DIS), GRA, NOD, BON, OR ?gra r6,6 c10,1 16,dis

CREATING STRUCTURES

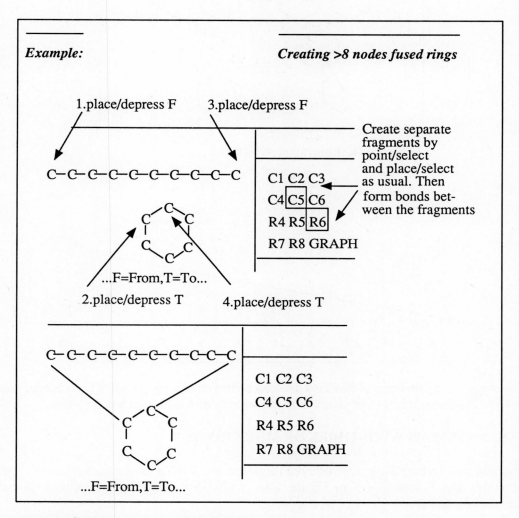

Figure 1/40. Creating large fused rings

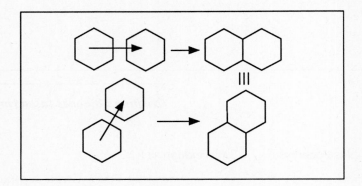

Figure 1/41. Equivalent structures

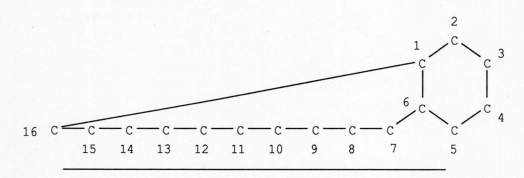

The shape of the structure is distorted and can be rearranged by the MOVE command (see p. 180). However, it is not necessary to shape the structure for use as a query.

FUSED SYSTEMS WITH THREE OR MORE RINGS

For the fused systems with three or more rings, the direction of fusion does affect the final structure. The fusion shown in Figure 1/42 results in two different structures of two distinct substances .
In the Registry File, the default (normal) direction of fusion is at the side directly across the side of previous fusion (Figure 1/43).

FUSED SYSTEMS WITH 3-8 MEMBERED RINGS

When we want to create fused ring systems and the fusion is going to be in the default (normal) direction, no indication of the direction of fusion is required in the GRAPH command.

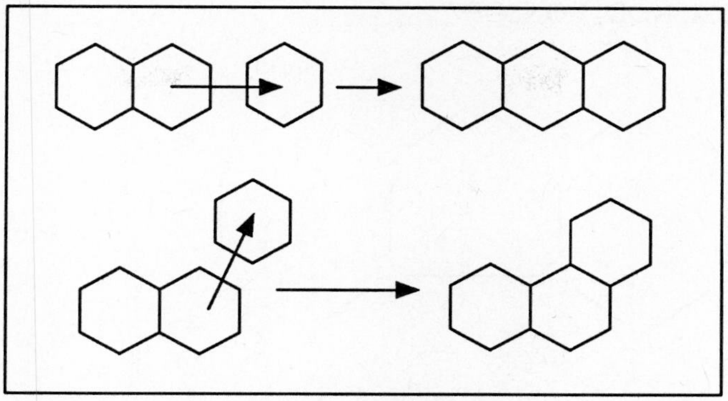

CREATING STRUCTURES

Figure 1/42. Nonequivalent structures

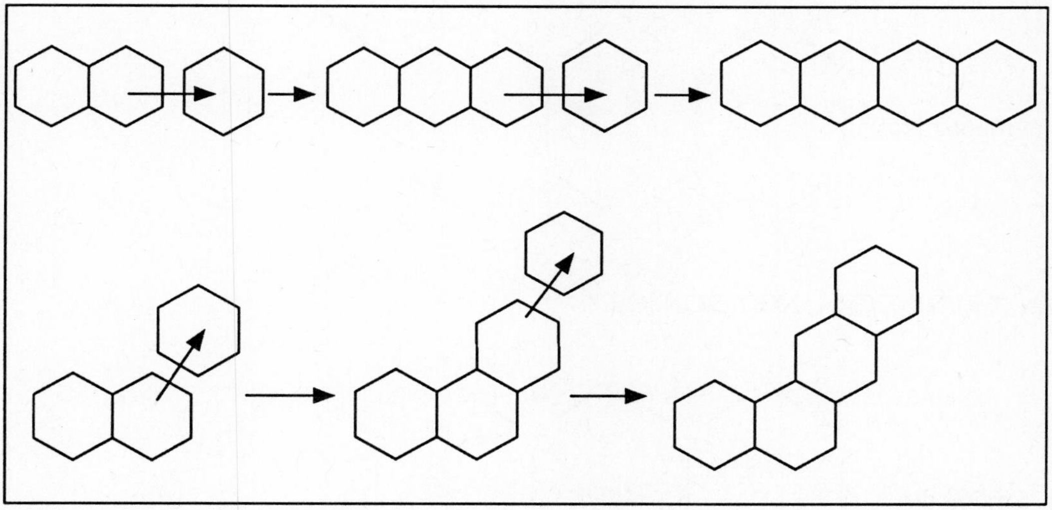

Figure 1/43. Normal direction of fusion

Example: Fusing three or more 3-8 membered rings

KEYBOARD GRAPH Rmno....z

Key:gra rmno.....z
r=ring
m,n,o,z=3-8

ENTER (DIS), GRA, NOD, BON, OR ?:

gra r656,dis

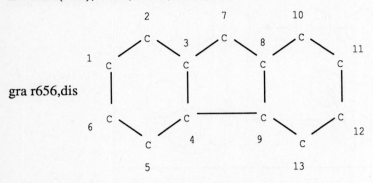

ENTER (DIS), GRA, NOD, BON, OR ?:

gra r6666,dis

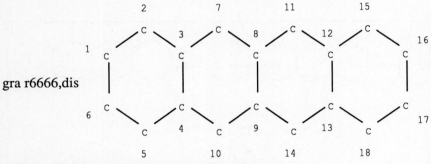

ENTER (DIS), GRA, NOD, BON, OR ?:

gra r6565,dis

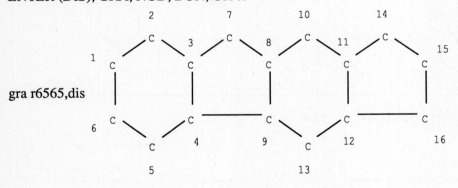

However, when a fused ring system is desired in which the fusion is in other than the default direction, we have to indicate whether the fusion is UP or DOWN from the previous side of fusion (Figure 1/44).

CREATING STRUCTURES

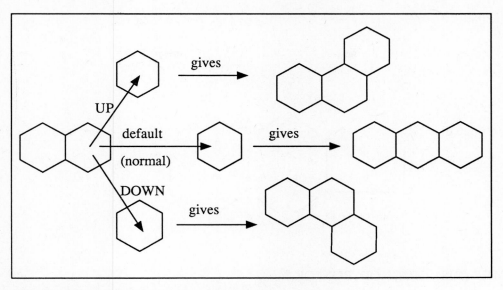

Figure 1/44. Specified direction of fusion

Example: *Fusing in nondefault direction*

KEYBOARD   GRAPH RmU(D)nU(D)....zU(D)

    Key: gra rmu(d)nu(d)...zu(d)
        r=ring
        m,n,z=3-8
        u=up
        d=down

ENTER (DIS), GRA, NOD, BON, OR ?:

gra r566u6,dis

ENTER (DIS), GRA, NOD, BON, OR ?:

gra r66u65,dis

ENTER (DIS), GRA, NOD, BON, OR ?:

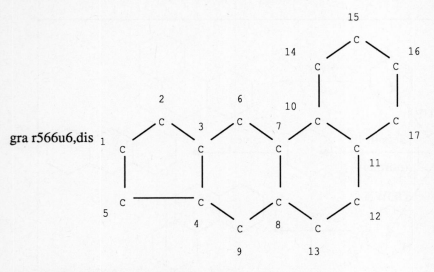

CREATING STRUCTURES

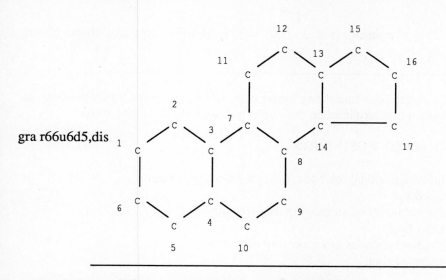

gra r66u6d5,dis

THREE OR MORE >8-MEMBERED FUSED RINGS

Similarly to the methods to create large single rings, or to fuse rings with more than 8 nodes, we follow the steps:
1. Attach a chain to a base ring.
2. Form a bond between the chain terminal node and the ring.

Example: Fusing 3 or more >8-membered rings

KEYBOARD GRAPH Rxy,m Cz,n o

    Key: gra rxy,m cz,n o
        r=ring
        x,y,z=number of nodes
        m=ring substitution locant
        n,o=locants of bond formed

ENTER (DIS), GRA, NOD, BON, OR ?:gra r66,8 c10,9 20dis

(Note that the MOVE Command (see p. 180) was used to achieve the shape of the displayed structure.)

The above methods create new fused ring systems in a one-step procedure. However, at times, we have already a ring structure on the screen and wish to fuse a ring to it.

MORE TECHNIQUES ON FUSING RINGS

The Registry File offers additional methods to create fused ring systems.
Fusion can be achieved by:
1. Incorporating adjacent nodes of an existing ring into the fused 3-8 membered ring.
2. Incorporating nonadjacent nodes of an existing ring into the fused 3-8 membered ring.

FUSION ONTO ADJACENT NODES OF A RING

When we have a ring fragment already formed, we can fuse a ring onto it so that two of the adjacent nodes of the existing ring become part of the new, fused ring.

Example: Fusing onto adjacent nodes of a ring

KEYBOARD GRAPH m n Rj

> Key: gra m n rj
> r=ring
> m,n=adjacent nodes locants
> j=3-8

Existing ring

ENTER (DIS), GRA, NOD, BON, OR ?:

CREATING STRUCTURES

gra 9 10 r5,dis

The same system, of course, can be created by fusing the 6-membered ring:

Existing ring

ENTER (DIS), GRA, NOD, BON, OR ?:

gra 7 8 r6,dis

FUSION ONTO NONADJACENT NODES OF A RING

When we have an existing ring and need to fuse a ring to it in such a way that two nonadjacent nodes of the existing ring become part of the fused ring, the fused ring will be formed through the shortest possible path along the nonadjacent nodes of the existing ring.

Example: *Fusing onto nonadjacent nodes*

KEYBOARD GRAPH m n Rj

Key: gra m n rj
r = ring
m,n = nonadjacent node locants
j = 3-8

Existing ring

ENTER (DIS), GRA, NOD, BON, OR ?:

gra 5 14 r5,dis

CREATING STRUCTURES

BRIDGED RINGS

Bridged ring systems are built by creating the base ring, substituting the base ring with a chain with the number of nodes to make the bridges and forming a bond between the terminal chain node and a ring node (Figures 1/45, 1/46).

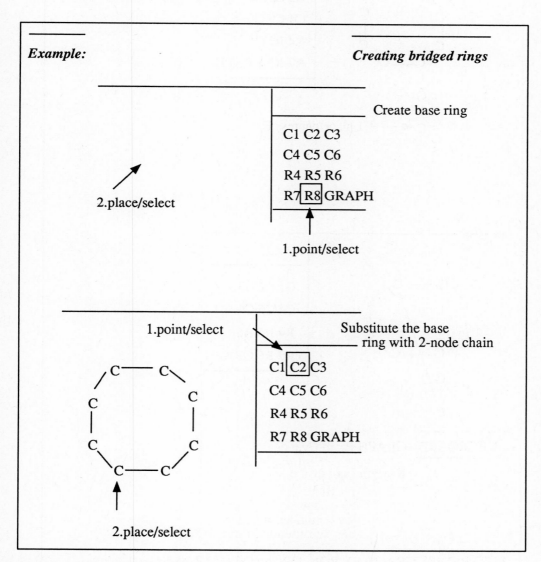

Figure 1/45. Creating bridged rings

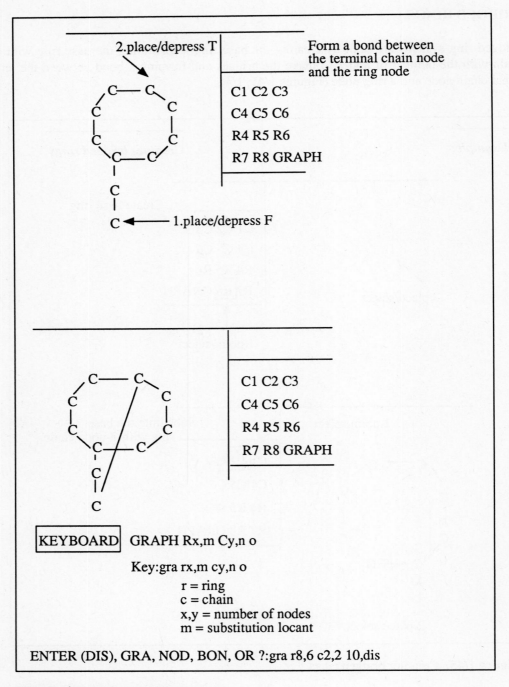

Figure 1/46. Creating bridged rings

CREATING STRUCTURES

To reshape the ring graph in Figure 1/46, use MOVE Command (see p. 180).

SPIRO RING SYSTEMS

These systems are created by selecting an R template, placing it in the Area B and placing the other ring, be it the same size ring or another size selected from the R template menu, on the node to be shared by the rings (Figures 1/47, 1/48).

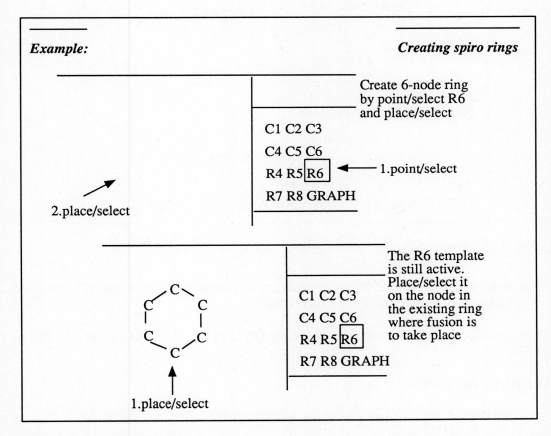

Figure 1/47. Creating spiro rings

HOW TO USE THE MENU GRAPH COMMAND

The Menu contains the STRUCTURE subcommand GRAPH in Area D.2 which can be activated to create any structure skeleton whose template is not available on the Menu.
We point/select GRAPH. The System responds with a prompt GRAPH: in Area A. The cursor disappears from the Menu and we type in Area A, after the prompt, any of the GRAPH specifications for structure skeletons, e.g., Cx, Rx, followed by carriage return.

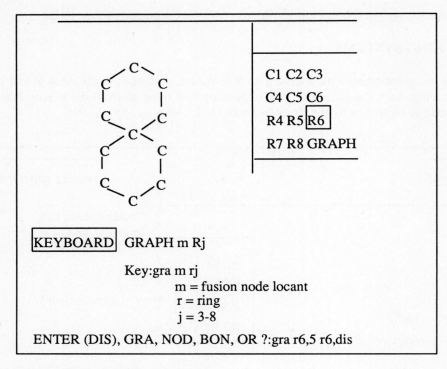

Figure 1/48. Creating spiro rings

The cursor reappears on the Menu. We place/select on Area B. The function is executed and displayed by the System (Figure 1/49, 1/50).

Alternatively, we can execute the Keyboard GRAPH Command in the Menu (Figure 1/51).

RING TEMPLATES

The building of ring structure graphs can be facilitated by using models which are stored in the Fragment File of the Registry System.
The Fragment File contains 38 predrawn single 5-12 membered rings, fused ring systems, full structures or structure skeletons of some widely used organic compounds, coordination compound fragments, and boron cages.

HOW TO USE RING TEMPLATES

Except where specifically shown in the fragment, all nodes are carbons, and all bonds have the value shown in the fragment by wavy, solid, or broken lines (see the section "Bonds", p. 90 for explanation of bond values and types). Some of the fragments show

CREATING STRUCTURES

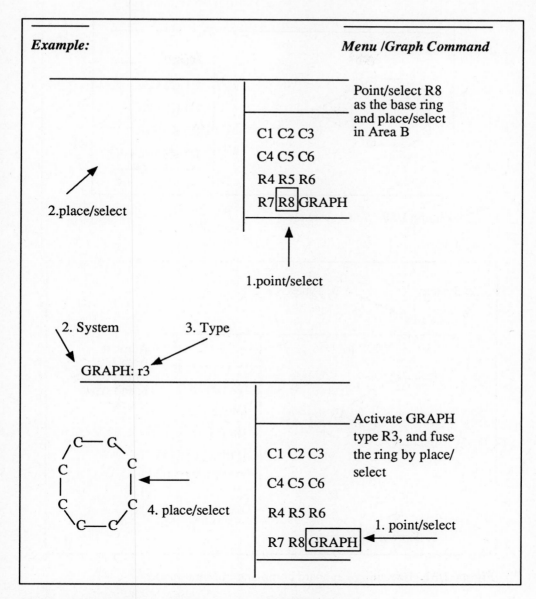

Figure 1/49. Using the GRAPH Command in Menu mode

hydrogens attached to nodes. If this kind of fragment is intended for SEARCH and a substitution is desired at the nodes showing hydrogen atoms, the hydrogen has to be deleted from the fragment (see section "Correcting/Modifying Structures", p. 168).
Each model in the Fragment File is assigned a numeric or a text code. These codes are used to recall the fragments from the Fragment File during the process of creating

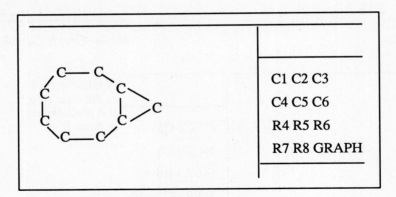

Figure 1/50. Spiro system

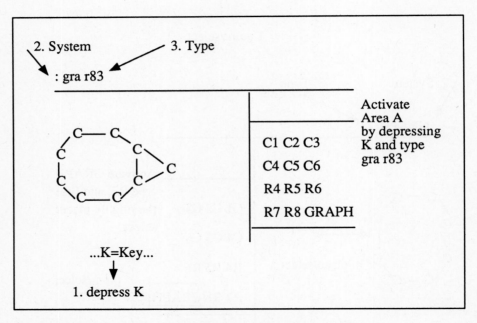

Figure 1/51. Executing keyboard command in Menu mode

structure graphs.
The codes can be displayed on the screen at the system prompt =>.

Example: *Displaying fragment codes*

Command:HELP STRUCTURE FRAGMENTS
=>help structure fragments

CREATING STRUCTURES

THE FRAGMENT FILE CONTAINS 38 RING SYSTEMS WHICH ARE COMMONLY USED IN STRUCTURE-BUILDING. TO USE ONE OF THESE AS A MODEL IN THE STRUCTURE COMMAND, ENTER ITS STRUCTURE CODE.
USE THE SAME METHOD (CALLING IT UP AS A MODEL) TO SEE THE STRUCTURE FOR A GIVEN CODE.

Structure Code	Gives basic (unsubstituted) structure of
5,6..12	5,6....12-membered ring
ACENAP	acenaphthene
ADAMAN	adamantane
ANTHRA	anthracene
BILINE	biline
CAROTE	carotene
CEPHAL	cephalosporins
FLUORN	fluorene
IBOGAM	ibogamines
INDENE	indene
MORPHN	morphine
NAPHTH	naphthalene
NORBRN	norbornane
PENICL	penicillins
PHENAN	phenanthrene
PHORBN	phorbine
PORPHN	porphines
PORPHY	porphyrazine
PROSTA	prostaglandins
PURINE	purine
SEQTER	sesquiterpenes
STEROD	steroids

Structure Codes for coordination centers:
COORD5,COORD6,COORD7,CO6LG3,CO6LG4
boron cages:
BO41,BO51,BO61,BO71

Example: *Selecting ring models*

If we are entering STRUCTURE Command:

=>structure
ENTER NAME OF STRUCTURE TO BE RECALLED (NONE):sterod

ENTER (DIS), GRA, NOD, BON, OR ?:menu vt
The Menu with the structure will be displayed (Figure 1/52).

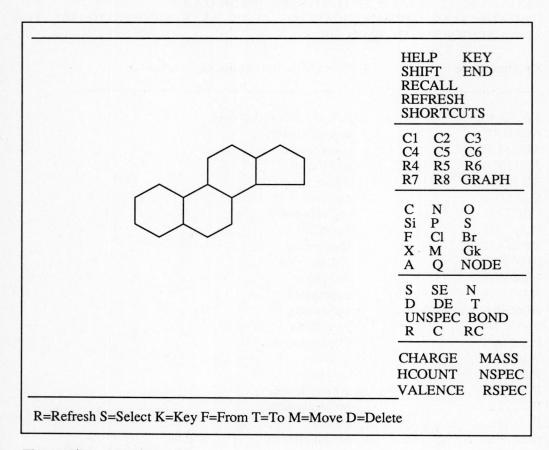

Figure 1/52. Selected model

If we already are in the Menu, we point/select RECALL in Area D.1 (Figure 1/53).

FREE-HAND DRAWING OF STRUCTURES

Structure skeletons can be created either entirely by free-hand drawing, or free-hand drawing can be combined with the Menu templates to build structure graphs.
In free-hand drawing, we use the single letter commands in Area C, F(From) and T(To) to go from a point to a point in Area B, thus creating nodes and bonds forming the graph we wish to create (Figure 1/54).
Free-hand drawing can also be invoked by the command DRAW at the System prompt

CREATING STRUCTURES

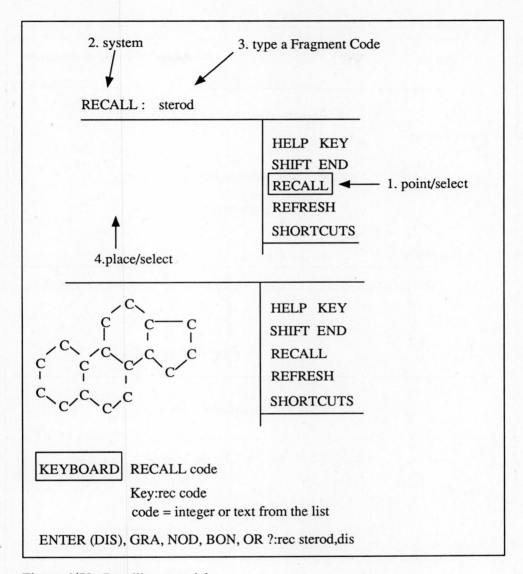

Figure 1/53. Recalling a model

ENTER(DIS), GRA, NOD, BON, OR ?:

To receive this prompt, we have to exit the Menu by depressing E on the keyboard. The Menu disappears and the message COPY AND CLEAR PAGE, PLEASE is issued by the System. At this message we have to shift the screen to graphics mode before the DRAW Command is entered (Figure 1/55).
Drawing by hand is shown in Figure 1/56.

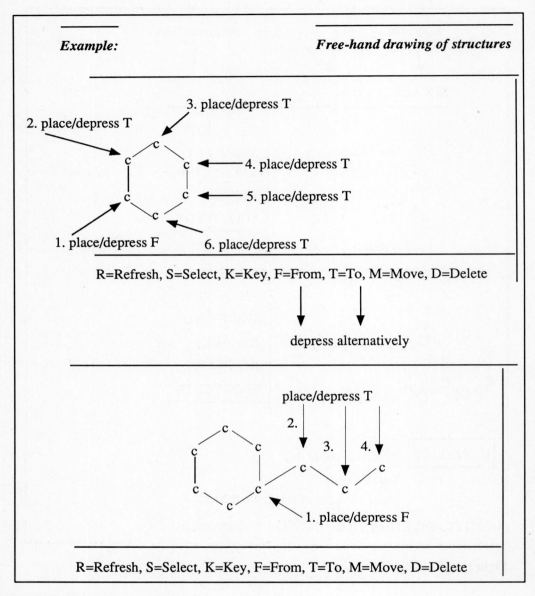

Figure 1/54. Free-hand drawing

Note: The shape of the skeleton in Figure 1/56 is the same as we have drawn it; the node numbering follows the path of drawing the skeleton.
To correct or modify the drawn skeletons we depress B. A slash will appear at each deleted node and bond (Figure 1/57).
To delete the entire structure and exit the DRAW mode, depress D.

CREATING STRUCTURES

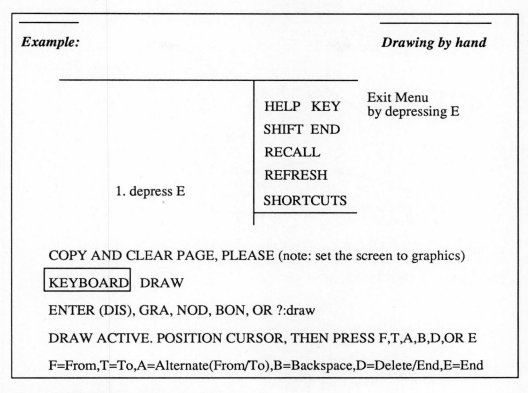

Figure 1/55. Initiating drawing by hand

The A key is used in place of F and T keys (Figure 1/58).

SPECIFICATION OF VALUES AT NODES

All nodes in the structure graphs we have created by using the structure skeleton templates in the Area D.-2. have the default carbon values. The Area D.-3. is used to select values for nodes in structure graphs.

VALUES AT SINGLE AND MULTIPLE NODES

The values (elements) in the Area D.3 are assigned to nodes by point/select a value followed by place/select it on the node or nodes which are to have the values (Figures 1/59, 1/60).

In some structures, we may want to assign so many heteroatoms of the same kind at multiple nodes that it would be faster to assign all the hetero values to all nodes and than reassign only a few of them back to carbon (Figures 1/61, 1/62).

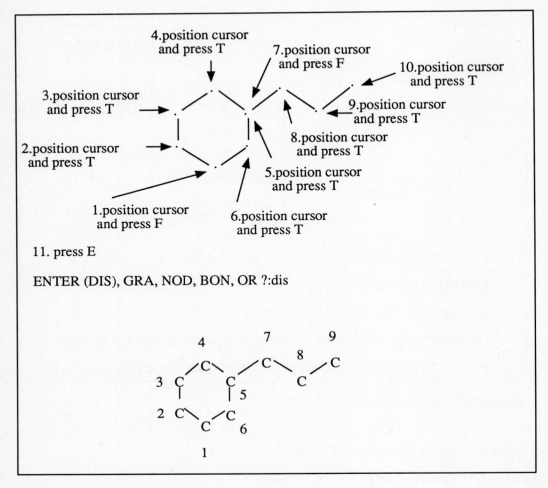

Figure 1/56. Drawing by hand

THE MENU NODE TEMPLATE

Area D.3 contains the NODE Command which can be activated to assign node values that are not on the Menu (Figure 1/63).

COMMON SUBSTITUENTS AND FUNCTIONAL GROUPS

Alkyl/aryl substituents and functional groups are assemblies of nodes with atom values joined by lines representing bonds.
These groups are always the same and appear repeatedly in structure graphs. Similarly to the Fragment File containing ring graphs, the Registry File comprises another file con-

CREATING STRUCTURES

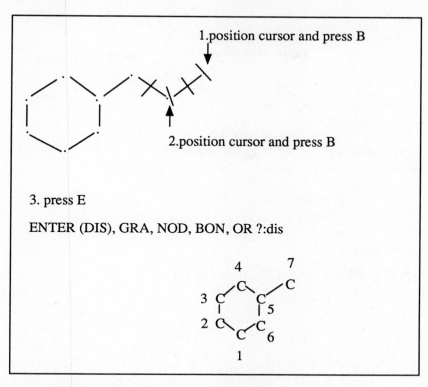

Figure 1/57. Deleting bonds in hand draw

taining preformed groups of this kind, called shortcuts, that can be attached to a structural fragment.
The alkyl/aryl substituents and functional groups available to the User are shown in the table.

SHORTCUTS FOR ALKYL/ARYL SUBSTITUENTS

Symbol	Description	Symbol	Description
CH	methyne	I-BUO	isobutoxy
CH	2methylene	T-BUO	*tert*-butoxy
ME	methyl	CF2	difluoromethylene
ET	ethyl	CBR2	dibromomethylene
N-PR	*n*-propyl	CCL2	dichloromethylene
I-PR	isopropyl	CI2	diiodomethylene
N-BU	*n*-butyl	CF3	trifluoromethyl
S-BU	*sec*-butyl	CBR3	tribromomethyl
I-BU	isobutyl	CCL3	trichloromethyl
T-BU	*tert*-butyl	CI3	triiodomethyl

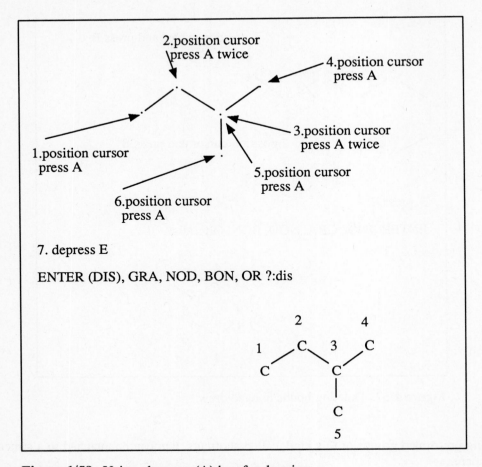

Figure 1/58. Using alternate (A) key for drawing

MEO	methoxy	PH	phenyl
ETO	ethoxy	O-C6H4	*o*-phenylene
N-PRO	*n*-propoxy	M-C6H4	*m*-phenylene
I-PRO	isopropoxy	P-C6H4	*p*-phenylene
N-BUO	*n*-butoxy	PHO	phenoxy
S-BUO	*sec*-butoxy		

SHORTCUTS FOR FUNCTIONAL GROUPS

Symbol	Description
C	Ncyano
CHO	formyl

CREATING STRUCTURES

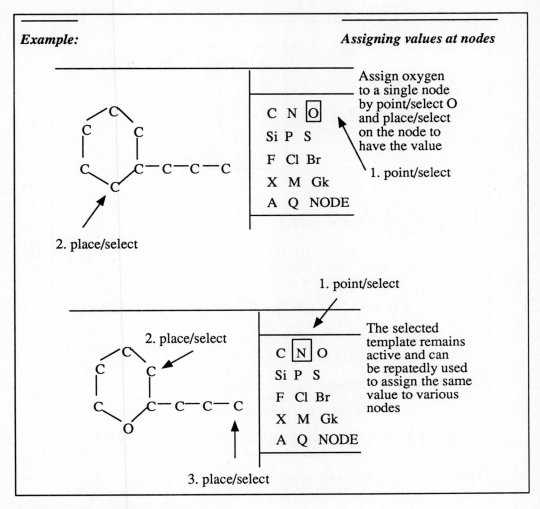

Figure 1/59. Assigning nodes values

CO2H	carboxy
C(O)CH3	acetyl
COSH	thiocarboxy
COS2H	dithiocarboxy
NH	imino
NH2	amino
NO2	nitro
OH	hydroxy
SH	mercapto
SO2	sulfonyl

```
              ┌─────────────────────────────────┐
              │   C        │ C  N  O            │
              │  / \       │                    │
              │ C   N      │ Si P  S            │
              │ │   │      │                    │
              │ C   C—C—C—N│ F  Cl Br           │
              │  \ /       │                    │
              │   O        │ X  M  Gk           │
              │            │ A  Q  NODE         │
              └─────────────────────────────────┘
```

| KEYBOARD | NODE m symbol |

    Key:nod m symbol
    m = node locant
    symbol = any element

ENTER (DIS), GRA, NOD, BON, OR ?:gra r6,4 c3,nod 5 O,dis

| KEYBOARD | NODE m n...y z symbol |

    Key:nod m n...y z symbol
    m,n,z = node locant
    symbol = any element

ENTER (DIS), GRA, NOD, BON, OR ?:gra r6,nod 5 O,3 9 N,dis

Figure 1/60. Assigning node values

    SO3H   sulfo
    OSO3H   sulfate
    PO3H2   phosphono
    OPO3H2   phosphate

When the structure graphs created by using the shortcuts are transferred to SEARCH, the System expands the shortcuts to structural fragments having all the atom values and bond types and values. The N-PR group, for example, is expanded to -CH2-CH2-CH3 with a minimum of hydrogen atoms at the attachment node, and an exact number of hydrogen atoms at all other nodes (see HCOUNT for detailed discussion of hydrogen atoms in structure graphs). Consequently, no substitution is allowed in the shortcut fragment.

The CO2H group is also expanded and the bonds are designated as normalized (see

CREATING STRUCTURES

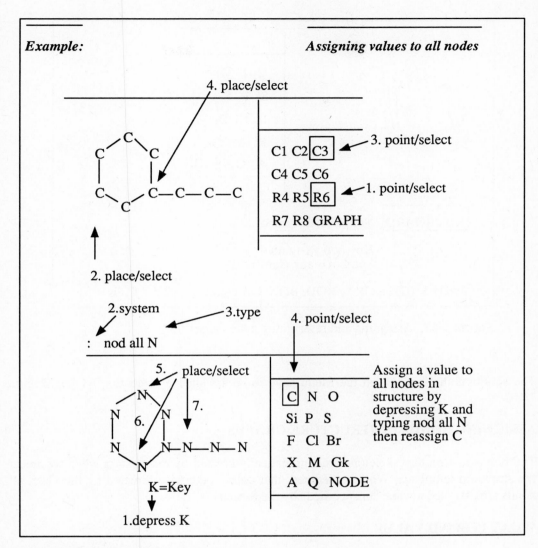

Figure 1/61. Assigning value to all nodes

BONDS for discussion of normalized bonds).
The bond values in the shortcuts cannot be changed by the User.

ASSIGNMENT OF SYSTEM SHORTCUTS

The System Shortcuts can be assigned at nodes by activating SHORTCUTS in Area D.1 followed by point/select the shortcut template (Figures 1/64, 1/65).

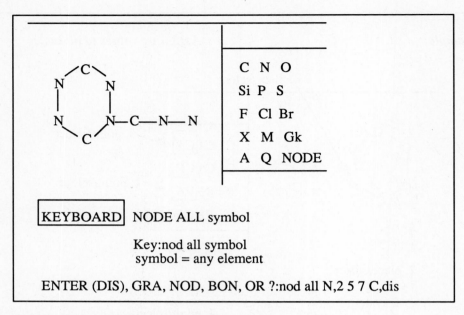

Figure 1/62. Assigning and reassigning node values

The shortcuts that are not on the Menu have to be assigned by selecting NODE (Figure 1/66).

ASSIGNING BONDS IN STRUCTURE GRAPHS

The bonds in structure skeletons are automatically created by the System when we select the structural templates. We have learned, that unless otherwise directed by the User, all bonds are assigned a value "unspecified" by the System.

WHAT IS BOND VALUE

Bond value is the kind of connectivity between nodes/atoms specified by the chemist, which meets the valency of the connected atoms and the desire of the chemist as to the nature of the structure being created.
The Registry File recognizes bond values whose definitions, symbols, and graphic representations are shown in Figure 1/67.

NORMALIZED BONDS IN REGISTRY FILE

The Registry System allows for registration of a chemical substance by only one structure. However, structures of many chemical compounds can be drawn by different chemists in a different way which results in multiple structures for the same compound. The structures

CREATING STRUCTURES

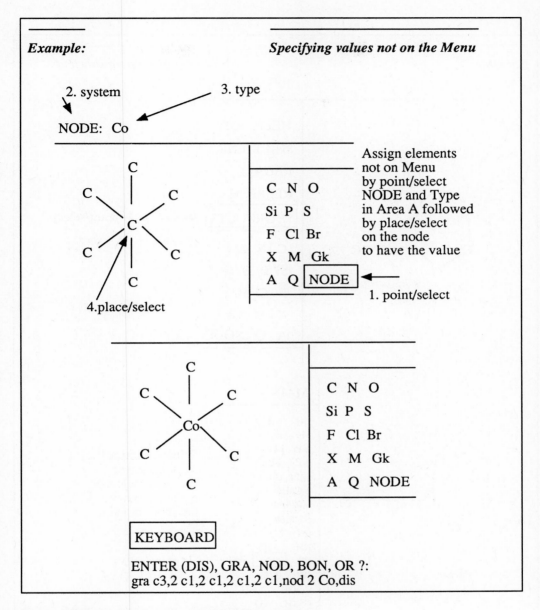

Figure 1/63. Using the Menu NODE Command

may differ in bond values, and/or position of charges, or hydrogen atoms.
To make it possible to store these kinds of structures and to prevent false hits on searching them, the concept of Bond Normalization has been developed and used in the registration of multiple structures of the same substance. The structures are assigned Normalized

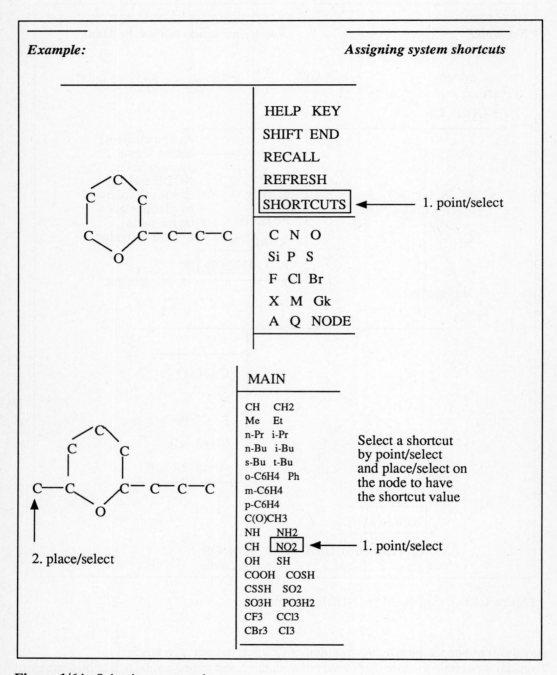

Figure 1/64. Selecting system shortcut

CREATING STRUCTURES

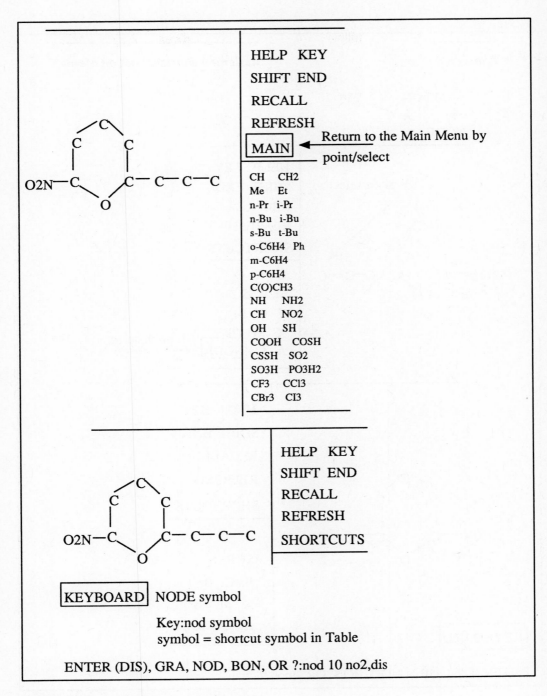

Figure 1/65. Assigning system shortcuts

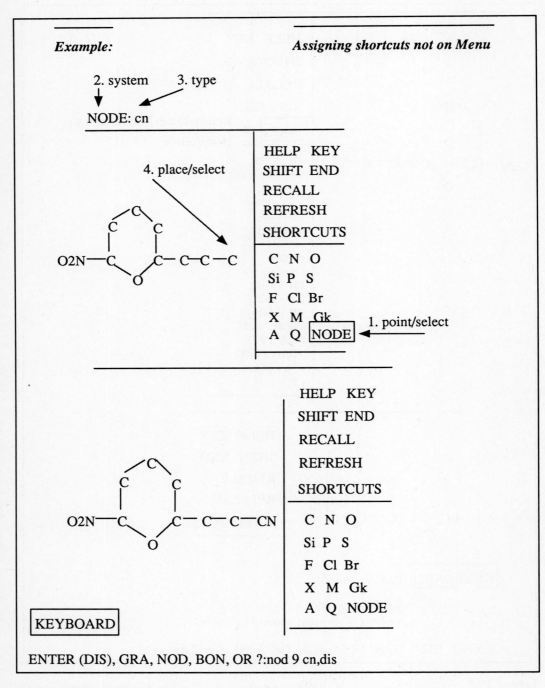

Figure 1/66. Assigning shortcuts not on the Menu

CREATING STRUCTURES

SYMBOL	BOND VALUE DEFINITION	DISPLAY
SE	Single exact	———
DE	Double exact	═══
T	Triple	≡≡≡
N	Normalized	·········

Figure 1/67. Graphics bonds

Bonds which allows representation of multiple structures by only one structure, thus ensuring that a substance will be retrieved no matter what the position of charges, bonds, and/or hydrogen atoms in the structure may be.
The Normalized Bond specification is used for:
1. Alternating single and double bonds in rings.
2. Bonds in tautomeric structures.
3. Delocalized bonds.

NORMALIZED BONDS IN RINGS AND RING SYSTEMS

The classical example of a substance that can be represented by more than one structure is benzene (Figure 1/68).

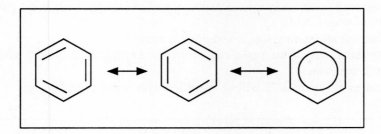

Figure 1/68. Resonance structures

For these structures to be recognized as being of an identical substance, the bonds are normalized.
The bonds of a single ring are normalized in the Registry System when:
1. The ring has an even number of nodes, and
2. the single and double bonds alternate around the ring.
Figure 1/69 shows examples of ring structures represented by normalized bonds.
However, ring structures that have single and double bonds, but the bonds do not alternate

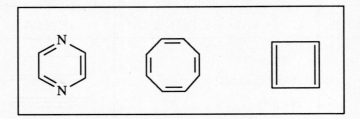

Figure 1/69. Candidates for normalized bonds

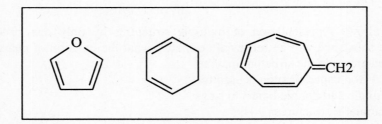

Figure 1/70. Candidates for exact bonds

around the ring or rings with an odd number of nodes, will not be specified as having normalized bonds but exact single and double bonds like the examples in Figure 1/70.

Fused ring systems, similarly, are specified with normalized bonds when:

1. The bonds are in a completely cyclic pathway and the rings in the bond pathway have an even number of nodes, and
2. the path is through alternating single and double bonds.

In fused ring systems, one of the fused rings may have normalized bonds, while the other may have exact bond values.

Thus, the structures in Figure 1/71 will be specified with normalized bonds.

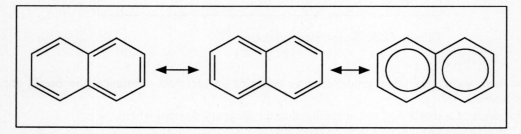

Figure 1/71. Candidate for normalized bonds

On the other hand, in the structure in Figure 1/72, the benzene ring has normalized bonds and the 5-membered ring has exact single and double bonds.

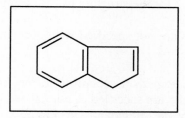

Figure 1/72

Note that the concept of "alternating single and double bonds" is not equivalent to the chemical definition of aromaticity or conjugation.

NORMALIZED BONDS IN TAUTOMERIC SYSTEMS

Tautomerism is a state of equilibrium between two or more structures that differ in the location of a mobile moiety. As a consequence of a change of location of the mobile moiety, the bond values for the structure containing the mobile moiety also change which results in the possibility of more than one structure for an identical compound.
The Registry File limits the many forms of tautomerism recognized in chemistry to that one which involves a three-node sequence which is generically expressed as is illustrated in Figure 1/73.

$$H - 1 - 2 = 3 \quad \rightleftarrows \quad 1 = 2 - 3 - H$$

Figure 1/73. Tautomeric system

The following requirements must be met in order for a structure to be classified as tautomeric in the Registry File:
The central node **2** is:
 C, N, P, S, As, Sb, Se, Te, Cl, Br, or **I**
The nodes **1** and **3** is:
 N, O, S, Se, or **Te**
A double bond is present between the central node and the node 3 or 1.
The node **3** must be either connected to hydrogen (or one of its isotopes) or have a

negative charge.

Carbon is not allowed as a value for the terminal node **1** and/or **3**. Carbon may be the value only for the central node **2** for the structure to be recognized as tautomeric in the Registry System.
For example, enols or enolates are not specified with normalized bonds, only with single exact and double exact bonds.

When the requirements for tautomerism according to the Registry System are met, the bonds in the tautomeric fragment are specified as normalized bonds.
For example, structures with the fragments in Figure 1/74 will be candidates for normalized bonds in those fragments.

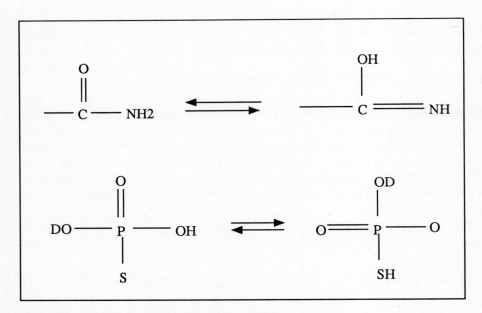

Figure 1/74. Fragments with normalized bonds

Ring structures with normalized (tautomeric) bonds are shown in Figure 1/75.
In the structure shown in Figure 1/76, we can see that the tautomerism is not limited to one three-node group and that the normalized bonds could cover the entire molecule.

NORMALIZED BONDS IN RING TAUTOMERIC SYSTEMS

Ring tautomeric systems may have overlaping tautomeric and alternating bonds.
A classical example is purine which can be represented by many structures (Figure 1/77).
The purine system has an even-numbered ring with alternating single and double bonds,

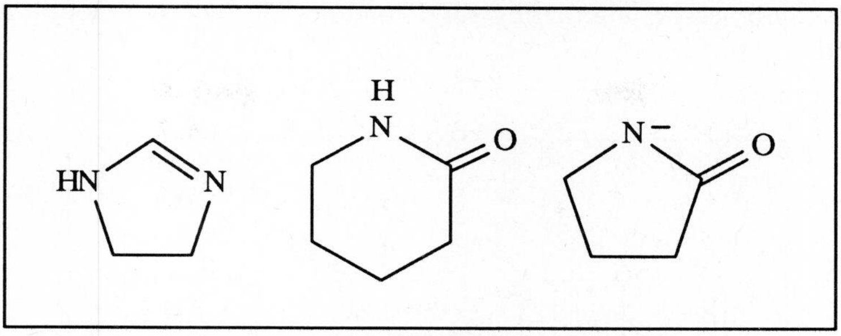

CREATING STRUCTURES

Figure 1/75. Ring structures with normalized bonds

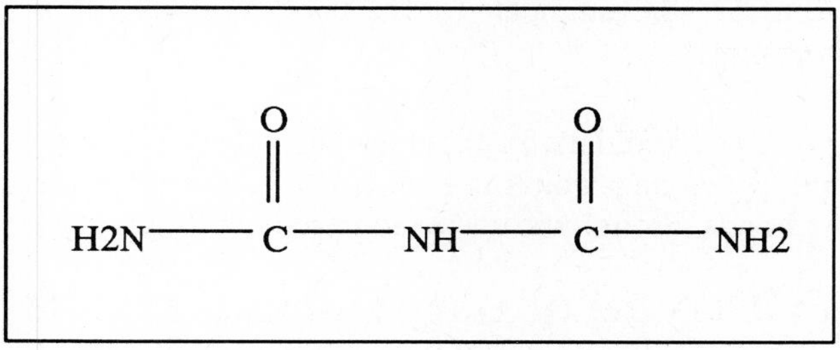

Figure 1/76. Candidate for normalized bonds

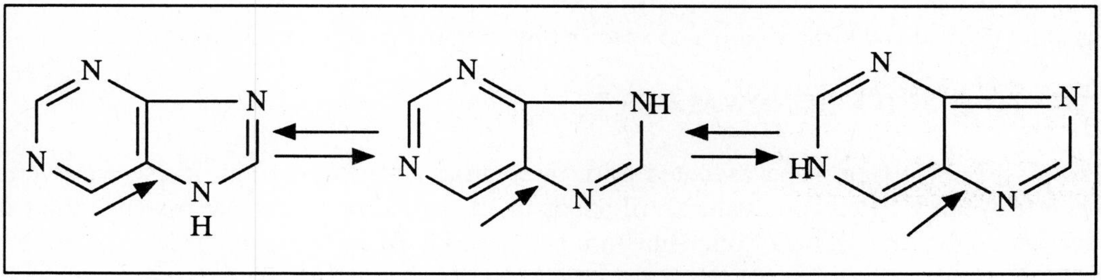

Figure 1/77. Overlaping tautomeric and alternating bonds

and tautomeric groupings in the five-membered ring. Consequently, all bonds will be specified as normalized except the arrow-marked bond which will be single exact.
The overlaping tautomeric and ring bonds have to be considered in totality before a decision is made whether the ring has alternating single and double bonds. The six-membered ring A in the structure in Figure 1/78 does not seem to have, without

considering the tautomerism, alternating single and double bonds, although it has an even number of nodes.

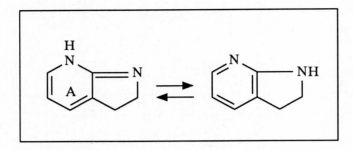

Figure 1/78. Candidate for normalized bonds

By itself, it would not be a candidate for the normalized bonds. However, the tautomerism brings a double bond in the ring which is then considered as the missing double bond to meet the "alternating bonds through the cycle" requirement.

NORMALIZED BONDS IN DELOCALIZED CHARGE SYSTEMS

Structures that contain a delocalized charge distributed over two or more nodes have delocalized bonds connecting these nodes. These bonds are specified in the Registry System as normalized bonds. Examples of structures of this kind are shown in Figure 1/79.

Note: if cyclopentadiene is in coordination compounds, all of the bonds are single and double exact.
Partially delocalized systems are also specified with normalized bonds (Figure 1/80).

HOW TO SPECIFY BOND VALUES

The SET BOND Command executed prior to creating a structure graph is used to override the default unspecified bond values. All bonds in all created structures are assigned the value specified in the SET BOND Command until another SET BOND (Figure 1/81).

BOND EXACT VALUES

To assign or reassign bond values in the course of creating structure graphs, we select the templates from Area D.4 and place/select them on the bond in the graph to have the exact bond value(s) (Figure 1/82). Assignment of bond exact results in retrieved structures having the exact specified bond values. We shall see later in describing generic structures, that more than one value can be assigned to a bond.

CREATING STRUCTURES

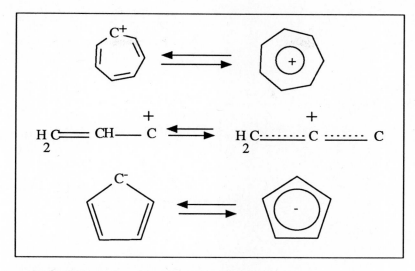

Figure 1/79. Delocalized charge systems

Figure 1/80. Delocalized charge system

NORMALIZED BOND VALUE

To specify normalized bonds we point/select N and place/select successively on the bonds (Figure 1/83). However, using the keyboard command, we can specify normalized value for all bonds in a ring in one step (Figures 1/84, 1/85).

The SET BOND Command sets bonds for all structures we are creating (Figure 1/81). If we do not use this command and wish to assign the same bond values to all bonds in a graph, we can use the Keyboard Command (Figure 1/86).

Example: *Overriding default bond value*

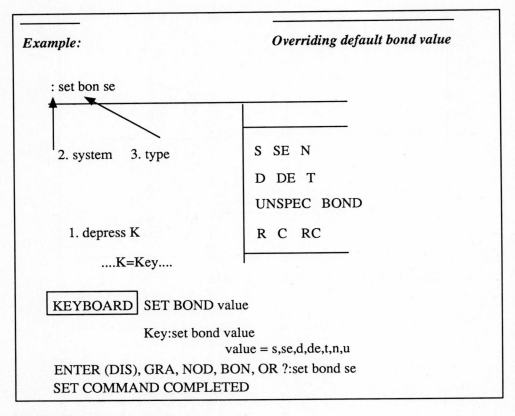

Figure 1/81. Setting bond value

Example: Assigning all bond values in single rings

| KEYBOARD | BOND R m n value

    Key:bon r m n value
        r=ring
        m,n=adjacent nodes locants in ring
        value=se,de,t

To change bond values only in this ring:

CREATING STRUCTURES

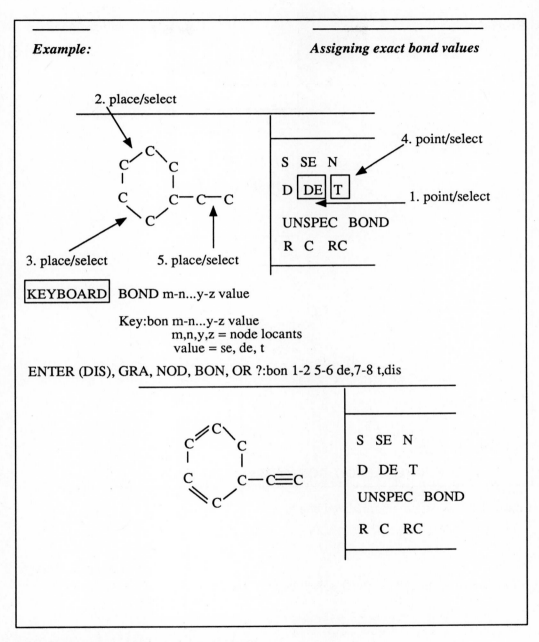

Figure 1/82. Assigning exact bond values

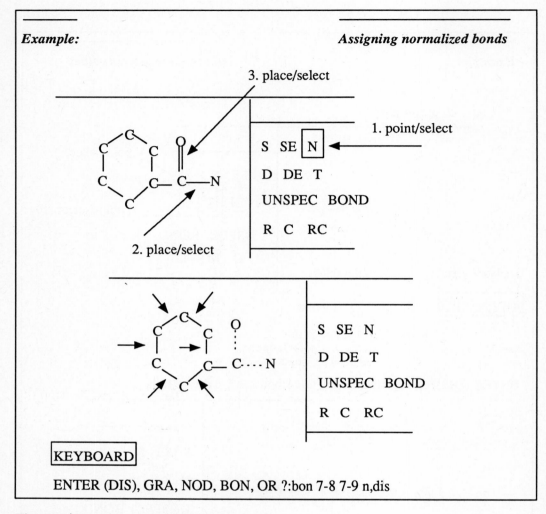

Figure 1/83. Assigning normalized bonds

ENTER (DIS), GRA, NOD, BON, OR ?: bon r 4 5 se,dis

In *fused ring systems,* the result of the all bond value depends on whether we specify:
1. Nonshared pair of adjacent nodes
The ring with the specified nonshared nodes will be assigned the all bond values.

CREATING STRUCTURES

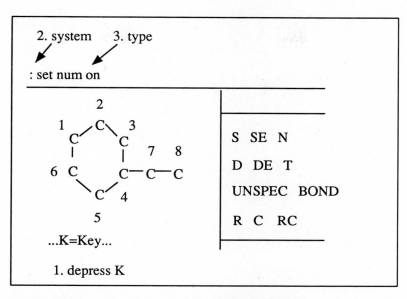

Figure 1/84. Assigning normalized bonds in rings

2. Shared pair of adjacent nodes
When the fused rings are of different size, the *smallest ring* will be assigned the all bond values.
When the fused rings are of the same size, the all bond values may be assigned to *either* of the fused rings.
Use *only nonshared pair of nodes* to specify a ring when the fused rings are of the same size and when differentiation of the rings matters.

Example: Assigning all bond values in fused rings

KEYBOARD | BOND R m n value

    Key:bon r m n value
        r=ring
        m,n=adjacent nodes
        value=se,de,t,n

1. Specification of nonshared adjacent nodes:

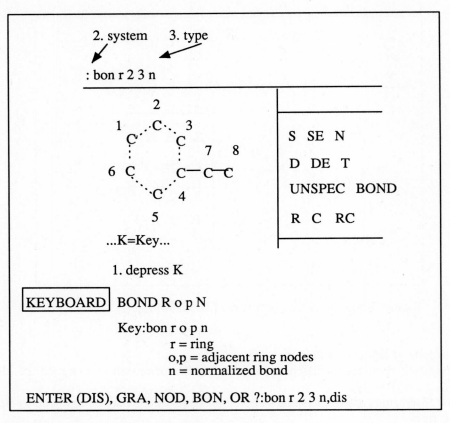

Figure 1/85. Assigning normalized ring bonds

ENTER (DIS), GRA, NOD, BON, OR ?:gra r46,dis

CREATING STRUCTURES

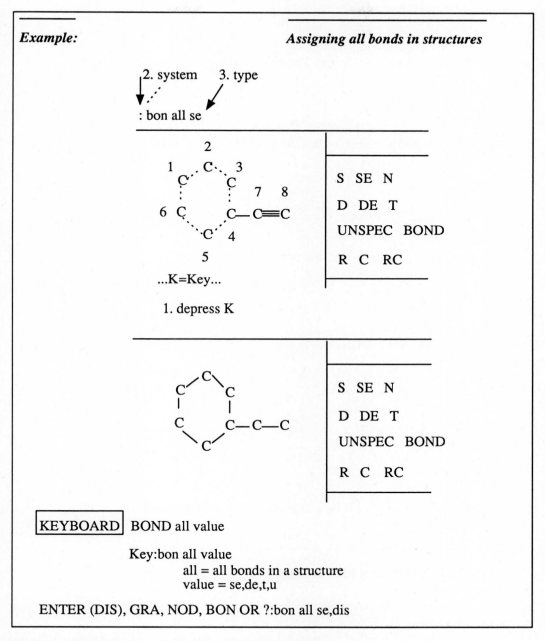

Figure 1/86. Assigning all same bonds in a structure

ENTER (DIS), GRA, NOD, BON, OR ?:bon r 1 4 se, dis

The smallest ring would also be assigned the single bonds by: bon r 2 3 se

2. Nonshared adjacent nodes for the following ring system:

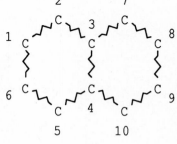

ENTER (DIS), GRA, NOD, BON, OR ?:

bon r 1 6 se,dis

However, either of the rings will be assigned the single bond value by specifying r 3 4 se (i.e., shared pair of nodes).

CREATING STRUCTURES FROM MODELS

Building of structures can be facilitated by using the shortcuts (see p. 84) or the ring fragments (see p. 76).

We can use yet another method to greatly speed up the creation of structures, by using models.

The models are structures or structural fragments that are registered in the Registry File or that the User builds and stores in a file for future use. The models can be recalled,

CREATING STRUCTURES

displayed, and manipulated by all the techniques we have so far learned or will learn later in this discussion.

REGISTRY SYSTEM MODELS

Each structure in the Registry File is assigned a Registry Number at the time the structure is entered into the System. The Registry Number is used to recall the structure, and modify it to arrive at the desired structure.
Since the CAS Registry Numbers are not compatible with the BRN in BEILSTEIN, we cannot use this technique in the Beilstein File. However, if we recall a model in the Registry File by the CAS Registry Number and receive the Ln for the structure query, the Ln (i.e. the structure) can be searched in the Beilstein File. Similarly, a structure created in the Beilstein File can be searched in the Registry File by its Ln.

HOW TO RECALL REGISTRY FILE MODELS

The models can be recalled at the time we activate the STRUCTURE Command. We are given an option to respond to the System prompt by the name of the structure to be recalled (if none is to be recalled we select the default response (NONE). To use a System model, we enter the Registry Number of the structure we want to recall.

Example: *Recalling system models*
At STRUCTURE Command Initiation

=>structure
ENTER NAME OF STRUCTURE TO BE RECALLED (NONE):150-13-0
ENTER (DIS), GRA, NOD, BON OR ?:menu vt
see (Figure 1/87).

If we already are in the Menu, we select RECALL (Figure 1/88).

The recalled structure graphs are displayed. When we turn on the node numbering by SET NUM ON, the node numbering may not be the same as in the structure built by the User. The bond and node values are those of the full registered structure. When more than one Registry Number (separated by a comma) are entered at the System prompt, the displayed graphs may overlap on the screen. Point/select SHIFT to rearrange the graphs on the screen (see page. 175).

LIMITS ON USING REGISTRY NUMBERS AS MODELS

Some Registry Numbers cannot be used to recall the graphs for modeling:

110 CAS REGISTRY

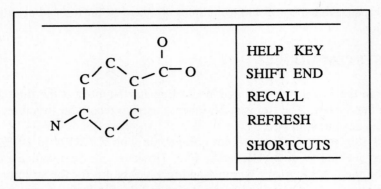

Figure 1/87. Recalled structure model

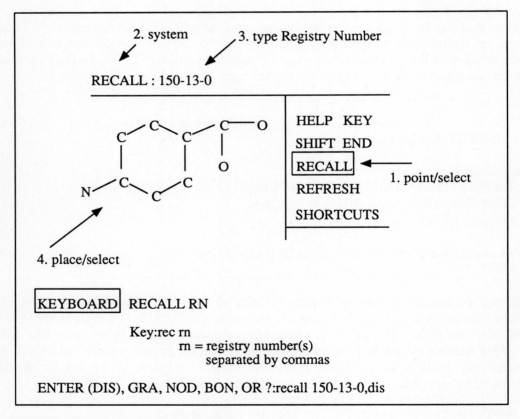

Figure 1/88. Recalling System model

CREATING STRUCTURES

1. Those substances which have been assigned Registry Numbers but have no connection table associated with them cannot be recalled as structure graphs.
2. Structures of multicomponent substances (see p. 27) cannot be recalled as whole structures by their Registry Numbers. However, the single components of the structures can be used.
3. Structures that are composed of more than one structural fragment, e.g., salts of acids, and one of the fragments is a single node fragment, are displayed without the single node fragment.
4. Stereoisomeric structures showing the stereobonds or stereocenters are not comprised in the Registry File. On recalling stereoisomers by their Registry Number, only structures with no stereo bonds are displayed.
5. Registry Numbers that have been deleted and replaced by CAS with another Registry Number can be used to recall a structure. The System automatically substitutes the valid Registry Number.

If any of the above described Registry Numbers is entered at the System prompt, the System issues a warning.

Example: Registry Numbers with no connection table

recall 9004-10-8
'9004-10-8' MAY NOT BE USED AS A MODEL

Registry Numbers of multicomponent substances

recall 8067-28-5
'8067-28-5 MAY NOT BE USED AS A MODEL
COMPONENTS
57-68-1
68-35-9 .....

However, the component Registry Numbers can be used to recall the structures.

Registry Numbers of multifragment substances

recall 127-09-3,dis
WARNING. SINGLE ATOM FRAGMENTS NOT INCLUDED IN MODEL

Registry Numbers of stereoisomers

ENTER NAME OF STRUCTURE TO BE RECALLED (NONE):22839--47-0
WARNING. STEREO DATA NOT INCLUDED IN MODEL. NOT SEARCHABLE

Deleted Registry Numbers as models

ENTER NAME OF STRUCTURE TO BE RECALLED (NONE):59217-40-2
REGISTRY NUMBER HAS BEEN REPLACED BY '826-36-8'

The structure graph which we requested is displayed but we are informed that it now has a new Registry Number.

USER MODELS

We can create any structure graph, very simple or very complicated, and store it in a User File for later recall. The structure graphs may be either full structures or they can be structural fragments that we can use as building blocks at any time in the structure building process. We can, in effect, supplement the System Fragment File (see p. 76) by our own fragment file.
To create User models, we proceed in building the desired structure graph as usual. After our graph is finished, we have to save it (store it).

HOW TO SAVE USER MODEL

User models can be saved:
1. for the current session in STN (Figure 1/89)
2. temporarily for a period of up to five days
3. permanently until canceled

1. Saving a model for the current STN session makes it possible to reenter the STRUCTURE Command and recall the saved structure graph for its modification or use it for building another structure.
The Ln assigned by the System for the structure is the specication needed to recall the graph.
All Ln assignments are erased, unless saved, at disconnecting from STN.
2. Saving the model temporarily for a period of up to 5 days is free of charge and absolves the User of the need to cancel the saved structure graph.
All temporary saves are erased by the System automatically over the weekend.
To save the model temporarily, the Ln must be assigned a name.

Example: *Saving User model temporarily*
Command:SAVE Ln TEMP name/q

    Key:save Ln temp name/Q (title)
        L=logical expression
        n=sequential number

CREATING STRUCTURES

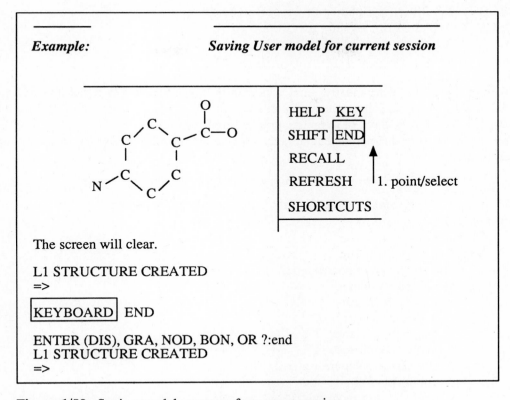

Figure 1/89. Saving model structure for current session

        temp=temporarily (erased over weekend)
        name=name for Ln
           name must:
           begin with a letter
           have 1-12 characters
           contain only letters a-z(A-Z) and
           numbers 0-9
           end with /Q (Q=query)
           not already be in use elswhere in
           the System
           not be:END,SAV,SAVE,SAVED,or
           L-number
        title=optional (memo) up to 40 characters

ENTER (DIS), GRA, NOD, BON OR ?:end
L1   STRUCTURE CREATED
=>save L1 temp benzene/q title

ENTER TITLE OR (NONE):structure for project Secret
QUERY 'L1' HAS BEEN SAVED AS 'BENZENE/Q'
=>

---

3. Saving the model permanently, until canceled by the User, makes it possible to run repeated searches of the same structure after each update of the Registry File (see Current Awareness and SDI, p. 259), or to build a private fragment file.

There is a charge for this permanent storage. To save the model permanently, the Ln must be assigned a name.

Example: *Saving User model permanently*

Command:SAVE Ln name/q

    Key:save Ln name/q title
        L=logical expression (see p...)
        n=sequential number
        name=name for Ln
          name must:
          begin with a letter
          have 1-12 characters
          contain only letters a-z(A-Z) and
           numbers 0-9
          end with /q (q=query)
          not be already in use elswhere in
           the System
          not be end,sav,save,saved,or
           L-number

ENTER (DIS), GRA, NOD, BON OR ?:end
L1   STRUCTURE CREATED
=>save L1 title
ENTER NAME(END):benzene/q
ENTER TITLE (NONE):structure for project Secret
QUERY'L1' HAS BEEN SAVED AS 'BENZENE/Q'
=>

HOW TO RECALL USER MODELS

The structures that have been saved either for current session, temporarily, or permanently and are recalled, are copied by the System on recalling them so that the original graph always remains intact even if we modified it or used it as a building block for another

CREATING STRUCTURES

structure graph. The structures that have been saved in current session and are being recalled in that session are recalled by their Ln designation.

Example: *Recalling User models by Ln*

KEYBOARD RECALL Ln

    Key:rec Ln
        L=logical expression
        n=sequential number

ENTER (DIS), GRA, NOD, BON OR ?:rec L1,dis
The graph will be displayed.
ENTER (DIS), GRA, NOD, BON OR ?:rec L1,menu vt
Menu and structure graph will be displayed.

If we have created a structure for which an Ln has been assigned and are reentering the STRUCTURE Command:
=>structure
ENTER NAME OF STRUCTURE TO BE RECALLED (NONE):L1
ENTER(DIS), GRA, NOD, BON OR?:dis (menu vt)
The graph will be displayed (Menu and graph will be displayed).

The structures that have been saved temporarily or permanently under a name have to be brought back into the STRUCTURE Command under the Ln designation. This is done by converting the name that we assigned to the Ln at the time of saving the structure back to the Ln designation. The conversion is accomplished by the command ACTIVATE at the System prompt =>.

Example: *Recalling User models by name*

Command: ACTIVATE name/q

    Key:activate name/q
        name=name given to Ln to save
        q=query
=>activate benzene/q
TITLE:STRUCTURE FOR PROJECT SECRET
L1   STR

The system assigns an Ln to the structure that has been saved under the name benzene/q. Should there be more than one structure graph saved under the name benzene, each graph will receive an Ln.

Now that we have the Ln, we can enter the STRUCTURE Command and display the structure(s).
=>structure
ENTER NAME OF STRUCTURE TO BE RECALLED (NONE):L1

FILE OF USER MODELS

The names of structure graphs that have been saved temporarily or permanently are stored by the System in the User's File from which they are recalled. We can locate the names of the graphs we have saved by displaying the content of the User's File with the use of the command DISPLAY at the System prompt =>.

Example: *Displaying User model names*
Command:DISPLAY SAVED/q

=>display saved/q
NAME   CREATED   NOTES/TITLE
BENZENE/Q   DEC 09 87   STRUCTURE FOR PROJECT SECRET

The saved structure graphs can be erased from the file. The graphs are, of course, lost for further recall. We use the command DELETE to erase saved names of graphs.

Example: *Deleting User models from file*
Command: DELETE name/q

   Key:delete name/q
        name=name for saved graph
        q=query
=>delete
ENTER NAME OF ITEM TO BE DELETED OR (?):benzene/q
DELETE BENZENE/Q? (Y)/N:y
BENZENE/Q DELETED

SUMMARY OF STRUCTURE BUILDING AT THIS POINT

We have learned how to build structure graphs by creating structure skeletons and specifying node and bond values, how to utilize the shortcuts and the ring fragments, and how to use models recalled either from the System File or from User File.
We have now enough knowledge and experience to build a structure query and submit it to

CREATING STRUCTURES

the Exact, Family, or Substructure Search (SSS) in the SEARCH subsystem (see p. 207).

So far, however, we have created structure queries in which we specified *single values* - atoms, bonds, substituents, or functional groups at specific nodes.

To utilize the most valuable capability of the Registry File, i.e., searching for generic structures, we have to expand our structure building know-how to creating generic structure queries.

9 CREATING GENERIC STRUCTURES

Generic structures are the graphs in which we specify *sets of variable values* at specific locants of the structure.

On searching generic structures in the Substructure Search (SSS) (see p. 218) we retrieve structures, if any, with at least one set of the variable values specified in the generic structure.

In creating generic structure graphs, we can specify the following variables:

1. a set of single atoms
2. negated atom or a limited set of negated atoms
3. multiatom structural fragments
4. bond values
5. bond types
6. bond values/bond types
7. chain node
8. ring node
9. chain/or ring node
10. ring specific (isolated or embedded)

HOW TO DEFINE VARIABLES IN GENERIC STRUCTURES

The Registry File has predefined some of the sets of variables. The sets have been assigned symbols that are used in assigning these sets in the generic structures. In addition, the System offers the User methods for defining any set of variable values that may be needed in building generic structures.

SYSTEM DEFINED VARIABLE ATOMS

The Registry File defines four symbols to specify a set of variable atom values that can be assigned to any node in the generic structure. The symbols are described in the table (Figure 1/90).

SYSTEM DEFINED GENERIC RING FRAGMENTS

The Registry File defines three symbols to specify generic ring fragments in structure graphs. The symbols are described in the Table (Figure 1/91).

See also p. 262-A1 for AK (alkyl) variable and Generic Group Category (GGC) specification. Any combination of these symbols can be specified in a structure.

CAS REGISTRY

SYMBOL	DEFINITION
A	Any element except hydrogen
M	Any metal*
Q	Any element except hydrogen and carbon
X	Any halogen

* Metal is defined as all elements except: Ar,As,At,B,Br,C, Cl,F,H,He,I,Kr,N,Ne,O,P,Rn,S,Se,Si,Te,Xe

Figure 1/90. System variables

SYMBOL	DEFINITION
CY	any kind of cyclic fragment
CB	any cyclic fragment with carbons only
HY	any cyclic fragment with at least one noncarbon atom

Figure 1/91. System variables

are selected and placed on nodes (Figure 1/92).
The variables in Figure 1/91 are assigned by the NODE Command (Figure 1/93).

USER DEFINED VARIABLES

The definition of variables by the User is accomplished by assigning a User symbol representing the set of variable values to nodes in the generic structures similarly to assigning the symbols of the System defined variables.
The User symbol for the variables is *G*. There can be up to 20 *G*s in a generic structure;

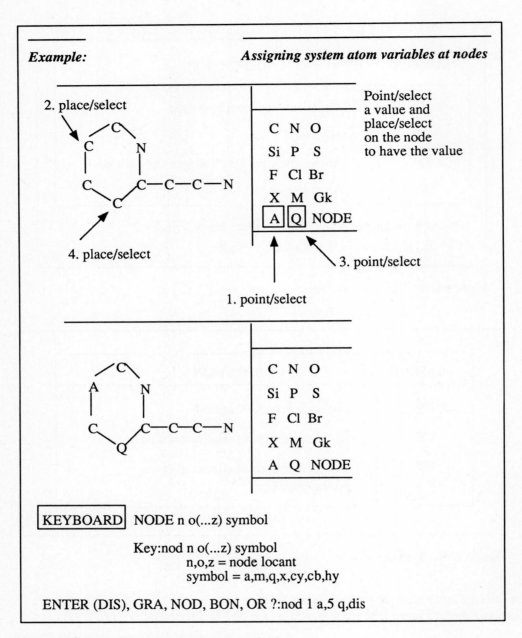

Figure 1/92. Assigning system atom variables

however, the structure must contain at least one node that is not *G*. Each *G* symbol can represent 2-20 different variables.

Figure 1/94 shows the composition of a structure with *G* symbols.

CREATING GENERIC STRUCTURES

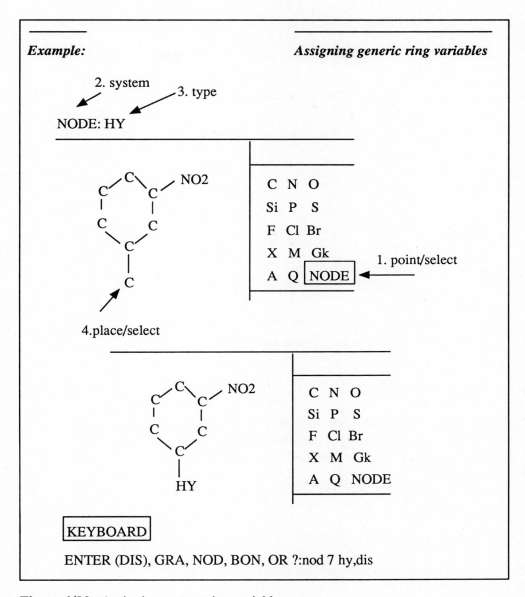

Figure 1/93. Assigning system ring variables

The *G* symbols (which are generically designated as Gk groups) are assigned to nodes by selecting Gk in Area D.3 (Figure 1/95, 1/96).
The values in Gk groups are defined by the command VARIABLE and/or REPEATING.

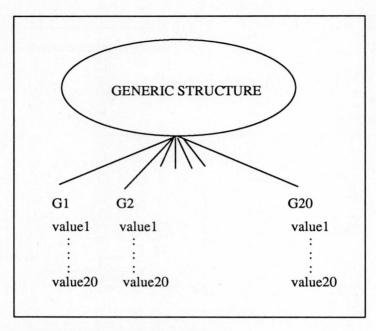

Figure 1/94. User variable groups Gk

THE COMMAND VARIABLE

The VARIABLE Command defines the values of Gk groups for sets of:
1. single atoms including hydrogen *(negated atoms cannot be used)* (see p. 124)
2. shortcut symbols *except o-, m-, and p-C6H4*
3. functional groups
4. generic ring symbols CB, CY, HY
5. variable symbols A, Q, X, M

VARIABLE ATOMS IN Gk GROUPS

The standard element symbols are used to define the values for Gk groups in the generic structures (Figure 1/97).

DISPLAY OF Gk GROUPS

The definition of Gk groups is displayed with the structure graph. We can request various formats of the Gk display (Figure 1/98), or we can turn the display off.

CREATING GENERIC STRUCTURES

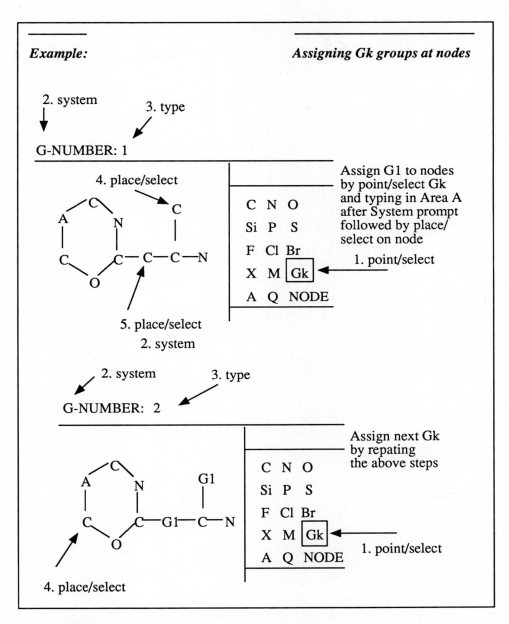

Figure 1/95. Assigning Gk groups

VARIABLE SHORTCUTS IN Gk GROUPS

All the symbols for shortcuts (see p. 84) can be used as values for the G groups (Figure 1/99), *except o-, m-, and/or p-C6H4.*

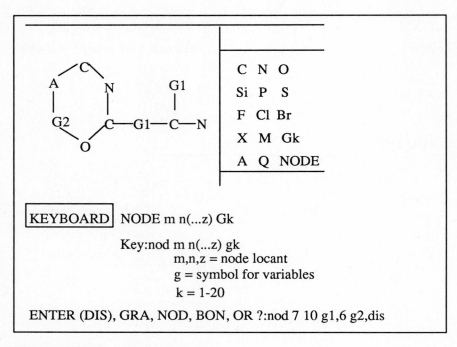

Figure 1/96. Assigning Gk groups

GENERIC RINGS IN Gk GROUPS

All the System symbols for generic rings (see p. 119) can be used as values in Gk groups (Figure 1/100).

NODE NEGATION

We can request that a specific atom value be excluded for a specified node in a graph (Figure 1/101). This is called node negation. When we exclude an atom as the value for a node, all other possible atom values are allowed as the values for the node. Consequently, the structures will be retrieved that have all the possible atom values but not the negated one.

The structure graph will show the negated value. At all other nodes, however, substitution by the value can take place. Similarly, the negated values may be the System symbols for a set of atoms, *except A and/or Q*. The table (Figure 1/102) shows the negated symbols for a set of elements. Gk groups cannot be negated.

STRUCTURAL FRAGMENTS IN Gk GROUPS

The definition of the Gk groups that we have discussed involved either single atom values

CREATING GENERIC STRUCTURES

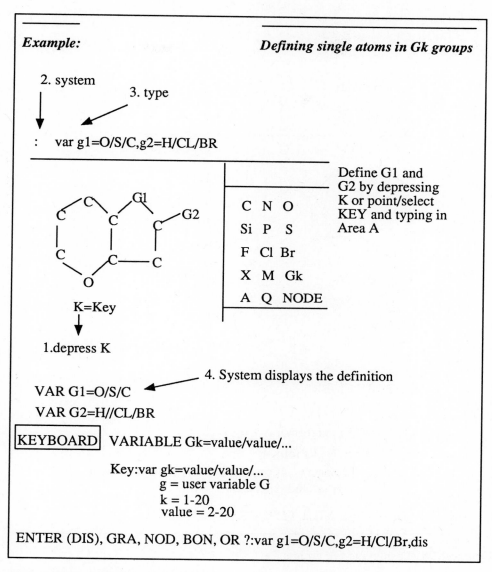

Figure 1/97. Defining Gk groups

or system predefined structural fragments, i.e., the shortcuts, functional groups, and the generic rings.
We can define any multiatom fragment of our choice as the value for Gk groups. The method of doing so consists of:
1. Creating the parent structure.
2. Specifying Gk groups in the parent structure.

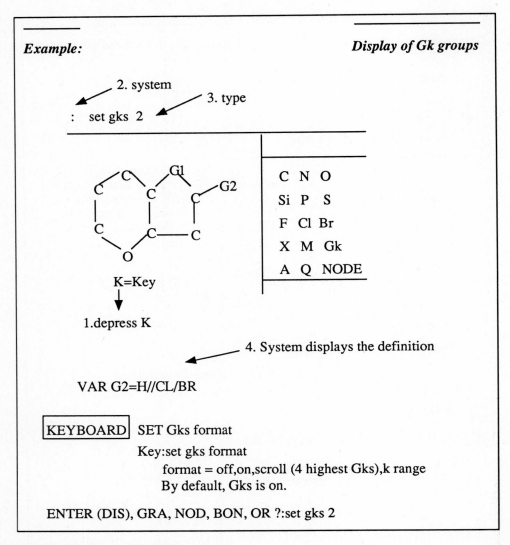

Figure 1/98. Displaying Gk groups

3. Creating the multiatom fragment in the same STRUCTURE Command (see p. 44).
4. Defining the Gk groups by the VARIABLE Command and citing the locant of the multiatom fragment attachment to the parent structure in the Gk group definition.

For one-node attachment, we cite only the locant in the parent (Figure 1/103). For multiple node attachment, we have to cite the locant both in the fragment and the parent (Figure 1/104).

The fragment in Figure 1/103 is bonded at the node 6 (indicated in the display by @ sign) to the parent at the G2 locant.

CREATING GENERIC STRUCTURES

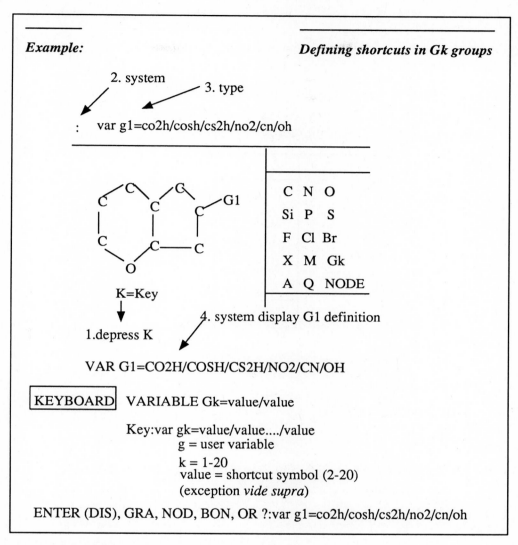

Figure 1/99. Defining shortcuts in Gk groups

Definition of a fragment with multiple points of attachment is shown in Figure 1/104. The fragment in Figure 1/104 is bonded by node 15 to the parent at node 3 and by node 16 to node 8 (indicated by the @ in the display).
Although hydrogens can be defined in Gk groups, a careful consideration has to be given

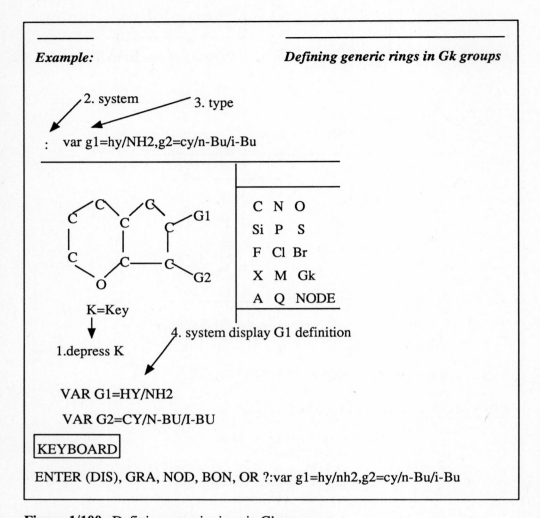

Figure 1/100. Defining generic rings in Gk groups

to define Gk group in a structure containing the following fragment:

To search for structures where R=hydrogen or any other substituent, the hydrogen that is already in the structure fragment has to be built-in in the fragment to be defined in the Gk group. We would build the following query:

CREATING GENERIC STRUCTURES

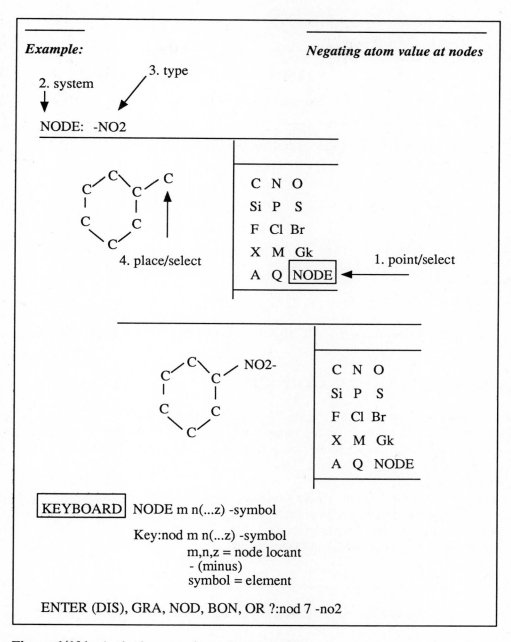

Figure 1/101. Assigning negative values at nodes

To retrieve structures where R=hydrogen, the definition of the G1 group will be CH2. For structures where, for example, R=PH,ET,NO2, etc., we have to build the following

SYMBOL	DEFINITION
-X	No halogen, no hydrogen
-M	No metals, no hydrogen

Figure 1/102. Negated system variables

fragments: CH——PH   CH——ET   CH——NO2
            1   2   3

and define them in the G1 group as follows:

KEYBOARD

ENTER (DIS), GRA, NOD, BON OR ?:var g1=ch2/1/2/3

For another method to formulate this kind of structure query, see **CONNECT** Command (p. 152).

USING Gk GROUPS WITHIN Gk GROUPS

We can build very comprehensive generic structures by incorporating Gk groups within Gk groups. There is no limit on the number of Gk groups embedded in one Gk group, except that a Gk group cannot be the point of attachment to the parent structure.
We proceed as follows:
1. Build the parent structure.
2. Assign the Gk group(s) in the parent structure.
3. Build the fragment(s) for the Gk group(s) in the parent.
4. Assign the Gk group(s) in the fragment(s).
5. Define Gk in the parent and Gk in the fragment.

Example: *Using Gk groups within Gk groups*

KEYBOARD VARIABLE

  Key:var gk(1)=value/../value/n,gk(2)=value/../value
     gk(1)=user symbol for variables in parent
     gk(2)=user symbol for variables in fragment

CREATING GENERIC STRUCTURES

      value=any of the values (*vide supra*)
      n=attachment node locant of fragment with one
        point of attachment (if more points of
        attachment, use m-n o-p)(*vide infra*)

1. Create Parent:
ENTER (DIS), GRA, NOD, BON OR ?:
gra r66,8 c2,11 c1,bon all n,8-11 se,11-13 d,11-12 s,nod 13 O
2. Assign Gk in parent:
ENTER (DIS), GRA, NOD, BON OR ?:nod 12 g1
3. Create Fragment in the same STRUCTURE Command:
ENTER (DIS), GRA, NOD, BON OR ?:gra c2,bon 14-15 se,nod 14 O
4. Define Gk in fragment:
ENTER (DIS), GRA, NOD, BON OR ?:nod 15 g2
5. Define Gk in parent:
ENTER (DIS), GRA, NOD, BON OR ?:var g1=ME/PH/OH/NH2/14
6. Define Gk in fragment:
ENTER (DIS), GRA, NOD, BON OR ?:var g2=ME/ET/N-PR,dis (Figure 1/105)
The fragment in Figure 1/105 is bonded by the node 14 to the parent

---

THE COMMAND REPEATING

Frequently, we are confronted with building a generic structure graph in which a fragment is a node or group of nodes that repeat x times.
In the chains below, the CH2 group repeats 1-4 times:

CH3CH2CH3 CH3CH2CH2CH3 CH3CH2CH2CH2CH3 CH3CH2CH2CH2CH2CH3

The structures may be represented as follows: $CH3(CH2)_n CH3$ n=1-4
Cyclic examples of this kind of structures are shown in Figure 1/106.
The structures in Figure 1/106 can be represented by the structure shown in Figure 1/107.
We can specify the repeating group of nodes as a Gk group and define it by the REPEATING Command.
The repeating group in the fragment may be:
1. single element
2. system shortcuts that have two points of attachment
3. system defined variable atoms
4. user defined multiatom fragments
The following requirements must be met for a fragment to be declared the Gk group:
1. The repeating group cannot contain Gk groups.
2. The repeating group must be attached to two nodes in the parent structure.

The repeating group can be attached to the parent by one node or by multiple nodes.

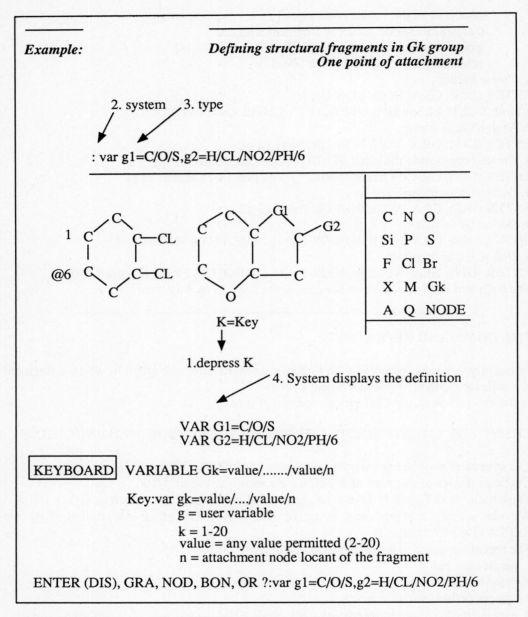

Figure 1/103. Structural fragment (one attachment point) in G group

REPEATING SINGLE ATOMS AND SHORTCUTS - SINGLE ATTACHMENT

Single attachment must be by **one node** of the repeating group to **two nodes** of the parent.

CREATING GENERIC STRUCTURES

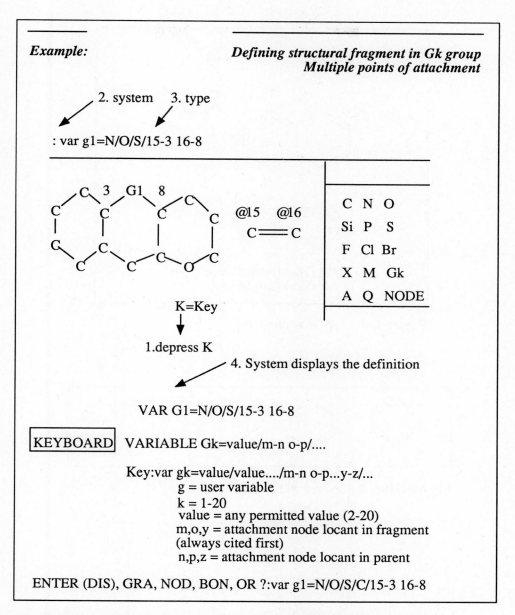

Figure 1/104. Structural fragment (multiple attachment) in G group

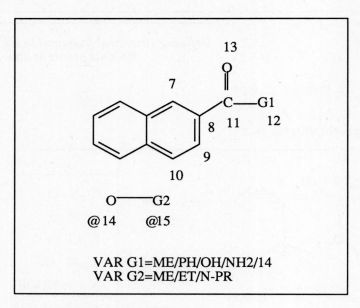

Figure 1/105. A Gk within a Gk

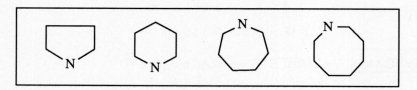

Figure 1/106. Structures with repeating units

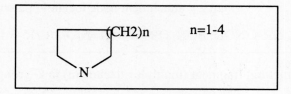

Figure 1/107.

CREATING GENERIC STRUCTURES

Example: *Repeating single atoms*

KEYBOARD REPEATING Gk=(x-y) symbol

Key:rep gk=(x-y) symbol
g=user symbol for variables
k=1-20
x-y=repeat range (1-20)
symbol=element

In the generic structure shown in Figure 1/108, the sulfur atom repeats 1-8 times; therefore, we can specify it as a Gk group and define the Gk group with the REPEATING Command.

Figure 1/108

To create the structure in Figure 1/108: 1. Create graph and assign G1:

ENTER (DIS), GRA, NOD, BON, OR ?:gra r3,nod 1 N,2 S,3 g1,dis

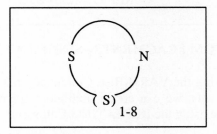

2. Define G1:
ENTER (DIS), GRA, NOD, BON, OR ?:rep g1=(1-8) S

Some of the system defined shortcuts (see p. 84) which have two points of attachement can be used in the definition of a Gk group.

Example: *Repeating system shortcuts*

KEYBOARD REPEATING Gk=(x-y) symbol

Key:rep gk=(x-y) symbol
     g=user symbol for variables
     k=1-20
     x-y=repeat range (1-20)
     symbol=system shortcut symbol

ENTER (DIS), GRA, NOD, BON, OR ?:

gra r5,nod 3 g1,rep g1=(2-5) ch2,dis

REP G1=(2-5) CH2

REPEATING MULTIATOM FRAGMENTS - SINGLE ATTACHMENT

As we have seen in discussing the VARIABLE Command, in addition to using the system defined symbols and shortcuts, we can create multiatom fragments and use them in Gk groups. We can do the same using the REPEATING Command.

Example: *Repeating multiatom fragments*
               One point of attachment

KEYBOARD REPEATING Gk=(x-y) n

   Key:rep gk=(x-y) n
      g=user symbol for variables
      k=1-20
      x-y=repeat range (1-20)
      n=attachment node locant

To search for the structure in Figure 1/109:
1. Create parent:
ENTER (DIS), GRA, NOD, BON, OR ?:gra c3,nod 1 3 CO2H
2. Assign G1 group:
ENTER (DIS), GRA, NOD, BON, OR ?:nod 2 g1
3. Create fragment:
ENTER (DIS), GRA, NOD, BON, OR ?:gra c2,nod 4 ch,5 oh
4. Define G1 group:
ENTER (DIS), GRA, NOD, BON, OR ?:rep g1=(3-10) 4,dis (Figure 1/110).

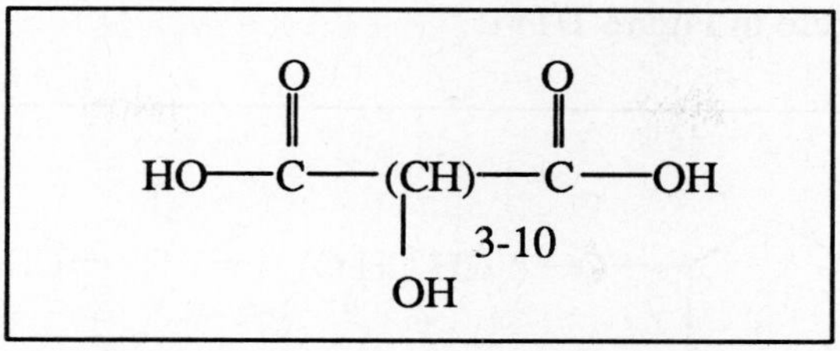

CREATING GENERIC STRUCTURES

$$HO-\underset{\underset{O}{\|}}{C}-(CH)-\underset{\underset{O}{\|}}{C}-OH$$
$$\underset{OH}{|}\ 3\text{-}10$$

Figure 1/109

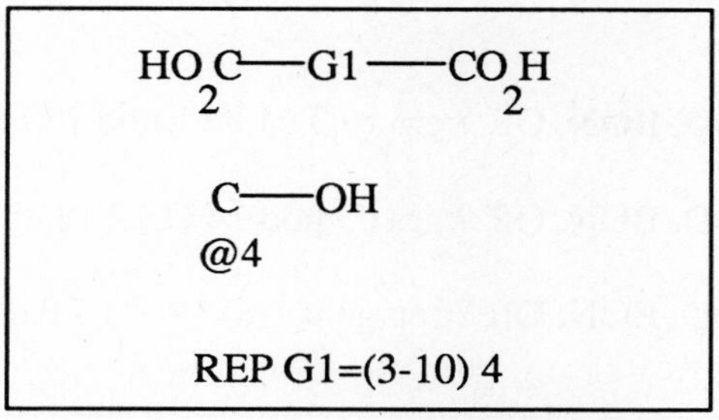

$$HO_2C-G1-CO_2H$$
$$C-OH$$
$$@4$$

REP G1=(3-10) 4

Figure 1/110

REPEATING MULTIATOM FRAGMENTS - MULTIPLE ATTACHMENTS

The multiple attachment is by **two nodes** of the repeating unit to **two nodes** of the parent. We have to cite the node locants of the attachments both for the fragment and the parent.

Example: *Repeating multiatom fragments*
Multiple points of attachment

KEYBOARD REPEATING Gk=(x-y) m-n o-p

    Key:rep gk=(x-y) m-n o-p
        g=user symbol for variables
        k=1-20
        x,y = repeat range
        m,o=attachment node locants of fragment
        n,p=attachment node locants of parent

To search for the structure in Figure 1/111:

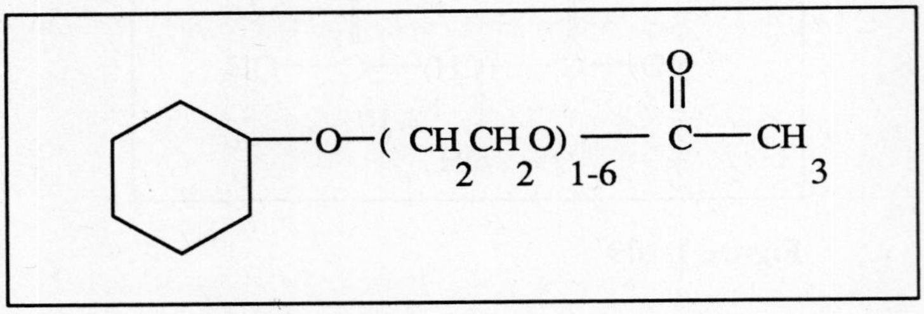

[Figure: cyclohexyl—O—(CH₂CH₂O)₁₋₆—C(=O)—CH₃]

Figure 1/111

1. Create parent:
ENTER (DIS), GRA, NOD, BON, OR ?:gra r6,3 c4,9 c1,nod 7 O,8 G1,11 O,bon 9-11 de
2. Create fragment:
ENTER (DIS), GRA, NOD, BON, OR ?:gra c3,nod 14 O,12 13 ch2
3. Define G1:
ENTER (DIS), GRA, NOD, BON, OR ?:rep g1=(1-6) 12-7 14-9,dis (Figure 1/112).

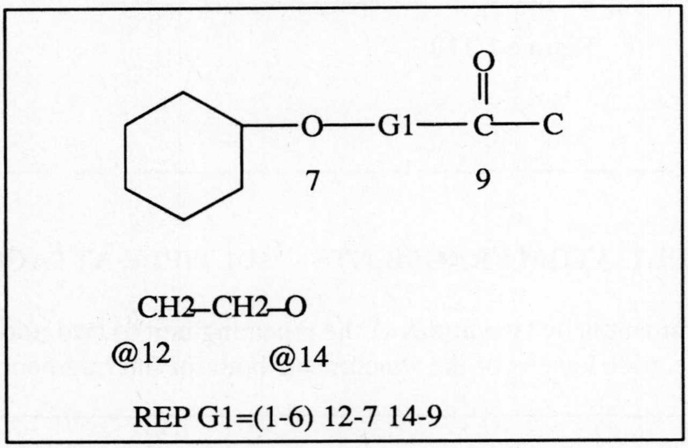

[Figure: cyclohexyl—O—G1—C(=O)—C, with positions 7 and 9 labeled; fragment CH2–CH2–O with @12 and @14]

REP G1=(1-6) 12-7 14-9

Figure 1/112

VARIABLE BOND VALUES IN GENERIC STRUCTURES

We have learned that the Bond Normalization concept (p. 90) allows for assigning a Normalized Bond in certain structures. In this process, in fact, the bond loses its exact character and becomes an "either or" bond. In generic structure queries, we extend this

CREATING GENERIC STRUCTURES

"either or" character of bonds to many more variables.
Variable bond values of a limited scope can be assigned by using the symbols for bond values. The table in Figure 1/113 shows the bond symbols and the values they represent.

SYMBOL	BOND VALUE DEFINITION	DISPLAY
S	Single exact or normalized	-----
D	Double exact or normalized	=====
U	Unspecified (any value)	∧∧∧

Figure 1/113. Bonds and graphics

By using the specifications in Figure 1/113, we can force the System to selectively search for structures with the specified bond values.

VARIABLE BOND EXACT OR NORMALIZED

The specification "single exact or normalized" signifies that the bond that was assigned this value may assume either the value of a single exact bond or a normalized bond. This specification is widely used in creating ring structure graphs and tautomeric structure graphs.

If we have the ring structure: we can specify the bond values single exact

(see p. 100). This specification does not allow for any other bond values and the search returns only the structure with exact single bonds.
On the other hand, if we specify the bonds as "single exact or normalized," we allow for search and retrieval of the structures in which the bonds are single, aromatic, or delocalized (Figure 1/114).
The structures in Figure 1/115 would not be retrieved.
By specifying the bond value "single exact or normalized" in a generic ring structure we also allow for retrieval of structures, if any, that have the query ring embedded in fused ring systems.
Similarly, in the example (Figure 1/116) where R=hydrogen or any substitutent, if the C=O double bond was specified as double exact (DE) and the C-N bond as single exact (SE), only N-unsubstituted structures would be retrieved because no tautomerism is recognized in this structure. However, the specification double or normalized (D) and sin-

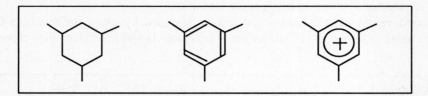

Figure 1/114

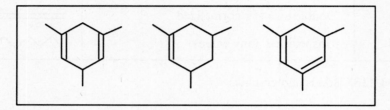

Figure 1/115

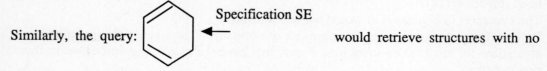

Figure 1/116

gle or normalized (S) or unspecified (U) for these bonds would retrieve both N-unsubstituted (R=H) and N-substituted (R=any substituent) structures because tautomerism is recognized by allowing single, double, normalized, or any bond value for the bonds. If the ring bonds were specified as single exact (SE), only cyclohexane derivatives would be retrieved. If the ring bonds were specified as single or normalized (S) or unspecified (U), cyclohexane, benzene, and fused ring systems would be retrieved.

Similarly, the query: [structure] ← Specification SE   would retrieve structures with no

fusion at the specified bond, while the specification single or normalized (S) would allow for fusion at that ring side.

For another method to specify isolated or fused ring systems, see the CONNECT Command (p. 152, 157).

CREATING GENERIC STRUCTURES

Figure 1/117 shows assignment of variable bond values.

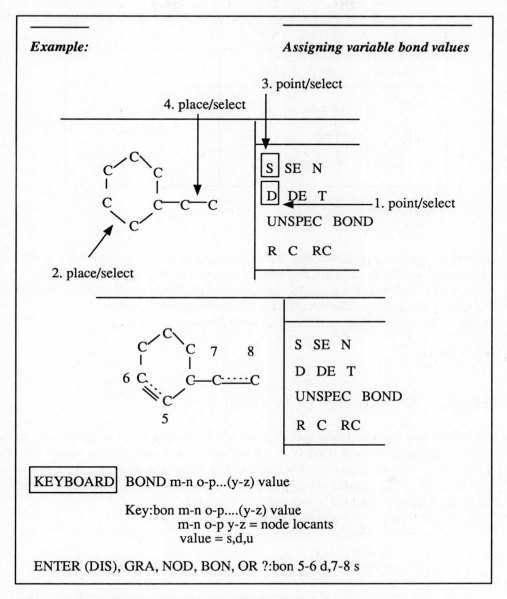

Figure 1/117. Assigning variable bond values

BOND TYPES IN GENERIC STRUCTURES

Along with variable bond values, we can assign bond types in generic structure graphs. The Registry File recognizes symbols for bond types which are described in the table (Figure 1/118).

SYMBOL	BOND TYPE	DISPLAY
C	Chain	———
R	Ring	—⊖—
RC	Ring or chain	—⊗—

Figure 1/118

By default, the bonds in rings are ring bonds and cannot be changed to chain bonds. All other bonds that are not in rings are chain bonds unless specified otherwise.
By respecifying the default values for bond types, we force the System to search and retrieve selectively only specific structures while others are excluded.
For example, (no SET BOND used), the two-node fragment created by:

KEYBOARD

ENTER (DIS), GRA, NOD, BON, OR ?:gra c2,dis C ∿ C

with no bond value and bond type specification will be assigned by the System the default bond value "unspecified" and bond type "chain" since the fragment is a chain. On searching this query, only structures will be retrieved in which the two-node fragment is part of chain structures (any substitution at both nodes is allowed). Ring structures in which the two-node fragment is embedded in rings will be excluded.
The structures in Figure 1/119 will be retrieved.

$$\diagdown_{\diagup}\!C\!-\!C\!\diagup_{\diagdown} \quad \diagdown_{\diagup}\!C\!=\!C\!\diagup \quad -C\!\equiv\!C-$$

Figure 1/119

The structures in Figure 1/120 will not be retrieved.

CREATING GENERIC STRUCTURES

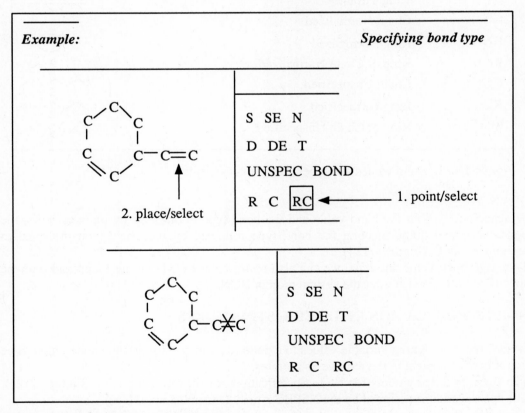

Figure 1/120

However, if we specified the bond type between the nodes as ring (R), only ring structures would be retrieved with the two nodes embedded in ring structures. If the same bond was specified as ring or chain (RC), all structures in which the two nodes are part of chain or ring structures would be retrieved. Specification of bond types is shown in Figure 1/121.

Figure 1/121. Bond type specification

It is possible to assign a combination of bond values and bond types in a generic structure.

The tables (Figures 1/122, 1/123) show the combinations.

SYMBOL	BOND DEFINITION	DISPLAY
CSE	Chain Single Exact	
RSE	Ring Single Exact	
RCSE	Ring or Chain Single Exact	
CDE	Chain Double Exact	
RDE	Ring Double Exact	
RCDE	Ring or Chain Double Exact	
CT	Chain Triple	
RT	Ring Triple	
CN	Chain Normalized	
RN	Ring Normalized	
RCN	Ring or Chain Normalized	
CU	Chain Unspecified	
RU	Ring Unspecified	
RCU	Ring or Chain Unspecified	

Figure 1/122. Bond values and types and graphics display

We can specify both the bond value and the bond type in one step by point/selecting the bond value and point/selecting the bond type followed by place/select on the bond to assign the values (Figure 1/124).

To assign bond value and type, we can also point/select BOND on the Menu and type the specification in Area A after the system prompt BOND:.

NODE SPECIFICATION IN GENERIC STRUCTURES

In addition to assigning variable values to a node, we can specify that the node ought to be in a ring or in a chain in the retrieved structures.

The Registry File symbols for Node Specification are shown in the table (Figure 1/125). The specification is assigned by the NSPEC Attribute Command (*vide infra*).

The nodes that are in rings are ring nodes and they cannot be respecified to chain nodes. All other nodes that are not in rings are chain nodes and can be specified as C, R, or RC.

Node Specification is dependent on Bond Type.

The values for a pair of nodes A----B in relation to the Bond Type by which they are

CREATING GENERIC STRUCTURES

SYMBOL	BOND DEFINITION	DISPLAY
CS	Chain Single Exact or Normalized	
RS	Ring Single Exact or Normalized	
RCS	Ring Single Exact or Normalized or Chain Single Exact or Normalized	
CD	Chain Double Exact or Normalized	
RD	Ring Double Exact or Normalized	
RCD	Ring Double Exact or Normalized or Chain Double Exact or Normalized	

Figure 1/123. Bond values and types and graphics display

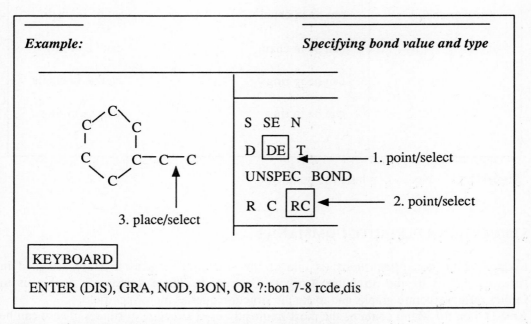

Figure 1/124. Specification of bond value and type in one step

connected are shown in Figure 1/126.

SYMBOL	DEFINITION
C	Node is in chain
R	Node is in ring
RC	Node can be in chain or ring

Figure 1/125. Node specifications

BOND TYPE	NODE A		B NODE
C(chain)	is in chain		is in chain
R(ring)	is in ring		is in ring
RC(ring or chain)	can be in chain		can be in chain
		or	
	can be in chain		can be in ring
		or	
	can be in ring		can be in chain
		or	
	can be in ring		can be in ring

Figure 1/126. Nodes in relation to bond type

STRUCTURE ATTRIBUTE COMMANDS

Node Specification (*vide supra*) is one of the structural features that are assigned in structure graphs by the Structure Attribute Commands. These commands enable us to further refine structure graphs and make the structure queries more specific.

Area D.5 on the Menu contains the most frequently used structure attributes. These can be point/selected and place/selected on the node to have the structure attribute. Those attributes that are not on the Menu have to be specified by activating Area A by depressing K or point/selecting KEY and typing the attribute followed by place/select on the graph.

CREATING GENERIC STRUCTURES

NODE SPECIFICATION

Node Specification in generic structure graphs is assigned by the Structure Attribute Command NSPEC.

Example: Assigning node specification

In the structure query (Figure 1/127), the C of the ester group is in a chain and, by default, its NSPEC is chain (C). Only alkyl esters would be retrieved.
To retrieve, selectively, aryl esters we specify that the C of the ester group must be in a ring not in a chain (Figure 1/127).
NSPEC Attribute is not displayed with the structure. To display it see the section "How to Display Structure Graphs" (p. 165).

RING SPECIFICATION

For structure graphs containing rings, we can specify the rings to be isolated, i.e., no fusion is allowed at the rings. For single rings, this specification results in isolation of the single rings. For fused ring systems, the entire fused ring structure will be isolated. If we wish to prevent fusion at individual rings in fused ring systems, we have to use the CONNECT Command (see p. 152) or HCOUNT (*vide infra*). Figure 1/128 shows assignment of ring specific.
To isolate the phenyl ring in Figure 1/129:

KEYBOARD

ENTER (DIS), GRA, NOD, BON, OR ?:rsp 11

To isolate the fused ring system in Figure 1/129:
ENTER (DIS), GRA, NOD, BON, OR ?:rsp 1

To isolate all rings in the structure in Figure 1/129:
ENTER (DIS), GRA, NOD, BON, OR ?:rsp

Ring specification is not displayed with the graph. To display it see the section "How to Display Structure Graphs" (p. 165).

HYDROGEN COUNT

In certain structures, the number of hydrogens at nodes is implied or has a default value. For example, the System Shortcuts (see p. 84) have a default number of hydrogens at nodes. The structure graphs that are submitted to Exact or Family Search (see p. 218) have an implied number of hydrogens at the nodes that do not show any substitution by other

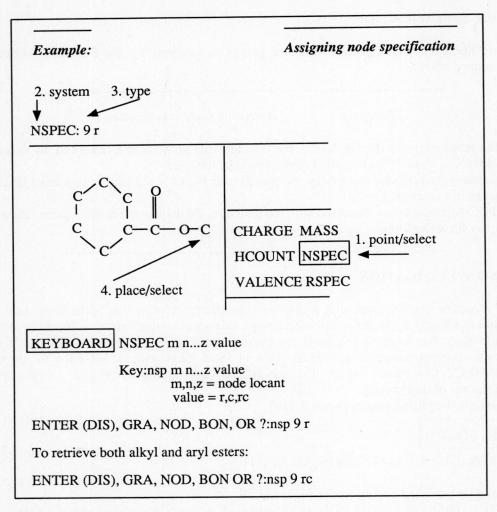

Figure 1/127. Assignment of node specific

elements or structural fragments. No substitution is allowed at these nodes.
However, when we submit a structure graph to Substructure Search (see p. 218), any substitution is allowed at all nodes to meet their valencies except when we prevent the substitution by explicit hydrogens or by specifying the number of hydrogens at a node by the HCOUNT Command.
We can specify exact (Figure 1/130), or minimum number of hydrogens at nodes.
The Hydrogen Count Attribute is displayed with the graph.

CREATING GENERIC STRUCTURES

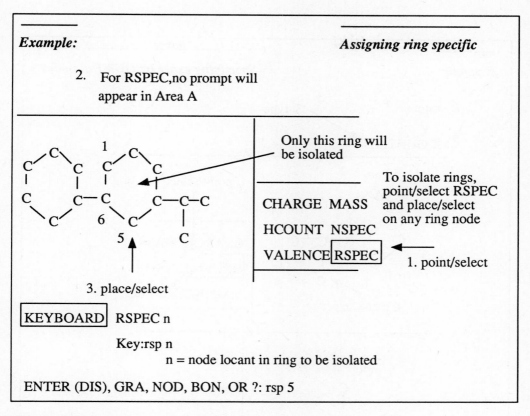

Figure 1/128. Ring specific assignment

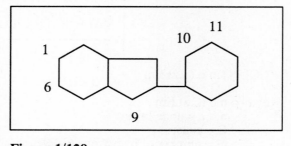

Figure 1/129

Example: **Specifying minimum HCOUNT at nodes**

To search for the substituted structure (Figure 1/131), we specify that the node 3 must have at least one hydrogen, but may have two hydrogens:

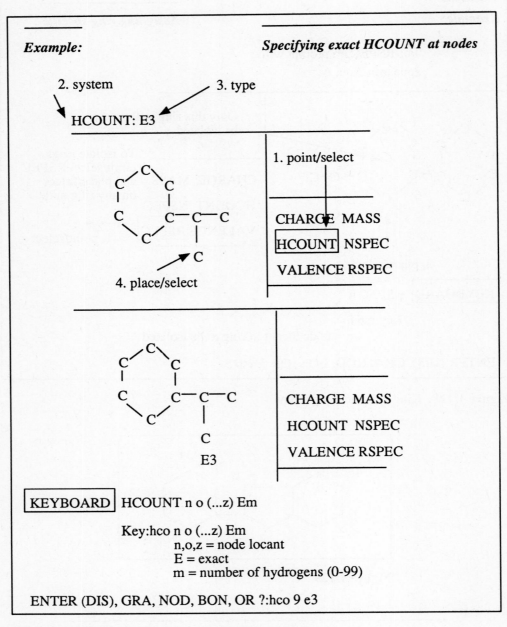

Figure 1/130. Assignment of exact hydrogen count

CREATING GENERIC STRUCTURES

[Figure showing a chemical structure with CH₃ groups, an H atom, R₁ substituent at node 3, and an O-O-R peroxide group at node 6. Legend: R,R1=anything including H]

Figure 1/131

KEYBOARD

1. Create structure query:
ENTER (DIS), GRA, NOD, BON, OR ?:gra c6,2 c1,2 c1,4 c1,4 c1,nod 5 6 O
2. Specify exact number of hydrogens:
ENTER(DIS), GRA, NOD, BON, OR ?:hco 1 7 8 9 10 e3
3. Specify minimum hydrogen count for node 3:
ENTER (DIS), GRA, NOD, BON, OR ?:hco 3 m1

We specified that the node 3 (Figure 1/131) has minimum number of hydrogens equal to 1. It means that there must be at least one hydrogen in the retrieved structure, i.e., one substituent must be a hydrogen. The other substituent to meet the valency of the node 3 can be anything, including hydrogen by default.
The node 6 does not have any HCOUNT specification, therefore, by default, there can be a hydrogen or any other element or structural fragment.
If we want to prevent substitution at the node 3, i.e., no other elements or structural fragments are allowed except hydrogens, we specify the exact HCOUNT for the node 3:

ENTER (DIS), GRA, NOD, BON, OR ?:hco 3 E2

It means that there must be two hydrogens at the node 3 in the retrieved structures; consequently, no substitution is allowed by other element or structural fragment. The node 6, again, has no HCOUNT specification, by default there can be a hydrogen or any element or structural fragment.

Example: **Specifying zero HCOUNT at nodes**

KEYBOARD
HCOUNT n o(...z) E0

Key:hco n o(...z) E0
   n,o,z=node locant
   E0=exactly zero

If we wish to retrieve structures with all possible substituents at the node 3, excluding hydrogens, we specify that there must be no hydrogen at node 3:

ENTER (DIS), GRA, NOD, BON, OR ?:hco 3 E0

It should be noted that in many cases the use of the HCOUNT Command is not necessary. The System Shortcuts have a minimum number of hydrogens at the attachment node and a maximum number of hydrogens at the remaining nodes. Thus, the structure in Figure 1/131 could be built using the shortcuts and no HCOUNT specifications:

KEYBOARD

ENTER (DIS), GRA, NOD, BON, OR ?:
gra c6,2 c1,2 c1,4 c1,4 c1,nod 5 6 O,1 7 8 9 ME,3 CH2,dis

The default number of hydrogens would prevent any substitution at the node 3 as well as in the ME groups. Substitution on the oxygen is allowed.

THE CONNECT COMMAND

Similarly to specifying the number of hydrogen attachments at a node by the HCOUNT Command, we can specify the number of nonhydrogen attachments at a node by the CONNECT Command. This command enables us to assign a degree of substitution and a type of substitution at nodes in generic structures.

THE DEGREE OF SUBSTITUTION AT NODES

The degree of substitution for a node is expressed as the number of nonhydrogen attachments at the node.
We can specify exact, minimum, or maximum number of nonhydrogen attachments at a node. If no degree of substitution is assigned to a node, all possible substitution at the node is allowed.

Example: ***Specifying exact number of substituents***

KEYBOARD CONNECT n o (...z) Em
   Key:con n o (...z) Em

CREATING GENERIC STRUCTURES

n,o,z=node locant
E=exact
m=number of nonhydrogen attachments (0-16)

Note: The number of non-hydrogen attachments "m" is the sum of nonhydrogen attachments already present in the structure graph plus the number of nonhydrogen attachments we wish to assign.

By creating the structure graph (Figure 1/132), we can search for derivatives selectively substituted at the nodes 1, 3, 4, and 6 and request no substitution at the node 5.

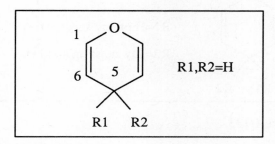

Figure 1/132

KEYBOARD

1. Create structure graph:
ENTER (DIS), GRA, NOD, BON, OR ?:gra r6,nod 2 O,bon 1-6 3-4 de
2. Specify degree of substitution:
ENTER (DIS), GRA, NOD, BON, OR ?:con 5 E2

We specified the exact number of nonhydrogen attachments equal to 2. Since there are already two nonhydrogen attachments in the graph, the specification is satisfied and no further substitution is allowed. There will be a CH2 fragment in the 5 position in the retrieved structures. Substitution at the other nodes is allowed.
To display CONNECT Attribute with the query, see p. 165.

Example: Specifying minimum number of substituents

KEYBOARD CONNECT n o (...z) Mm

Key:con n o (...z) Mm
n,o,z=node locants

M=minimum
m=number of nonhydrogen attachments (0-16)

Minimum signifies that there must be at least the number of nonhydrogen attachments expressed by the "m" but there may also be more than the minimum specified by the "m." The "m" is the sum of nonhydrogen attachments already present in the graph plus the number of attachments we wish to assign.

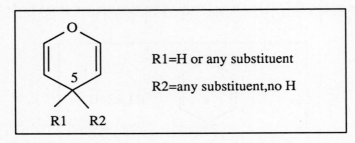

Figure 1/133

If in the structure (Figure 1/133) R1 can be H or any substituent, and R2 cannot be H but can be any substituent, we specify:

KEYBOARD

ENTER (DIS), GRA, NOD, BON, OR ?:con 5 M3

This specification indicates that there must be at least three nonhydrogen attachments at the node 5. There are already two nonhydrogen attachments in the graph so that to satisfy the requirement, one additional nonhydrogen attachment is required. The fourth attachment to fill the valency of the node may be either H or any substituent.

Example: Specifying maximum number of sustituents

KEYBOARD CONNECT n o (...z) Xm

    Key:con n o (...z) Xm
    n,o,z=node locant
    X=maximum
    m=number of nonhydrogen attachments (0-16)

Maximum signifies that the number of nonhydrogen attachments can be up to the maximum expressed by the "m," but there can also be less than the maximum specified by

CREATING GENERIC STRUCTURES

the "m."
The "m" is the sum of nonhydrogen attachments already present in the graph plus the number of nonhydrogen attachments we wish to assign.

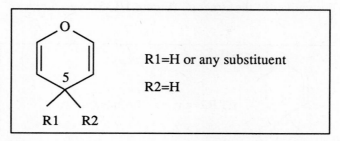

Figure 1/134

To retrieve derivatives of this type (Figure 1/134), we specify:

| KEYBOARD |

ENTER (DIS), GRA, NOD, BON, OR ?:con 5 X3

This specification indicates that up to three nonhydrogen attachments at the node 5 are allowed. There are already two nonhydrogen attachments in the graph. If there is a third nonhydrogen attachment (R1) (up to the maximum), the fourth must be hydrogen (R2). If there is a hydrogen (R2), the fourth may also be a hydrogen (R1) (less than maximum).

THE TYPE OF SUBSTITUTION

In addition to specifying the degree of substitution, we can direct the System to search for derivatives of our query structure having either chain substituents, ring substituents, or both of these. We specify a Bond Type (ring, chain, ring or chain) of the substituent. If no Bond Type is specified, the default is ring or chain. If the structure graph has a Node Specification, the NSPEC must agree with the specified Bond Type. If it does not, the Node Specification must be respecified.

Example: *Specifying type of substitution*

| KEYBOARD | CONNECT n o (...z) E(M,X)m symbol

  Key:con n o (...z) E(M,X)m symbol
     n,o,z=node locant
     E=exact

M=minimum
X=maximum
symbol=c,r,rc

To search for the derivatives of the structure (Figure 1/135), we specify:

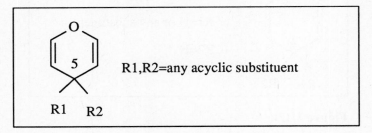

Figure 1/135

| KEYBOARD |

ENTER (DIS), GRA, NOD, BON, OR ?:con 5 E2 c

This means that there must be exactly two nonhydrogen attachments at the node 5 and they must be by the chain bond. We do not have to take into account the two already present nonhydrogen attachments since no more substitution is possible due to the valency of carbon (node 5). We would retrive structures like those in Figure 1/136.
Note: The Bond Type assignment does not affect the bonds in the substituents; therefore, rings may be in the alkyl substituents. However, the substituent itself may be a ring while the bond of the substitution is the chain bond.
If we specified:

| KEYBOARD |

ENTER (DIS), GRA, NOD, BON, OR ?:con 5 E4 r

This would mean that there must be exactly four nonhydrogen attachments at the node 5 and they must have ring bonds. The spiro ring systems like those in Figure 1/137 would be retrieved.
To retrieve both chain and ring substituents, we would specify:

| KEYBOARD |

ENTER (DIS), GRA, NOD, BON, OR ?:con 5 E4 rc

CREATING GENERIC STRUCTURES

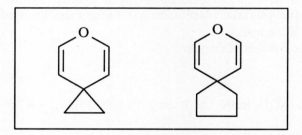

Figure 1/136

Figure 1/137

SELECTIVE FUSION OF RINGS BY THE CONNECT COMMAND

To prevent fusion at individual rings in a fused ring system, the RSP Structure Attribute (see p. 147) cannot be used. However, the CONNECT Command enables us to direct fusion at a specific ring in fused rings. To fuse a ring at the ring B in the structure in Figure 1/138 and prevent fusion at the ring A:

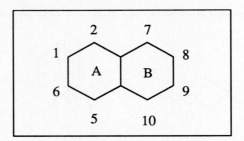

Figure 1/138

Example: *Selective fusion of rings*

KEYBOARD

1. Create the graph:
ENTER (DIS), GRA, NOD, BON, OR ?:gra r66,bon all n
2. Specify nonhydrogen connection:
ENTER (DIS), GRA, NOD, BON, OR ?:con 1 2 5 6 E2 r

The nodes 1 2 5 6 have been specified to have exactly 2 ring type nonhydrogen connections. No fusion can take place at ring A since there are already exactly 2 ring connections at each of the nodes.
To force fusion at the 1-6 side of ring A:

KEYBOARD

ENTER (DIS), GRA, NOD, BON, OR ?:con 2 5 7 8 9 10 E2 r,1 6 E3 r

THE CHARGE COMMAND

As one of the Structure Attributes, we can specify a charge on one or more nodes in the structure graphs. By selecting the command CHARGE we can assign a positive, negative, or zero charge (Figure 1/139). If no charge is specified, both charged and uncharged substance structures will be retrieved.

The assignment of charges is dependent on the Registry System convention which is complicated. It is recommended not to use this command in ambiguous cases and rather retrieve both charged and uncharged substance structures.

CREATING GENERIC STRUCTURES

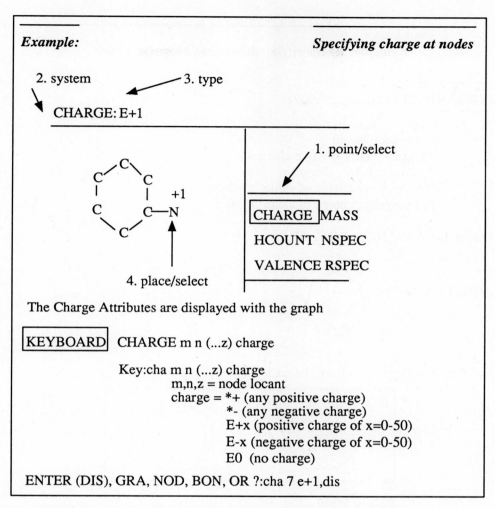

Figure 1/139. Specification of charges at nodes

THE DELOCALIZED CHARGE COMMAND

In the structure graphs in which a charge may be delocalized we specify the charge by the Delocalized Charge Command. If no specification is given, all substances with delocalized or with no delocalized charge will be retrieved.

The assignment of delocalized charges is dependent on the conventions used in the Registry System. The theory is complicated. It is recommended not to use this command and retrieve both charged and uncharged substance structures.

Example: **Specifying delocalized charge at nodes**

KEYBOARD DLOC m n (...z) charge
    Key:dlo m n (...z) charge
    m,n,z=node locant
    charge=*+ (any positive charge)
    *- (any negative charge)
    E+n (positive charge of value "n"=0-5)
    E-n (negative charge of value "n"=0-5)

ENTER (DIS), GRA, NOD, BON, OR ?:

gra r7,dlo 1 2 3 E+1,dis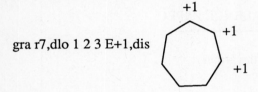

The structures in Figure 1/140 will be retrieved:

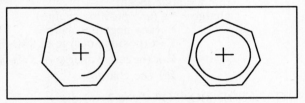

Figure 1/140

THE HYDROGEN MASS COMMAND

The HCOUNT Command (*vide supra*) assigns a specified number of hydrogens of normal (1) mass value. The abnormal mass hydrogens can be assigned by the HMASS Command. If no HMASS values are specified, structures with hydrogens of all masses will be retrieved.

CREATING GENERIC STRUCTURES

Example: **Specifying abnormal hydrogen mass**

KEYBOARD HMASS m n (...z) symbol

Key:hma m n (...z) symbol
m,n,z=node locant
symbol=D (deuterium)
T (tritium or higher isotope)
* (any abnormal hydrogen mass)

The values D, T, and * signify that at least one abnormal mass hydrogen must be present at the node.

KEYBOARD

ENTER (DIS), GRA, NOD, BON, OR ?:

gra c3,2 c1,hma 2 D,dis

```
        C
        |
   C — C — C
```

(The HMASS Attribute is not displayed with the graph. To display it, see STRUCTURE DISPLAY (p. 165)).

The following structure will be retrieved:

$$\begin{array}{c} CH_3 \\ | \\ H_3C - C - CH_3 \\ | \\ D \end{array}$$

THE MASS COMMAND

The MASS Command is used to specify any value of abnormal mass at nodes (Figure 1/141).
If the MASS Structure Attribute is not specified, structures of substances with elements of normal and abnormal masses will be retrieved.
The value "j" represents exactly the required mass. By specifying a normal mass at a node, the abnormal masses are excluded at the node. By specifying an abnormal mass, only ele-

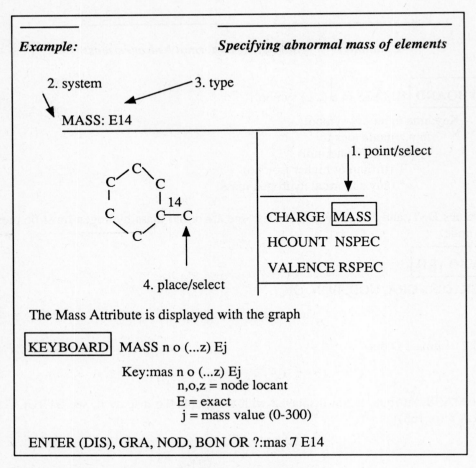

Figure 1/141. Assigning abnormal mass of elements

ments of that exact abnormal mass will be retrieved.

THE VALENCE COMMAND

The Valence Attribute Command assigns a specific or any valence to a node, be it a normal valence or abnormal valence. If no valence is assigned, nodes with all possible valences will be contained in the retrieved structures.

This command can be used, similarly to the CONNECT Command *(vide supra)*, to assign a degree of substitution at a node. The System will be forced to search and retrieve only those structures with the specified valences.

CREATING GENERIC STRUCTURES

Example: *Specifying valence at nodes*

KEYBOARD VALENCE n o (...z) Ej

　　Key: val n o (...z) Ej
　　　　n,o,z=node locant
　　　　E=exact
　　　　j=valence value (1-16)
　　　　val n o (...z) * =any abnormal valence

ENTER (DIS), GRA, NOD, BON, OR ?:

gra c4,nod 3 S,val 3 E2,dis C∽C∽S∽C　(no SET BOND used)

The structure (Figure 1/142) will be retrieved in which the sulfur atom is singly bonded to two substituents only, chain or ring (the other nodes can have any substitution), while the structures with multiple bonds at the sulfur will not be retrieved.

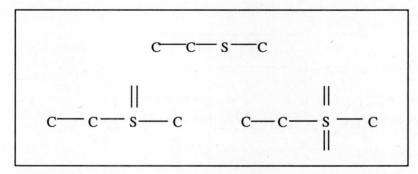

Figure 1/142

STRUCTURE ATTRIBUTES IN Gk GROUPS

The Structure Attributes cannot be assigned directly to Gk groups.
If we want to specify a Structure Attribute for one of the values defined in a Gk group, we have to build the value as a separate fragment and assign the Structure Attribute in the fragment (Figure 1/143).

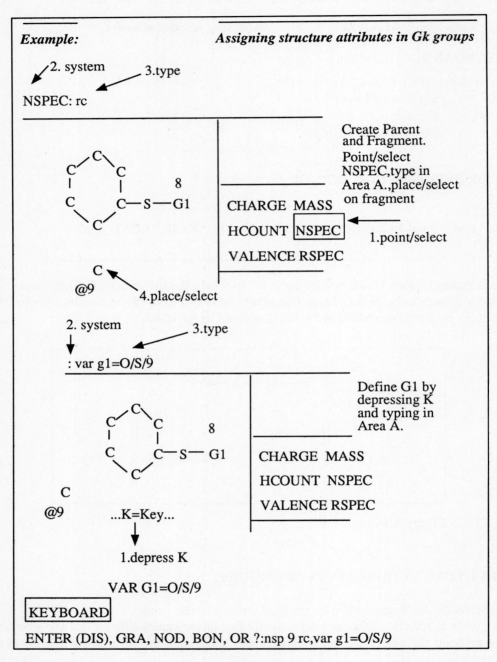

Figure 1/143. Assigning structure attributes to Gk groups

CREATING GENERIC STRUCTURES

HOW TO DISPLAY STRUCTURE GRAPHS

In the Keyboard Commands mode, the DISPLAY Command is the default in the line of options:
ENTER (DIS), GRA, NOD, BON, OR ?:
When we wish to activate the default (in parentheses) we enter a "." followed by carriage return. The graph is displayed on the screen. If no structure graph is available for display, the System repeats the prompt:
ENTER (DIS), GRA, NOD, BON, OR ?:

The DISPLAY Command is used to display, in addition to the Structure Image (SIM), other data related to structure graphs. We have seen, for example, that some of the Structure Attributes display in the structure graph, while some of them do not. We can display them by the DISPLAY Command.

The following is a list of items that can be displayed by the DISPLAY Command:
1. Structure Image (SIM)
2. Structure Attributes (SAT)
3. Structure Image and Attributes (SIA)
4. Structure Connection Table (SCT)
5. Structure Data (SDA)

In Menu mode, structure image is always displayed. Only 2 and 4 are the valid options.

Example: *Displaying structure attributes*

| KEYBOARD | DISPLAY SAT

   Key:dis sat

ENTER (DIS), GRA, NOD, BON, OR ?:

gra r6,6 c1,hco 1 2 3 4 5 E1,6 E0,nsp 7 rc,rsp,dis

ENTER (DIS), GRA, NOD, BON OR ?:dis sat

NODE ATTRIBUTES:
HCOUNT IS E1 AT 1

HCOUNT IS E1 AT 2
HCOUNT IS E1 AT 3
HCOUNT IS E1 AT 4
HCOUNT IS E1 AT 5
HCOUNT IS E0 AT 6
NSPEC IS RC AT 7

GRAPH ATTRIBUTES:
RSPEC I
NUMBER OF NODES IS 7

We can display attributes for a single node only:

| KEYBOARD | DISPLAY m SAT

or for a set of nodes:

| KEYBOARD | DISPLAY m-n SAT

    Key:dis m sat
      dis m-n sat
        m,n=node locant

Example: Displaying structure image and attributes

| KEYBOARD | DISPLAY SIA

    Key:dis sia

ENTER (DIS), GRA, NOD, BON, OR ?:

gra r6,6 c1,hco 1 2 3 4 5 E1,6 E0,nsp 7 rc,rsp,dis sia

NODE ATTRIBUTES:
HCOUNT IS E1 AT 1
HCOUNT IS E1 AT 2
HCOUNT IS E1 AT 3
HCOUNT IS E1 AT 4

CREATING GENERIC STRUCTURES

HCOUNT IS E1 AT 5
HCOUNT IS E0 AT 6
NSPEC IS RC AT 7

GRAPH ATTRIBUTES:
RSPEC I
NUMBER OF NODES IS 7

Example: *Displaying structure connection table*

KEYBOARD DISPLAY SCT
    Key: dis sct

ENTER (DIS), GRA, NOD, BON, OR ?:dis sct

(See p. 31 for displayed connection table.)

We can display connection table for only a single node:

KEYBOARD DISPLAY m SCT

or for a set of nodes:

KEYBOARD DISPLAY m-n SCT

Example: *Displaying structure data*

KEYBOARD DISPLAY SDA
    Key:dis sda

ENTER (DIS), GRA, NOD, BON, OR ?:dis sda

All data, i.e., structure image, attributes, and connection table are displayed.

---

10 CORRECTING/MODIFYING STRUCTURES

In the process of creating structure graphs we may make errors, occasionally, resulting in a structure graph that is not the one we wanted to create. The errors might be simply due to typing errors that we might miss correcting before the graph is finished.
At times, we will finish a structure graph which is correct but which, on searching it, does not retrieve exactly the structures we were hoping to retrieve.
We need to correct and modify structure graphs after they are completed and displayed.
We also need to modify models recalled from the System File or the User File.

Modification and correction of structure graphs is performed by the DELETE Command to erase:
1. Single or multiple nodes
2. Single or multiple bonds
3. Single or multiple structure attributes
4. Entire chains
5. Entire rings
6. Entire structures

HOW TO DELETE NODES

All bonds from deleted node(s) will be deleted. Consequently, deletion of nodes gives rise to different results depending on the nature of the deleted node(s).
If the node is a terminal chain node, we obtain a shorter chain. If the node is an internal chain node or is part of a ring, we obtain structural fragments resulting from deleting the node(s) and the bonds by which the node(s) was connected. In some cases this may be undesirable but in other cases we may be able to create, in one stroke, structural fragments ready to search (Figure 1/144).
Note: When we delete a node or nodes from a ring and obtain chain fragments, the Node Specification and Bond Type have changed from ring to chain.

Example: *Deleting nodes*

| KEYBOARD | DELETE m (..z)

    Key:del m (..z)
        m,z=node locant
ENTER (DIS), GRA, NOD, BON, OR ?:

CAS REGISTRY

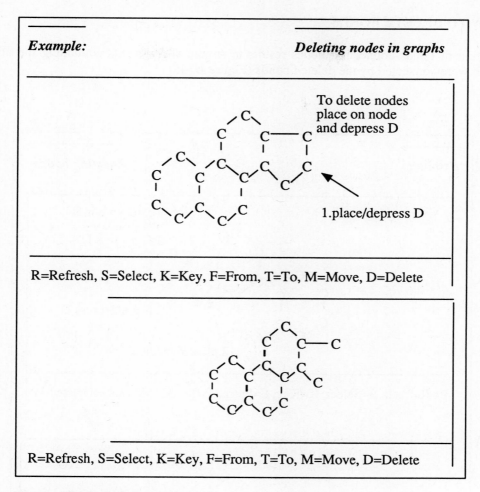

Figure 1/144. Deleting nodes

```
              1     2     3     4     5     6
gra c6,dis
              C — C — C — C — C — C
ENTER (DIS), GRA, NOD, BON, OR ?:

              2     3           5     6
del 1 4,dis
              C — C           C — C
```

HOW TO DELETE BONDS

Deletion of bonds connecting nodes results in structural fragments containing the nodes which were connected by the deleted bonds (Figure 1/145).

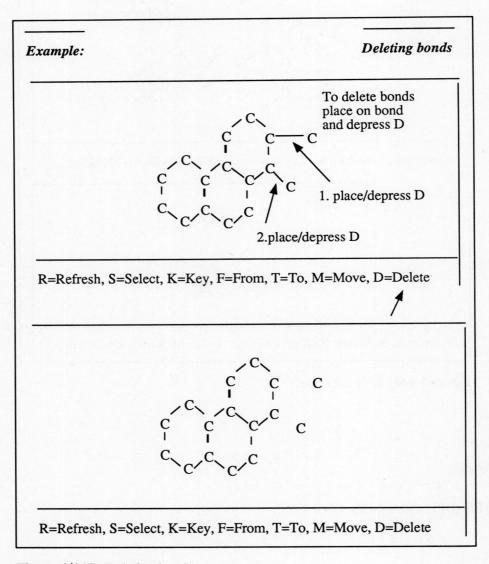

Figure 1/145. Deleting bonds

CORRECTING/MODIFYING STRUCTURES

Example: *Deleting bonds*

Note: When we delete bonds in rings and obtain chain fragments, the Bond Type and Node Specification have changed from ring to chain.

KEYBOARD DELETE m-n o-p...(y-z)

   Key: del m-n o-p...(y-z)
      m,n,o,p,y,z = node locant of bond

Existing chain:
```
        1    2    3    4    5
        C——C——C——C——C
```

ENTER (DIS), GRA, NOD, BON, OR ?:

del 1-2 3-4,dis
```
        1    2    3    4    5
        C    C——C    C——C
```

HOW TO DELETE CHAINS

We can delete isolated chain structure graphs or chains that are connected to rings.

Example: *Deleting entire chain*

KEYBOARD DELETE C n

   Key: del c n
      c = chain
      n = node locant in chain to be deleted

Existing structure: hexagonal ring with nodes 1,2,3,4,5,6 connected to chain C-C-C-C at positions 7,8,9,10 attached at node 3.

ENTER (DIS), GRA, NOD, BON, OR ?:del c 10,dis

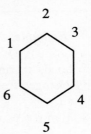

HOW TO DELETE RINGS

Single isolated rings or rings which are connected to chains or to ring systems by acyclic bonds can be selectively deleted. However, individual rings in fused ring systems have to be deleted node by node (Figure 1/146).

Example: *Deleting entire single ring*

KEYBOARD DELETE R n

   Key:del r n
      r=ring
      n=node locant of ring to be deleted

Existing structure:

```
       2
    1 /―\ 3   7   8   9   10
     |   |―― C ― C ― C ― C
    6 \_/ 4
       5
```

ENTER (DIS), GRA, NOD, BON, OR ?:del r 2

```
  7    8    9    10
  C ―― C ―― C ―― C
```

CORRECTING/MODIFYING STRUCTURES

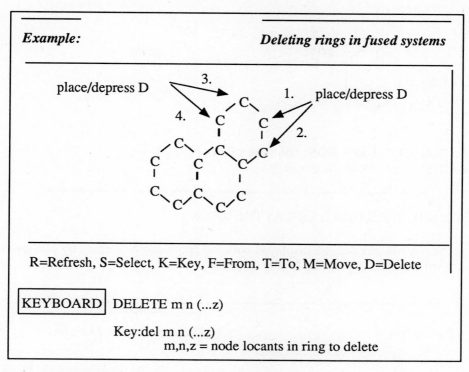

Figure 1/146. Deleting rings in fused systems

HOW TO DELETE STRUCTURE GRAPH

The entire structure can be deleted from display. The structure is lost unless previously saved. When we have separate structures on the screen, we can selectively remove any of them.
The structures are automatically deleted, unless saved, on disconnecting from STN.

Example: *Deleting entire structure*

KEYBOARD DELETE S n

Key:del s n
  s=structure
  n=any node locant in structure to be deleted

ENTER (DIS), GRA, NOD, BON, OR ?:

gra r6,3 c4,dis

```
      2    7   8   9   10
    ╱─╲3
   1    ├──C───C───C───C
    ╲─╱
   6    4
      5
```

ENTER (DIS), GRA, NOD, BON, OR ?:del s 9,dis
ENTER (DIS), GRA, NOD, BON OR ?:

HOW TO DELETE STRUCTURE ATTRIBUTES

The Structure Attributes that are displayed with the graph can be deleted by placing on the Structure Attribute and depressing D (Figure 1/147).

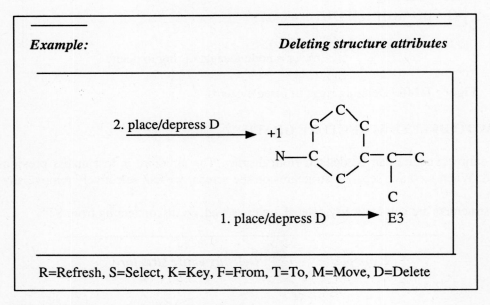

Figure 1/147. Deleting structure attributes

Those Structure Attributes that are not displayed with the graph have to be deleted by DELETE Command.

| **KEYBOARD** | DELETE n o(...z) symbol

CORRECTING/MODIFYING STRUCTURES

>   Key:del n o(...z) symbol
>   n,o,z=node locant
>   symbol=see Table
>   del all symbol- deletes all attributes
>   represented by symbol in graph

The following symbols are used for deleting structure attributes:

Attribute	Symbol
Charge	CHA
Delocalized charge	DLO
Connect	CON
Hydrogen count	HCO
Hydrogen mass	HMA
Mass	MAS
Node specification	NSP
Ring specification	RSP
Valence	VAL

Figure 1/148 shows deletion of NSPEC and RSPEC.

RESPECIFICATION OF NODE AND BOND VALUES

When we want to change the values that have been assigned to nodes and bonds, we simply point/select the new value on the Menu and place/select it on the node or bond that are to have the new value (Figure 1/149).

CHANGE IN CHAIN STRUCTURES

The length of a chain can be changed either by deleting terminal nodes to make the chain shorter or by adding chain fragments to make it longer.
Straight chains can be converted to branched chains by adding a chain fragment.
Node and bond values in a chain can be reassigned.

CHANGE IN RING STRUCTURES

The size of rings cannot be reassigned once the ring has been created. All other values in ring structures can be changed except Node Specification and Bond Type.

HOW TO MOVE STRUCTURE GRAPHS ON SCREEN

The structure graphs displayed in Area B can be manipulated as to the position of the entire graph, or parts of the graph, on the screen (Figure 1/150).

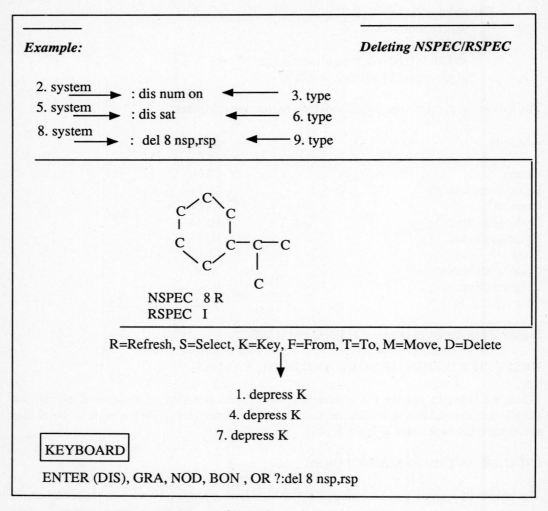

Figure 1/148. Deleting NSPEC and RSPEC

Note: If we are in the Keyboard Command mode there must be at least two structure graphs on the screen in order to move them about the screen.

KEYBOARD MOVE S n Um(Dm,Lm,Rm)

    Key:move s n Um(Dm,Lm,Rm)
      s=structure
      n=any node locant in structure
      m=number of spaces (terminal dependent)

CORRECTING/MODIFYING STRUCTURES

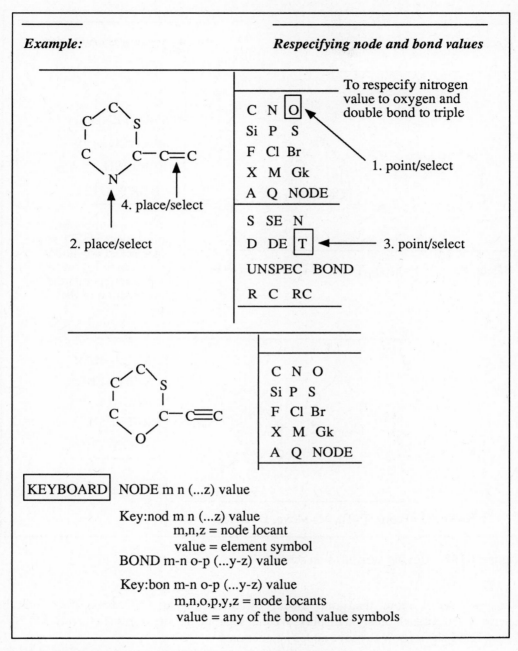

Figure 1/149. Respecification of node and bond values

U=up
D=down   L=left   R=right

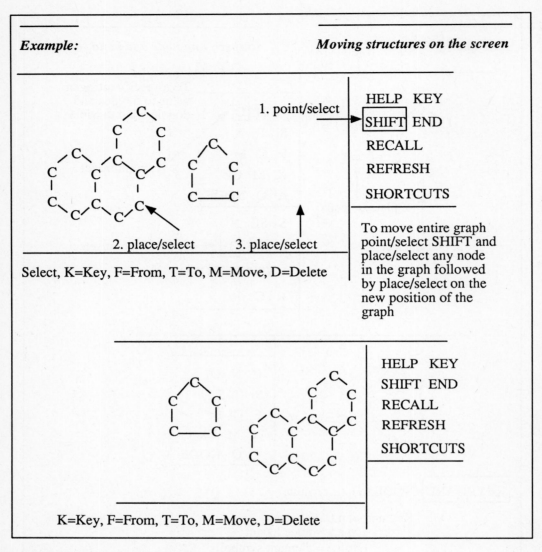

Figure 1/150. Moving structures on screen

When we recall more than one model by their Registry Numbers, they may be superimposed on each other on display. We can separate the structures (Figure 1/151).
Figure 1/152 shows how to move nodes in structures.
Those manipulations of the position of structure graphs that cannot be performed by the Menu templates can be accomplished by depressing K or point/select Key and entering Keyboard Commands.

CORRECTING/MODIFYING STRUCTURES

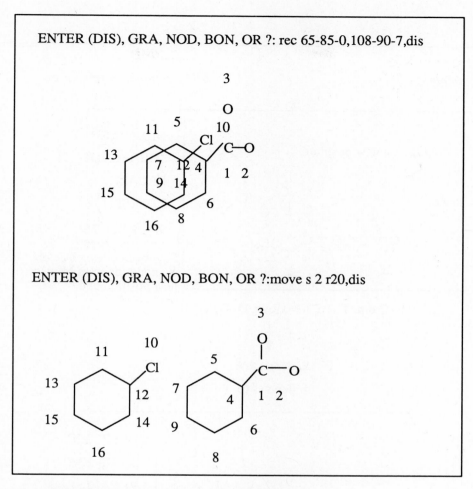

Figure 1/151. Moving structures on the screen

KEYBOARD

We have learned that the screen of a terminal or a microcomputer represents a page and that the System periodically issues a prompt CLEAR AND COPY PAGE PLEASE to prepare the screen for additional display. The first structure we create is always displayed in the left upper corner of the screen. When we continue in building and displaying structure graphs, the System, without our intervention, proportionally displays the graphs when there is more than one graph on the screen.

We may get graphs on display which overlap or substituents in separate graphs may overlap. At times, we may create a structure graph whose shape is distorted especially when we create large rings from chains.

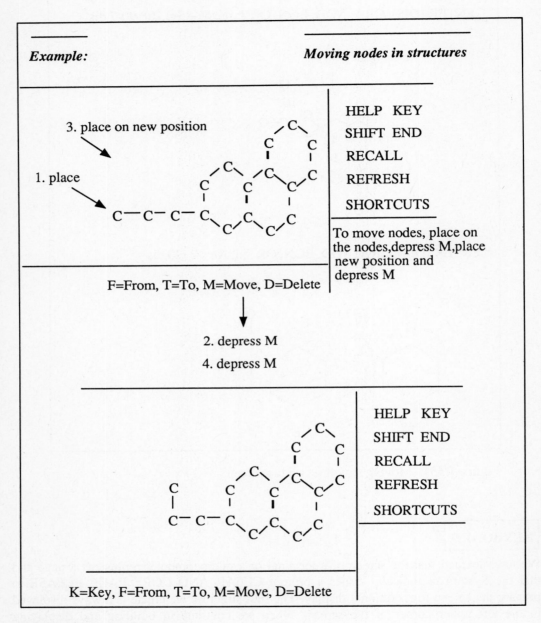

Figure 1/152. Moving nodes

We are able to manipulate structure graphs displayed on the screen and reposition and/or reshape them by the MOVE Command.

CORRECTING/MODIFYING STRUCTURES

HOW TO RESHAPE DISTORTED STRUCTURE GRAPHS

Reshaping of structures is done by moving nodes in the structure.
For example, when we create rings or ring systems with more than 8 nodes from chains, the resulting graph is not a ring but a chain or a distorted ring. When we create a graph that is too long or too wide to fit on the screen, the graph is displayed in segments. We can reshape the graph and bring it on the screen.

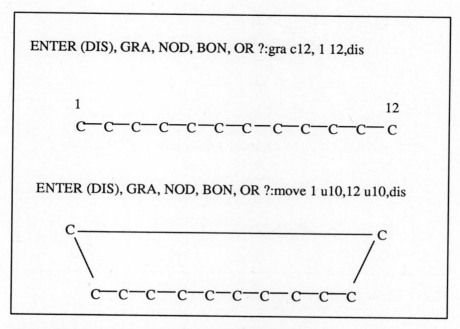

Figure 1/153. Moving nodes and reshaping structure

Example: *Moving nodes in structure graphs*

KEYBOARD MOVE n (...z) Um(Dm,Lm,Rm)

    Key:mov n (...z) Um(Dm,Lm,Rm)
     n,z=node locant to be moved
     U=up
     D=down
     L=left
     R=right
     m=number of spaces to move (terminal

ENTER (DIS), GRA, NOD, BON, OR ?:gra r66,r5,10 11,dis

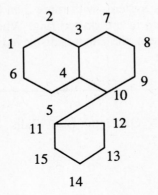

ENTER (DIS), GRA, NOD, BON, OR ?:move r 11 r20u10,dis

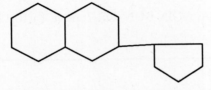

Figure 1/154. Moving fragment in structure

dependent)
(Any combination of U,D,L,R is allowed)
Figure 1/153 shows how to reshape a structure.
We can move rings or chains within a structure graph or we can move separate structural graphs.

Example: *Moving fragments in graphs*

KEYBOARD MOVE X n Um(Dm,Lm,Rm)

Key:mov X n Um(Dm,Lm,Rm)
X=r(ring), c(chain)
n=node locant in ring or chain to be moved
U=up

CORRECTING/MODIFYING STRUCTURES

D=down
L=left
R=right
m=number of spaces to move
(Any combination of U,D,L,R allowed).
Figure 1/154 shows how to move fragments in structures.

11 STRUCTURES BY STN EXPRESS

In the sections 7-10 of Part One we have described the techniques of building graphic structures online, i.e., while we were connected to CAS ONLINE - Registry File. In this section 11, we discuss the techniques of building graphic structures offline, i.e., before we connect with STN-CAS ONLINE. The graphic structures built offline can be uploaded to the Registry File after we connect with STN and can be thus used as structure queries for searching the Registry File.
Graphic structures are created offline in STN Express.

WHAT IS STN EXPRESS

We have listed STN Express as one of the Emulation Softwares (see p. 14) which is needed to utilize a microcomputer as a terminal for connecting and communicating with STN or, for that matter, with other online systems. However, STN Express is not only the Emulation Software. It is a package of softwares which offers many other functions, especially in running the STN system of databases. It is not within the scope of this book to fully describe STN Express. The function we will be concerned with here is the capability of creating query structures offline and uploading them for search in the Registry File when we connect to STN-CAS ONLINE.

LIMITS ON USING STN EXPRESS GRAPHICS

While the function of STN Express to build structures offline is very advantageous because we can create queries with no clock running in our account with STN, the structures created in STN Express cannot be edited/modified by the techniques of structure building in the Registry File while we are searching with the uploaded query and, *vice versa*, the query structures created in Registry File cannot be downloaded and manipulated in STN Express. It is possible to edit/modify the query structure prepared in STN Express or to create a new one while we are online, by switching to STN Express. The connection, however, will be still in effect and the connect charges will apply for the time we are, temporarily, in STN Express working with the query.
It should be noted, however, that while query structures created in Registry File cannot be downloaded to STN Express, the result of our search, i.e., the retrieved structures including the text data can be dowloaded from Registry File and captured on disk through STN Express for later printing and manipulating the search results offline.

OPERATION OF STN EXPRESS

STN Express is a system which is menu driven. We will find that understanding the

CAS REGISTRY

concept of structure building which we have discussed in previous sections is necessary for creating structures in STN Express.

Most of the operations of STN Express are self-explanatory and/or the menu will show which keys are operational and the system will display messages to guide us through the operations.

Part Two describes in detail ChemBase, the system for graphic structures operated through menus. The reader will find a detailed discussion in Part Two and will find many similarities in these operations. Therefore, we will only minimally describe the operations in STN Express. However, the definition of Gk groups will be discussed in detail since it is a more involved procedure.

STN menus are operated with the use of the mouse and the keyboard keys.

F1 calls Help system.

Moving the mouse, or typing the first letter of the commands on top of the menu, or pressing the keyboard arrows will move us about the STN Express Main menu and will highlight the commands. Pull-down menus will be activated from which we can select the appropriate options.

Selecting from the pull-down menus is accomplished by:
mouse - left button (selects)
<cr> - selects

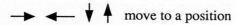 move to a position

To cancel pull-down menu:
click left button outside of menu or press ESC.

CREATING STRUCTURES

The discussion of the nature of query structures in previous sections of Part One is the prerequisite of using STN Express for structure building.

Structures are created in the Query mode which is entered from the Main Menu (Figure 1/155).

The Query Menu is then displayed (Figure 1/156).

FREE-HAND DRAWING

Drawing with the mouse pointer is always active. The bottom of the menu shows the current atom and bond for drawing. Drawing is continuous by clicking the mouse left button along the path of the structure. To discontinue at a point, click the point again and move to another point to start from there.

In selecting from pull-down Menus:
left mouse button selects for single use
right mouse button selects for multiple use
In creating structures:
left mouse button click creates new value
right mouse button click modifies existing value

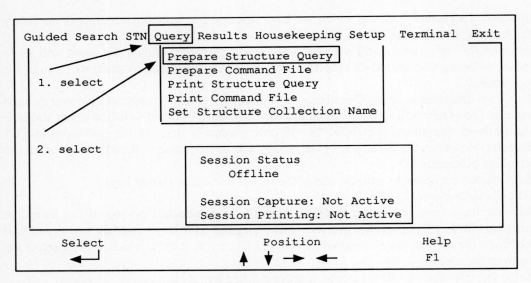

Figure 1/155. STN Express Main Menu (status offline)

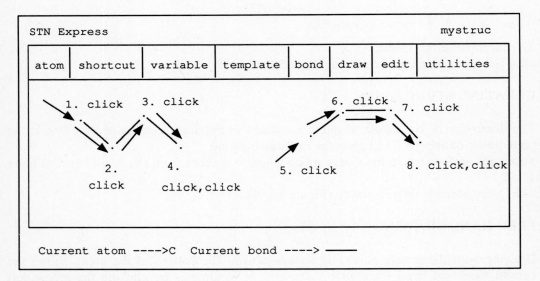

Figure 1/156. Query Menu

Pressing keyboard "d" erases at the pointer location.
Pressing space bar restores default carbon and single bond values for drawing.

STRUCTURES BY STN EXPRESS

SPECIFYING ATOMS

Select ATOM in the Menu (Figure 1/157) and select values from the displayed table by pointing the atom and clicking the mouse (*vide supra*).

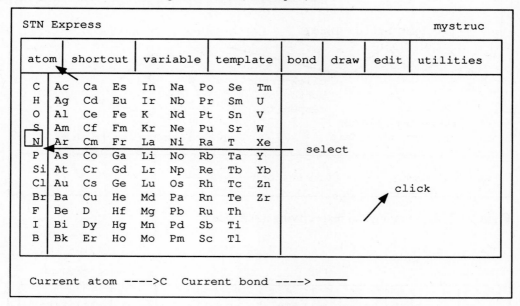

Figure 1/157. Atom Menu

The left button is loaded with the atom and when we draw, the atom is the current value for the first atom in the drawing.
The right button is loaded with the atom and is clicked on the atom(s) which is to be respecified with the new value (Figure 1/158).

SPECIFYING SHORTCUTS

Select SHORTCUT in the menu (Figure 1/159).
Select the value from the table as with the atom selection (*vide supra*).

SPECIFYING VARIABLES

Select VARIABLE from the menu (Figure 1/160) and select values from the table as with atom selection (*vide supra*). The technique of defining generic groups using "* point of attachment" and "G1 User definable" will be discussed later (see p. 195).

USING TEMPLATES

Select TEMPLATE (Figure 1/161) followed by selecting from the displayed Menu. When

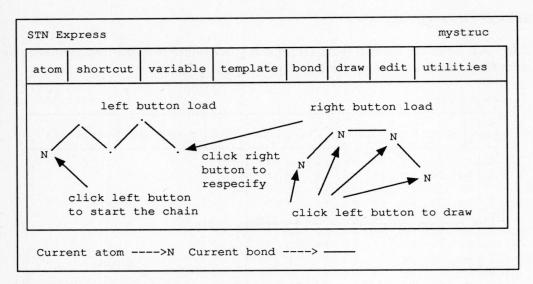

Figure 1/158. Drawing and respecifying atom values

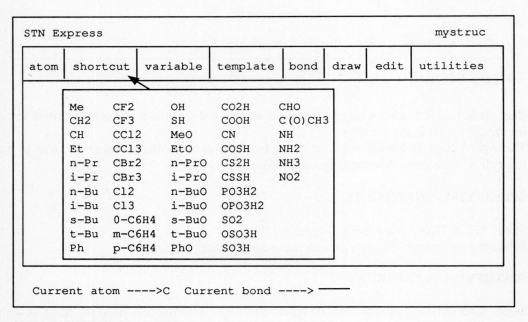

Figure 1/159. Shortcuts Menu

we select, e.g., aromatics, the available templates will be displayed (Figure 1/162). We select by clicking an atom or bond of the template. The remaining templates are displayed by clicking "next" in the table. Templates are automatically available for fusion.

STRUCTURES BY STN EXPRESS

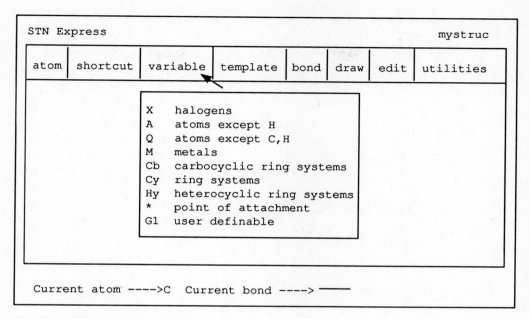

Figure 1/160. Variables Menu

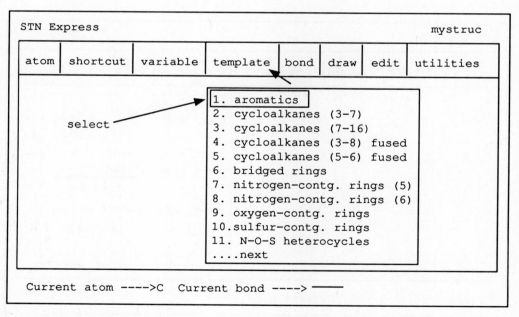

Figure 1/161. Template Menu

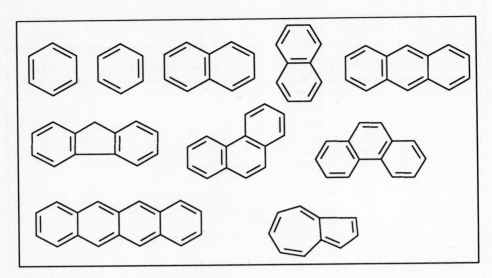

Figure 1/162. Aromatics templates

DRAWING BONDS

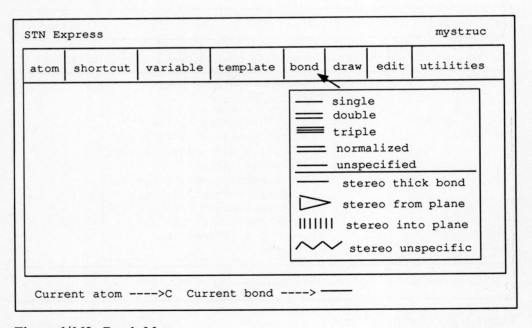

Figure 1/163. Bonds Menu

STRUCTURES BY STN EXPRESS

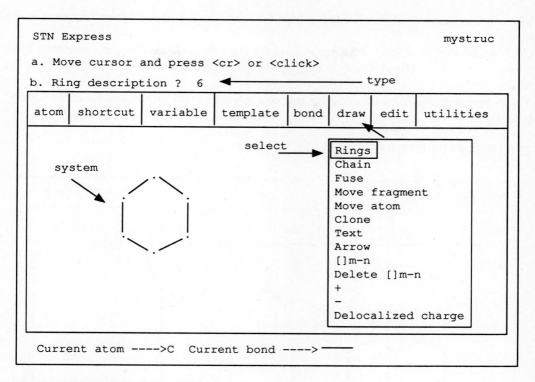

Figure 1/164. Quick Draw Menu

The bond table is displayed upon selecting BOND (Figure 1/163). When we select a bond from the table, the selected bond becomes the current bond for drawing or modifying existing bonds by the technique described for atoms (*vide supra*).
STN Express automatically recognizes aromatic bonds and tautomeric bonds and marks them as normalized in the query for a search.

QUICK DRAW

In addition to the free-hand drawing, we can employ the DRAW Command to specify chains and rings to be drawn by STN Express (Figure 1/164).
Selecting Rings or Chain from the table results in the prompt "a" at the top of the Menu to position the pointer and prompt "b" to input the size of the ring (3-15) or the length of the chain (1-15) we wish to be drawn (Figure 1/164).
The FUSE Command in the Menu allows for fusing separate existing fragments which we have free-hand drawn or which are the templates (*vide supra*). Pointing an atom in one fragment will fuse at the atom we select in the second fragment (note that the first pointed atom overlays the second one), pointing a bond fuses at the bonds. The keyboard fusion commands (see p. 49) can be used to construct fused ring systems.

Moving fragments or atoms is accomplished by the Move atom or Move fragment in the menu.
Copying is done by Clone. Descriptive text for the structures is added by selecting Text. Charges and Delocalized Charges can be specified in structures by selecting from the Menu.
The Arrow option allows for creating reaction schemes (see "Uploading Reaction Query" p. 204).

EDITING STRUCTURES

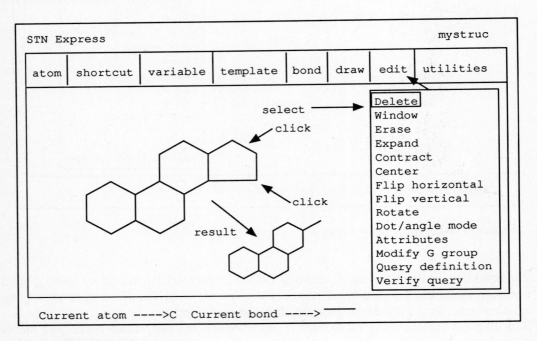

Figure 1/165. Edit Menu

As we have learned, atoms and bonds can be deleted by pointing to the item and pressing "d." The same effect is achieved by selecting Delete from the EDIT Menu (Figure 1/165) followed by clicking the item to be deleted.
If we want to delete a group of atoms and bonds (a fragment) we select Window and draw a box around the group to be deleted (or otherwise manipulated) as a unit.
To clear the screen of all structures, select Erase.
Expand, Contract, Center, Flip horizontal (i.e., an atom), Flip vertical (i.e. an atom), and Rotate in the Edit Menu allow for manipulating the structures or fragments in the structures as to their position on the screen. Dot/angle mode changes the appearance of structures to that one with no dots between bonds, just angles. Selecting Attributes displays a Menu from which we select the attribute by clicking a line (Figure 1/166).

STRUCTURES BY STN EXPRESS

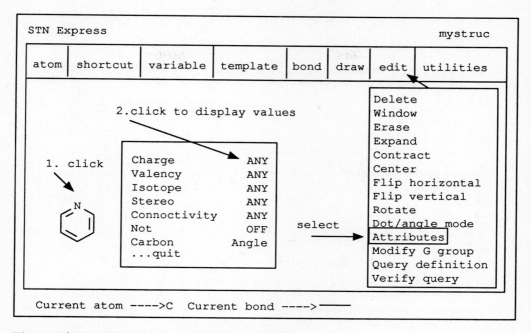

Figure 1/166. Edit Menu - Structure Attributes

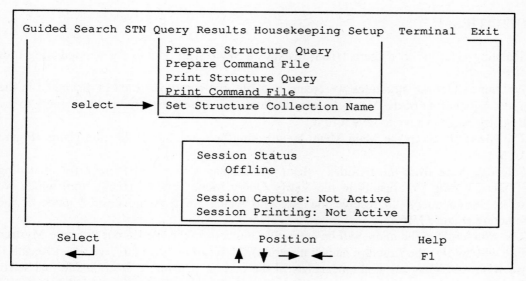

Figure 1/167. Creating file for structures

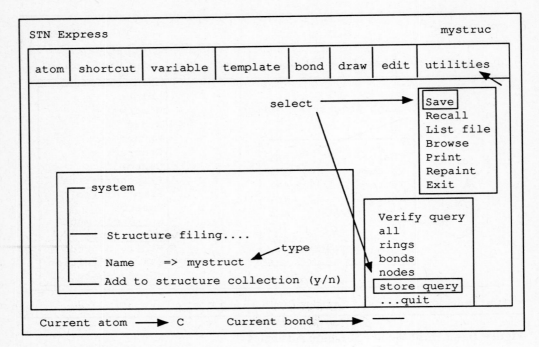

Figure 1/168. Utilities Menu

UTILITIES

Selecting UTILITIES (Figure 1/168) displays a Menu with options for manipulating files of structures.

We can collect the structures we create in a file, the name of which is displayed in the upper right corner of the Menu. In fact, if we wish to upload the structure as a query to the Registry File, we have to save it first.

The file is set up in the Main Menu by selecting Set Structure Collection Name (Figure 1/167).

Selecting Save from the Utilities Menu will display a message: Prepare for uploading <Y/N>. Typing "Y" results in the Verify Query Menu (Figure 1/168) from which we select "Store query" to receive the box into which we type the desired name of the structure (Figure 1/168).

The structure and the name will be entered as a record in the file (in our example Mystruc File will contain the structure named mystruct). To recall the structure, we select Recall. If we wish to display all structure names in the current structure collection file, we select List file.

Browse allows for browsing in the structure file.

Print will print the structure on the screen.

Repaint redraws the screen.

Exit returns us to the Main Menu.

STRUCTURES BY STN EXPRESS

DEFINING GENERIC GROUPS

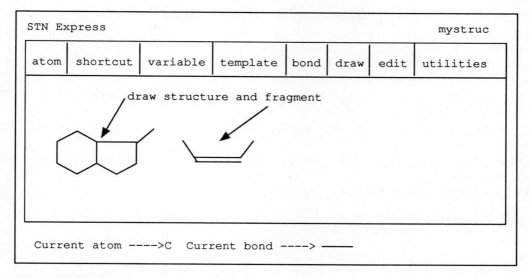

Figure 1/169. Parent and fragment for Gk

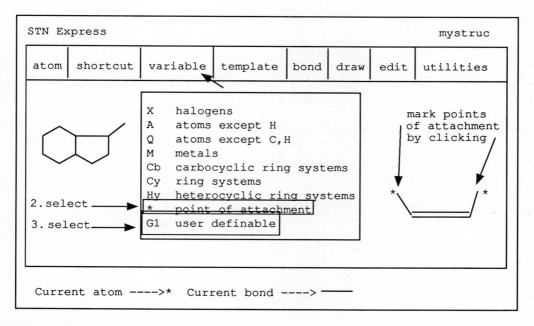

Figure 1/170. Specifying attachments

Generic groups are defined by using the options in the VARIABLE and EDIT pull-down menus.

The basics for defining a Gk (k=1-20) are as follows: we select from atoms, shortcuts, and variables that will be in the Gk. If we wish to have a fragment in the Gk, the fragment must be drawn and the point of attachment(s) must be marked before we select a Gk to define from the VARIABLE Menu. The defined Gk thus becomes a unit which can be inserted into the parent at the desired locant similarly as we insert an atom. The defined Gk is displayed in the VARIABLE pull-down menu from which it is selected.

The steps for defining Gk groups are:
1. Create the parent as usual and the fragment (if we want a fragment in the Gk group)(Figure 1/169).
2. Mark the points of attachment in the fragment by selecting * point of attachment in the pull-down menu (Figure 1/170).
3. Select G1 User definable (Figure 1/170).
4. Select atoms (Figure 1/171).

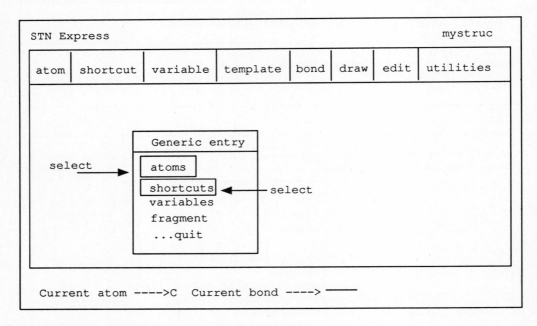

Figure 1/171. Defining atoms and shortcuts in Gk

5. Select from the displayed table (Figure 1/172):
   a. select atom 1
   b. select atom 2
   c. etc. (each atom will be marked by *, to deselect click the atom again)
6. Click outside the table (Figure 1/172) to return to the Menu (Figure 1/171).
7. Select Shortcuts (Figure 1/171), select the shortcut; click outside (Figure 1/173).

STRUCTURES BY STN EXPRESS

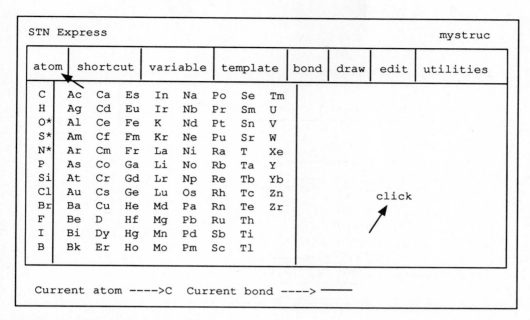

Figure 1/172. Selecting atoms for Gk

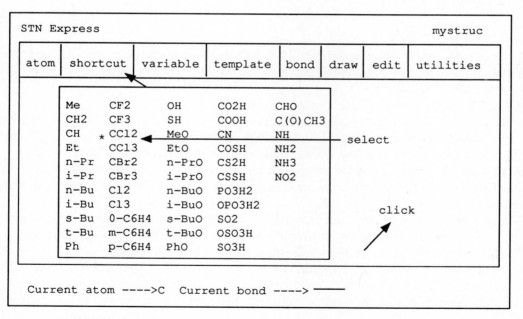

Figure 1/173. Selecting shortcut for Gk

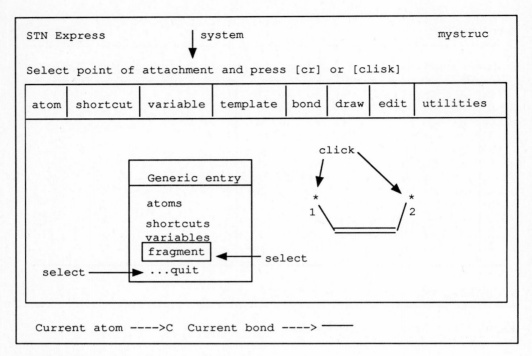

Figure 1/174. Defining fragment for Gk

8. Select Fragment (Figure 1/174). The system will prompt for selecting point of attachments. Click the star-marked points which will be numbered 1 and 2 by the system (Figure 1/174) and click END.
9. To end defining the Gk, select ...quit from the menu (Figure 1/174).
The Gk (e.g., G1) has been defined and is now displayable in the VARIABLE pull-down menu (Figure 1/175). To insert G1 in the parent, select it and click in the parent as though G1 were an atom (it is, actually, shown in the Menu as the Current atom (Figure 1/175).
If we have a fragment in the Gk group with multiple points of attachment, like the [*1-*2] fragment, we specify the attachments to the parent by selecting EDIT and Query definition (Figure 1/176).
The system will prompt us by the message "a" (Figure 1/176) and, upon our acceptance by "Y," will highlight the points of attachment in the parent and issue the message "b" (Figure 1/176). We click the atom in the fragment which is supposed to be attached to the highlighted node in the parent (Figure 1/176) and repeat for the second attachment. The highlighting will then be removed. We are returned to Verify Query Menu; we ....quit.
The attachments can be verified by selecting EDIT/Verify Query and Generic (Figure 1/177).
The parent and fragment attachment points will be shown highlighted. Upon the message "b" (Figure 1/177) we are returned to the Verify Query Menu (Figure 1/177) and we can

STRUCTURES BY STN EXPRESS

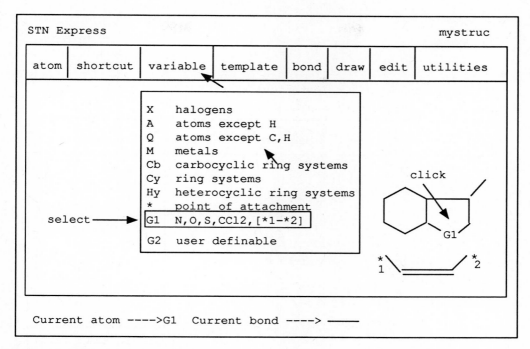

Figure 1/175. Inserting Gk in parent

either quit or proceed in verifying the other query definitions in the Verify Query Menu.
To define the second Gk (e.g., G2):
1. Draw the fragment (if we wish to have a fragment in the Gk) and mark the point of attachment (Figure 1/178) as described above. The defined G2 will be displayed in the VARIABLE Menu as G2 [*3].
2. Insert the G2 at the locant (Figure 1/178).

MODIFYING Gk GROUPS

We can make changes in the definition of Gk groups. (We can also delete all defined values in Gk group. If we do so, the Gk remains in the VARIABLE menu with no values but is not usable any more. However, if we define a Gk, we do not have to use it. To delete Gk from the parent, delete as with an atom.)
For example, to add atom values to G2:
1. Select Modify G group (Figure 1/179).
2. Select G2 (Figure 1/179).
3. Proceed in defining atom values as we have shown for G1.
The added values will be displayed in the VARIABLE Menu, e.g., G2 Cl,Br,[*3].

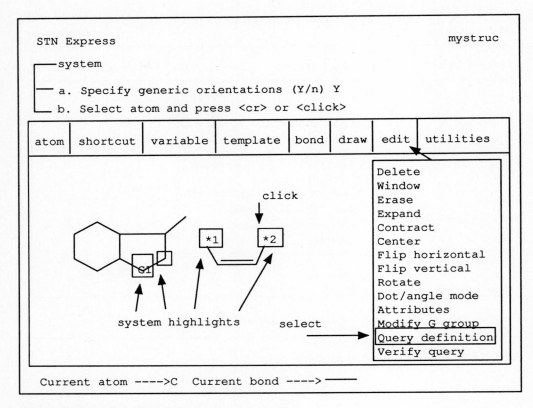

Figure 1/176. Specifying attachments to parent

DEFINING REPEATING GROUPS

Note that in STN Express the repeating groups are not defined as Gk groups (compare p. 131); however, they count in the limit of 20 groups per structure.
To define a repeating groups, we use the DRAW option (Figure 1/180).
The following steps define a repeating group:
1. Prepare the structure with the group.
2. Select [ ]m-n from the Menu (Figure 1/180).
3. Click the bonds at each end of the group (Figure 1/180) at System prompt.
4. Type the repeating range at the System prompt "from to" at the top of the Menu
5. The range will be displayed (Figure 1/180).

The repeating group can be deleted by selecting Delete []m-n from the menu and clicking the group.

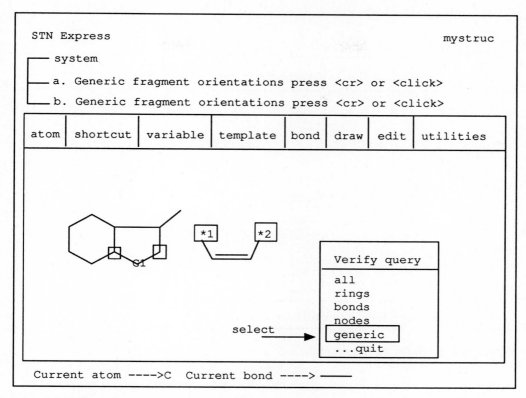

Figure 1/177. Verifying attachment points

UPLOADING QUERY STRUCTURE TO REGISTRY FILE

When we connect to STN-CAS ONLINE and enter the Registry File, we are ready to upload the query structure at the => prompt. Pressing F2 will bring up the STN Express Main Menu (Figure 1/181).

Note that three additional options are available in the menu when we are online. Select Upload Structure Query to receive the box (Figure 1/182).

If we elect to display all structure names in the current Structure Collection File by <cr> or clicking [End] the Menu shown in Figure 1/183 is displayed and we select from the names in the Menu.

As a shortcut, we can also press F8 to receive the prompt at the bottom of the Menu for query name (without having to go through the Main Menu).

When the upload is completed, we receive the Ln for the query and are ready to search.

To draw or modify query structure while online, press F2 to access the Main Menu and proceed from there as usual. Note that the limit for staying in STN Express Query mode online is 20 minutes before the system disconnects.

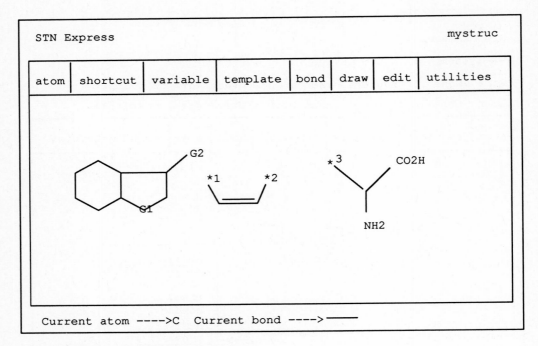

Figure 1/178. Defining fragment for G2

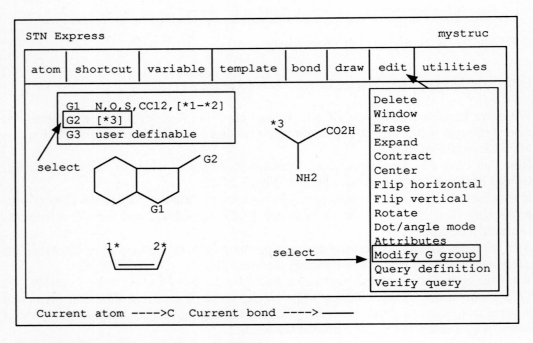

Figure 1/179. Adding value to existing Gk

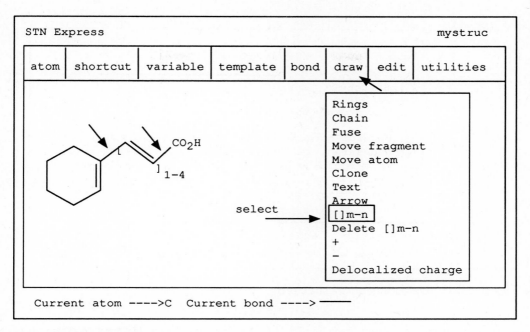

Figure 1/180. Defining repeating group

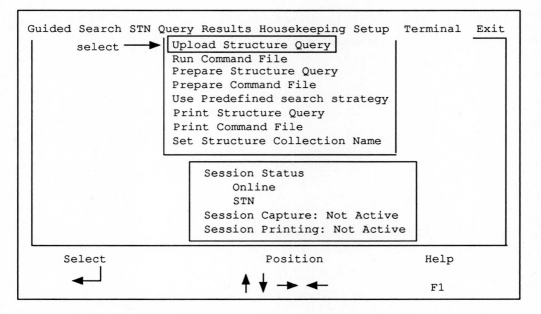

Figure 1/181. Main Menu Online

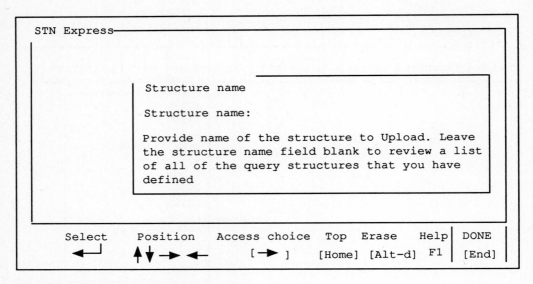

Figure 1/182. Uploading query structure

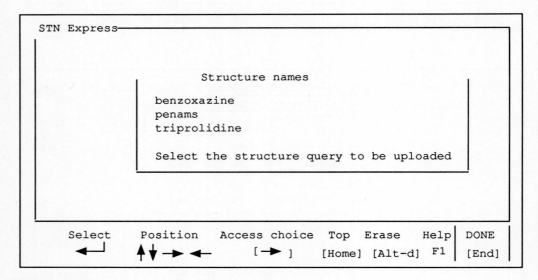

Figure 1/183. Selecting query name to upload

UPLOADING REACTION QUERY FOR CASREACT

The reaction query is prepared using the Arrow option in the DRAW Menu (Figure 1/184).
The role of structures in reactions is specified in the EDIT/Verify query (Figure 1/185).

STRUCTURES BY STN EXPRESS

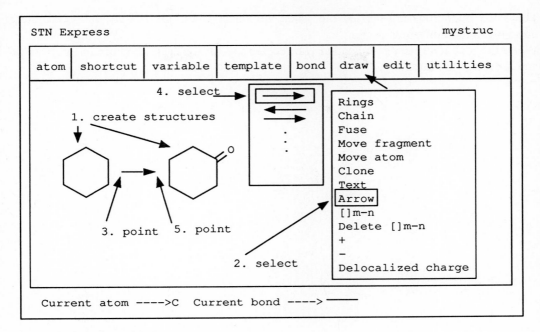

Figure 1/184. Creating reaction query

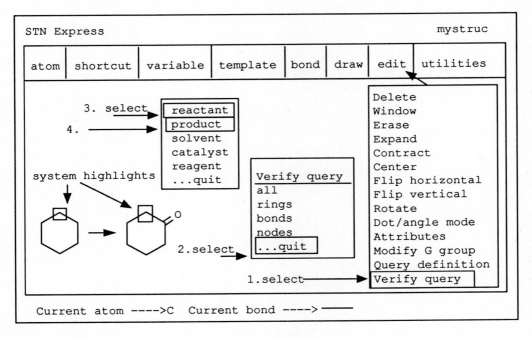

Figure 1/185. Specifying role of reaction structures

The same operations are carried out when we select UTILITIES/Save and select Store query from the pull-down Menu.
When the reaction query is uploaded, we receive the Ln for each structure in the reaction and are ready to search.

12 STRUCTURE SEARCHING

The Registry File can be searched using graphic structure queries composed of:
1. single or multiple structures
2. structures combined by Boolean Operators
3. structures combined with screens by Boolean Operators

Text terms are also searchable in the Registry File by the text searching techniques. However, we shall limit our subsequent discussion to structure searching.

The structures that we create in the Menu mode, in the Keyboard Commands mode, in STN Express, or ChemTalk Plus (see Part Five), are transferred to the SEARCH Subsystem of the Registry File to search and retrieve structures that comply with the structural requirements specified in the structure queries.
The structure query is submitted to the SEARCH subsystem as a logical expression. The simplest form of a logical expression is the Ln assigned by the System to structure queries when we END the Structure Command (see p. 40). Complex logical statements are defined by the User with the use of the Boolean Operators (see p. 247).
The SEARCH subsystem is activated at the System prompt => by the command SEARCH. (See p. 262-A 4 for the new search option CSS.)

Example: *Initiating SEARCH*
                                                        Command:SEARCH Ln

    Key:search Ln
       Ln=logical expression

=>search L1
ENTER TYPE OF SEARCH: (SSS), FAMILY, OR EXACT:.
ENTER SCOPE OF SEARCH: (SAMPLE), FULL, RANGE:

In addition to specifying the criteria for our search in the structure query itself, we are prompted by the System to define additional conditions as to how the query is to be searched when we invoke the SEARCH subsystem, i.e., we have to specify TYPE OF SEARCH, and SCOPE OF SEARCH.

SCOPE OF SEARCH IN THE REGISTRY FILE

The Registry File is a large file containing over 9 million structures and growing by 10 to

15 thousand structures per week. The System offers the User a flexibility in how to search the File.
We can specify the following Scope of Search:
1. Sample Search - to test our query in a File segment
2. Full Search - a search in the entire File
3. Range Search - a search in Registry Number range
4. Batch Search - an offline search to override search limits

SAMPLE SEARCH

The Sample Search is a search of a file segment which represents 5% of the full Registry File. The 5% file segment has been selected statistically from the full Registry File to represent all the substances in the full file. The purpose of the Sample Search is to test effectiveness of our structure query to retrieve the desired structures and to find out whether the System Limits (*vide infra*) will not be exceeded. The Sample Search results in a projection upon which we are able to see whether the query can be run in the Full File, whether the Full Search will run to completion, and how many answers we can expect.
The Sample Search has a limit of 1,000 iterations and 50 answers (the answers can be displayed when the search is completed). However, if more than 1,000 structures are located in the screen search (see p. 11), the exact number of structures found is reported but only 1,000 of them are subjected to iteration (see p. 11). The projection for the Full File search is given based on the actual number of structures found in the screen search. Similarly, if more than 50 answers is the result of iteration, the exact number of answers is reported. Upon reviewing the answers, we can decide whether to modify our query structure or use it as it is for the Full Search.
The Sample Search is the System default because it is the search most commonly used to test the correctness of our query. No fee is charged for the Sample Search.

Example: *Running SAMPLE SEARCH*

We will search the REG File for dithianes of the substructure in Figure 1/186 where R can be any substituent but the ring cannot be embedded in larger fused systems. We create the structure query shown in Figure 1/186 as usual.
=>file reg
=>structure
ENTER NAME OF STRUCTURE TO BE RECALLED (NONE): none (or ".")
ENTER (DIS), GRA, NOD, BON, OR ?:
gra r6,6 c1,6 c1, nod 1 5 S,hco 7 8 e2,rsp,bon all se
ENTER (DIS), GRA, NOD, BON, OR ?:end
L1 STRUCTURE CREATED
=>search L1
ENTER TYPE OF SEARCH: (SSS), FAMILY, OR EXACT:sss
ENTER SCOPE OF SEARCH: (SAMPLE), FULL, OR RANGE:sample

STRUCTURE SEARCHING

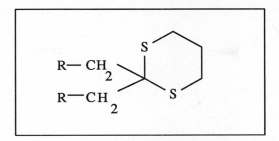

Figure 1/186. Query structure

```
SAMPLE SEARCH INITIATED 11:11:20
SAMPLE SCREEN SEARCH COMPLETED - 148 SUBSTANCES TO ITERATE
100.0% PROCESSED         148 ITERATIONS    18 ANSWERS
SEARCH TIME: 00.00.09

FULL FILE PROJECTIONS:  ONLINE  **COMPLETE**
             BATCH   **COMPLETE**
PROJECTED ITERATIONS:    2231 TO    3689
PROJECTED ANSWERS:        106 TO     614

L2      18 SEA SSS SAM L1
=>
```

SEARCH STATISTICS AND SEARCH PROJECTION

The Sample Search produces statistical data that we can use to monitor the progress of the search and to evaluate the effectiveness of our structure query. In the Sample Search, we are able to decide whether to proceed to the Full Search, or to modify the query and run another Sample Search.
The final statistics is displayed when the search is completed (for short searches) or in increments while the search is in progress (for long searches).
The last line of the search statistics is shown in Figure 1/187.
A repeated Sample Search with the same structure query returns identical results.
The System gives a FULL FILE PROJECTION which can be COMPLETE, INCOMPLETE, or UNCERTAIN. These projections are based on the actual number of iterations and answers produced by the Sample Search of our structure query.
Figure 1/188 shows the limits for Full Search and the actions the user has to take when the limits are reached.
The command:
HELP SEARCH PROJECTIONS
will result in instructions how to proceed if the projection is INCOMPLETE or

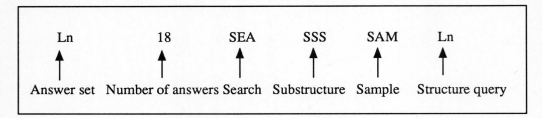

Figure 1/187. Substructure Search (SSS) statistics

UNCERTAIN.
The answers retrieved in the Sample Search can be displayed (see p. 226).

FULL SEARCH

Full Search is run in the entire Registry File. The File contains over 9 million structures and is updated every week by 10-15 thousand structures.
When we wish to perform a Substructure Search (SSS), the Full Search should not be run without prior Sample Search. We will be certain that our search will not exceed the System limits and that we do not have a structure query that returns false hits. However, when we intend to perform an EXACT or FAMILY Search, it is necessary to run the Full Search even if the Sample Search produces statistics showing 0 answers. This is because the Sample Search is a search of 5% of the File and the exact match to our structure query may not be in that portion of the File. However, the correctness of the structure query is tested in the Sample Search even for EXACT and FAMILY Search.

Example: *Running FULL SEARCH*

We will use the query L1 as in the above Sample Search to run a Full Search in the REG File.
=> search L1
ENTER TYPE OF SEARCH: (SSS), FAMILY, OR EXACT:sss
ENTER SCOPE OF SEARCH: (SAMPLE), FULL, OR RANGE:full
FULL SEARCH INITIATED 11:11:57
FULL SCREEN SEARCH COMPLETED -   2900 SUBSTANCES TO ITERATE
   83.0% PROCESSED   2406 ITERATIONS   218 ANSWERS
  100.0% PROCESSED   2900 ITERATIONS   268 ANSWERS
SEARCH TIME: 00.00.30
L3   268 SEA SSS FUL L1
=>

STRUCTURE SEARCHING

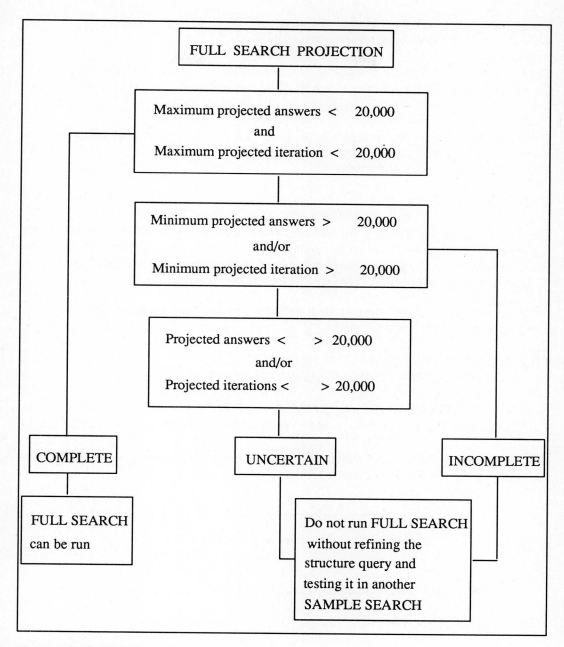

Figure 1/188. Full Search projection

The answers in the set L3 can be displayed (see p. 226).
We can turn off the display of search statistics by:
=>set status off
Then, only the last, i.e., the 100% line will be shown with full projection information.
To display statistics of a completed search if the "set status off" has been employed, we use:
=>display query Ln
To outline the difference in the results in the SAMPLE and FULL search of our query L1:
SAMPLE SEARCH..........18 answers
FULL SEARCH...............268 answers

RANGE SEARCH

The structures in the Registry System are assigned Registry Numbers when they are registered. The Registry Numbers are indicative of the time when the structure was entered into the File. The highest number is associated with a structure registered last. An update on Registry Numbers is given when we enter the Registry File:
=> file registry
COST IN U.S. DOLLARS   SINCE FILE ENTRY   TOTAL ENTRY SESSION
FULL ESTIMATED COST   0.35   0.35

FILE 'REGISTRY' ENTERED AT 12:10:29 ON 12 MAY 89
COPYRIGHT (C) 1989 AMERICAN CHEMICAL SOCIETY

STRUCTURE FILE UPDATES: HIGHEST RN 120520-41-4
DICTIONARY FILE UPDATES: 6 MAY 89 (890506/ED) HIGHEST RN 120496-10-8

When we already are in the Registry File, the update on Registry Numbers can be requested by the NEWS FILE Command
at the => prompt:

Example: *Update on Registry Numbers*
                    Command: NEWS FILE

=>news file
NEWS FILE   May 6 Recent updates to the Registry File

Recent Updates Highest Registry Number

May 6, 1989   120520-41-4
Apr 30, 1989   120408-07-3
Apr 22, 1989   120292-57-1
Apr 15, 1989   120142-33-8
Apr 9, 1989   120019-16-1

STRUCTURE SEARCHING

Apr 1, 1989   119903-92-3
Mar 25, 1989   119784-94-0
Mar 18, 1989   119677-03-1
Mar 11, 1989   119564-98-6
Mar 4, 1989   119432-84-7
Feb 25, 1989   119322-50-8
Feb 18, 1989   119180-30-2

The Registry File contains 9,675,394 Registry Numbers. All can be displayed. The Registry File is updated weekly. References are current through CA Volume 110, Issue 18.

TSCA flag in the LC field reflects the status of substances on the non-confidential Toxic Substances Control Act Inventory as of April, 1988.
=>

We can use RANGE SEARCH to limit our search to a range of Registry Numbers, e.g., to the structures registered at a certain date. For example, we have a structure published in 1980 and its Registry Number. We wish to find related substances published since then. These structures will have higher Registry Numbers. We perform a Range Search with a range of Registry Numbers consisting of the Registry Number of our structure and the highest one in the Registry File at the time of the search.

The following are the formats to specify ranges of Registry Numbers (RN):

To retrieve structures of	Specify RN format
Substances with RNs lower than and including a given RN	(,XXXX-XX-X)
Substances with RNs within a given range	(XXXX-XX-X, YYYY-YY-Y)
Substances with RNs higher than and including a given RN	(XXXX-XX-X,)

The Range Search can be also employed when we run a search and the System limits are exceeded. A search run in the Full File is always complete from the highest Registry Number down to the Registry Number when the search stops and an answer set is created. Therefore, searching for lower Registry Numbers than the lowest one in the answer set by a Range Search will continue the search in the Full File from the lowest Registry Number in the answer set to the lowest Registry number indicated in the Range Search. The following example is of a search with the projection INCOMPLETE:

FULL SCREEN SEARCH COMPLETED - 27811 SUBSTANCES TO ITERATE
  72% PROCESSED   20000 ITERATIONS   57 ANSWERS
INCOMPLETE SEARCH (SYSTEM LIMITS EXCEEDED)

FULL FILE PROJECTIONS:	ONLINE	**INCOMPLETE**
	BATCH	**COMPLETE**
PROJECTED ITERATIONS:		27811
PROJECTED ANSWERS	57 TO	105

L5 57 SEA SSS FUL L4
=>

Since the limit is 20,000 iterations, the search stops when there are still 7,811 screened structures to iterate. The answer set L5 contains 57 answers which have the Registry Numbers in descending order. If we display the last, i.e. the lowest Registry Number, we can then create a Range in which to continue the search.
To display the lowest Registry Number (see p. 226):
=>display rn 57
L5 ANSWER 57 OF 57
RN 54774-89-9
This Registry Number is used to search from it down to the lowest in the Full File by incorporating it in the search range.

Example: *Running RANGE SEARCH*

=>search L4

ENTER TYPE OF SEARCH: (SSS), FAMILY, OR EXACT:sss
ENTER SCOPE OF SEARCH: (SAMPLE), FULL, OR RANGE:range
ENTER RANGE OR (ALL): (,54774-89-9)
RANGE MORE THAN 100,000 SUBSTANCES. WILL BE BILLED AS A FULL FILE SEARCH.(Note: This is a System warning. If the number of substances is less than 100,000 the charge is less.)
SEARCH INITIATED 22:00:09
SCREENING
RANGE SCREEN SEARCH COMPLETED - 7967 SUBSTANCES TO ITERATE
    10.9% PROCESSED 867   ITERATIONS   2 ANSWERS
    68.6% PROCESSED 5469   ITERATIONS   12 ANSWERS
    98.3% PROCESSED 7832   ITERATIONS   20 ANSWERS
  100.0% PROCESSED 7967   ITERATIONS   21 ANSWERS

L6 21 SEA RAN=(,54774-89-9) SSS L4
=>

STRUCTURE SEARCHING

If we now combine (see p. 247) the answer sets, we arrive at a set containing answers to our query from the entire Registry File:
=>s L5 or L6
L7   77 L5 or L6
=>
The L7 set can be now displayed (see p. 226) or printed (see p. 242).

BATCH SEARCH

When we expect to exceed the System limits or when a Sample Search produces statistics that include a warning that the ONLINE Search exceeds the System limits, we can perform a BATCH Search. The search projection includes information for BATCH Search.
The BATCH Search is a search for which we submit the query to the Search subsystem while we are online, i.e., connected to the host computer. However, the search itself will be performed offline, overnight, by the System and the answers will be ready to display online the next day, or they can be requested as offline prints.

Example: **Running BATCH SEARCH**
               Command: BATCH

The following query L9:

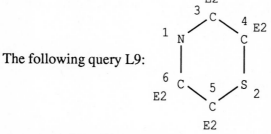

was searched in the REG file.
:dis sat
NODE ATTRIBUTES:
HCOUNT IS E2   AT   3
HCOUNT IS E2   AT   4
HCOUNT IS E2   AT   5
HCOUNT IS E2   AT   6

GRAPH ATTRIBUTES:
RSPEC I
NUMBER OF NODES IS 6
:end
L9 STRUCTURE CREATED

=>search L9

ENTER TYPE OF SEARCH: (SSS), FAMILY, OR EXACT:sss
ENTER SCOPE OF SEARCH: (SAMPLE), FULL, OR RANGE:sample
SAMPLE SEARCH INITIATED 11:21:59
SAMPLE SCREEN SEARCH COMPLETED - 2168 SUBSTANCES TO ITERATE

32.6% PROCESSED   706 ITERATIONS   50 ANSWERS
46.1% PROCESSED   1000 ITERATIONS   50 ANSWERS

15 MAY 89 11:22:40   STN INTERNATIONAL   P0006
INCOMPLETE SEARCH (SYSTEM LIMIT EXCEEDED)
SEARCH TIME: 00.00.36

FULL FILE PROJECTIONS: ONLINE   **INCOMPLETE**
                              BATCH   **COMPLETE**
PROJECTED ITERATIONS: 40573 TO 46147
PROJECTED ANSWERS: 2949 TO 4595

L10   50 SEA SSS SAM L9
=>

The Sample Search statistics revealed that the System limits will be exceeded for the FULL ONLINE Search of this query. However, the BATCH Search is projected as complete.
The BATCH Search limits are:
50,000 iterations
50,000 answers
We can therefore perform a BATCH Search:
=>batch

ENTER QUERY L# FOR BATCH REQUEST:L9
ENTER BATCH REQUEST NAME (END):hetero/b (Note: The name must end with /b.)
ENTER TYPE OF SEARCH: (SSS), FAMILY, OR EXACT:sss
ENTER SCOPE OF SEARCH: (FULL) OR RANGE:full
QUERY 'L9' HAS BEEN SAVED AS BATCH REQUEST 'HETERO/B'
=>

The System performs the search offline and creates an Answer File of the same name (ending with /a) as our Batch Name. The Answer File is ready to display the next day. We have to activate the file to display the answers (Note: To display the answers we have to be in the Registry File.)

STRUCTURE SEARCHING

Figure 1/190. Structure to search

ST 3:(+ -)

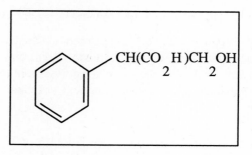

ST 1:S

[Structure: benzene ring—CH(CO$_2$H)CH$_2$OH]

ST 1:R

Example: *Running FAMILY SEARCH*

Create a structure query (we will use the same one as for the EXACT Search above) and END the STRUCTURE Command to obtain an Ln for the query, and to be returned to the System prompt =>.

=>search L1
ENTER TYPE OF SEARCH: (SSS), FAMILY, OR EXACT:family
ENTER SCOPE OF SEARCH: (SAMPLE), FULL, OR RANGE:full
SEARCH INITIATED 18:42:17

FULL FILE SEARCH COMPLETE
L3   11 SEA FAM FUL L1

The FAMILY SEARCH returned 11 answers which are the records of the substances registered in the Registry File whose structures match our query. Five of these substances are represented by the same structures as in the EXACT Search shown above. In addition, six substances are multicomponent substances and a sodium salt of the structures shown below:

Component 1   Component 2

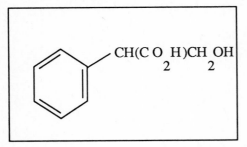

Component 1

H$_2$NCH$_2$CH$_2$SCH$_2$CH$_2$NHAc

Component 2

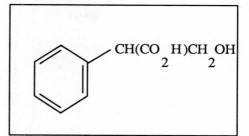

Component 1

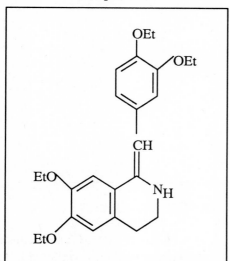

Component 2

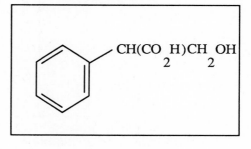

Component 1

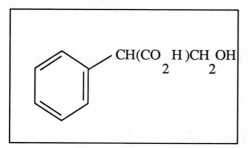

Component 2

Component 1

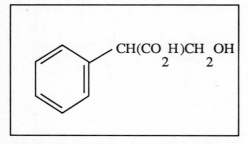

Component 2

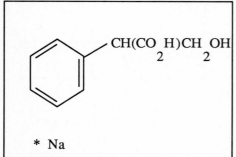

* Na

Example: **Running SUBSTRUCTURE SEARCH**

Create a substructure query (we will use the same query shown in the EXACT Search) and END the STRUCTURE Command to obtain an Ln for the query, and to be returned to the System prompt =>

=>search L1
ENTER TYPE OF SEARCH: (SSS), FAMILY, OR EXACT:sss
ENTER SCOPE OF SEARCH: (SAMPLE), FULL, OR RANGE:sample
SAMPLE SEARCH INITIATED 18:42:17

SAMPLE SCREEN SEARCH COMPLETED - 342 SUBSTANCES TO ITERATE
100.0% PROCESSED 342 ITERATIONS   44 ANSWERS
SEARCH TIME: 00.00.04

FULL FILE PROJECTION: ONLINE **COMPLETE**
                           BATCH **COMPLETE**
PROJECTED ITERATIONS   5523   TO 7659
PROJECTED ANSWERS   464 TO 1230t

L4   44 SEA SSS SAM L1

STRUCTURE SEARCHING

The Sample Search returned 44 answers (50 is the maximum of answers for a Sample Search) and the File Projection permits us to perform the Full Search.

=>search L1
ENTER TYPE OF SEARCH:(SSS), FAMILY, OR EXACT:sss
ENTER SCOPE OF SEARCH:(SAMPLE), FULL, OR RANGE:full

SEARCH INITIATED 11:03:20

FULL FILE SEARCH COMPLETE
L5   1187 SEA SSS FUL L1

The SUBSTRUCTURE FULL SEARCH returned 1187 answers which are the records of the substances registered in the Registry File whose structures match our query. Among the retrieved substances are those which have been retrieved by the EXACT FAMILY SEARCH shown above. The other retrieved substances have in their structures the structural fragment of our query substituted by any possible substituent and/or embedded in larger fused systems. Two of the retrieved answers are shown in Figure 1/191 and Figure 1/192.

Figure 1/191. Retrieved structure

As we can see, a variety of structures will be retrieved in the SUBSTRUCTURE SEARCH (any substitution is allowed). The effect of recognizing tautomers by the System (see p. 97) can be seen in the retrieved structures whose fragment can be represented as shown in Figure 1/193.
The differences between the EXACT, FAMILY, and SUBSTRUCTURE SEARCH can be outlined by comparing the number of the retrieved answers:
EXACT SEARCH...............5 substances
FAMILY SEARCH............11 substances
SUBSTRUCTURE SEARCH....1187 substances

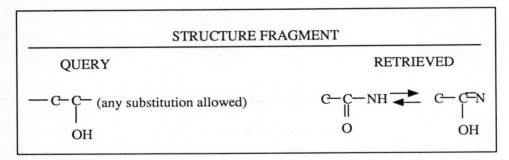

Figure 1/192. Retrieved structure

Figure 1/193. Tautomerism in structure fragment

LIMITS ON SAMPLE, FULL, AND BATCH SEARCH

The Registry File imposes limits on how many structures will be subjected to search in the Registry File (Figure 1/194).

LIMITS ON SUBSTRUCTURE, FAMILY, AND EXACT SEARCH

The Registry File imposes limits on using the variable values in structure graphs as shown in Figure 1/195.

DISPLAY OF ANSWERS TO A STRUCTURE SEARCH

When a search is completed and we receive the answer set Ln, the answers are ready to display. The following is a description of a variety of specifications we can use to display the answers at the System prompt =>.

STRUCTURE SEARCHING

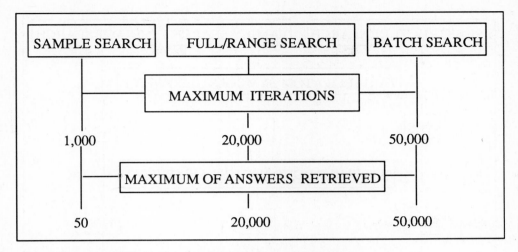

Figure 1/194. Limits on number of retrieved structures

SPECIFICATION	DISPLAY
DIS	the next Answer (Answer 1 if no Answer yet displayed) in the default format (SUB)
An answer number	a specific answer in the default format (SUB)
Several answer numbers	those answers in the default format (SUB)
A range of numbers	a range of answers in the default format (SUB)
A format	the last answer in this new format
*format	in a specified format different from the default format (SUB)
DIS SCAN	review answers in the SAMPLE SEARCH (no charge for the display)
DIS BROWSE	answers in random order and in specified format

An answer number is any number of the total of retrieved answers in the Answer Set Ln.
A format is a set of data fields from the record (answer) we wish to display.
We can display information related to substances only, to documents only, or both. The format may be System defined or User defined.

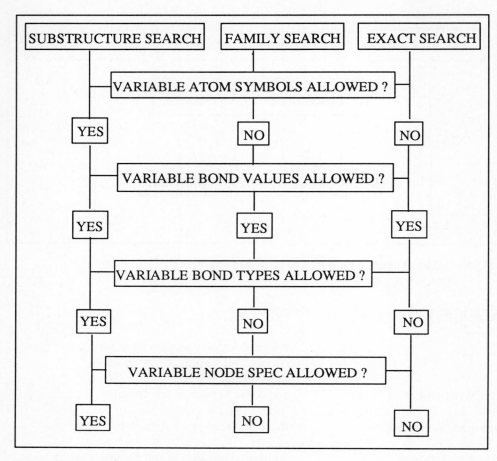

Figure 1/195. Limits on variables in queries

DISPLAY OF SUBSTANCE INFORMATION

The System predefined formats are:

FORMAT	DISPLAY
REG	CAS Registry Number(s)
SUB	Substance Data (includes CAS Registry Number, Molecular Formula, Names, and Structure Image)
SAMPLE	CA Index Name, Molecular Formula, and Structure Image, Class Identifier, RegNo Locator, Source of Registration, Reference Indicator (No Registry Number and Name Synonyms)

STRUCTURE SEARCHING

SCAN   CA Index Name, Molecular Formula, Structure Image, Reference Indicator

Answer number and format (except SCAN) may be combined.
Examples are:

SPECIFICATION	DISPLAY
5-8 REG	Answers 5 through 8 in REG format
9 SUB	Answer 9 in SUB format
3,7 REG	Answers 3 and 7 in REG format

The User defined formats are composed of codes of the data fields of the records in the Registry File. The codes are listed below:

CODE		DATA FIELD
RN		CAS Registry Number
IN		CA Index Name
SY		Synonym
DR		Deleted Registry Number
RR		Replacing Registry Number
PR		Preferred Registry Number
AR		Alternate Registry Number
MF		Molecular Formula
AF		Alternate Molecular Formula
CI		Substance Class Identifier
	AYS	Alloys
	CCS	Coordination compounds
	CTS	Registered concept
	GRS	Generic registration
	IDS	Incompletely defined substance
	MAN	Manually registered substance
	MNS	Mineral
	MXS	Mixture
	PMS	Polymer
	RIS	Radical ion
	COM	Component
	RPS	Ring parent
SR		Source of Registration
LC		Registry Number Locator
IL		Isotope at Unknown Location
ST		Stereochemistry Text Descriptor
CM		Component Number

STR   Structure Diagram
REF   Number of References in CA and CAOLD Files

The codes (except CM) can be specified individually or in any combination. Examples are:

FORMAT	DISPLAY
STR	graphics structure of retrieved substance
STR ST	graphics structure and stereodescriptors
IN SY RN	index name, synonyms, and Registry Number

DISPLAY OF DOCUMENT INFORMATION

We can request display of the bibliographic information, abstract, and(or) indexing information for the 10 most recent documents citing the retrieved substance. To do this, enter individual document information field codes, predefined document formats, or a combination of these. A request for document information must be combined with either the RN field or the REG or SUB format of the substance information. The following fixed, system predefined combinations of fields can be used:
(AN = CA Number)

FORMAT	DISPLAY
BIB	AN, plus Bibliographic Data
IND	Index Data
ABS	AN, plus Abstract
ALL	combination of SUB BIB ABS IND

These document fixed formats may be combined with the substance fixed formats or with the substance data field codes (substance data must be cited first in this combination). Examples are:

SUB BIB
RN ST BIB
REG IN AN TI AU

The data field codes for a document are given below:

CODE	DATA FIELD
AN	Chemical Abstracts Number
TI	Title of Document
AU	Author or Patent Inventor

STRUCTURE SEARCHING

CS	Corporate Source or Patent Assignee
LO	Corporate Source or Patent Assignee Location
PI	Patent Information
AI	Patent Application/Priority Information
CL	Patent Classification
SO	Source (Name of Journal, Volume, Issue, Pages)
SC	Chemical Abstracts Section Code
SX	Chemical Abstracts Section Cross-Reference Code
DT	Document Type
CO	CODEN
IS	ISSN (International Standard Serial Number)
PY	Publication Year of Original Document
LA	Language of Original Document
AB	Abstract Text
KW	Keywords
IT	Index Entries (corresponds to CA Volume Indexes)
RN	Registry Number

When the search is completed we are returned to the => prompt. To display the answers at this prompt for the last answer set, we just use DIS. To display answers from any other previous set, we have to include the Ln answer set in our display command followed by the above described specifications:

L15   89 SEA SSS FUL L1
=>dis 1-10 (displays answers from the last set L15)

=>display L10 all 2-5 (displays answers from the set L10 in the ALL format))

Example: *Displaying answers*

The following examples will show the display of the records in various formats of our full search on dithianes (see p. 210).
We have just completed the search, therefore, DIS will show the answers of the last answer set.

SCAN format

=>d scan
L3 268 ANSWERS

IN 1,3-Dithiane, 2-butyl-2-(2,2-diethoxyethyl)- (9CI)
MF C14 H28 O2 S2

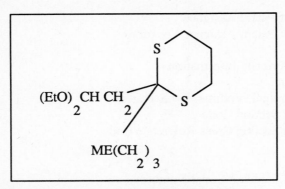

1 REFERENCES IN FILE CA (1967 TO DATE)

HOW MANY MORE ANSWERS DO YOU WISH TO SCAN? (1):

(We can enter any number, or NONE to exit the SCAN mode.)

Default SUB format

=>dis
ANSWER 1 OF 268

RN 38292-96-5
IN 1,3-Dithiane, 2-butyl-2-(2,2-diethoxyethyl)- (9CI)
MF C14 H28 O2 S2

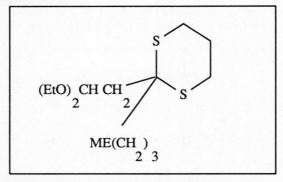

1 REFERENCES IN FILE CA (1967 TO DATE)

REG format

=>dis L3 2 reg

2   38292-97-6

Range of answers in SUB format

=>dis 3-5 sub

L3 ANSWER 3 OF 268

RN 38292-98-7
IN 1,3-Dithiane-2-ethanol, 2-butyl-.alpha.-(2-methyl-1,3-dithian-2-yl)- (9CI)
MF C15 H28 O S4

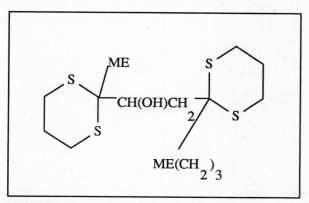

1 REFERENCES IN FILE CA (1967 TO DATE)

17:24:01 COPY AND CLEAR PAGE, PLEASE

L3 ANSWER 4 OF 268

RN 38293-03-7
IN 1,3-Dithiane-2-acetaldehyde, 2-pentyl- (9CI)
MF C11 H20 O S2

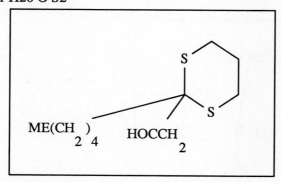

17:24:09 COPY AND CLEAR PAGE, PLEASE

L3 ANSWER 5 OF 268

RN 83313-13-7
IN 1,3-Dithiane, 2,2-bis(1-naphthalenylmethyl)- (9CI)
MF C26 H24 S2

CI COM

[Structure: 1,3-dithiane with two naphthylmethyl substituents at the 2-position]

3 REFERENCES IN FILE CA (1967 TO DATE)

Separate answers in SAMPLE format

=>dis 8,10 sample

L3 ANSWER 8 OF 268

IN 1,3-Dithiane-2-ethanol,.alpha.-methyl-2-[(2,2,6-trimethyl-1,3-dioxan-4-yl)methyl]-(9CI)
MF C15 H28 O3 S2

[Structure of the named compound]

2 REFERENCES IN FILE CA (1967 TO DATE)

STRUCTURE SEARCHING

17:24:49 COPY AND CLEAR PAGE, PLEASE

L3 ANSWER 10 OF 268

IN 1,3-Dithiane, 2-[3-(1-ethoxyethoxy)propyl]-2-(2-oxiranylethyl)-
   (9CI)
MF C15 H28 O3 S2

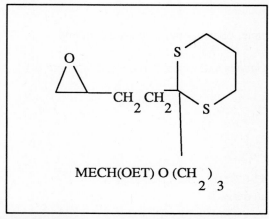

1 REFERENCES IN FILE CA (1967 TO DATE)

17:24:57 COPY AND CLEAR PAGE, PLEASE

BIBLIOGRAPHIC format

=>dis 15 reg bib

15   76495-04-0

REFERENCE 1

AN CA94(11):83910m
TI Synthesis of optically active 2-methyl- and 2-ethyl-1,6-dioxaspiro[4.4]nonane and
    -[4.5]decane pheromones from a common chiral precursor
AU Hungerbuehler, Ernst; Naef, Reto; Wasmuth, Daniel; Seebach, Dieter;Loosli, Hans
    Rudolf; Wehrli, Adolf
CS Lab. Org. Chem., Eidg. Tech. Hochsch.
LO Zurich CH-8092, Switz.
SO Helv. Chim. Acta, 63(7), 1960-70
SC 27-21 (Heterocyclic Compounds (One Hetero Atom))
DT J
CO HCACAV
IS 0018-019X

PY 1980
LA Ger

INDEX TERMS format

=>dis 15 reg ind

15   76495-04-0

REFERENCE 1

KW dioxaspirononane; dioxaspirodecane; pheromone; dioxaspirononane
IT Pheromones
   (of Pityogenes species, dioxaspirononanes and dioxaspirodecanes, prepn. of)
IT 627-30-5   691-84-9   2203-35-2
   (addn. of, with Et vinyl ether)
IT 109-92-2
   (addn. of, with di-Et malate)
IT 505-23-7
   (alkylation of)
IT 30502-41-1P   52500-29-5P   61847-07-2P
   (prepn. and alkylation by, of dithiane deriv.)
IT 76494-96-7P
   (prepn. and bromination of)
IT 64028-90-6P
   (prepn. and conversion of, to epoxide)
IT 76494-99-0P   76495-00-6P
   (prepn. and epoxide ring cleavage of)
IT 76495-01-7P   76495-02-8P   76495-03-9P   76495-04-0P
   (prepn. and hydrolysis of)
IT 76495-05-1P   76495-06-2P   76495-07-3P   76495-08-4P
   (prepn. and hydrolysis of, dioxaspiro compd. from)
IT 76494-97-8P
   (prepn. and reaction of, with bromoethyl oxirane)
IT 76494-95-6P
   (prepn. and redn. of)
IT 76494-98-9P
   (prepn. and redn. of, with bromoethyl oxirane)
IT 69744-43-0P
   (prepn. and resoln. of)
IT 72229-31-3P
   (prepn. and tosylation of)
IT 69744-44-1P   76041-91-3P   76041-92-4P   76042-02-9P
   76119-65-8P   76495-09-5P
   (prepn. of)

STRUCTURE SEARCHING

ABSTRACT format

=>dis 15 reg abs
15   76495-04-0

REFERENCE 1

AN CA94(11):83910m
AB Four title compds. (I; R = Me, Et; n = 1,2) were prepd. from the bromoepoxide (S)-(-)-II, readily available from malic acid. Alkylation of 1,3-dithiane, first with Cl(CH2)nOCHMeOEt (n = 3,4) and then with II, followed by oxirane cleavage with LiBHEt3 or Li dimethylcuprate gave III (m = 3,4; R = H, Me; R1 = H, CHMeOEt) in 60-80% overall yields. Acetal and thioacetal hydrolyses of III gave I as 3:2 diastereomeric mixts.; theE/Z-epimers of I (R = Me, n=1) were sepd. by preparative gas chromatog. I (R = Et,n =1) is the aggregating pheromone of Pityogenes chalcographus. For diagram(s), see printed CA Issue.

ALL format

=>dis 15 all

L3 ANSWER 15 OF 268

RN 76495-04-0
IN 1,3-Dithiane-2-propanol,2-[4-(1-ethoxyethoxy)butyl]-.alpha.-e (9CI)
MF C17 H34 O3 S2

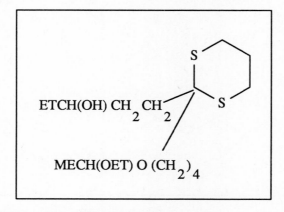

1 REFERENCES IN FILE CA (1967 TO DATE)

REFERENCE 1

AN CA94(11):83910m
TI Synthesis of optically active 2-methyl- and 2-ethyl-1,6-dioxaspiro[4.4]nonane and -[4.5]decane pheromones from a common chiral precursor

AU Hungerbuehler, Ernst; Naef, Reto; Wasmuth, Daniel; Seebach, Dieter; Loosli, Hans Rudolf; Wehrli, Adolf
CS Lab. Org. Chem., Eidg. Tech. Hochsch.
LO Zurich CH-8092, Switz.
SO Helv. Chim. Acta, 63(7), 1960-70
SC 27-21 (Heterocyclic Compounds (One Hetero Atom))
DT J
CO HCACAV
IS 0018-019X
PY 1980
LA Ger
AB Four title compds. (I; R = Me, Et; n = 1,2) were prepd. from the bromoepoxide (S)-(-)-II, readily available from malic acid.
Alkylation of 1,3-dithiane, first with Cl(CH2)nOCHMeOEt (n = 3,4) and then with II, followed by oxirane cleavage with LiBHEt3 or Li dimethylcuprate gave III (m = 3,4; R = H, Me; R1 = H, CHMeOEt) in 60-80% overall yields. Acetal and thioacetal hydrolyses of III gave I as 3:2 diastereomeric mixts.; the E/Z-epimers of I (R = Me, n = 1) were sepd. by preparative gas chromatog. I (R = Et, n = 1) is the aggregating pheromone of Pityogenes chalcographus. For diagram(s), see printed CA Issue.

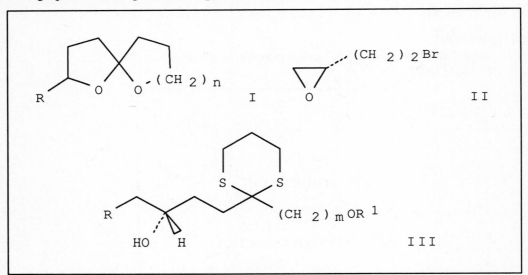

KW dioxaspirononane; dioxaspirodecane; pheromone dioxaspirononane
IT Pheromones (of Pityogenes species, dioxaspirononanes and dioxaspirodecanes,prepn. of)
IT 627-30-5   691-84-9   2203-35-2
   (addn. of, with Et vinyl ether)

IT 109-92-2
   (addn. of, with di-Et malate)

STRUCTURE SEARCHING

IT 505-23-7
   (alkylation of)
IT 30502-41-1P   52500-29-5P   61847-07-2P
   (prepn. and alkylation by, of dithiane deriv.)
IT 76494-96-7P
   (prepn. and bromination of)
IT 64028-90-6P
   (prepn. and conversion of, to epoxide)
IT 76494-99-0P   76495-00-6P
   (prepn. and epoxide ring cleavage of)
IT 76495-01-7P   76495-02-8P   76495-03-9P   76495-04-0P

17:28:18 COPY AND CLEAR PAGE, PLEASE

L3 ANSWER 15 OF 268
RN 76495-04-0
   (prepn. and hydrolysis of)
IT 76495-05-1P   76495-06-2P   76495-07-3P   76495-08-4P
   (prepn. and hydrolysis of, dioxaspiro compd. from)
IT 76494-97-8P
   (prepn. and reaction of, with bromoethyl oxirane)
IT 76494-95-6P
   (prepn. and redn. of)
IT 76494-98-9P
   (prepn. and redn. of, with bromoethyl oxirane)
IT 69744-43-0P
   (prepn. and resoln. of)
IT 72229-31-3P
   (prepn. and tosylation of)
IT 69744-44-1P   76041-91-3P   76041-92-4P   76042-02-9P
   76119-65-8P 76495-09-5P
   (prepn. of)

Display BROWSE

=>display browse

ENTER (L3) OR L#:14
ENTER (DIS), ANSWER NUMBER(S), FORMAT(S), OR END:

Note that in the Display Browse mode, the System gives us choices:
(dis)- default SUB format
answer number(s)- enter any number of the answer set
format(s)- enter any desired format (*vide supra*).

HOW TO SAVE ANSWER SET

The Answer Set generated in our search can be saved in a User File for later recall. To save the Answer Set we assign a name to it.

Example: *Saving answer set*
                       Command: SAVE Ln name

Key:save Ln name (title) (temp)
   Ln=answer set assigned by the System
   name=any name which must:
      begin with a letter
      have 1-12 characters
      contain only letters and numbers
      end with /A
      not be already in use in saved files
      not be end, sav, save, saved, Ln
   title=optional, up to 40 characters
   temp=temporary storage (erased automatically
   over the weekend)

To save our Answer Set L3, we proceed at the System prompt:
=>save L3 title
ENTER NAME (END):test/a
ENTER TITLE (NONE):for project LEARN
ANSWER SET 'L3' HAS BEEN SAVED AS 'TEST/A'
=>

HOW TO RECALL SAVED ANSWER SET

To display the answers in a saved Answer Set, we have to be in the database where the answers have been retrieved, i.e., for recall of an Answer Set generated in the Registry File, we have to be in the Registry File. We use the ACTIVATE Command to recall an Answer File, and to receive an Ln for the Answer Set which is then specified in the DISPLAY Command.

Example: *Recalling saved answer set*
                       Command: ACTIVATE name/A

=>activate test/a
TITLE: FOR PROJECT LEARN

STRUCTURE SEARCHING

```
L1         STR
L3         268 SEA SSS SAMPLE L1
```

L1 can be displayed in the STRUCTURE Command as a model for structure building, if desired. To see the actual structure, we use the DISPLAY QUERY Command.

HOW TO DISPLAY SAVED ANSWER FILE NAMES

When we have too many Answer Files saved, chances are we wish to recall the names under which the Answer Sets have been saved. This can be done in the HOME File or any other File we are in, at the System prompt.

Example: *Displaying saved answer file name*
Command: DISPLAY SAVED/A

=>display saved/a

NAME	CREATED	NOTES/TITLE
STEROID/A	12 NOV 87	25 ANSWERS IN FILE REGISTRY
TEST/A	15 MAY 89	268 ANSWERS IN FILE REGISTRY

If we know the saved Answer File name, we can request directly a display of the specific File:
=>display test/a

NAME	CREATED	NOTES/TITLE
TEST/A	15 MAY 89	268 ANSWERS IN FILE REGISTRY

HOW TO DELETE SAVED ANSWER FILE

Permanently stored Answer Files can be deleted in the HOME File or in any File we are in by the DELETE Command.

Example: *Deleting saved answer file*
Command: DELETE name/a

Key:delete name/a
  name=name of saved Ln answer set

=>delete test/a
DELETE TEST/A? (Y)/N:y
TEST/A DELETED

PRINTING SEARCH ANSWERS

When we display the answers to a search, we can get a hard copy of the structures and text of the records by printing on our printer that is connected to the terminal or microcomputer we are using. The technique of printing differs and depends on the hardware we are using. The terminals usually have a key on the keyboard which is pressed to obtain a screen dump, i.e., whatever is on the screen is printed on paper. The microcomputer Emulation Softwares afford techniques of dumping the screen similarly to the terminal technique. We can also download the entire session or part of it, i.e., save it on disk as a file and print it later when we disconnect from the host computer. The manuals for the hardware and software should be consulted for how to proceed.

We can also request offline printing on the high-speed laser printers that are connected to the host computer at the STN Service Center. The prints are of high quality. They are sent to the requester by mail.

To request offline prints of the answers of our current session, we use the PRINT Command at the System prompt =>.

To request offline prints of the answers that have been saved, we have to be in the file in which the answer set has been created, and have an active Ln for the answer set, i.e., the saved answer set must first be activated to receive an Ln for the set.

Example: *Printing answers offline*
                    Command: PRINT

=>print

ENTER (Ln), L#, OR ACCESSION NUMBER:

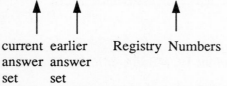

current   earlier   Registry Numbers
answer   answer
set   set

We have an option to enter current answer set, any answer set of the current session, activated answer set, or specific Registry Number(s).

ENTER (Ln), L#, OR ACCESSION NUMBER:L3

L3 CONTAINS 268 ANSWERS CREATED ON 15 MAY 89 AT 11:15:12

STRUCTURE SEARCHING

MAILING ADDRESS = (your ID address will appear here)
CHANGE MAILING ADDRESS? (N)/Y:y

We have an option to change the standard mailing address related to the ID to another mailing address at which we wish to send the prints. If we exercise the option, the System prompts as follows:

ENTER (END) OR NAME: John Doe
ENTER (END) OR ADDRESS: 0000 Any Street
ENTER (END) OR ADDRESS: Any City, Any State Zip
ENTER (END) OR ADDRESS:end
MAILING ADDRESS=(the new mailing address entered at the prompts will appear here)
CHANGE MAILING ADDRESS? (N)/Y:n
ENTER PRINT FORMAT (SUB) OR ?:

We have an option to use the standard SUB or any of the formats we have discussed in displaying answers to structure search (see p. 226).

PRINT ENTIRE ANSWER SET? (Y)/N:y

We have an option to answer Yes or No to print all answers in the set. If we enter N, we are prompted for answer numbers.

215 ANSWERS PRINTED FOR REQUEST NUMBER XXXXXX

The System assigns a number to our print request as a reference for all enquires and for deleting the print request.

HOW TO DELETE PRINT REQUEST

Print requests can be canceled only during the current session. After LOGOFF, the requests cannot be canceled (a call to the STN staff will help).

         Example: *Canceling print request*
                                Command: DELETE PXXXXXX

    Key:delete PXXXXXX
        PXXXXXX=print request number

=>delete PXXXXXXX

HOW TO DISPLAY PRINT REQUESTS

Print requests are stored in the User File created by the System. They can be displayed in the HOME File or in any other File.

Example: *Displaying print requests*
Command:DISPLAY PRINT

=>display print

REQUEST	ANSWERS SET	ANSWERS	STATUS
Pxxxxx	Ln	XX	ACTIVE
PXXXXX	Ln	XXXX	DELETED
.	.	.	.
.	.	.	.

13 SEARCH STRATEGY

A search strategy for searching the Registry File is an important prerequisite to retrieving structures that meet our goals. It is always possible to retrieve some structures from the Registry File even without any planning. However, many of the retrieved structures may be false hits or we may not retrieve all of the structures in the File that would satisfy our search requirements. A search strategy should be set up before the actual terminal session is started. Otherwise, waste of time and money may result. No search should be run as Full Search without first running a Sample Search.

The first stage of a strategy planning is the creation of our structure query. We have to carefully define our goals as to what kind of structures we want to retrieve and, in building the structure query to, appropriately, specify:
1. nodes and/or node variables
2. bonds and/or bond variables
3. normalized bonds in tautomers
4. normalized bonds in rings

For example, the structure query in Figure 1/196. in which the ring has unspecified and

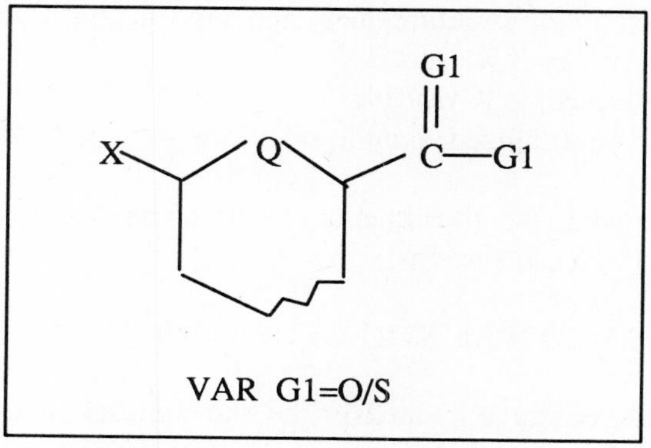

Figure 1/196. Query structure

single bonds and the chains have exact single and double bonds, will retrieve structures containing any halogen (X), any atom except carbon and hydrogen (Q), and either S or O atoms in the position of the G1 groups. Any possible substitution is allowed at S and O and at the ring nodes. The unspecified bond value will allow retrieval of single and/or double bond in the ring. Since no ring specification has been assigned, the ring can be in

fused systems. Since no normalized bonds have been specified in the (thio)carboxyl group, no free acids or salts will be retrieved.
See the discussion of Substructure Searching in Part Two, p. 380, for a detailed analysis of a Substructure Search.

The second stage of strategy planning involves a decision whether to search single or multiple structure query and whether the structure graphs are to be created in the same or in separate Structure Commands.

SEARCHING FOR ONE-COMPONENT STRUCTURES

To search for one-component structures, we formulate queries with:
1. single structure, or
2. multiple structures created in the same Structure Command

SEARCHING WITH A SINGLE STRUCTURE QUERY

The search of a single structure in the query will retrieve:

1. In Exact or Family Search:
An exact structure matching the query structure. No substitution is allowed in the retrieved structure(s). Query structure ring fragments will not be embedded in fused ring systems.

2. In Substructure Search:
An exact structure matching the structure query and any substituted derivatives.
Structural analogs of the query if we specify:
a. System and/or User defined node variables
b. unspecified, single or normalized, double or normalized, tautomeric and/or ring bond variables.
Any substitution is allowed (the substituents can be of any kind); rings in the structure query can be embedded in fused systems.

SEARCHING WITH MULTIPLE-STRUCTURE QUERY

The search of multiple structures created in the same Structure Command will retrieve structures in which:

1. The structural fragments can be connected by any possible bonds either directly or through any kind of nodes.
2. The structural fragments cannot be in separate components (dot-disconnected formulas).
3. The structural fragments will not overlap.

In Substructure Search, any substitution is allowed. The Exact or Family Search do not allow for any substitution.

SEARCH STRATEGY

BOOLEAN OPERATORS IN STRUCTURE SEARCH

Complex queries can be defined by using the Boolean Operators.
The Registry File recognizes three Boolean Operators. Their functions are shown in the Venn diagrams in which the shaded areas show quantitatively the term A and the term B in the answer to the search statement (Figure 1/197).

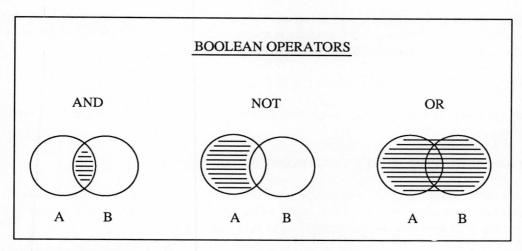

Figure 1/197. Boolean operators

When we have two terms - A, B - in a search statement:

The AND operator requires that both A and B must be present in the answer to the search statement (most restrictive).
The NOT operator requires that the NOTed item be excluded in the answer to the search statement (less restrictive).
The OR operator requires that either A or B or both be in the answer to the search statement (no restriction).

The Registry File allows for complex logical statements in which more than one Boolean Operator can be used.
In a complex logical statement, the Boolean Operators are processed in the order AND or NOT first (reading from left to right) and OR last. Therefore, in complex logical statements where ambiguity may arise, it is necessary to place the logical statements in parentheses (to nest them) in order for the System to process the statement as was the intention.
For example, a statement A OR B AND C OR D will be searched by processing B AND C first followed by combining the result of B AND C by OR with A or D which was not intended by the search statement.

The correct search statement is the one nested in parentheses: (A OR B) AND (C OR D). The nested A OR B, C OR D will be processed first and the results will be combined by AND, which was originally intended.

The Boolean Operators will have the following effect on the search and the nature of retrieved structures:

AND Operator: Structure A and Structure B can be in separate components, i.e., the substance will be represented by structure A and structure B. They can also be in the same substructure but with overlap of structural features.

NOT Operator: Structure B will be excluded from the retrieved structures.

OR Operator: either Structure A or Structure B or both will be among the retrieved structures.

SEARCHING FOR MULTICOMPONENT STRUCTURES

To search, selectively, for multicomponent structures, we create structure queries in separate Structure Commands. Each query will be assigned an Ln. The Lns are combined with the AND Boolean Operator to retrieve structures in which:

1. The structural fragments may be connected by any
   possible bonds in one component or they can be
   in separate components.
2. The structural fragments may or may not overlap.

In Exact Search, no substitution in the structure query is allowed. In Substructure Search, any substitution is allowed.

Example: *Searching with BOOLEAN OPERATORS*

To selectively retrieve the addition compound that was among the substances retrieved in our FAMILY SEARCH (see p. 222) we use the Boolean Operator AND in the EXACT SEARCH of the query:
=>structure
ENTER NAME OF STRUCTURE TO BE RECALLED (NONE):.
ENTER (DIS), GRA, NOD, BON, OR?:
gra r6,3 c3,7 c1,nod 9 O,10 co2h,bon r 1 2 n,dis

SEARCH STRATEGY

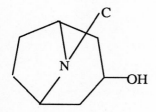

ENTER (DIS), GRA, NOD, BON, OR?:end
L1   STRUCTURE CREATED
=>structure
ENTER NAME OF STRUCTURE TO BE RECALLED (NONE):.
ENTER (DIS), GRA, NOD, BON, OR?:
gra r7,1 c2,3 c1,5 8,nod 8 N,10 OH,move 9 d10,11 r5d5,dis

ENTER (DIS), GRA, NOD, BON, OR?:end
L2   STRUCTURE CREATED

AND

=>search L1 and L2

ENTER TYPE OF SEARCH: (SSS), FAMILY, OR EXACT:exact
ENTER SCOPE OF SEARCH: (SAMPLE), FULL, OR RANGE:sample
SAMPLE SEARCH INITIATED 21:13:22
SAMPLE SCREEN SEARCH COMPLETED - 9 SUBSTANCES TO ITERATE
100.0% PROCESSED   9 ITERATIONS
SEARCH TIME 00.00.10
FULL FILE PROJECTIONS: ONLINE   **COMPLETE**
                            BATCH   **COMPLETE**
PROJECTED ITERATIONS: 1 TO 9
PROJECTED ANSWERS: 1 TO 5
L3   0 SEA EXA SAM L1 AND L2

=>search L1 and L2
ENTER TYPE OF SEARCH: (SSS), FAMILY, OR EXACT:exact
ENTER SCOPE OF SEARCH: (SAMPLE), FULL, OR RANGE:full
FULL SEARCH INITIATED 21:15:21
SCREENING

FULL SCREEN SEARCH COMPLETED - 9 SUBSTANCES TO ITERATE
100.0% PROCESSED   9 ITERATIONS
SEARCH TIME 00.00.08
L3   1 SEA EXACT FUL L1 AND L2
=>display

ANSWER 1 OF 1

RN 6164-69-8
IN Benzeneacetic acid, .alpha.-(hydroxymethyl)-, compd. with
    exo-8-methyl-8-azabicyclo[3.2.1]octan-3-ol (1:1) (9CI)
SY 1.alpha.H,5.alpha.H-Tropan-3.beta.-ol, tropate (salt)
    (8CI)
SY Tropic acid, compd. with pseudotropine (1:1)
SY Pseudotropine tropic acid salt
SY 8-Azabicyclo[3.2.1]octan-3-ol, 8-methyl-, exo-,
    .alpha.-(hydroxymethyl)benzeneacetate (salt) (9CI)
MF C9 H10 O3 . C8 H15 N O
RN 6164-69-8

CM 1

RN 529-64-6
MF C9 H10 O3
CI COM

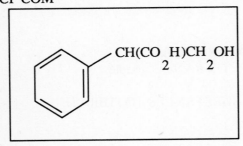

CM 2

RN 135-97-7
MF C8 H15 N O
CI COM
ST 2:EXO

SEARCH STRATEGY

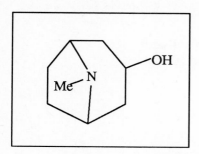

0 REFERENCES IN FILE CA (1967 TO DATE)

=>

To search for the isomeric structures (Figure 1/198) we combine the structure queries by the Boolean OR, or use the VPA specification (see p. 262-A 2)

Figure 1/198. Query structures

Structure A = L1
Structure B = L2
Structure C = L3

=>structure
ENTER NAME OF STRUCTURE TO BE RECALLED (NONE):.
ENTER (DIS), GRA, NOD, BON OR ?:gra r6,1 c1,3 c1,4 c1,nod 7 x,8 co2h,9 oh,bon r 1 2 n,end
L1   STRUCTURE CREATED
=>structure
ENTER NAME OF STRUCTURE TO BE RECALLED (NONE):L1
ENTER (DIS), GRA, NOD, BON OR ?:del 7,gra 6 c1,nod 10 x,end

L2 STRUCTURE CREATED
=>
ENTER NAME OF STRUCTURE TO BE RECALLED (NONE):L1
ENTER (DIS), GRA, NOD, BON OR ?:del 7,gra 5 c1,nod 10 x,end
L3 STRUCTURE CREATED

OR

=>search L1 or L2 or L3

ENTER TYPE OF SEARCH: (SSS), FAMILY, OR EXACT:sss
ENTER SCOPE OF SEARCH: (SAMPLE), FULL, OR RANGE:sample
SAMPLE SEARCH INITIATED 22:09:57
.
.
.
=>display L4 3

L4 ANSWER 3 OF 22

RN 67127-63-3
IN Benzoic acid, 3-acetyl-4-chloro-2-hydroxy- (9CI)
MF C9 H7 Cl O4

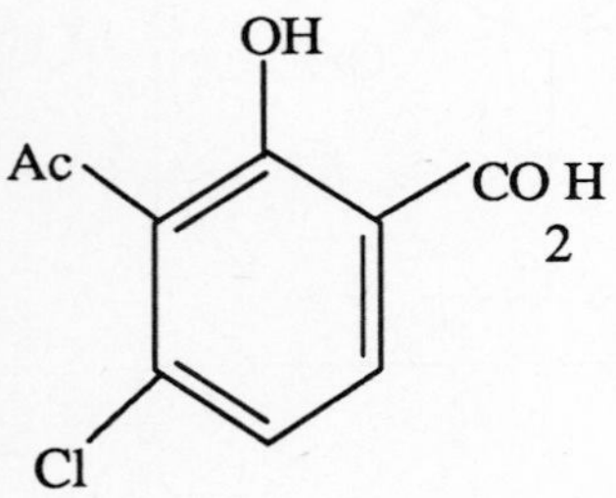

1 REFERENCES IN FILE CA (1967 TO DATE)

=>dis 4
L4 ANSWER 4 OF 22

RN 67127-79-1
IN Benzoic acid, 3-acetyl-5-fluoro-2-hydroxy- (9CI)
MF C9 H7 F O4

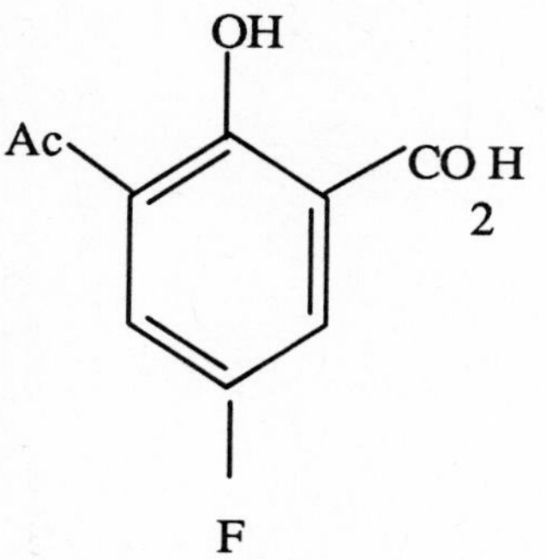

SEARCH STRATEGY

2 REFERENCES IN FILE CA (1967 TO DATE)

Similarly, the Boolean Operator NOT would be used if we wished to search for structure A (Figure 1/199) but wanted to exclude the structure B (Figure 1/199).

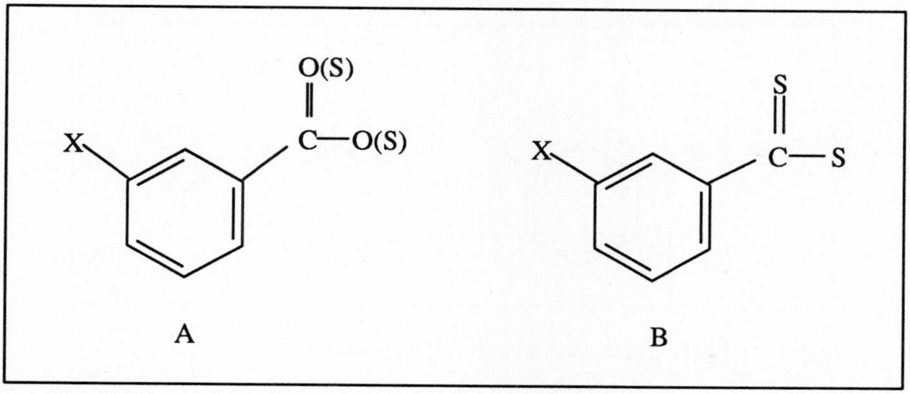

Figure 1/199. Query structures

We would build the structure query B (Figure 1/199) and the generic structure query C (Figure 1/200):
We would receive L1 for the query B and L2 for the query C. The queries are then combined for search as follows:

NOT

=>search L2 not L1

Structures containing the fragment B will be excluded from the answers to our query.
We have to realize that the NOT operator is absolute and that it blocks the inclusion of the NOTed structure, which in some cases may not be desirable. In our case, for example, structures like the one shown in Figure 1/201 will be also excluded although a fragment of

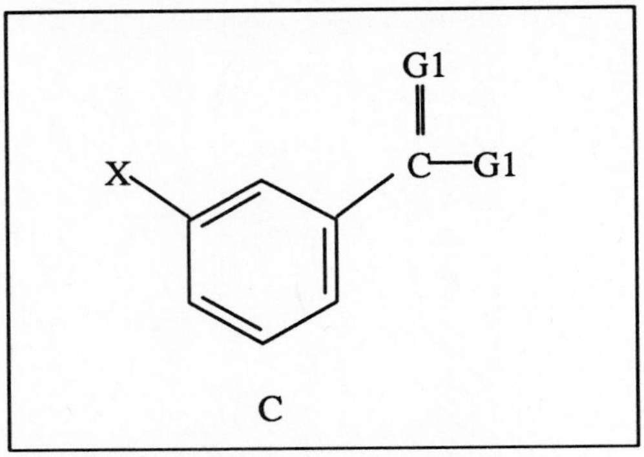

Figure 1/200. Query structure

the structure was of interest.

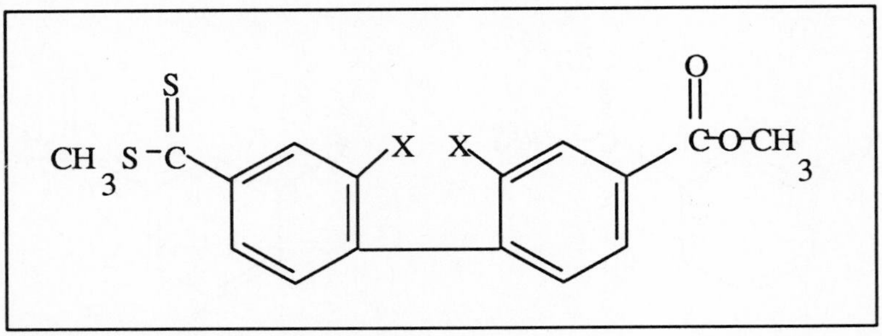

Figure 1/201. Excluded structure from search

SEARCHING STRUCTURE QUERIES COMBINED WITH SCREENS

As we have discussed (see p. 8), screens can still be used for searching structures in the Registry File. It would not be efficient, however, to code the entire structure in screens. The System generates screens automatically for us from the query structure in the course of a search. But in certain cases it is advantageous, or it may be necessary, to include some of the screens in the search statement to achieve a total recall of the desired structures. It is not within the scope of this book to describe the screen system. Those readers interested in learning the system will contact the Chemical Abstracts Service. However, there are screens, called Graph Modifier Screens, which are not automatically generated by the System. We shall discuss them since they cannot be ascribed to structures by the techniques of building structure graphs. They may prove to be useful for some searches.

SEARCH STRATEGY

GRAPH MODIFIER SCREENS

Graph Modifier Screens are divided into three categories:
1. screens describing structural features
2. screens for multicomponent substances
3. screens for class identifiers

Their descriptions are given below:

GRAPH MODIFIER SCREENS

SCREEN	DESCRIPTION
	Structural feature screens
2039	abnormal mass - all isotopes
2045	deuterium
2046	tritium and higher H isotopes
2047	isotope at unknown locant
2041	abnormal valence
2040	charges: fixed,tautomeric,delocalized
2042	delocalized charge only
2076	tautomer
	Multicomponent substance screens
2127	2 or more components
2077	3 or more components
2078	4 or more components
2079	single atom fragment

SUBSTANCE CLASS IDENTIFIER SCREENS

2050	alloy
2049	coordination compound
2048	incompletely defined substance (ID)
2071	ID - unknown structure
2072	ID - unknown point of attachment
2073	ID - ester
2074	ID - hydrogen (bond)
2053	manual registration
2052	mineral

2051	substance named "mixture with" in CA index
2043	polymer (general category)
2067	homopolymers and copolymers (A)x,(A.B)x
2068	polymers defined as structural repeating units (SRU)
2069	SRU with end groups X-(-Y-)n- Z
2070	SRU without end groups -(-Y-)n-
2054	radical ion

HOW TO SPECIFY SCREENS

Screens are created by the SCREEN Command and are assigned by the System an Ln. The Ln is then combined with an Ln for a structure graph by a Boolean Operator.

Example: *Specifying screens*
Command: SCREEN screen

Key:SCREEN screen
screen=screen numerical code

To search for any polymers of the structure:

$$CH2=CH-C(R)=CH2 \quad R=\text{any substituent}$$

1. Build the structure as usual.
2. End STRUCTURE Command.

L1 STRUCTURE CREATED
=>
3. Create screen query.
=>screen
ENTER SCREEN EXPRESSION OR (END):2043
L2 SCREEN CREATED
=>search L1 and L2

The retrieved substances will be only polymers as stipulated by the screen 2043 and will be of the structure defined in the structure query L1. Substances that are not polymers will be excluded in the retrieved structures since the Boolean Operator AND requires that the screen 2043 specification for polymers must be in the retrieved record-structure. Substances of the structure L1 that are not polymers do not have the screen 2043 associated with them in the record of the Registry File, and therefore, would not meet the logical statement requirement.

SEARCH STRATEGY

We can specify more than one screen in our query.
To search for the polymers of the above structure but labeled with tritium, we generate the following screen query:
=>screen
ENTER SCREEN EXPRESSION OR (END):2043 and 2046
L3 SCREEN CREATED
=>search L1 and L3

We can use any of the Boolean Operators both in the screen query generation and in the search logical statement:

=>search L1 and L2 not L3

The retrieved substances will be polymers of the above structure but will not be labeled with tritium.

=>screen
ENTER SCREEN EXPRESSION OR (END):2039 not 2046
L4 SCREEN CREATED
=>search L1 and L2 and L4

The retrieved substances will be polymers of the above structure and will have any isotope except tritium.

CROSSOVER INTO OTHER STN FILES

The result of a search in the Registry File is a list of Registry Numbers associated with the structures registered in the File. The Registry Numbers are, in fact, the accession numbers for each record in the File which contains all the data and information related to the substances represented by the structures.

CA AND CAOLD FILES

As we have seen (see "Display of Answers to Structure Sesarch", p. 226), some of the data and information related to both the substance and the document in which it was published can be displayed in the Registry File. However, the Registry File affords display of only the 10 most recent references to documents. Others can be only displayed, after being searched for, in the CA File and CAOLD File of the CAS ONLINE System. We may also be interested in conducting a selective search for substances related to some specific chemical, physical, biological, or other data and information in the other STN Files and databases which have Registry Number data fields in their records.
We can crossover with the list of Registry Numbers from the Registry File to perform selective searches and display answers in these files if they contain CAS Registry

Numbers as searchable fields.

Example: *Crossover into CA Files*

We have performed a structure search and the L2 answer set has been created:

L2   450 SEA SSS FUL L1
=>file CA (or file CAOLD)
FILE 'CA' ENTERED AT 14:05:45 ON 15 NOV 87
COPYRIGHT 1987 BY THE AMERICAN CHEMICAL SOCIETY
=>search L2

The list of Registry Numbers in L2 is now searched in the CA File. In this simple search, all references to the publications that described the substances of the structures registered under the Registry Numbers since 1967 would be retrieved. However, it is possible to search in the CA Files by combining the set of Registry Numbers with the subject index terms in the CA Files using the Boolean Operators.
For example, we can selectively search for biological properties of the retrieved structures (substances) in the Registry File by the following query:

=>search L2 and toxicity

The answer to this query will be references to publications describing the toxicity of all substances we have retrieved in the answer set L2.
The search and display techniques in the CA Files are not within the scope of this book. The interested readers will contact Chemical Abstracts Service for further information.

CASREACT

CASREACT is the file in the CAS ONLINE System which contains chemical synthetic reaction information from over 100 journals from 1985. The file can be searched, however, only by the text-search techniques using Registry Numbers of the structures involved in the reactions as reactants, products, reagents, catalysts, and solvents.
Let us assume that we have performed a search in the Registry File for the structures of the reactant and product of the reaction in Figure 1/202 and have the L1 and L2 comprising the Registry Numbers of the structures.
We crossover with these RegNos to the CASREACT File:
=>file CASREACT
and search for the reaction:
=>s L1/rct (L) L2/pro
We retrieve the reaction which is displayed as shown in Figure 1/203.

SEARCH STRATEGY

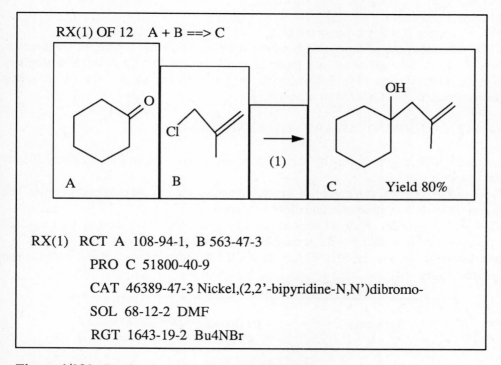

Figure 1/202. Reaction search in CASREACT

Figure 1/203. Retrived reaction from CASREACT

CURRENT-AWARENESS SEARCHES

Current-awareness searches are those that are run at regular time intervals, after a comprehensive retrospective search has been performed, to retrieve data that have been added to the Registry File since our retrospective search in regular, weekly updates. Thus, we can keep our data and information related to a certain project up to date.

The Registry File offers the User two methods to perform these searches:
1. Online search (current-awareness)
2. Offline search (selective dissemination of information - SDI)

ONLINE CURRENT-AWARENESS SEARCH

An online current-awareness search is essentially a Range Search of SSS, EXACT, or FAMILY type of search carried out by the User while connected to the host. The range is limited to 80,000 substances (Registry Numbers) for the search to qualify as a current-awareness search for billing purpose. We have already discussed the Range Search (p. 212) and Query Saving (p. 112) which are the techniques necessary for running the current-awareness search.
The search can be run at any time by the User.
It is to be noted that the newest records in the Registry File may not be, in some cases, complete. This is because when a substance is registered, the structure and Registry Number are entered first. The bibliographic and other data of the record including substance name are added later at an unspecified time.

AUTOMATIC CURRENT-AWARENESS SEARCH (SDI)

An automatic SDI search is run by the System every two weeks (the Registry File is updated weekly).
SDI profiles, i.e., queries, can comprise structures and/or screens, or dictionary terms. The profile cannot have both structures and dictionary terms.
To define an SDI profile, create a query as usual to obtain an Ln for the query. If a stored query is to be used as an SDI profile, it must be first activated.
The prints with answers to an SDI search are mailed to the requestor, or the answers can be displayed online.

Example: *Requesting SDI search*
Command:SDI

Key:sdi (title)
  title=optional up to 40 characters

=>sdi
ENTER QUERY L# FOR SDI REQUEST:L1
ENTER SDI REQUEST NAME (END):thiazoles/s
(Note: Observe name requirements as in Saving Answer Set; append "/s" to the name.)
ENTER TYPE OF SEARCH:(SSS), FAMILY, OR EXACT:family
WARNING - QUERIES FOR FAMILY SEARCH ARE SUBJECT TO

SEARCH STRATEGY

RESTRICTIONS. ENTER ? AT THE NEXT PROMPT FOR AN EXPLANATION
ENTER METHOD OF DELIVERY:(MAIL), ONLINE, OR BOTH:both
MAILING ADDRESS = (address for the ID will appear)
CHANGE MAILING ADDRESS? (N)/Y:n
ENTER PRINT FORMAT (SUB) OR ?:sub
ENTER MAXIMUM NUMBER OF HITS TO BE PRINTED PER RUN (100):.
System: QUERY 'L1' HAS BEEN SAVED AS SDI REQUEST 'THIAZOLES/S'

DISPLAYING ANSWERS TO SDI SEARCH

If we specify in the SDI request the Method of Delivery ONLINE, the System will create and store the Answer Set of the SDI search in a file with the same name as the SDI request, appended with a consequtive number of the run. In our case, the created file would be named THIAZOLE01/A for the first run, THIAZOLE02/A for the second run, etc.

To find out whether we have any answers for our SDI searches, we use the DISPLAY Command for Answer Set:

Example: *SDI for online display*

=>display saved/a

NAME	CREATED	NOTES/TITLE
AZACYTIDINE/A	14 JAN 86	20 ANSWERS IN FILE REGISTRY
SPIRORINGS/A	11 JUN 87	56 ANSWERS IN FILE REGISTRY
.	.	.
.	.	.
THIAZOLES26/A	09 DEC 87	12 ANSWERS IN FILE REGISTRY
THIAZOLES01/A	12 JAN 88	0 ANSWERS IN FILE REGISTRY

To display the answers, activate the stored answer set and use the assigned Ln in the DISPLAY Command.

Example: *Displaying SDI answers online*

=>activate thiazoles01/a

L1   STR

L2   12 SEA SSS FUL L1
=>dis L2 sub 1-12
The answers 1-12 will be displayed in the SUB format.

Displaying SDI searches

To display a list of all SDI searches:
=>display saved/s
To display a specific SDI search:
=>display thiazoles/s

Deleting SDI

To delete an SDI search:
=>delete thiazoles/s

ADDENDUM

New features have been implemented in the CAS Registry File in September 1989. They are covered in this Addendum.

NEW SYSTEM VARIABLE SYMBOL

A new system variable symbol has been added to the system variables (see Figure 1/91, p. 119):

AK = any carbon chain fragment (alkyl)

The AK variable can be attached to any ring node or to a chain hetero node. It cannot be attached to a chain carbon node.

To specify the AK, use the Menu NODE (see Figure 1/101, p. 129) or:

|KEYBOARD| NODE m AK

    Key: nod m ak
       m = node locant

Example: gra r66,8 c1,nod 8 N,11 ak,bon all se,dis

NEW STRUCTURE SUBCOMMANDS

New subcommands for building generic structures are available within the STRUCTURE command at the : prompt (see p. 22).

1. Generic Group Category (GGC)

The system variables AK, CY, CB, and HY (*vide supra*) can be further specified by the definitions listed in the table:

Symbol	Category	Definition
LIN	Linear	Group must be linear chain
BRA	Branch	Group must be branched chain
SAT	Saturated	Only single bonds allowed

UNS	Unsaturated	At least one nonsingle bond required
LOC	Low Carbon	Must contain 6 or fewer carbons
HIC	High Carbon	Must contain more than 6 carbons
LOQ	Low Hetero	Must contain only 1 noncarbon atom
HIQ	High Hetero	Must contain more than 1 noncarbon atom
MCY	Monocycle	Must be monocyclic
PCY	Polycycle	Must be polycyclic

The GGC can be used for the above variables only. It cannot be used for elements, Gk groups, or system defined variables A, Q. It is specified as follows:

> **KEYBOARD** GGC symbol
>
> Key:ggc symbol
> symbol = any symbol from table
> any combination allowed unless
> mutually exclusive

Example (the above structure is the query):

ENTER(DIS), GRA, NOD, BON, OR ?:ggc 11 bra sat hic

The AK will be a branched, saturated carbon chain with more than 6 carbons.

2. Variable Point of Attachment (VPA)

The VPA allows for defining alternative points of attachment of a substituent to a **ring** or **ring** system.

This subcommand allows for defining queries with variable substitution locant thus eliminating the need to build separate structure queries and search them with the Boolean operator OR to retrieve positional isomers (see Figure 1/198, p. 251). We can specify this query by the VPA.

The VPA is specified as follows:

> **KEYBOARD** VPA n-a/b/c...... bond value
>
> Key:vpa n-a/b/c.... bond value
> n = locant of substituent (either chain or ring);
> can be element, shortcut, Gk, generic symbol or a multinode fragment
> a,b,c....**ring** locants of attachment (up to 20)
> cannot be generic rings or Gk
> bond value = any bond value

Note that using hydrogen as the substituent may lead to false drops.

Example:

ENTER(DIS), GRA, NOD, BON, OR ?:
gra r6,3 c1,4 c1,c1,nod 7 co2h,8 oh,9 x,bon r 1 2 n,3-7,4-8 se,vpa 9-1/5/6 se,dis

```
           2
       ╱═══╲    CO₂H
    @1╱     ╲ ╱
            ║ 3
    @6╲     ╱
       ╲═══╱
           ╲
        @5  OH
            8

         @9X
```

The variable points of attachment are shown by the @ for the substituent X (which was build as a separate fragment).
Note: proper bond values (exact or normalized) must be specified in the parent depending on the effect on bonds by the substituent.

We can build more than one substituent and specify the points of attachment for each one thus forcing the search to find multisubstituted structures with the specified attachments.

Note: in using a Gk as the substituent, H is not to be defined in the groups. If a multiatom fragment is the substituent, only one node of the fragment can be specified as the attachment.

NEW TYPE OF SEARCH

We have learned that there are three types of search we can perform in the Registry or Beilstein Files:
1. Exact search
2. Family search
3. Substructure search

The Exact and Family searches retrieve the structures exactly as they were drawn in the query with no substitution allowed. Substructure search, however, allows, by default, for any substitution or fusion at the locants where they can take place (see p. 217; see also p. 380 for a discussion of Substructure search).
We now have a choice to perform a Closed Substructure Search (CSS) which is an "exact" substructure search, i.e., a substructure search in which no substitution or fusion is allowed, unless we specify the locants at which substitution or fusion is desired. The hits of CSS will be structures which will contain the query structure fragment exactly as it was drawn and will be substituted or fused only at the locants we specified in the query.

Closed Substructure Search (CSS)

CSS is available for all the Scopes of search, i.e., Sample, Full, Range, and Batch search (see p. 207) and SDI search mode (see p. 260).
The CSS search is now one of the options in the prompt we receive when we start a search:

=>search
ENTER LOGIC EXPRESSION OR QUERY NAME (END):L1
ENTER TYPE OF SEARCH: (SSS), CSS, FAMILY, OR EXACT:css

We have used the query in Figure 1/190 (see p. 221) in Substructure search (SSS) and retrieved 1187 structures (see p. 224). The SSS allowed for any substitution or fusion at the locants indicated by the arrows

If we wanted to prevent substitution or fusion at all of these positions (and thus also narrow the result of the search), we would simply submit the query L1 to CSS with no further specifications:
=>search
ENTER LOGIC EXPRESSION OR QUERY NAME (END):L1
ENTER TYPE OF SEARCH: (SSS), CSS, FAMILY, OR EXACT:css
ENTER SCOPE OF SEARCH: (SAMPLE), FULL, OR RANGE:sample

Note that "no substitution" in CSS is absolute, i.e., it applies to all units of the query. If we have the system variables, shortcuts or fragments from the Fragment File, or Gk in a query, they are all subject to this "no substitution".
The "no substitution" requirement in CSS is achieved by automatic assignment to each of the nodes in the query the CONNECT values. For example, the above query is automatically assigned by CSS the CONNECT value E2 for the position 8 so that no substitution is possible at that locant (there are already two nonhydrogen attachments).
If we wanted to allow for substitution at some of the positions in the query, we would open the positions for substitution by the CONNECT command (see p. 152).
For example, to allow substitution in CSS of the above query at the position 8, excluding hydrogen, we would specify:

ADDENDUM

ENTER (DIS), GRA, NOD, BON, OR ?:con 8 e4

Note that in queries for CSS we cannot use HCOUNT (see p. 147) or VALENCE (see p. 162) to control substitution; we must use CONNECT although we can combine CONNECT with HCOUNT or VALENCE.

For example, to allow hydrogen as one of the substituents at position 8 in the above qeury:

ENTER (DIS), GRA, NOD, BON, OR ?:con 8 m3,hco 8 e1

To allow fusion at the A side and substitution at positions 7 and 8 in the above query, we would specify:

ENTER (DIS), GRA, NOD, BON, OR ?:con 1 6 e3,7 e4,8 m3,hco 8 e1

Similarly, we would use CONNECT for opening substitution of the A and Q variables, and, for Gk groups, we would build the Gk as a separate fragment and assign CONNECT to the fragment.

If a bond type is not specified for the substitution, the default "unspecified" is assumed. Or, we can specify R, C, RC by using the NSPEC Attribute (see p. 147).

PART TWO

CHEMBASE

1 SYSTEM BACKGROUND

In Part One we have discussed CAS ONLINE-Registry File, the system for search and retrieval of chemical structures.
In Part Three and Part Four we will discuss other systems for graphic molecular structures and chemical reactions, MACCS-II and REACCS. These systems run on mini/mainframe computers and can be accessed through personal computers, i.e., microcomputers with graphics capability.
In this Part Two, we describe and discuss a system for graphic molecular structures and chemical reactions which is run as a standalone system on IBM and compatible personal computers, i.e., microcomputers with graphics capability. In turn, this microcomputer system can be interfaced with the mini/mainframe systems to form a complete data and information management system for chemist to handle both internally and externally generated data and information composed of chemical structures and associated numeric and textual data.

CPSS - Chemist's Personal Software Series[R]

CPSS is a series of software for personal computers (IBM and compatibles) developed and marketed by Molecular Design Ltd.
CPSS comprises the following programs:

ChemBase[R]	- A chemical database system
ChemText[R]	- A wordprocessing system
ChemTalk[R]	- A communication and graphics terminal emulation system
ChemHost[R]	- A communication manager system that resides on the mini/mainframe and operates in conjunction with ChemTalk

The microcomputer-mini/mainframe systems interface is shown in Figure 2/1.

MOLECULAR DESIGN LTD.

Molecular Design Limited (MDL) was founded in 1978 by Dr. Stuart Marson and Professor W. Todd Wipke to provide guidance and services for management of large files of chemical structures for the chemical and pharmaceutical industry. The consultation services soon developed into providing complete computer graphics systems for chemical structure storage and retrieval with associated numeric and textual data. The objective had been to enable chemists to run these programs themselves and to develop corporate and

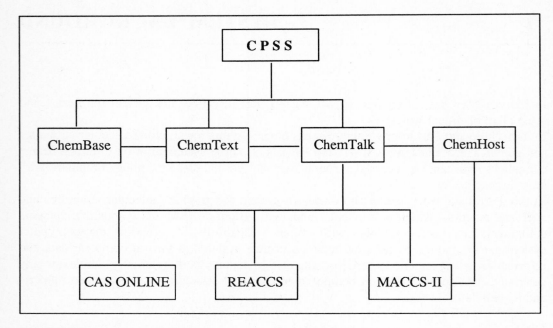

Figure 2/1. Systems interface

departmental databases.

One of the first programs, the Molecular Access System (MACCS), was developed as a system for low-cost minicomputers which, at the same time, could be adapted for the large mainframes.

It was a natural second step to develop a system which could handle chemical reactions and, in 1981, the Reaction Access System (REACCS) was born.

Recognizing the need for generating by computer reports and manuscripts comprising both chemical structures and numeric and text data, MDL also developed programs to prepare formatted forms and transfer data to these forms from MACCS and REACCS, i.e., The Data Access System (DATACCS) which was later integrated with MACCS in MACCS-II.

These systems formed the first comprehensive Database Management System (DBMS) for chemical information processing. In addition to these chemical information management systems, MDL has developed a number of programs for molecular design/ modeling, e.g. PRXBLD and DISP, ADAPT, etc.

In 1986, following the trend of integrating microcomputers into the mainframe and minicomputers network, MDL offered the Chemist's Personal Software Series (CPSS), including ChemBase, a subset of the MACCS-II and REACCS systems for the IBM PC and compatibles in addition to complete chemical wordprocessing (ChemText) and communication system (ChemTalk).

MDL has grown from the consulting service beginnings to an enterprise with over 150 people. The headquarters are located in San Leandro, California with regional offices in

SYSTEM BACKGROUND

New Jersey, Illinois, England, and Switzerland, and with distributors in Japan.
In 1987, MDL was purchased by the Maxwell Communication Corporation of England.

2 SYSTEM DESIGN

CHEMBASE

ChemBase is a microcomputer system for storing, searching, and retrieving molecular structures, chemical reactions, and associated numeric and textual data.
A collection of molecules, reactions, and data, contained in a single disk file, forms a database.
In ChemBase, a database can be used for only molecules or only reactions, or for both molecules and reactions. If a database contains both molecules and reactions, they are kept functionally separate although they are manipulated from the same database. Numeric and textual data can either be part of the structural databases or we can have a database with only numeric and textual data with no structures.

STRUCTURES

Storage, search, retrieval, and display of structures is based on the same principles as discussed in Part One. A connection table (see p. 29) for a structure contains all atoms and bonds in the structure along with any unique specifications, e.g., charges, isotopic labels, and abnormal valences. The encoding of structures in form of the connection tables is performed automatically by the System when we create a structure. We are able to create files containing the encoded structure data and save them in SDfiles, RDfiles, Molfiles, or Rxnfiles (*vide infra*) and transfer these files into ChemBase databases or to the mainframe MDL Systems, MACCS-II and REACCS.
The System automatically calculates molecular formulas, molecular weights, and elemental compositions from the data of stored structures and reactions. These, in turn, can be searched and displayed.
ChemBase recognizes aromaticity for six-membered rings drawn with alternating single and double bonds (see also the discussion of aromaticity, p. 95) and will store and retrieve these aromatic resonance structures no matter how they have been input in the System.
In ChemBase structures, the carbon atoms are implied at bond junctions or at the end of a bond. When we create a structure, by default, all atoms are carbons unless we specify other atom values. All drawn bonds are single bonds unless we specify otherwise.
Hydrogens are implied at atoms where no other substitution is specifically assigned to fill the atom valences. By default, hydrogens are not displayed in structures. We can, however, turn the display of these implicit hydrogens on/off. In structure searching, explicit hydrogens are assigned in query structure to prevent substitution by any other element or fragment.

STEREOISOMERIC STRUCTURES

ChemBase Version 1.3 recognizes stereochemical structures both for registration (entry) in a database and searching.

Structures can be drawn with solid or dotted lines, solid wedged or striped hashed stereobonds. In ChemBase Version 1.2, however, the stereochemical designations are used for display of structures only. The search subsystem of ChemBase Version 1.2 does not recognize stereoisomerism. Structure queries with stereobonds return all of the stereoisomers that have been registered. However, when we save stereochemical structures from ChemBase Version 1.2 or 1.3 in SDfiles and RDfiles, the stereochemical data are retained and can be transferred into MACCS-II and REACCS which recognize stereoisomerism in searches.

For that purpose, the CHIRAL label can be assigned to a structure in ChemBase. SDfiles and RDfiles store the three-dimensional coordinates with a structure if these are available. It is to be noted that the three-dimensional coordinates in a file can be read into ChemBase and they are retained with the structure even if we modify the structure in ChemBase. This makes it possible to use the structure for manipulation in a modeling program.

REACTIONS

Chemical reactions in ChemBase are composed of molecules labeled as reactants, intermediates, and/or products. The reaction schemes are stored and displayed in the customary format with plus and arrow signs indicating the role of each of the molecules in the reaction. Although the reactions are composed of individual molecules, the molecular structures are not automatically stored separately, in contrast to REACCS.

Reactions can be searched with queries composed of individual structures or of entire reaction schemes. In searching, the System treats intermediates as both reactants and products.

ChemBase offers a unique capability to search for chemical transformations within a molecule, i.e., for reaction transforms of a fragment within a molecule.

NUMERIC AND TEXTUAL DATA

Numeric and textual data can be stored with each of the molecules and reactions or we can create a database containing only numeric and textual data with no structures.

The data associated with molecules and reactions can be simply a text description of a molecule or a reaction, or the data can be results of chemical, physical, biological, pharmacological, or clinical research of chemical substances. Note that the special text descriptions attached to a molecule or a reaction structures are not recognized by MACCS-II and REACCS if we transfer them to these systems from ChemBase, unlike the regular data fields which can be transferred.

DATABASE STRUCTURE

Molecules, reactions, and numeric and textual data are registered in the database as records (entries). One molecule or one reaction along with numeric and textual data form one record.

Each record is composed of data fields. The records (entries) are identified by ID numbers. The data fields are identified by field names. Figure 2/2 shows the database structure.

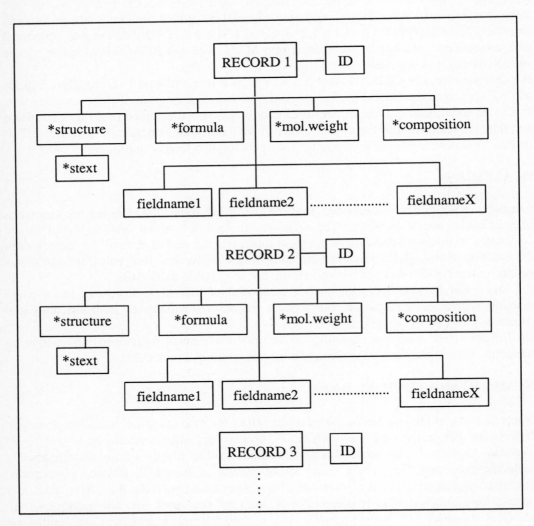

Figure 2/2. Database structure

SYSTEM DESIGN

ID - THE RECORD IDENTIFIER

Each record in the database is assigned a unique identifier (ID) which fills the ID data field. The ID is the first data field in the definition of a database. Molecules have separate IDs from the reactions IDs. The field type for the ID must be either of the field types:
1. an integer
2. a fixed-length text string

If we use the option 1, the ID numbers do not have to follow sequentially. We can assign an entry the ID number 20 following the entry with ID number 10 and later register data in between the ID 10 and 20.

If we use option 2, we can use any text string containing numbers and/or letters (e.g. CA Registry Numbers which we retrieve in the CPSS-CAS ONLINE interface, see p. 771) or company codes for substances. Upper- and lowercase letters are equivalent in fixed-length text IDs.

SPECIAL DATA FIELDS

The data fields that begin with an asterisk in Figure 2/2 are special ChemBase data fields. With the exception of the *stext field (see p. 311), we do not have any control over these data in editing or displaying them or in updating a database. The *formula, *mol. weight, and *composition are automatically generated by ChemBase for each of the registered molecules and cannot be edited. However, except for the *composition field, all the special fields can be searched by data queries (see p. 403).

The special data fields can be used as data field names (*vide infra*). The asterisk has to be included.

If we want to calculate our own molecular formulas, molecular weights, or elemental compositions (the System calculates only elemental composition for the entire molecule) we have to define separate regular data fields and name them when we set up the database.

DATA FIELD NAMES

The names of the regular data fields (fieldname1 to fieldnameX in Figure 2/2) are specified according to our choice when we define the database structure (see p. 340).

The names must be unique within a database. The first character of the name must be a letter.

The names may:
1. be up to 30 characters long
2. consist of upper- and lowercase letters or numbers, spaces, and the special characters -,_,/,#

DATA FIELD TYPES

Data field type specifies the nature of the data a data field will hold. The data field type is specified at the time we define the database structure (see p. 340).

The field type can be any of the following values:

1. Real - a single real number with decimal point and optional exponent and sign. The number size may be up to $\pm 7 \times 10^{302}$.
2. Integer - a single integer with no decimal point or exponent.
3. Range - a pair of real numbers separated by a hyphen, usually representing minimum and maximum data.
4. Date - any date from March 1900 to 2079.
5. Fixed text - a fixed-length string of a length specified when the field is defined. Space for the number of characters in the field is reserved in the database whether data for the field is registered or not. Therefore, this field should be used only for regularly entered data in order not to waste space in the database. It is the field type that has to be used if ID (*vide infra*) is a text string.
6. Variable text - a string of text of variable length. The total text length is limited only by the available memory of the User system. Space for variable text is not reserved in the database unless the data are actually registered in the database.

CHEMBASE FILES

In ChemBase we can read and write a variety of files that are used to store data within ChemBase or for transfer to other systems.

Figure 2/18 (p. 290) shows the type of files we encounter in CPSS System. We now describe and discuss the files that are specific for ChemBase.

CPSS programs write and read difrent type of files for different data. The following are the files that are valid in ChemBase (see also p. 287 for discussion of the Clipboard).

Molecule Files and Reaction Files

A Molecule File (Molfile) and a Reaction File (Rxnfile) have the file name extension .MOL and .RXN, respectively.

These files store the connection table describing one molecule (Molfiles) or one reaction (Rxnfiles), along with any of the structure text (*stext) we may have attached to a molecule or a reaction. No other numeric or textual data associated with a molecule or a reaction in the database can be stored in Molfiles or Rxnfiles.

Molfiles and Rxnfiles can be read, of course, by ChemBase and also by MACCS-II and REACCS. They can be used to transfer data from database to database in ChemBase or from ChemBase via ChemTalk to the mainframe systems. However, REACCS does not recognize intermediates in reactions, so it disregards them if they are specified in a ChemBase reaction stored in Rxnfile.

SYSTEM DESIGN

SDfiles and RDfiles

SDfiles (Structure/Data Files) are the files for storing molecules, RDfiles (Reaction/Data Files) are the files for storing reactions.
The names of these files have the file name extension .SDF and .RDF, respectively.
While Molfiles and Rxnfiles contain the connection table data describing only one molecule or reaction, the SDfiles and RDfiles can be used to store information for any number of molecules or reactions including numeric and textual data associated with the molecules or reactions. SDfiles and RDfiles are written and read by ChemBase, MACCS-II, and REACCS and are used to transfer data between the Systems via ChemTalk/ChemHost.

Data Files

Data File names have the extension .ASC. The Data File stores only numeric and textual data from ChemBase database. No structures can be stored in Data Files. The data are written or read in a table format, a line of data for each record according to the Table Definition (see p. 356). The Data File is an ASCII File used to transfer data from/to ChemBase and any system which can use ASCII Files.
Note that the ChemBase Data Files are not equivalent to datfiles in MACCS-II.

List Files

List Files have the file name extension .LST. The List Files store the IDs of the records that we have located in a database by search or retrieve operations. No structures, or numeric or textual data can be stored in the List Files. The list of IDs is merely a reference list which is tied with the database from which the IDs have been retrieved. Therefore, a List File is only meaningful in the database from which it has been created and saved. MACCS-II and REACCS List Files are not compatible with ChemBase List Files.

Form Files

Form Files have the file name extension .FRM. They are used to store the display format FORM (see p. 346). We can save either the FORM definition including box labels, or we can save the FORM with current data, including structures (see p. 353). The Form Files are useful to save data for later display or for transfer to ChemText.

Table Files

Table Files have the file name extension .TBL. They are used to store the display format TABLE (see p. 349). Contrary to saving the Form, we can save only the TABLE definition including the column labels, but no data.

Template Files

Template Files have the file name extension .TPL. These files store the User templates (see p. 305) for drawing molecular structures.

PostScript Files

PostScript Files have the file name extension .PS. These are the files for ChemBase output to be printed on PostScript printers such as the Apple LaserWriter or incorporated into desktop publishing programs that accept encapsulated PostScript files.

Database Files

Note DO NOT (p. 287).

Database Files have the file name extension .DB. They store all the data registered in a database. Although Database Files are files stored in the Clipboard directory, they behave differently.
If we have a Database File Name entered in the SETTINGS (see p. 284), that database is the default database and is opened automatically by the System when we start ChemBase, becoming the current database. We do not have to read or write Database Files like the regular files. We carry out operations in the current database. When we call on another database (by USE(Database) on the Main Menu) or exit ChemBase, the current database is closed and exited automatically by the System.
It is a good practice to back up Database Files periodically and frequently.
ChemBase Database File size is limited to 8 megabytes. Depending on the size of structures and the amount of data associated with the structures, we can create databases containing several thousands of molecules and/or reactions.

INPUT/OUTPUT SYSTEM

The input and output of data in ChemBase databases is channeled through a system of formatted windows, i.e., FORMS (see p. 346) and TABLES (see p. 346) and is based on the concept of VIEWing data.
Molecule View/Form/Table I/O channels data related to molecules in the format of the FORM or the TABLE.
Reaction View/Form/Table I/O channels data related to reactions in the format of the FORM or the TABLE.

DATABASE OUTPUT

Search (retrieval) of the data from a database is based on the following conditions:

1. "Which data do I want to see?"

SYSTEM DESIGN

ChemBase offers the following options:
a. molecular structures with the associated numeric and textual data
b. structural reactions with the associated numeric and textual data

2. "How do I want to see the data?"

ChemBase offers the following options:
a. a FORM with boxes for each of the data fields of interest for one of the retrieved molecule or reaction records
b. a TABLE with columns for each of the data fields of interest for all the retrieved molecule or reaction records, one line per record

Figure 2/3 illustrates the options we have in ChemBase for displaying data.

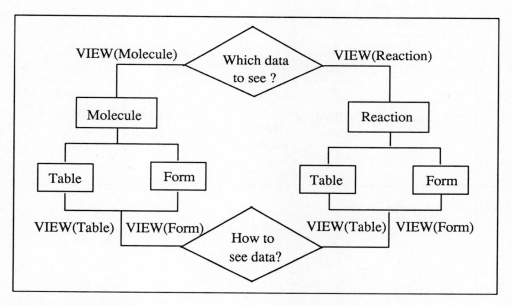

Figure 2/3. Output system

Molecule Forms and Tables are different and separate from Reaction Forms and Tables.
If we are in Molecule View, the System automatically uses the Molecule Form/Table. If we are in Reaction View, the System automatically uses the Reaction Form/Table.
If we are in Molecule Form/Table, the System is automatically in Molecule View. If we are in Reaction Form/Table, the System is automatically in the Reaction View.

DATABASE INPUT

Registration (entry) of new or updated data into a ChemBase database is accomplished through the same system (Figure 2/4).
The FORM is used to enter, edit, or delete one record in a database. The TABLE is used to automatically enter multiple records from an SDfile, RDfile, or Data File.

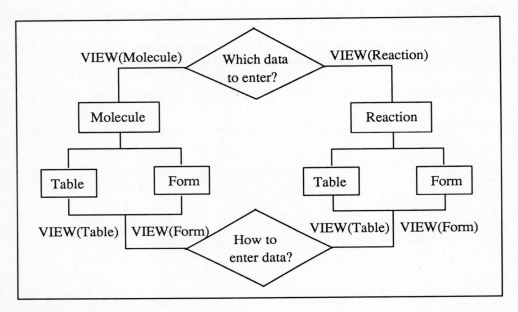

Figure 2/4. Input system

3   SYSTEM OPERATION

ChemBase is operated through Menus which are displayed on the screen. The Menus contain commands for instructing the System to perform a specific action. The system of Menus is built around the Main Menu whose functions are the central part of the ChemBase program. The Main Menu is linked to four separate Menus which control the Editors. There is a Text Editor available throughout the system which is used in any of the program operations. The system of the Main Menu and Editors is illustrated in Figure 2/5.

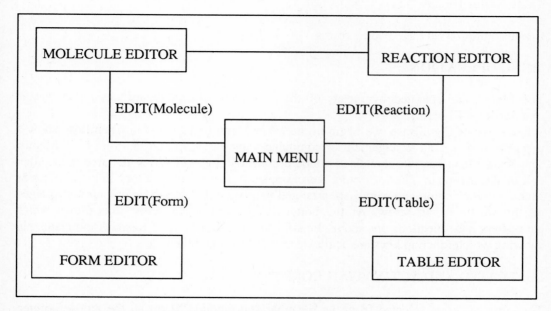

Figure 2/5.   System of Main Menu and Editors

Main Menu (Figures 2/6, 2/7)

Creates and updates databases
Manages files
Searches, retrieves, displays, and prints data
Sets parameters for running the system

Molecule Editor (Figure 2/8)

Creates and edits molecules and creates reactions

Reaction Editor (Figure 2/9)

Manages and edits reaction schemes

Form Editor (Figure 2/10)

Creates box-format data display forms

Table Editor (Figure 2/11)

Creates table-format data display forms

MENU OPERATION

The Menus are operated by means of the mouse, by the keyboard keys, and by the function keys F1-F10.
When we start ChemBase, we begin in the Main Menu. All Menus in ChemBase show a menubar which is an arrangement of command names in a line on the top of the Menus. The Main Menu has two menubars which are switched by activating the arrows at either end of the menubar. The Molecule, Reaction, Form, and Table Editors show also a side Menu in which command names are arranged in a column. The mouse pointer is displayed on the screen as an arrow. At the bottom of the screen, a status line shows which parameters and selections are active. Pressing the function key F5 displays the status bar showing which function keys are available in that Menu and their uses.

SELECTING AND ACTIVATING COMMANDS

The commands are selected from the menubar and the side Menu by the mouse pointer. We will use the following terms to describe the Menu operations:

Point - move the mouse pointer to and over a command name so that the name becomes highlighted.
Click - depress and release the mouse button.
Drag - point, depress the mouse button and, while holding the button, move the pointer to a new position and release the button.
Place - move the mouse pointer to a blank position on the Menu display area or to a position in a structure in the display area.
When we point/click to activate a command, the command will remain highlighted as long as it is active.
Many of the commands have options which are displayed on a pull-down Menu. The

SYSTEM OPERATION

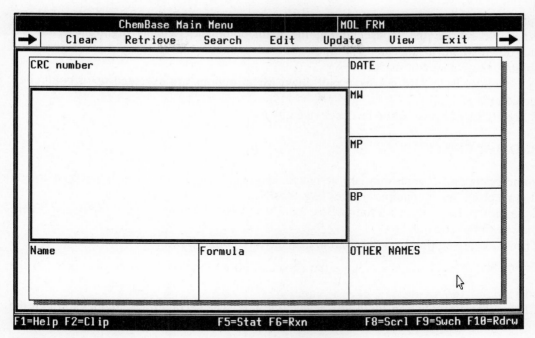

Figure 2/6. Main Menu (First Menubar) in FORM View

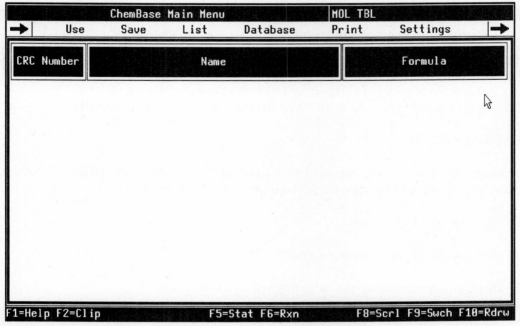

Figure 2/7. Main Menu (Second Menubar) in TABLE View

options are activated when we Drag the mouse pointer until the option is highlighted. When we release the mouse button, the main command remains highlighted but the pull-down Menu disappears. The option is active until we point/drag another option or point/click another command.
The commands that do not have pull-down Menu options are executed upon point/click the command.
A highlighted command can be deactivated by the ESC key.

PROMPT BOXES

Some commands, when they are activated, return a Prompt Box which acts as a toggle (toggle box), i.e., it contains choices (e.g. Yes/No).
Other commands return a Prompt Box (edit box) into which we have to input text. There may be more than one box.
The following editing keys are used to manage the Prompt Boxes:

→ / ← The arrow keys toggle the choices in a toggle box or move the cursor in the edit box.

Click - The first click inside a box selects the box. The second click toggles through the choices available.

↓ / ↑ The arrow keys switch from previous to next box.

⏎ Carriage return accepts the content of a box and switches to the next one. If there are no other boxes, it executes the option.

F9 - Undoes any changes made since start of editing.
F10 - Accepts the contents of all boxes and executes the command.
ESC - Ignores any changes and removes Prompt Box, canceling the command.

FUNCTION KEYS

The keyboard function keys are used to activate commands that are not shown on the menus. The F1, F2, and F5 keys are active in any part of ChemBase.

F1 - Activates the Help System.

F2 - Displays the Clipboard in browse mode.

F5 - Displays alternate status line indicating function of the keys.

The function of the keys is shown on the status line (F5 switches it on/off) on each of the menus.

SYSTEM OPERATION

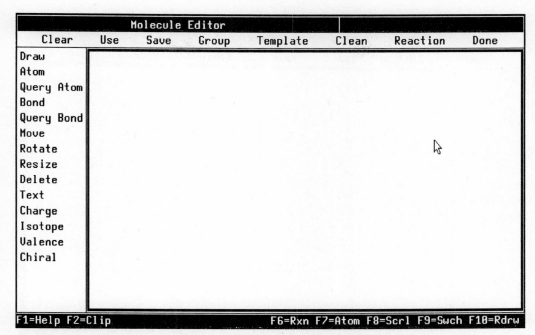

Figure 2/8. Molecule Editor

Figure 2/9. Reaction Editor

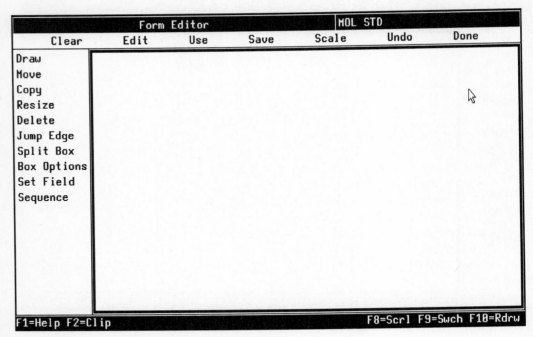

Figure 2/10. Form Editor

Figure 2/11. Table Editor

SYSTEM OPERATION

CURSOR CONTROL KEYS

HOME, END, PGUP, PGDN, and these combined with CTRL key, and the arrow keys ← → ↓ ↑ move the text cursor — .

ESCAPE

The ESC key on the keyboard is used to cancel and abandon any operation which is in progress or any command which is active.
The following are examples of interrupting an operation with the ESC key:
-a search or a lengthy output operation (e.g., printing).
-editing a block of text in a box on a display form. Any editing changes are disregarded and the previous content of the box is valid.
-to unload from the mouse pointer a side-bar option which is active.
-to clear a highlight.
ESC is also used in drawing structure skeletons to interrupt the Continuous Draw mode (see p....).

TEXT EDITING KEYS

TAB,SHIFT/TAB - moves cursor from box to box in FORM or TABLE
F5 - word wrap on/off
F6 - reformats text
F7 - clears box of text
F9 - restores text in box, undoing any made changes
F10 - accepts text, terminates editing
ESC - does not accept text, terminates editing

Cursor control:

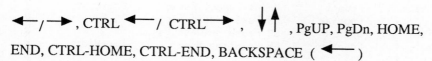

←/→ , CTRL ←/ CTRL→ , ↓↑ , PgUP, PgDn, HOME, END, CTRL-HOME, CTRL-END, BACKSPACE (←)

DEL - deletes character under cursor
INS - switches to replace character under cursor or to insert character before cursor
F3 - deletes word from current cursor position
F4 - deletes to the end of the current line from current cursor position

ERROR MESSAGES

The System issues two kinds of error messages:

Notification messages are displayed at the bottom of the screen. They can be ignored by the User or they can be taken into account for further action.

Recoverable messages are displayed in a box in the center of the screen. They must be acted upon to prevent an error in running the System. Click, or press (cr) or ESC to remove the box and take the required action.

SYSTEM SETTINGS

When we load the ChemBase program into our micro, we have to input certain parameters in the Setup mode concerning our hardware which have to match those required by ChemBase to be able to run the software.

To run the ChemBase programs in a manner matching our needs, the System offers the User choices that are specified in the Settings Window which is displayed upon activating the SETTINGS Command on the Main Menu.

The Settings Window is shown in Figure 2/12.

The values in this window are the default values which have been set up when we installed ChemBase. We can edit the values to the settings of our choice.

```
┌─────────────────────────────────────────────────────────────────────────────┐
│              ChemBase Main Menu                    MOL TBL        INS       │
│                                Settings                                     │
│ Clipboard:           C:\CPSS\CLIP                                           │
│ Molecule Form:       TUTORMOL           Hydrogen Labels:       No           │
│ Molecule Table:      TUTORMOL           Stereo Bonds:          Standard     │
│ Reaction Form:       TUTORRXN           Atom Numbers:          No           │
│ Reaction Table:      TUTORRXN           Stext/Abbreviations:   Yes          │
│ Date Format:         mm/dd/yy           Bond Spacing:          Normal       │
│ Time Format:         12 Hour            Scaling Mode:          Fit Box      │
│ Printer:             Epson              Background:            Black        │
│ Printer Port:        LPT1:              Foreground:            White        │
│ Help Level:          Novice             Bond Length [1 - 30]:  10           │
│─────────────────────────────────────────────────────────────────────────────│
│ Database:            TUTORIAL           Forms Per Page:        1            │
│ Log Transactions:    No                 Print Orientation:     Page         │
│ ID Duplicate:        Prompt             Print Border:          Form Edge    │
│ Structure Duplicate: Prompt             Print Aspect Ratio:    Preserve     │
│ LW Text Font:        Courier            CB Page Length (in):   11.0         │
│ LW Text Font Size:   6                  CB Pause Between Pages: No          │
│ LW Structure Font:   Courier            CB Printing Mode:      Immediate    │
│ LW Structure Font Size: 6                                                   │
│─────────────────────────────────────────────────────────────────────────────│
│ F1=Help F2=Clip F3=Word F4=Line F5=Read F6=Save        F9=Undo F10=Done     │
└─────────────────────────────────────────────────────────────────────────────┘
```

Figure 2/12. Settings Window

The Settings Window is divided into two sections. The upper half of the window contains the **system-wide settings** that affect the entire CPSS system and can be changed from any of the CPSS programs. The bottom half of the window contains the **individual-program settings** that affect only a specific CPSS program and can be edited only in that program.

SYSTEM OPERATION

The editing is performed by the techniques described above (see "Prompt Boxes," "Function Keys," and "Cursor Control Keys").
The following is a brief description of the Settings:

Clipboard (*vide infra*)
Molecule Form: file name of Molecule Form (see p. 346)
Molecule Table: file name of Molecule Table (see p. 349)
Reaction Form: file name of Reaction Form
Reaction Table: file name of Reaction Table
 (click the line twice to display available files for the
 forms and tables)
Date Format: see options by clicking the line
Time Format: toggle 12, 24 hours
Printer: click the line for make and model of your printer
Printer Port: click for connection port for your printer
Help Level: toggle (Novice, Expert)
Hydrogen Labels: options No, Yes, Hetero (see p. 298)
Stereo Bonds: toggle No, Standard, Alternate
Atom Numbers: toggle No, Yes (see Figure 2/30)
Stext/Abbreviations toggle No, Yes (see p. 311)
Bond Spacing: options Small,Normal,Large
Scaling Mode: select an option by clicking the line:
 Fit Box - rescales size of structures from database or
 from Molecule or Reaction Editor to fit in a box in Form
 ignoring structures original size
 Std Bonds - rescales the size of structures according
 to a set average Bond Length
 As drawn - structures are displayed exactly as they were
 drawn
Foreground,Background: determines the screen color
Bond Length: set range 1-30 (for use with Std Bonds)
Database: click twice for name of database in the Clipboard
Log Transaction: toggle No,Yes - records changes in
 database and saves them in .JNL file
ID Duplicate: options Prompt,Overwrite,Skip manage
 registration of duplicate structures from SDfiles, RDfiles,
 and Data Files
Structure Duplicate: options Prompt,Overwrite,Skip
 manage registration of duplicate structures
LW: settings for the Apple LaserWriter printer
Forms Per Page: options 1,2,4,6,8,List manage number of
 forms printed on a page (used with Print List on the Main
 Menu)
Print Orientation: toggle Page,Rotated Page (Rotated Page

is valid only for Apple LaserWriter, HP LaserJet Plus and Toshiba)
Print Border: toggle Screen Edge,Form Edge specifies to print screen display only or entire form
Print Aspect Ratio: toggle Preserve,Fill Page specifies to print forms as shown on the screen or to fill available space on a page
CB Page Length: click to specify
CB Pause between Pages: toggle No,Yes for continuous or per page print
CB Printing Mode: click for options:
Background - to continue ChemBase operation while printing
To disk - to save a file to disk as Postscript (.PS) or dot-matrix (.DM) file
Immediate - to print a file at once

CONFIGURATION FILE

The default settings have been saved in the c:\cpss\CPSS.CFG file. When we start ChemBase, this file is always first read by the System. If we edit the default settings, we can save them in the same default file. To do so:
1. Edit the settings to new values.
2. Press F6.
A window, shown in Figure 2/13, is displayed.

> Write configuration file: c:\cpss\CPSS.CFG

Figure 2/13

To store the new settings in the default file whose name is shown in the box: press F10, or (cr), or click outside of the box
However, we can save as many settings as we wish with each window containing specific settings for our different tasks we want to accomplish with ChemBase. To save the settings in different than the default file:
1. Erase the name in the box.
2. Type a new filename with path.
3. Press (cr), or click, or press F10.

To have the System read different settings than the default file:
1. Press F5.
2. Erase the existing file name and type a new file name in the box (Figure 2/14).

SYSTEM OPERATION

> Read configuration file: c:\cpss\CPSS.CFG

Figure 2/14

THE CLIPBOARD

The Clipboard is a window which, when called, is superimposed on the current screen. The window shows a menubar and a list of files. Unless we have created and stored some files, the initial files are the default files that come with the ChemBase software. They were loaded into the Clipboard at the time of installing ChemBase.
The Clipboard can be called by:

1. Pressing F2:

The Clipboard is displayed with the menubar (Figure 2/15). The menubar commands are selected as usual. The function of the commands is as follows :
Change(Comments) - to edit comments (memos) for the files
Change(Clipboard) - to switch from the current to another Clipboard
Info - displays the total number of files in the current Clipboard and remaining disk capacity
Sort(Name) - displays files by name in alphabetical order
Sort(Type) - displays files alphabetically by file extension
Sort(Date/Time) - displays files sorted by date and time
Select(By Name/Type) - displays a subset of the files that match the specified name/type
Select(By Date) - displays a subset of the files that match the specified date range
Select(All) - cancels the subsets of files and restores the full list of files
Delete - **erases the file from disk** and removes the file name from the Clipboard
DOS - switches to DOS. Type Exit in DOS to return to ChemBase
  Note: **DO NOT copy an open ChemBase database through the Clipboard DOS window. Both the original and the copy will be unusable.**
Done - removes the Clipboard from the screen
The displayed files can be browsed but not selected.

2. Dragging USE:

When we activate an option from the pull-down menu of the USE Command, the Clipboard displays only the files related to the operation we are currently conducting. For example, when we are drawing a structure, the USE(Molecule) option in the Molecule Editor displays the Clipboard showing only the .MOL files (Figure 2/16).
The files can be selected.

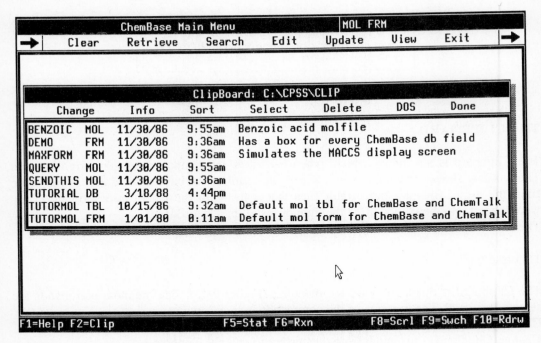

Figure 2/15. The Clipboard

The file extension is an important appendage to the file name since it determines the contents and use of the file. It is created automatically when we save a file in a CPSS program. When we transfer files from other systems into CPSS, we have to include the file extension in the file pathname. The filename extensions that we encounter in CPSS are shown in Figure 2/18.

3. Activating SETTINGS:

Clicking the lines in the SETTINGS Window (Figure 2/12) displays the Clipboard showing those files (Figure 2/17).

To select a file from the Clipboard, point/click the file name.
To scroll and browse in the Clipboard, use the keyboard keys up/down arrows, PgUp, PgDn, CtrlHome, CtrlEnd.

CREATING A CLIPBOARD

The default Clipboard has been named CLIP and it resides in the CPSS directory. It is, in fact, a subdirectory as shown in Figure 2/18.
We can have the Clipboard in any subdirectory that we create in our hiearchy of directories. Or, we can create separate Clipboards in any directory or subdirecotry, e.g., MYCLIP in Figure 2/18, and save files in any of the Clipboards.

SYSTEM OPERATION

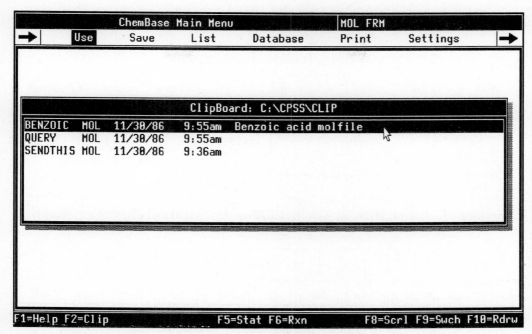

Figure 2/16. The Clipboard displaying .MOL files

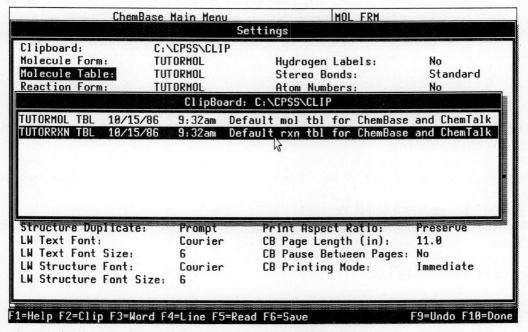

Figure 2/17. Selection from Settings displays Clipboard files

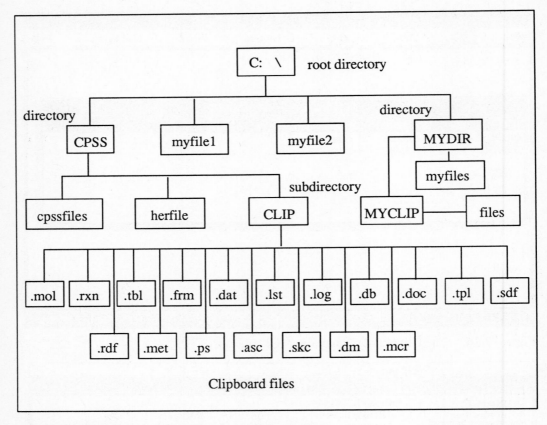

Figure 2/18. Directory system with Clipboard files

The Clipboards can be created in DOS or in the Clipboard Window by dragging the option CHANGE(Clipboard).
When we input the complete pathname of the new Clipboard, the System displays a message "Unable to open Clipboard." Click to remove the message. The System issues a prompt:
"Should it be created?." Select Yes.
To make a new Clipboard the default Clipboard, use SETTINGS Window to change the Clipboard name.

STARTING CHEMBASE

ChemBase program is invoked at the DOS prompt:

C> chembase (cr)

SYSTEM OPERATION

The Software License Agreement is displayed. Enter: (cr)

The program is loaded and the Main Menu with the default Molecule FORM is displayed (Figure 2/6). We are always in the Molecule View/Form at the start of ChemBase.

The most important commands on the Main Menu for activating the ChemBase operations are USE, VIEW, and EDIT.

THE USE COMMAND

USE(Database) - displays Clipboard with Database Files for selection of a database if we do not want to use the default database specified in SETTINGS.

THE VIEW COMMAND

VIEW(Reaction) - transfers operation from Molecules to Reactions. If Molecule FORM is current, switches to the Reaction FORM.
VIEW(Molecule) - transfers operation from Reactions to Molecules. If Reaction TABLE is current, switches to the Molecule TABLE.
VIEW(Form) - switches from TABLE to FORM, stays in current Molecules or Reactions.
VIEW(Table) - switches from FORM to TABLE, stays in current Molecules or Reactions.

Figures 2/3. and 2/4. illustrate these transfers.

THE EDIT COMMAND

The Edit Command is activated to enter the Editors from the Main Menu.
EDIT(Molecule) - displays the Molecule Editor Menu
EDIT(Reaction) - displays the Reaction Editor Menu
EDIT(Form) - displays the Form Editor
EDIT(Table) - displays the Table Editor

Figure 2/5 illustrates these transfers.

4   CREATING STRUCTURES

MOLECULE EDITOR

The Molecule Editor (Figure 2/8) is the center of action for creating molecular structures and chemical reactions. The Reaction Editor is the center of action for editing and manipulating chemical reaction schemes.
ChemBase Version 1.3 has additional commands and options in the Molecule Editor which we will describe later in the discussion.
We enter the Molecule Editor by point/drag EDIT(Molecule).

BUILDING STRUCTURES

We build structures in order to:
1. register them in a database
2. to formulate structural queries for searching
3. to retrieve text and numeric data associated with structures
4. to assemble reactions
5. to save them for transfer to the other MDL systems

DRAWING STRUCTURE SKELETONS

Structure skeletons are drawn by activating the DRAW option on the Menu. A pull-down menu will appear from which we can select Normal, Continuous, or Rubber Band Draw mode. All of these draw modes create a single bond. The difference is in the technique of drawing.
Normal Draw (Figure 2/19) requires that each atom be clicked twice, except the first and last one.
In Continuous Draw mode (Figure 2/20), we click only once for each node in the path of drawing. If we want to break the continuity of the path, we press ESC and place on the position where we want to start a new bond. If we do not ESC, the bond will be drawn between the last node and the new position we place on.
In Rubber Band mode, we place the mouse pointer on the position where the bond starts and drag the pointer in any direction. The bond follows the direction and becomes fixed in that direction at release of the mouse button (Figure 2/21).
At times, when we draw a structure, the shape of it is distorted. We can reshape structures to the proper shape by using the CLEAN option on the Menu. Drag CLEAN(All) to clean, i.e., to normalize bonds and angles for all structures on the screen (Figures 2/22, 2/23).
CLEAN(Fragment) cleans and redraws the fragment (i.e., a structural unit on the screen not connected to another structural unit) on which we place/click a bond or an atom within the fragment.

Example: *Normal Draw*

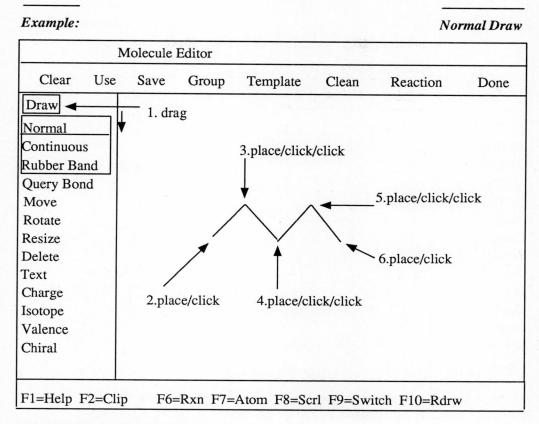

Figure 2/19. Normal Draw mode

ChemBase Version 1.3 has the option CLEAN(Undo) which reverses CLEAN.

SPECIFYING ATOMS

Specifying atoms in structure skeletons we have just drawn is, in fact, respecification of the default carbon value to another value. The same technique of respecification is used when we wish to respecify a heteroatom to another value. Activating ATOM on the Menu displays a pull-down menu of atoms. An atom is loaded into the mouse pointer by dragging the mouse to the desired atom symbol. We then place/click on the position in the structure skeletons where we want the atom (Figure 2/24). If we want to create an unconnected atom, we place/click on a position outside the structure. The atom value is active for repeated use until a new one is selected.

The atoms that are not on the pull-down menu are displayed in form of the periodic table by dragging ATOM(Other). The element symbols are selected from the periodic table by

Example: *Continuous Draw*

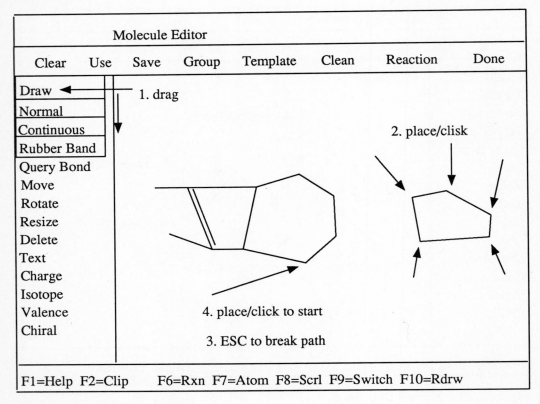

Figure 2/20. Continuous draw

point/click an element in the table followed by point/click DONE on the menubar (Figure 2/25) which returns us to the structure. Place/click on the position in the structure where the atom is desired. The menubar contains also the option SPECIAL. Point/click SPECIAL displays a pull-down menu with R, X, Deuterium, Tritium, and amino acids symbols that can be selected to insert into the structure (Figure 2/26).
Note, however, that the special symbols R, X, and the amino acids R groups have no structural significance. However, the T (tritium) and D (deuterium) labels have structural meaning when inserted in structures.
Figure 2/24 also shows the steps 7 and 8 to specify, if desired, ATOM(Value) as real number or a range of real numbers including exponents. After we place/click on the oxygen atom, a prompt box will appear:
Enter value or range of values:

The specified value becomes attached to the atom. It can be deleted by the steps 7 and 8

CREATING STRUCTURES

Example: *Rubber Band Draw*

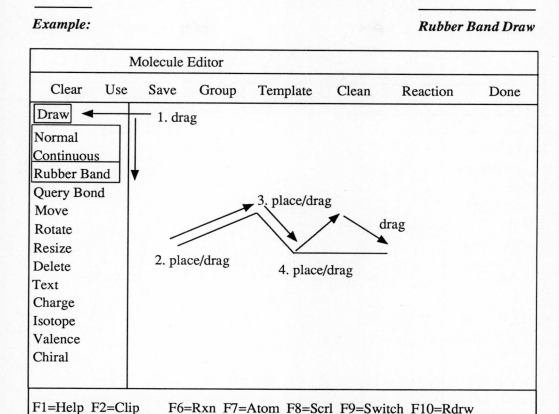

Figure 2/21. Rubber Band mode

and deleting the value from the prompt box by F4. The atoms with specified values can be searched for by including the atom values in the query for SSS, RSS, TSS search (see section "Molecule Searching" and "Reaction Searching").

ChemBase Version 1.3 has a toggle in the F9 window (see p. 298) which allows for turning on/off the display of atom values. We can also use the SETTINGS:

Atom Value Display Yes/No....turns on/off display of atom values

SPECIFYING BONDS

To assign bond values, we respecify the default single bond value. Similarly, any bond value in the structure can be respecified (Figures 2/27, 2/28).

The BOND(Highlight) option can be used to highlight reaction centers in reaction queries.

Example: *Reshaping structure skeletons*

```
┌─────────────────────────────────────────────────────────────────────┐
│                      Molecule Editor                                │
├─────────────────────────────────────────────────────────────────────┤
│  Clear    Use   Save   Group   Template  │ Clean │ ↘Reaction  Done │
│  Draw                                    ┌───────┐   ↓             │
│  Atom                                    │  All  │      drag       │
│  Query Atom                              │Fragment│                │
│  Bond                                    └───────┘                 │
│  Query Bond                                                        │
│  Move                                                              │
│  Rotate                                                            │
│  Resize                                                            │
│  Delete                                                            │
│  Text                                                              │
│  Charge                                                            │
│  Isotope                                                           │
│  Valence                                                           │
│  Chiral                                                            │
├─────────────────────────────────────────────────────────────────────┤
│  F1=Help F2=Clip    F6=Rxn F7=Atom F8=Scrl F9=Switch F10=Rdrw     │
└─────────────────────────────────────────────────────────────────────┘
```

Figure 2/22. Cleaning structures

Although the highlight is ignored by ChemBase in searching with these queries, the query can be used in REACCS for searching reaction centers (see p. 682). Molfiles and Rxnfiles retain the specified bond highlight.

STEREOCHEMISTRY IN CHEMBASE VERSION 1.3

ChemBase 1.3 recognizes stereochemical structures in searching ChemBase 1.3 databases by:
RETRIEVE(Current)(p. 374), SEARCH(SSS/RSS)(p. 380, 396), and SEARCH(TSS)(p. 398).
The Molecule Editor in ChemBase 1.3 has two options for specifying stereocenters and stereobonds:
QUERY ATOM(Stereo)....marks stereocenter with a box
QUERY BOND(Stereo)....marks atoms connected with the bond by boxes

CREATING STRUCTURES

```
┌─────────────────────────────────────────────────────────────────────┐
│                          Molecule Editor                             │
│   Clear    Use    Save    Group    Template    Clean   Reaction  Done│
│  Draw    │                                                           │
│  Atom    │                                                           │
│  Query Atom                                                          │
│  Bond                                                                │
│  Query Bond                                                          │
│  Move                                                                │
│  Rotate                                                              │
│  Resize                                                              │
│  Delete                                                              │
│  Text                                                                │
│  Charge                                                              │
│  Isotope                                                             │
│  Valence                                                             │
│  Chiral                                                              │
│                                                                      │
│  F1=Help F2=Clip    F6=Rxn F7=Atom F8=Scrl F9=Switch F10=Rdrw        │
└─────────────────────────────────────────────────────────────────────┘
```

Figure 2/23. Cleaned structures

See the discussion on stereochemical drawing rules and stereochemical searches in Part Three and Part Four. The discussion applies to ChemBase 1.3.
Note that in ChemBase 1.3 the following drawings do not produce equivalent structures (contrary to MACCS-II).

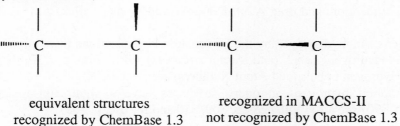

      equivalent structures   recognized in MACCS-II
  recognized by ChemBase 1.3 not recognized by ChemBase 1.3

See also p. 269 for Stereochemistry in ChemBase 1.2.

Example: *Specifying atoms*

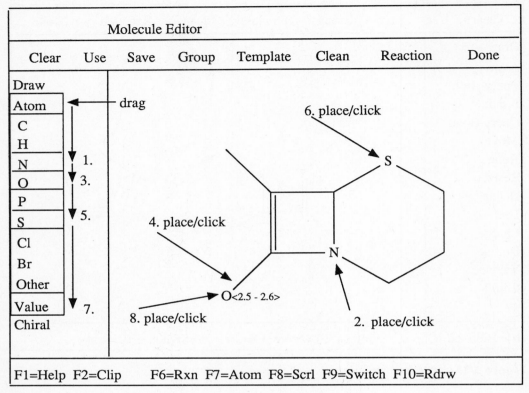

Figure 2/24. Specification of atoms

HOW TO SPECIFY HYDROGENS

Explicit hydrogens at atoms are assigned by:
drawing a bond at the atom and drag ATOM(H) followed by place/click at the end of the bond.
Implicit hydrogens can be displayed by the F9 function key selections (*vide infra*). Figure 2/29 shows the Hydrogen Label (and Atom Numbers) OFF. Figure 2/30 shows the structure if the Hydrogen Labels (and Atom Numbers) are turned ON.
Explicit and implicit hydrogens in query structures have a significant meaning in searching (see p. 390)(see also "Valence," p. 309, 310 and for use, p. 162).

USING FUNCTION KEY F9

The SETTINGS values (see p. 285) which affect the structure display can be overridden in

CREATING STRUCTURES

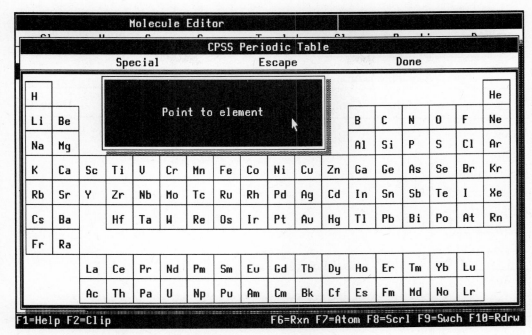

Figure 2/25. Selecting ATOM(Other)

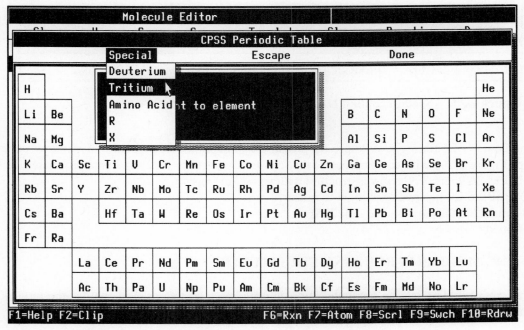

Figure 2/26. Selecting Special Atoms

Example: *Specifying bonds*

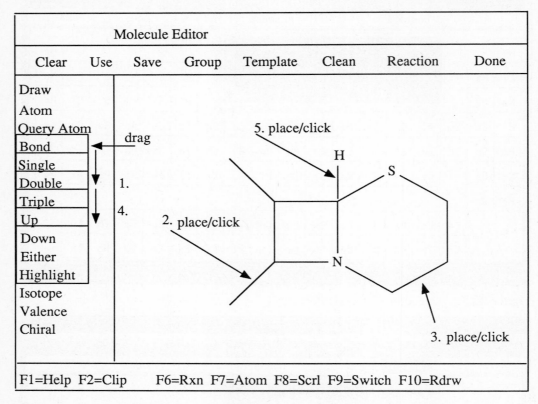

Figure 2/27. Specification of bonds

the course of building a structure by pressing F9 for molecule switches. A window pops up (Figure 2/29) with the parameters that can be changed by clicking the appropriate line. Figures 2/29 and 2/30 show a structure with some Settings on.
ChemBase 1.3 has two additional toggles:
Atom Value Display (see p. 294, 295)
Moltext Size (see p. 332, 334)

USING MODELS IN BUILDING STRUCTURES

Structure building can be facilitated by using predrawn structures. We can use the models that are included in ChemBase software or define our own models.

CREATING STRUCTURES

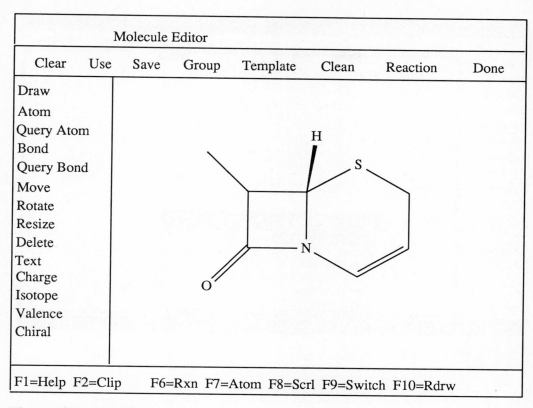

Figure 2/28. Specified bonds

SYSTEM MODELS

ChemBase System contains 57 models, i.e., single rings, polycyclics, chains, and common functional groups that can be recalled by activating the TEMPLATE option on the Menu and selecting one of the options on the pull-down menu. A table is then displayed from which we select a structural fragment (Figures 2/31-2/34).

HOW TO USE SYSTEM MODELS

The templates can be used as starting fragments to build a structure, or as substituents in an existing structure. The rings and polycyclics can also be used to build fused ring systems (Figures 2/35-2/37).
To select:
1. **template as a separate fragment**
We point/click **an atom or a bond** of a template in the table. We are then returned to the Menu. We place/click on the position where we want the template to display.

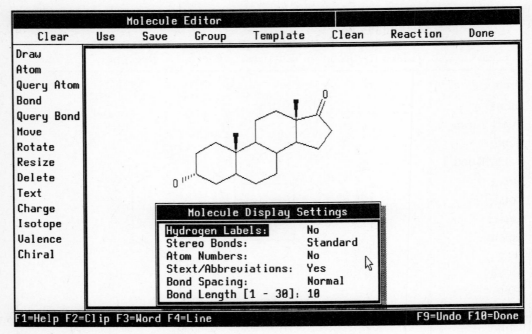

Figure 2/29. Function Key F9

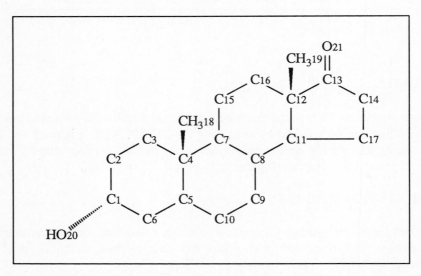

Figure 2/30. F9 Function Key - Atom Numbers and Hydrogen Labels ON

CREATING STRUCTURES

Example: *Activating templates*

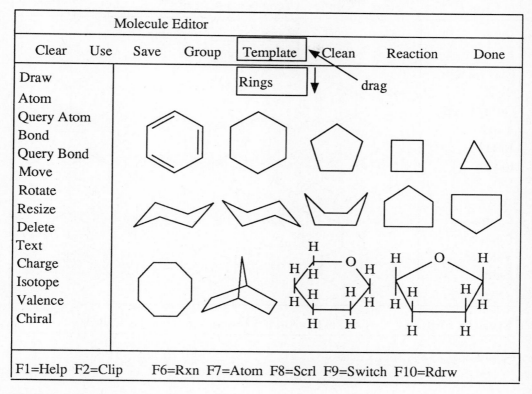

Figure 2/31. System models

2. template as a substitutent
We point/click **an atom** in a template in the table. We are then returned to the Menu. We place/click on the atom in the parent where the substituent is to be connected by a single bond.

3. ring template for fusion
We point/click **a bond** in the template. We are returned to the Menu. We place/click on the bond in a parent where the fusion is desired.

In addition to templates activated by the TEMPLATE option, the System contains 35 common substituents that are stored in Molfiles in the MOLS directory created at the time we install ChemBase.

The substituents are:

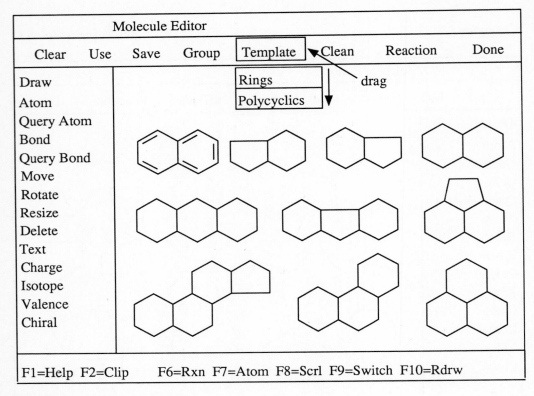

Figure 2/32. System models

SYSTEM SUBSTITUENTS

CCl3, CF3, CHO, CN, CNN, CNO, COCH3, CONH2, COOH, CSO, Et, iPr, N2, NC, NCO ,NMe2, NMe3, NNN, NO, NO2, NSO, NSS, OCN, OCOCH3, OEt, OMe, OPh, Ph, R3, R4, R5, R6, SO2, SO3H, tBu

These fragments are substituted in a parent by using the function key F7.

USING FUNCTION KEY F7

The function key F7 can be used for both atom specification and for substitution with a fragment (Figures 2/38, 2/39).
To circumvent dragging the ATOM option, we:
1. place on the atom in our structure where we wish to specify an atom or to attach one of the substituents from the MOLS directory, and press F7
The System displays a prompt box.
2. type =, or - followed by any of the element symbols

CREATING STRUCTURES

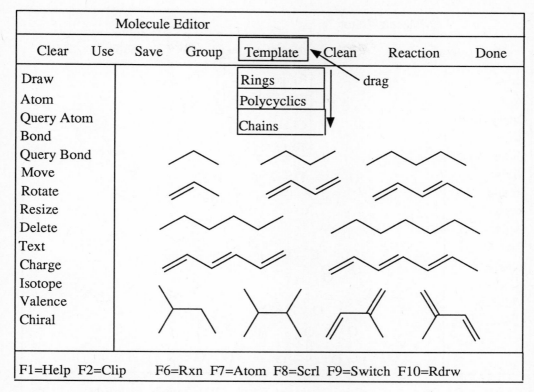

Figure 2/33. System models

followed by a carriage return to **replace** the atom we pointed to, or
3. type any of the element symbols or any of the substituents in MOLS directory followed by carriage return to **substitute** at the atom we pointed to in our structure

USER MODELS

We can create any substituent, structural fragment, or a complete structure and store these for recall.

Templates

User templates are created by drawing a structural fragment that we desire to use as the template by the usual drawing techniques. The fragment is saved by SAVE(Template) option on the Menu (Figure 2/40).
The fragment in Figure 2/40 has been saved under the name mytempl.tpl in the \cpss\clip directory. When we want to use this User template, we drag TEMPLATE(Other) on the Menu. The Clipboard will appear from which we select the template file.

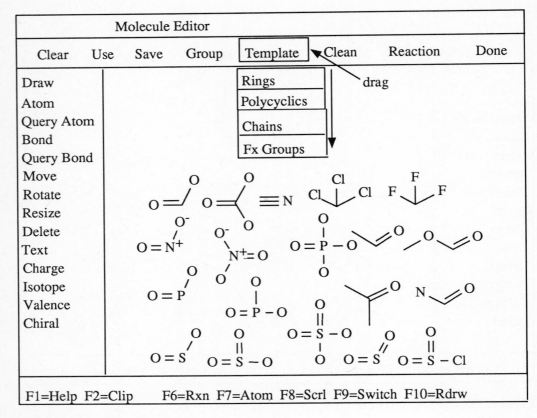

Figure 2/34. System models

The fragment will be displayed and can be used similarly as the System templates.

Substituents

The System substituents discussed above are structural fragments that have been predrawn and stored in Molfiles under the names CCl3.mol, CF3.mol, etc. They are loaded at the time of ChemBase installation into MOLS directory which is a subdirectory in the CPSS directory. For example, the path for the CCl3 file is \cpss\mols\ccl3.mol.
We can create our own substituents of any kind that will complement the System substituents. We draw the desired structural fragment as usual, **starting with the atom which we want to be the point of attachment as atom number 1,** and save it in a Molfile (Figure 2/41). The Molfile is then copied into the MOLS subdirectory by the DOS Command COPY. However, the MOLS subdirectory can be in other than the CPSS directory, e.g. \mydir\mols.
We have saved the fragment in \cpss\clip\mysub.mol file (Figure 2/41). To copy it to the MOLS subdirectory, we use, at the DOS level, the command: copy mysub.mol \cpss\mols.

CREATING STRUCTURES

Example: *Using System templates*

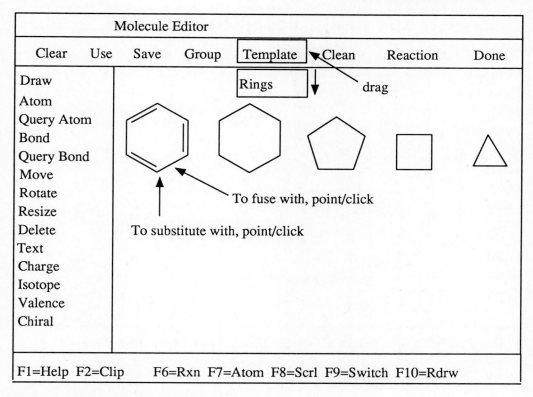

Figure 2/35. Using System templates

The fragment will be ready to be used with the F7 key. Unless we erase the copied file \cpss\clip\mysub.mol, we can use it as a regular Molfile. For example, if we did not copy the file into the MOLS subdirectory, the fragment can still be used as a substituent. We point to the atom in an existing structure and press F7. At the System prompt:

Atom or substituent:

we have to type the full file path, e.g., \cpss\clip\mysub.mol.
The F7 allows for bringing a fragment from the MOLS directory and displaying it as a Group Abbreviation (see p. 332). We use *molfilename (asterisked filename) at the F7 prompt.

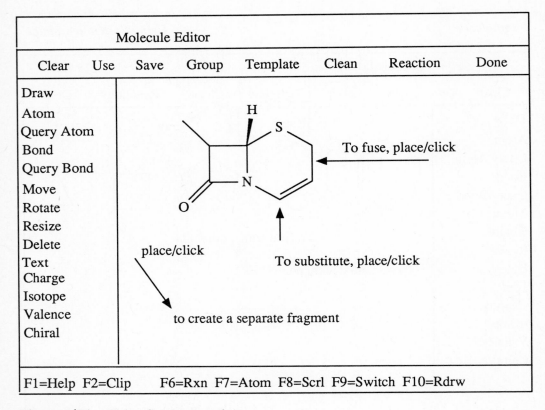

Figure 2/36. Using System templates

Complete Structures

Complete structures are drawn by the usual techniques and can be permanently stored in Molfiles in any directory by SAVE(Molfile). They can be recalled by USE(Molfile) option on the Menu followed by selecting the file from the displayed Clipboard. The structure is displayed and can be modified by any of the structure building methods. When we create a template and store it by SAVE(Template) on the Menu, the template can be recalled as a structure by selecting USE(Template). However, recalling a template through the USE(Template) command does not allow for using it with the F7 technique.

CHIRAL DESCRIPTOR

When we build structures, we can indicate stereobonds in the structures (see Figures 2/27, 2/28). However, in ChemBase 1.2 the stereobonds are disregarded in searching if we use such structures as a query. Nevertheless, we can mark these structures by the descriptor CHIRAL which is attached to a structure to prepare it for transfer to MACCS-II or REACCS and to insure that these systems will recognize the structure as one of the

CREATING STRUCTURES

```
┌─────────────────────────────────────────────────────────────────┐
│                        Molecule Editor                          │
│   Clear    Use    Save    Group   Template    Clean   Reaction   Done │
│ Draw                                                            │
│ Atom                                                            │
│ Query Atom                                                      │
│ Bond                                                            │
│ Query Bond                                                      │
│ Move                                                            │
│ Rotate                                                          │
│ Resize                                                          │
│ Delete                                                          │
│ Text                                                            │
│ Charge                                                          │
│ Isotope                                                         │
│ Valence                                                         │
│ Chiral                                                          │
│                                                                 │
│ F1=Help F2=Clip    F6=Rxn F7=Atom F8=Scrl F9=Switch F10=Rdrw    │
└─────────────────────────────────────────────────────────────────┘
```

Figure 2/37. Using System models

stereoisomers.
Point/click CHIRAL on the Menu turns the CHIRAL label on and off for the current structure.
ChemBase 1.3 uses query structures labeled CHIRAL for retrieving structures with the specified absolute configuration from ChemBase 1.3 databases. See the discussion on stereochemical searches in MACCS-II and REACCS (p. 511, 649).

CHARGE, ISOTOPE, AND VALENCE SPECIFICATION

The atoms in structures can be assigned negative, positive, free-radical charge, isotopic masses, and valences, where it is appropriate.

Charged Atoms and Free Radicals

Negative. positive, and free-radical specifications can be assigned to atoms by point/drag CHARGE(+,-,^) on the Menu. Charges are added in units from -3 to +3 by place/click on the atom to have the charge. A + sign negates a - sign and vice versa. The caret ^ indicates

Example: *Using function key F7*

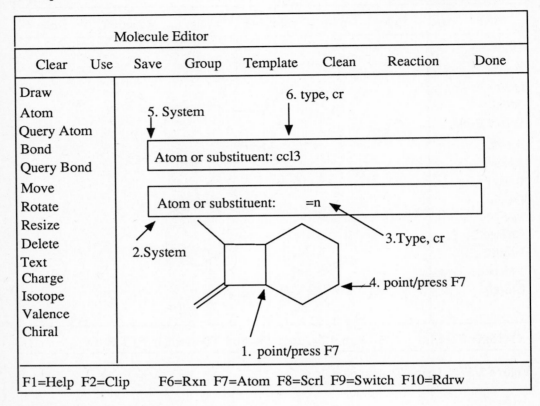

Figure 2/38. Using function key F7

a radical and can be canceled by another ^.

Isotopes

By default, all atoms in created structures are of normal mass. To change the mass value, we drag ISOTOPE(+,-) and place/click on the atom to have the abnormal mass. Each place/click on + increases the mass up to the limit for the atom, each place/click on - decreases the mass up to the normal mass for the atom.

Valence

When we are building a structure, the System automatically checks if the normal atom valence is met by the attached bonds. If we do not draw bond(s) at an atom which connect another specified atom(s), the valence is filled by implied hydrogen connections which is

CREATING STRUCTURES

	Molecule Editor							
Clear	Use	Save	Group	Template	Clean	Reaction	Done	

Draw
Atom
Query Atom
Bond
Query Bond
Move
Rotate
Resize
Delete
Text
Charge
Isotope
Valence
Chiral

F1=Help F2=Clip   F6=Rxn F7=Atom F8=Scrl F9=Switch F10=Rdrw

Figure 2/39. Structure created by F7 technique

reflected in the molecular weight and formula calculated by ChemBase. If a valence is exceeded, the System issues a warning to that effect. However, we are allowed to exceed a valence of an atom and continue, if desired, in building the structure. A specific valence can be assigned to an atom by point/click VALENCE followed by typing the valence value in the prompt box. Place/click on the atom to have the valence applies the valence to the atom. To restore the normal valence value, respecify the atom or point/click VALENCE and type the normal valence value in the prompt box followed by place/click on the atom.

TEXT DESCRIPTOR

Text descriptor is a text string that we may wish to attach to a structure we have created. The text string can "float" with the structure, i.e., it is not within the structure, or it can be "embedded," i.e., be within the structure. Both the floating and the embedded strings move with the structure once we exit the Molecule Editor. The floating string may be just a short trivial name for the structure or it may describe in some fashion the created structure and its purpose. The embedded string is termed "atom alias." Both the floating and embedded

Example:  *Storing User templates*

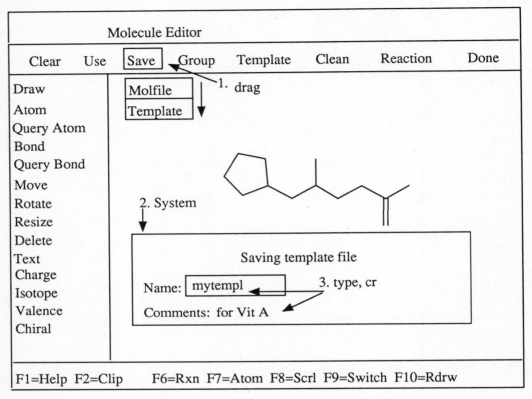

Figure 2/40. Saving User template

text string is registered with the structure. Structurally, the floating or the atom alias are meaningless. The underlying value is still structurally valid (see also Group Abbreviate, p. 332). In Exact, SSS, RSS, TSS searches, or on transfer to MACCS-II/REACCS, they are disregarded. However, we can search for them in Data Search by the *stext search technique (see p. 410).
The text string is entered by dragging TEXT(Edit,Move,Delete) on the Menu:

TEXT(Edit)....followed by place/click on the position where we wish to start the text (anchor the first character) invokes a prompt box into which we type one line of text followed by <cr>. More than one line of text can be input by repeating the procedure.
Note that the first character of the input string is highlighted. It is the anchor of the string. In atom aliasing we can make any character of the string the anchor by preceding the character by ^ (caret). The character will be the anchor. In searching, the ^ (caret) must be included in the *stext query. Stext/Abbreviation in the F9 switches turns the atom alias

CREATING STRUCTURES

Example: *Storing fragments for F7*

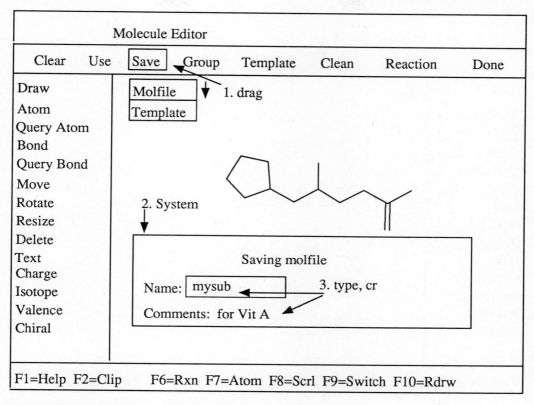

Figure 2/41. Saving a substituent for F7

display on/off.
Figures 2/43 and 2/44 show the use of floating and embedded strings.

TEXT(Move)....followed by place/clicking the anchor character and dragging the string moves the text to any new position. Note that only the floating text can be moved.

TEXT(Delete)..followed by place/click the anchor character deletes the text string.

The SETTINGS:

Stext/Abbreviations.....turns on/off display of text strings

Example: *Specifying variable atoms*

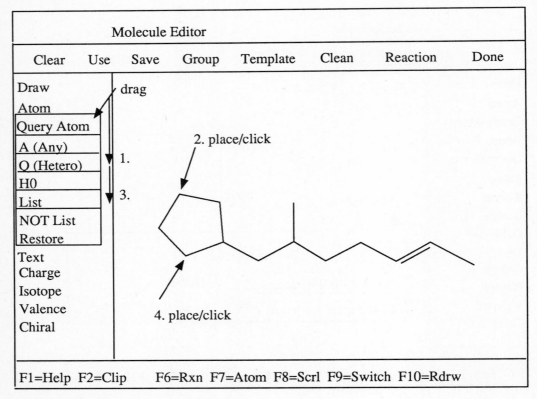

Figure 2/42. Specifying variable atoms

BUILDING GENERIC STRUCTURES

We have so far discussed drawing of structure skeletons and specifying single atom and bond values in the drawn structures. The structures are ready to register in a database or to use for a search. If we use these structures as queries, the search will retrieve only structures with the single specific atom or bond values of the query (see "Molecule Searching," p. 372).

To formulate structural queries in which the atoms and bonds have more than one value, we have to discuss how to build generic structures with variable atoms and bonds.

Generic structures are drawn by the same techniques we have used for creating structures. When we determine which of the atoms and bonds in the skeleton ought to have more than one value, we activate the options on the Menu:

1. FOR SPECIFYING VARIABLE ATOMS (Figures 2/42-2/44)

CREATING STRUCTURES

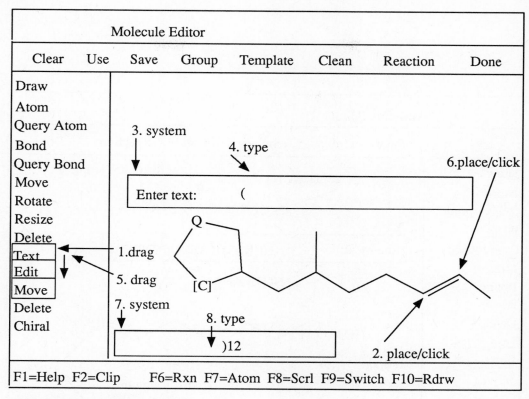

Figure 2/43. Variable atoms

QUERY ATOM(A, Q, H0, List, NOT List, Restore)
A   any atom except hydrogen
Q   any atom except carbon and hydrogen
H0   no hydrogens
List   atom values selected from ATOM options
NOT List   atom values to be excluded
Restore   restore to the initial single atom value

2. FOR SPECIFYING VARIABLE BONDS (Figure 2/45)

QUERY BOND(Any, Arom, S/D, Ring, Chain, Restore)
Any   any bond value
Arom   aromatic bond
S/D   single or double exact bond value
Ring   ring bond type (changes chain bond type)
Chain   chain bond type (changes ring bond type)
Restore   restore to the initial single bond value

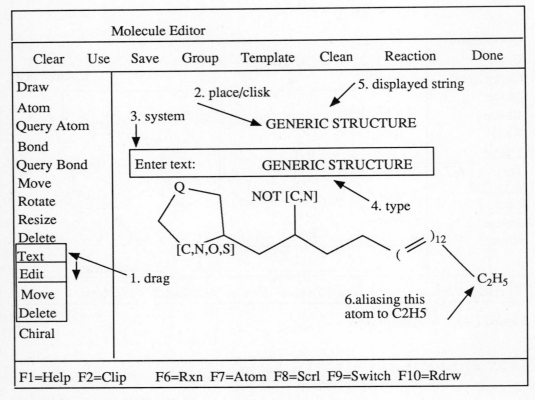

Figure 2/44. Variable atoms

CREATING STRUCTURES

Example: *Specifying variable bonds*

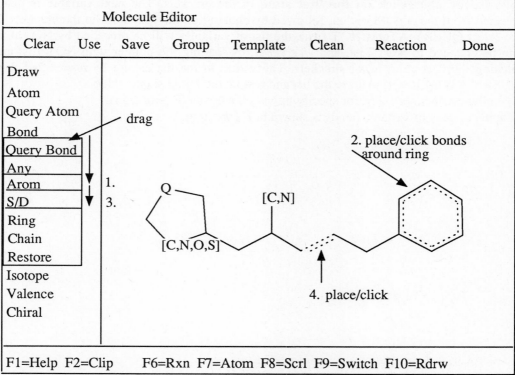

Figure 2/45. Variable bonds

Figure 2/46

When we select the LIST option, the first atom in the list of variables will be the current value as we specified it in drawing the structure. To assign our variables, we follow by activating the ATOM option and select an element symbol as usual (including the Other) followed by **place/click on the first atom** in the brackets. The next variable is again selected from the ATOM option, followed by clicking on the first atom in the brackets. If we select QUERY ATOM(NOT List), the list of variables will be preceded by NOT. List can be changed to NOT List, and *vice versa,* by loading the pointer with NOT List and clicking the first atom in the brackets. The values in the list cannot be respecified. Use QUERY ATOM(Reset) to erase the list and restore the initial single value.

A similar procedure is used for specifying variable bonds (Figure 2/45).

Graphic display of variable bonds is shown in Figure 2/46.

5 CREATING REACTIONS

Chemical reaction schemes are composed of structures of the molecules taking part in a reaction. The structures in a reaction scheme are interrelated by + and ⟶ signs. By convention, the structures in front of the ⟶ sign are reactants tied with the + sign; the structures after the ⟶ sign are intermediates or products which may also be tied with the + sign.

BUILDING REACTIONS

In ChemBase, reactions are constructed from molecular structures that we draw **in the Molecule Editor**. We label the drawn structures according to their function in the reaction scheme as:
1. reactant
2. intermediate
3. product
by activating the REACTION(Reactant,Intermediate,Product) option **on the Molecule Editor Menu.** When we do so, each structure retains this label in the System and is treated as such by the Reaction Editor until we respecify the structure reaction label. To monitor reaction building **in the Molecule Editor**, each of the labeled structures is assigned by the System a capital letter, A to Z, and a **Background Reaction Line** is displayed at the bottom of the Molecule Editor Menu, e.g.:

$$A + B + C \longrightarrow D + E \longrightarrow F + G$$

The Background Reaction Line can be displayed at any time in the Molecule Editor, if a current reaction exists, by pressing the F6 key.
Simultaneously, the **Reaction Editor** automatically places the labeled structures in the proper position in the reaction scheme **in the Reaction Editor** (at this point, we do not see it). We can imagine the Molecule Editor and the Reaction Editor working in parallel as is illustrated in Figures 2/47-2/50.
To see and edit the structural reaction scheme, we have to switch to the **Reaction Editor**.
Before we start to build a new reaction, it is a good practice to clear the Molecule and Reaction Editors from all previously used structures or reactions. The System retains the current structure or reaction even if they are not on display and if we did not clear them, we might get these mixed up with the new structures and reactions.
Point/drag CLEAR(Molecule,Reaction) on the Molecule Editor Menu or point/click CLEAR on the Reaction Editor Menu.
The concerted actions taking place in each of the editors are shown in Figure 2/47.

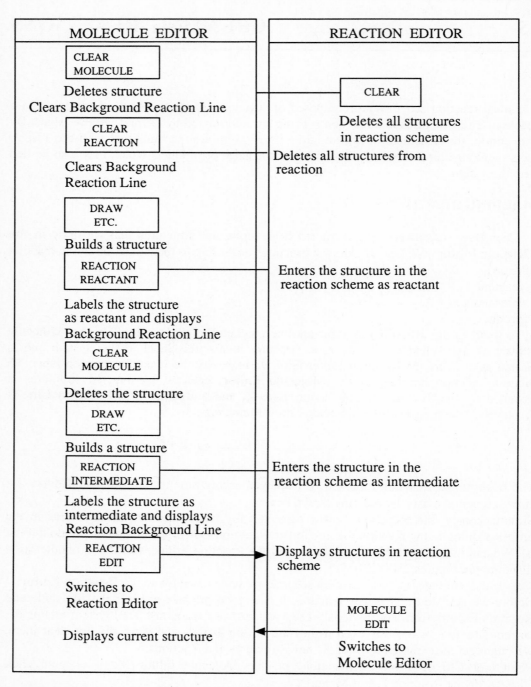

Figure 2/47. Dynamic interaction of Molecule/Reaction Editor

CREATING REACTIONS

REACTION EDITOR

The Reaction Editor is the center of action for manipulating reactions schemes. (The Molecule Editor is the center of actions for manipulating individual molecular structures.) The Reaction Editor Menu is shown in Figure 2/9 (p. 281).

When we construct a reaction and switch to the Reaction Editor, the reaction is displayed in a System defined, default layout. At times, we may desire to rearrange the default display format or to change/modify the molecules participating in the reaction.

EDITING REACTIONS

The Reaction Editor allows us to carry out the following changes:
1. rearrange the reaction layout
2. change the + sign to the ⟶ sign
3. change the structure reaction label
4. change/modify a structure in the scheme
5. add a structure to the scheme
6. delete a structure in the scheme

The mechanical changes 1-2 are described in the section "Correcting/Modifying Reactions" (p. 338).
The dynamic interaction of Molecule/Reaction Editor allows for making the changes 3-6.

Example: *Editing a reaction*
Command: MOLECULE(Select)

To **copy a structure** from a structural reaction scheme in the Reaction Editor and transfer it into the Molecule Editor, we use the MOLECULE(Select) option (Figure 2/48). The structure to be copied is selected by place/click an atom or a bond in the structure.

Command: MOLECULE(Change)

To move a structure from the Reaction Editor into the Molecule Editor for editing and thus effectively **delete a structure** from the reaction scheme, we use the MOLECULE (Change) option (Figure 2/49). The structure is selected similarly as in the MOLECULE(Select) operation (*vide supra*).

Command: DELETE

To **delete a structure** from a reaction scheme in the Reaction Editor **and abandon it** we use the option DELETE (Figure 2/50.)

Note that after deleting a structure, the reaction indicator line is automatically adjusted and the structures in the reaction scheme are assigned the appropriate letters.

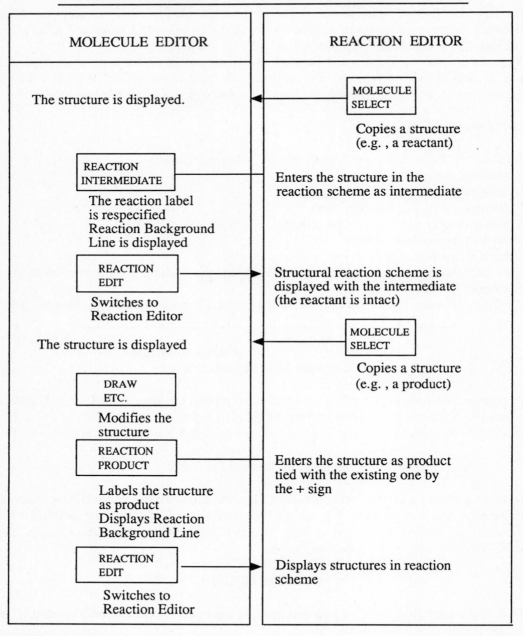

Figure 2/48. Editing a reaction

CREATING REACTIONS

DYNAMIC INTERACTION OF MOLECULE/REACTION EDITOR

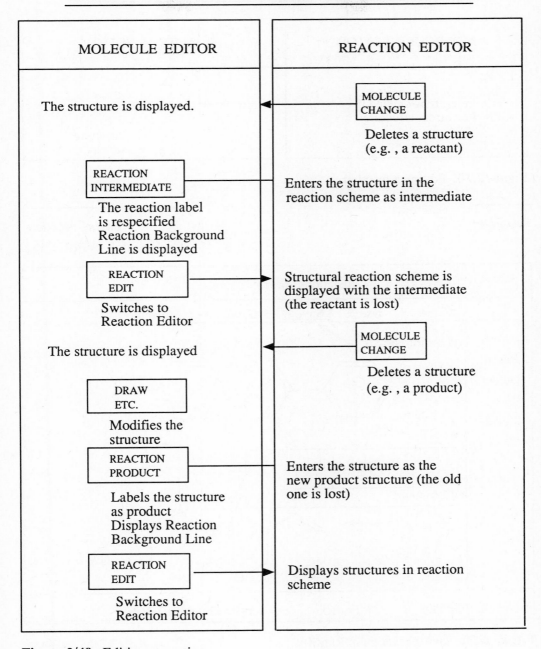

Figure 2/49. Editing a reaction

DYNAMIC INTERACTION OF MOLECULE/REACTION EDITOR

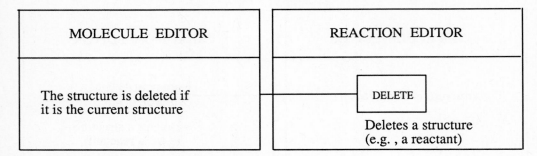

Figure 2/50. Editing a reaction

Example: *Saving reaction*

Command: SAVE(Rxnfile)

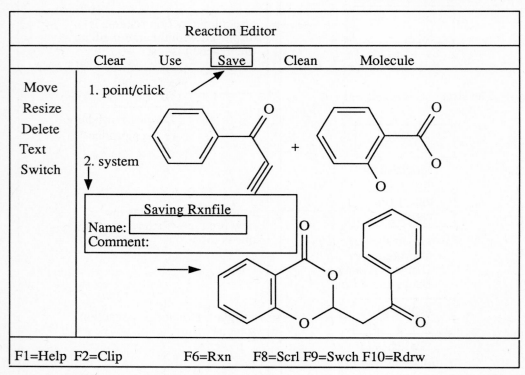

Figure 2/51. Saving reaction in Rxnfile

CREATING REACTIONS

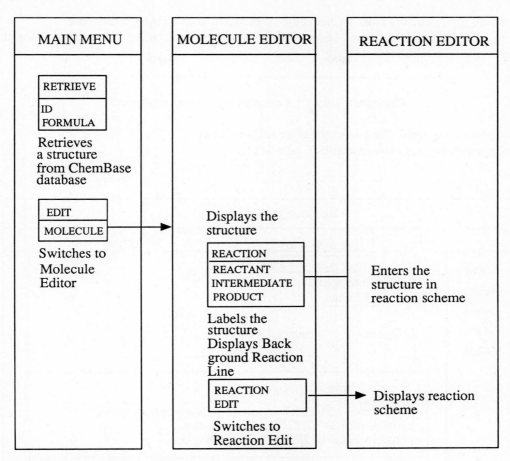

Figure 2/52. Using a retrieved structure as model

SAVING REACTIONS IN RXNFILE

The reactions we have built can be saved in a Rxnfile for later recall (Figure 2/51).

USING MODELS

Building of structures can be facilitated by using models of:
1. complete structures
2. structural fragments
3. complete structural reactions.

Complete Structures

We can use complete structures as models by:

a. retrieving a structure from a ChemBase or MACCS-II or REACCS database
b. recalling a structure from User molfile followed by labeling the structure for the reaction we are building.

Example: *Complete structures as models*

Model structure from ChemBase database (Figure 2/52).
Model structure from User molfile (Figure 2/53).

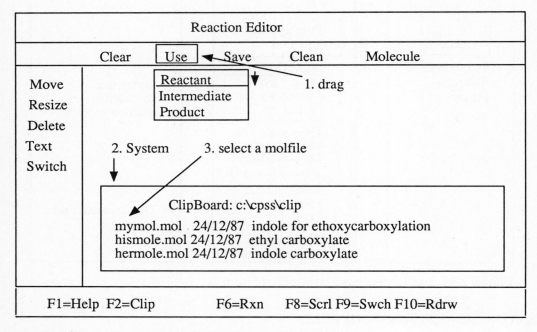

Figure 2/53. Using molfile as model structure for a reaction

The reactant is displayed (Figure 2/54)
Repeating the steps to recall additional Molfiles and applying the appropriate reaction label constructs the reaction in the Reaction Editor (Figure 2/55). If we wish, we can edit the reaction.

Structural Fragments

Structural fragments are recalled from the System or User Template File. They can be used intact or they can be modified by any of the structure building techniques to be used as the building blocks for creating new structures.

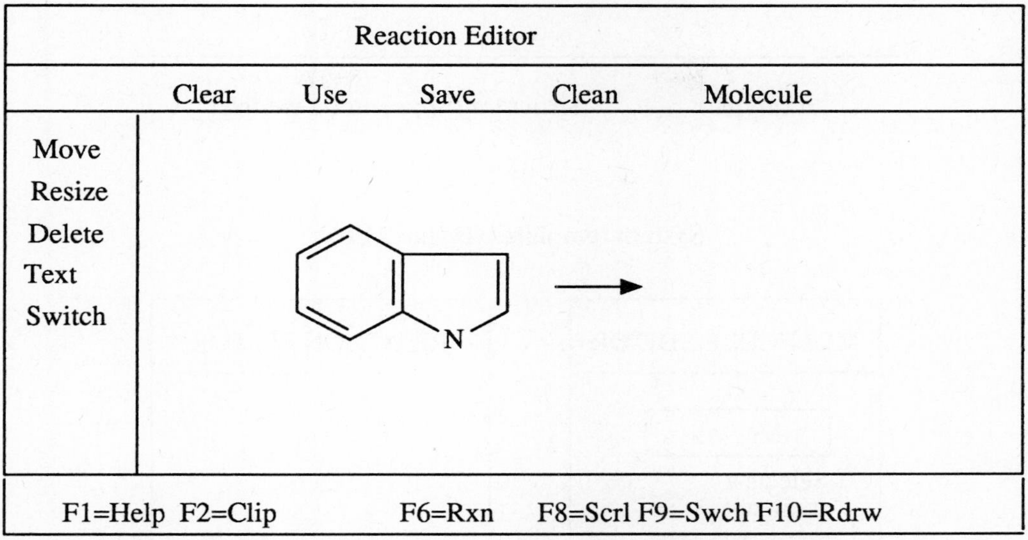

CREATING REACTIONS

	Reaction Editor				
	Clear	Use	Save	Clean	Molecule
Move Resize Delete Text Switch					
F1=Help F2=Clip		F6=Rxn	F8=Scrl F9=Swch F10=Rdrw		

Figure 2/54. Displaying reaction reactant

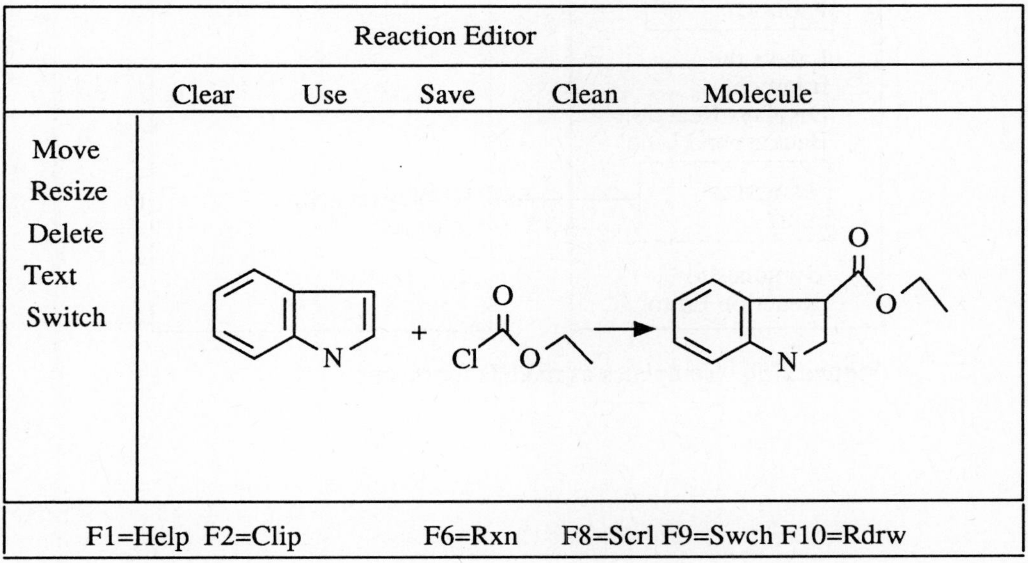

	Reaction Editor				
	Clear	Use	Save	Clean	Molecule
Move Resize Delete Text Switch					
F1=Help F2=Clip		F6=Rxn	F8=Scrl F9=Swch F10=Rdrw		

Figure 2/55. Constructing a reaction

Example: ***Structure fragments as models***

System templates (Figure 2/56)

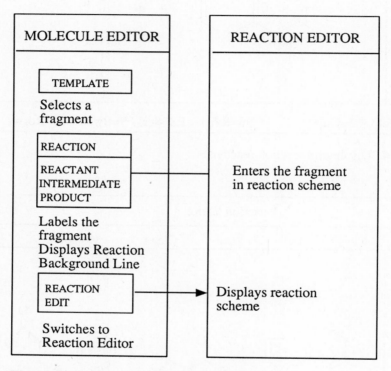

Figure 2/56. Templates as models for reaction

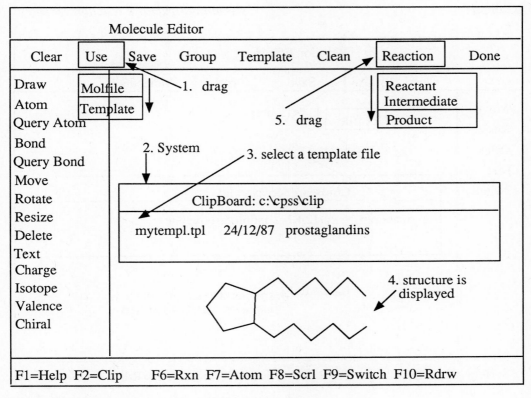

Figure 2/57. User templates as models for reaction

Example: *Recalling structural reaction schemes*

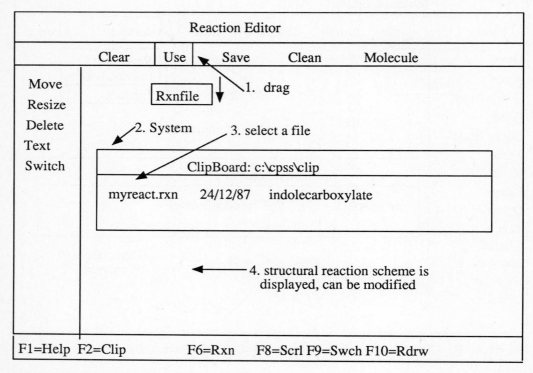

Figure 2/58. Saved reactions as queries

User Templates (Figure 2/57)

Complete Structural Reaction

Complete reactions can be brought into the Reaction Editor by:

a. retrieving a reaction from a ChemBase database (p. 394)
b. recalling a reaction from User RxnFile (Figure 2/58)
c. retrieving a reaction from REACCS and transferring it to Chembase (see p. 767)

TEXT DESCRIPTOR

Similarly to appending a text descriptor to molecular structures we can append lines of text to reactions.
The functions of the options TEXT(Edit, Move, Delete) on the Reaction Editor Menu are exactly those of the options in the Molecule Editor (see p. 311).

6   CORRECTING/MODIFYING STRUCTURES

In the course of building molecules and reactions, we might need to make corrections in the drawn structures, modify the current structures and reactions, and/or move the structures about the screen to create a format most suitable to our need. We also want to be able to alter structures that are recalled from Molfiles or Rxnfiles so that they can be used as models for building new structures.

ChemBase offers several techniques to make corrections or to modify structures and reactions.

DELETING ATOMS AND BONDS

DELETE(Atom) - deletes an atom and all the bonds the atom was connected with (Figures 2/59, 2/60)
DELETE(Bond) - deletes a bond and all atoms that were connected by the bond (Figures 2/59, 2/60)

DELETING FRAGMENTS

In ChemBase, a fragment is either an isolated, individual structure or it is a group of atoms within a structure.
To delete a fragment/structure, we drag:

DELETE(Fragment) - deletes an entire individual structure (place/click an atom in structure)

To delete a fragment/group, we have first to define the group. The group must be attached by an acyclic bond to the rest of the structure:

GROUP(Define) - defines the group within a structure (Figure 2/61.)
DELETE(Group) - deletes the defined group (Figure 2/61)
GROUP(Cancel) - cancels the group definition
To delete an entire structure, select DELETE(Fragment) and place/click on any atom or bond in the structure to be deleted. To delete a defined group, select DELETE(Group) and place/click on any atom in the group.

GROUP ABBREVIATION

When we define a group in a structure (*vide supra*), we can abbreviate the group to a string text which becomes attached to the specified position occupied by the defined

CHEMBASE

Example: *Deleting atoms/bonds*

```
Molecule Editor
Clear   Use   Save   Group   Template   Clean   Reaction   Done
Draw
Atom
Query Atom          4. place/click
Bond
Query Bond
Move
Rotate
Resize
Delete    ←—— drag
Atom    ↓ 1.
Bond    ↓ 3.          2. place/click
Group
Fragment
Chiral

F1=Help  F2=Clip   F6=Rxn  F7=Atom  F8=Scrl  F9=Switch  F10=Rdrw
```

Figure 2/59. Deletion of atoms and bonds

group. Groups abbreviated in this manner retain their structural significance and can be searched for. For example, to abbreviate the side chain in the structure in Figure 2/61 which was defined as a group:
1. Drag GROUP(Abbreviate).
2. Click the acyclic bond followed by an atom in the group.
3. In the prompt box: Enter group abbreviation: input (CH2)3CO2H.
The side chain will be replaced by the specified abbreviation.
To convert the abbreviation to the original side chain:
1. Drag GROUP(Expand).
2. Click the abbreviation.
To edit an existing abbreviation, use TEXT(Edit)(see p. 311.)
To search for the abbreviated groups, we use Data Search *stext query (see p. 410).
See also Text Descriptor and Atom Alias, p. 311.
Stext/Abbreviations SETTINGS turns on/off display of Group Abbreviations.
Structures with abbreviated groups can be saved in Molfiles and Rxnfiles.

Example: *Deleting atoms/bonds*

	Molecule Editor						
Clear	Use	Save	Group	Template	Clean	Reaction	Done

Draw
Atom
Query Atom
Bond
Query Bond
Move
Rotate
Resize
Delete
Text
Charge
Isotope
Valence
Chiral

F1=Help F2=Clip   F6=Rxn F7=Atom F8=Scrl F9=Switch F10=Rdrw

Figure 2/60.   Deletion of atoms and bonds

ChemBase 1.3 has an option to make Stext, Atom Symbols, Atom Aliases, and Group Abbreviations in various font sizes of the Helvetica typeface. Press F9 and choose the size by clicking Moltext Size.
In ChemBase 1.3 we can also insert the Greek letters α, β, and δ in floating text or atom aliases.
Press ALT and type, on the keypad: 224 (for alpha), 225 (for beta), and/or 235 (for delta).

RESIZING STRUCTURES

The size of the structures that are displayed on the Menu can be changed. We can resize the entire structure or, by defining a group, a fragment within a structure by point/drag:

RESIZE(Larger) - enlarges a structure or a group by 20% size increments
RESIZE(Smaller) - reduces a structure or a group by 20% size increments
RESIZE(Like Bond) - resizes the length of bonds taking one of the existing bonds as a

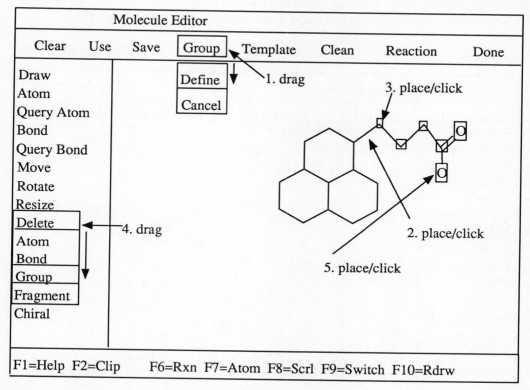

Figure 2/61. Deletion of a fragment/group

measure
RESIZE(Std Bonds) - resizes the average length of bonds accrding to the Settings (see p....)
RESIZE(Fit Screen) - resizes all structures on the screen to center the diagram on the screen

RESHAPING AND MOVING STRUCTURES ON THE SCREEN

The Molecule Editor has the option CLEAN(All, Fragment) which reshapes structures automatically without the User's intervention (see Figures 2/22, 2/23).
However, at times, we need to reshape structures according to our need and/or move them to the specific locations on the screen. We can reshape structures by point/drag:
MOVE(Atom) - moves an atom and all the attached bonds
MOVE(Bond) - moves a bond and the attached atoms

MOVE(Group) - moves a defined group in a structure
All of the above movements are analogous to the DRAW(Rubber Band) mode (see p. 292, 295) in that that the atom, bond, or group is anchored by the mouse pointer and is dragged to a new position on the screen. The shape of the structure changes accordingly.

Example: ***Rotating structures***

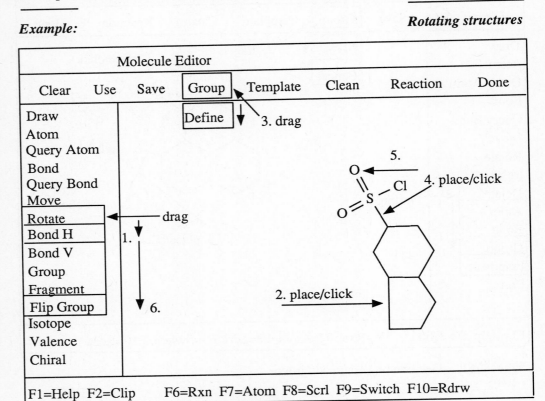

Figure 2/62. Rotation of structures

On the other hand, when we wish to move an entire structure, without changing its shape, we point/drag:
MOVE(Fragment) - moves an entire structure about the screen

ROTATING STRUCTURES ON THE SCREEN

If we want to manipulate the **entire structures** without changing the shape of them, we use the options (Figures 2/62, 2/63):
ROTATE(Bond H) - rotates the structure so that the selected bond becomes horizontal
ROTATE(Bond V) - rotates the structure so that the selected bond becomes vertical
If we want to manipulate a defined fragment/group or a fragment/structure, we point/click

CORRECTING/MODIFYING STRUCTURES

the options:

ROTATE(Group) - rotates the defined group around the end atom of the acyclic bond as the center of rotation
ROTATE(Fragment) - rotates the entire structure around its center
ROTATE(Flip Group) - flips the group by 180 degrees around the bond of attachment

Example: *Rotating structures*

Molecule Editor								
Clear	Use	Save	Group	Template	Clean	Reaction		Done
Draw Atom Query Atom Bond Query Bond Move Rotate Resize Delete Text Charge Isotope Valence Chiral								
F1=Help F2=Clip F6=Rxn F7=Atom F8=Scrl F9=Switch F10=Rdrw								

Figure 2/63. Rotation of structures

7   CORRECTING/MODIFYING REACTIONS

Editing individual structures in reactions has been discussed previously (see p. 321) since it is closely related to building reactions in the dynamic relationship of the Molecule/Reaction Editor.

However, we can also manipulate the structures in a reaction scheme as to their position and thus change the reaction layout in the Reaction Editor.

Example: *Rearranging reaction layout*

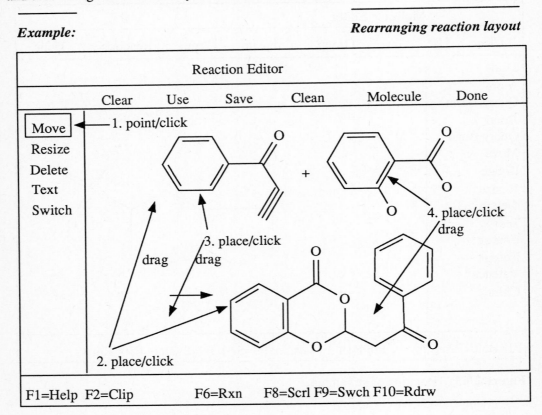

Figure 2/64.   Rearranging reactions

REARRANGING DEFAULT REACTION LAYOUT

We can move the structures in a reaction scheme about the screen (Figures 2/64, 2/65) to arrive at a format we desire. However, bear in mind that rearranging the position of structures in a reaction scheme does not change their reaction label and function assigned

CHEMBASE

to them in the Molecule Editor.

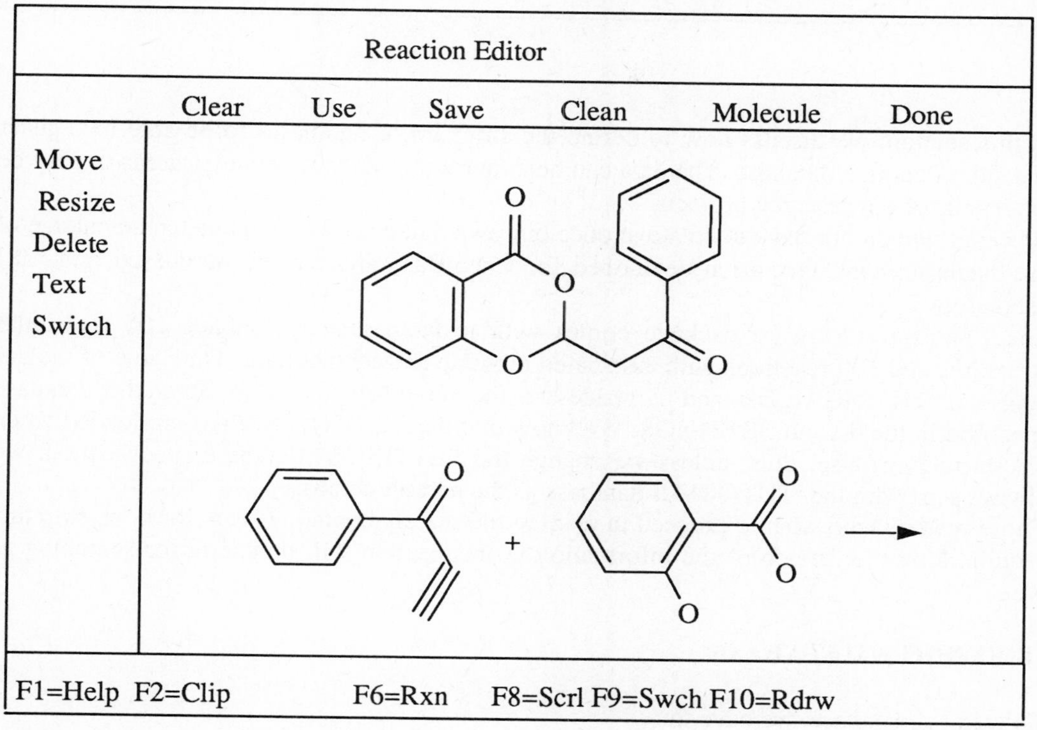

Figure 2/65. Rearranging reactions

The CLEAN option in the Reaction Editor has a different function than in the Molecule Editor (see Figures 2/22, 2/23). It restores the default layout of the reaction, if it has been changed, i.e., the option does not reshape the structures in reactions.

The + and ⟶ signs in reaction schemes can be moved by placing the mouse pointer over the sign and dragging it to a new position. The ⟶ sign can be also made shorter or longer by anchoring one of its ends by the pointer and dragging left or right. Again, the reaction labels of the structures do not change by moving the + and ⟶ signs.

The ⟶ between intermediates can be changed to the + sign by point/click the SWITCH option and place/click on ⟶ to be changed.

RESIZING STRUCTURES IN REACTIONS

The options RESIZE(Larger, Smaller, Like Bond, Std Bonds) in the Reaction Editor have the same functions as those in the Molecule Editor (see p. 334) and can be activated to change the size of the individual structures in reactions.

8 SETTING UP DATABASES

In this section, we discuss how to define the structure of databases to be able to register and store data in a database. The data can be either imported from other systems or may be the results of our research projects.

However, we do not have to create at once our own database. We can take an advantage of the databases that have been developed for ChemBase and can be purchased (e.g., ISI databases).

The ChemBase software package comes with a database that contains 100 molecular structures and 100 reactions with associated numeric and textual data. The name of the database is TUTORIAL.DB and it resides in the directory \cpss\clip. It is the database specified in the default SETTINGS. We know that the SETTINGS values are loaded when we start ChemBase, thus, unless we change the SETTINGS database specification, we always start with the TUTORIAL database as the default database.

Those readers who wish to proceed in reading the search techniques can, for now, skip this section. However, much of the information in this section will be useful for searching as well.

DEFINING A DATABASE

To create a database, we have to:
1. Define the data fields
2. Define the data types that we wish to have in the database. Careful planning is advisable at this stage. After the database definition is completed and saved, we cannot:
1. Change the field types.
2. Delete data fields.

The Main Menu is the starting gate to the database definition. We point/drag the DATABASE(Create) option (Figure 2/66). A window is overlayed on the Main Menu (Figure 2/67).

Example: ***Database definition***
Command: DATABASE(Create)

The window displays a list of the data fields and data types of the current, in our case the default TUTORIAL database.

CHEMBASE

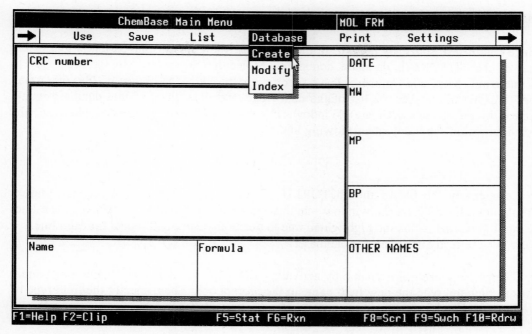

Figure 2/66. Creating a database

Figure 2/67. Defining database fields

MOLECULES DATABASE

Since ChemBase starts, by default, in the Molecule View, the displayed list is that of the part of the TUTORIAL database containing Molecular Structures and we are ready to define a new Molecule Database. If we specified in the SETTINGS (see p. 284) other than the TUTORIAL database, a list of the data fields and data types of that database would appear in the Database(Create) Window. If we do not have an opened database, the Window would show only the following line:

$$\text{ID} \qquad \text{Integer ID}$$

We can empty the list of the TUTORIAL database definition by deleting the entries. Point/click DELETE on the window menubar and move the mouse pointer over the lines to highlight a line. Click the highlighted line to delete the entry and repeat for each line. Now, we are ready to define our database:

1. Change the current field name by activating EDIT on the menubar (Figure 2/68); erase the existing and type the new field name in the prompt box. Then, specify the field type by clicking the Field Type line in the prompt box.
2. Add new data fields and data types by activating ADD on the menubar (Figure 2/69) and filling the prompt box with new specifications.

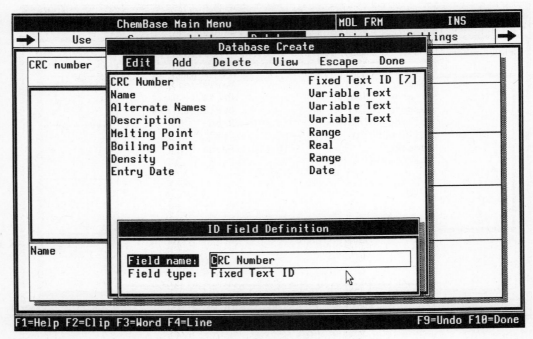

Figure 2/68. Editing Field Names and Field Types

SETTING UP DATABASES

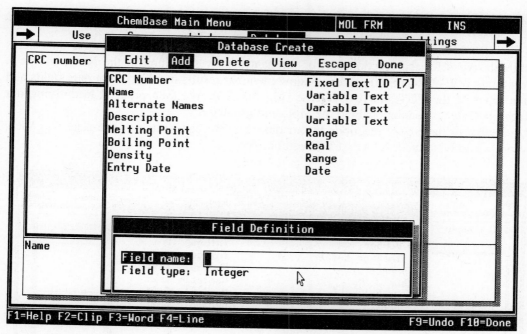

Figure 2/69. Defining a Field

We can selectively delete some of the data fields in the TUTORIAL database definition and leave the other in and/or selectively edit some of the existing data fields:

1. Select only a specific line in the definition by clicking DELETE on the menubar and click the line to be deleted.
2. Activate ADD and enter new data fields and data types.
3. Activate EDIT and change selectively a highlighted data field by editing the specification in the prompt box.

When we are satisfied with our database definition, we click DONE on the window menubar and enter the name of our database in the prompt box. Then, we are prompted for a password for the database. We can either enter a carriage return (no password) or a password.
The database structure is saved as a file in the Clipboard with the file name extension .DB. It is ready for input of data.
Although the database is a file in the Clipboard, when we work with the database we do not save the file as we would do with the ordinary files. A database is "opened" by the System when it is loaded through the SETTINGS specification or by the USE(Database) Command in the Main Menu and it is "closed" when we exit from it by another USE(Database) selection or by EXIT(DOS). All newly registered entries or edited existing entries are automatically saved.

REACTIONS DATABASE

If we wish to have both molecules and reactions in the database, we define the data fields and data types for the reaction part of the database by switching to the Reaction View. Point/drag VIEW(Reaction) on the window menubar results in a list of the data fields and data types of the TUTORIAL Database (Figure 2/70). We then proceed in defining our Reaction Database analogously to the Molecule Database.

If we wish to have only reactions in our database, we can switch to the Reaction View before we activate the DATABASE(Create) option.

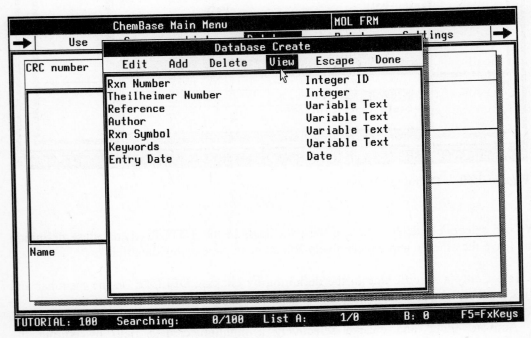

Figure 2/70. Switching to Reaction database

NUMERIC AND TEXTUAL DATABASE

We can build databases with no structures. The procedure for defining such database is the same as for the Molecule or Reaction database. However, the ID field has to be in. In registering data in this database, we simply do not draw any structures and register only numeric and text data and assign an ID for each of the records.

ESCAPE

If we, for any reason, wish to abandon the process of defining a new database at any point, we point/click ESCAPE on the window menubar. We are returned to the Main Menu. No

SETTING UP DATABASES

changes have been recorded and no new databases have been created.

MODIFYING DATABASES

Any changes in the database definition can be made before the database definition is saved.
However, as we have pointed out above, only limited changes can be made in the database definition after it has been completed and saved.
We are allowed to:

1. Change existing data field names.
2. Add new data fields and new data types.

The modifications are carried out by activating the command DATABASE(Modify) on the Main Menu menubar. The overlay window is the same as with the DATABASE(Create) option and the techniques in manipulating the entries are the same as in defining the database.
Note that if we add new data fields by DATABASE(Modify), they can be removed only before the modified database structure is saved.

PASSWORD CHANGE

If we wish to change the password that we have assigned to a database (*vide supra*), we open the database and point/drag DATABASE(Modify) to display the window. Then point/click DONE on the window menubar. The prompt box will appear with the password that can be erased and a new one typed in.

9 CREATING FORMS AND TABLES

We have learned (see p. 274) that the data input and output in a database is channeled through a system of Forms and Tables. We have discussed how the Forms and Tables are interchanged for viewing molecules and reactions. In this section, we will discuss how to create the Forms and Tables and how to prepare them for reception of data.

WHAT IS A FORM

In ChemBase, a Form is a formatted outline for data which can be:
1. Displayed on the screen.
2. Printed on a printer to obtain a hard copy.
3. Saved in a file for recall.

The Form consists of a **box or boxes** which can be separated from each other or which can have common borders. The borders of the boxes can be suppressed so that no visible box(es) will show; however, the data will be shown in the reserved box space.
The Form can channel **one data record** for display or registration in the database.
Usually, we have one Form for molecules and another Form for reactions.
A Form can have as many boxes as there are data fields in the database. Or, a Form can have only boxes for selected data fields in the database.

WHAT IS A BOX

A box in the Form is destined to hold data that correspond to the data field in the database to which the box has been linked on output or input operations.
A box can be a **data box** whose content changes according to the records we have retrieved from a database or we want to register. A box can be a **label box** whose content is permanent and always the same when the Form is used. A box in a Form can be both data and label box showing a permanent label and data that change according to the retrieved record. Boxes can have special features in their border outline. We can draw lines and/or arrows to link boxes in a Form.

CREATING FORMS

The Forms are created in the Form Editor which is shown in Figure 2/10 (p. 282).
When we start ChemBase, the default Form which is specified in the SETTINGS (see p. 284) will be displayed on the screen. We can modify it and save it under the same or different filename. If we wish to start from scratch, we point/drag CLEAR(Screen) on the menubar to have a new screen. The following examples show the major operations to create and manipulate boxes.

CHEMBASE

Operations in Figure 2/71:

Activate DRAW(Box) to draw box A, B and C.
Activate MOVE(Box) to move box B to the new location
Activate JUMP EDGE(Left) to join box A and C.

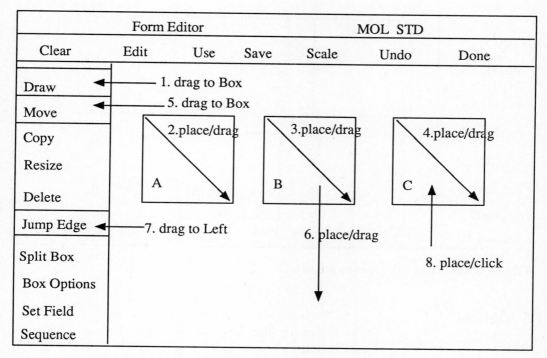

Figure 2/71. Drawing boxes in a Form

Figure 2/72 shows the results of the operations in Figure 2/71.
Operations in Figure 2/73:
Activate RESIZE to change the size of box B.
Activate SPLIT BOX(Horizontal) to divide box C. A new box D is formed.
Figure 2/74 shows the results of the operations in Figure 2/73.

MORE ON CREATING FORMS

ACTIVATE	TO

DRAW(Solid Line
    Dotted Line
    Arrow)
	draw lines or arrows anywhere in a form.(A line dividing a box **does not create** a new box.)

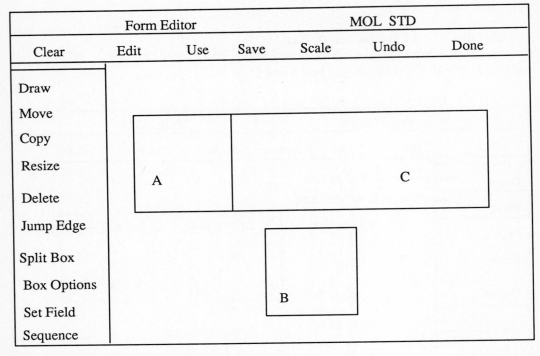

Figure 2/72. Drawing boxes in a Form

MOVE(Line)	move an existing line to a new location by place/drag the line.
COPY	copy a box by place/drag the box to a position on the form.
RESIZE	change size of a box by dragging a side or corner of the box.
DELETE	delete box or line by clicking the object.
JUMP EDGE(Right,Up,Down)	join boxes on right, up, or down. If no box next to the jump side, the box extends to the screen edge.
SPLIT BOX(Vertical)	divide box (**creates a new box**).
SEQUENCE	change the sequence of boxes which are in the order they were drawn.
CLEAR(Structure,Text)	erase structure, data in boxes (labels are not cleared).
EDIT	change box labels, press TAB to jump from box to box.
SCALE	switch from BIG/STD to display

CREATING FORMS AND TABLES

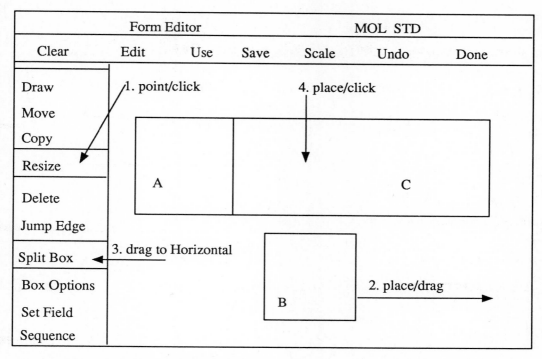

Figure 2/73. Resizing and splitting boxes

	entire or part of Form (F8 and left and right arrow keys scroll the Form in BIG scale).
UNDO	cancel and return to original/none Form.
DONE	switch to the Main Menu with the current Form (will be shown in BIG scale).

WHAT IS A TABLE

In ChemBase, a Table is a formatted outline for data. Similarly to the Form (*vide supra*), we can display the Table on the screen, print it, or save it in a file.
The Table consists of **columns** that can be separated or joined with each other. When we construct a Table in the Table Editor, the columns are separated by vertical lines. However, when we use the Table for display of data or when we print a Table with data, the lines do not show. Column titles are in the boxes at the top of the Table.
The Table channels a **list of records**. Similarly to the Form, we have one Table for molecules and a separate Table for reactions.

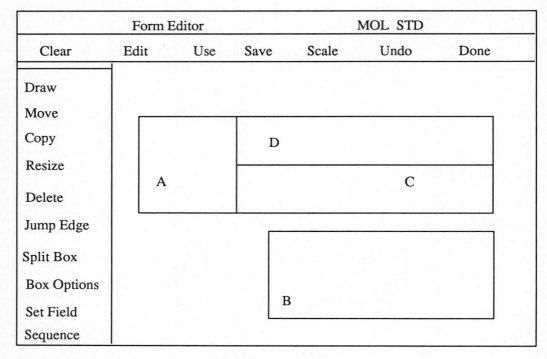

Figure 2/74. Resizing and splitting boxes

A table can have as many columns, limited by the size of the page, as we have data fields in the database. Or, a table can have only columns for selected data fields in the database.

WHAT IS A COLUMN

A column is destined to hold data of one specific data field for a list of records whereby the data are arranged, one line per record, in the column.
The columns in Tables are used:
1. To display data of many records that we retrieve from a database. This allows for browsing in the list to select specific records, one by one, that we may wish to see displayed in the Form. Note that the Columns in Tables do not display structures, they are displayed in the Form when we select an entry from the Table and switch to the Form.
2. To print a list of records (including structures).
3. To transfer large files of data (without structures) via ASCII data files.
4. To register the content of SDfiles and RDfiles.

CREATING FORMS AND TABLES

CREATING TABLES

Tables are created in the Table Editor which is shown in Figure 2/11 (p. 282).
If we have the default Table on the screen, we can modify it to arrive at our table. If we want to start from scratch, we clear the Table from the screen by CLEAR(Screen).

The major operations in the Table Editor are as follows.

Operations in Figure 2/75:

1. Activate DRAW to define columns in the Table.
2. Place/click successively on position 2, 3, 4.
Each place/click will result in a column of a width determined by the place/click position on the screen.
The final table of three columns is shown in Figure 2/76.

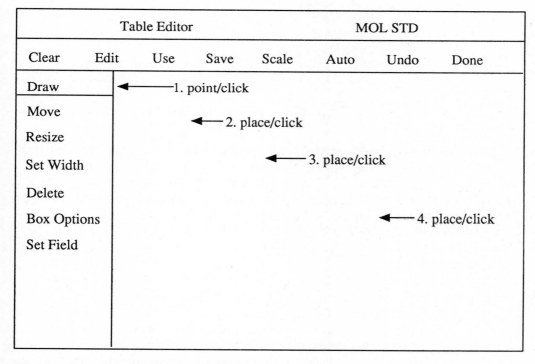

Figure 2/75. Drawing columns in Table

Operations in Figure 2/77:
To change column size, use the MOVE and RESIZE options:
Activate MOVE and drag the right edge of the column A. The column A becomes larger

Table Editor					MOL STD		
Clear	Edit	Use	Save	Scale	Auto	Undo	Done
Draw							
Move							
Resize							
Set Width							
Delete							
Box Options							
Set Field							

Figure 2/76. Drawing columns in Table

up to the dotted line on the account of the column B whose left edge is now the dotted line.

On the contrary, if we activate RESIZE and drag the right edge of the new column B, it becomes larger but the size of column C does not change; it is shifted full-size to the right. The result of the changes in Figure 2/77 are shown in Figure 2/78.

MORE ON CREATING TABLES

ACTIVATE	TO
SET WIDTH	specify size of column by number of characters. A prompt box will appear for the specification followed by place/click the column to have the width. The column size will adjust accordingly.
DELETE	delete column by clicking it.
EDIT	enter/edit column title boxes. Press TAB to jump from title box to title box, finish with F10. Data in

CREATING FORMS AND TABLES

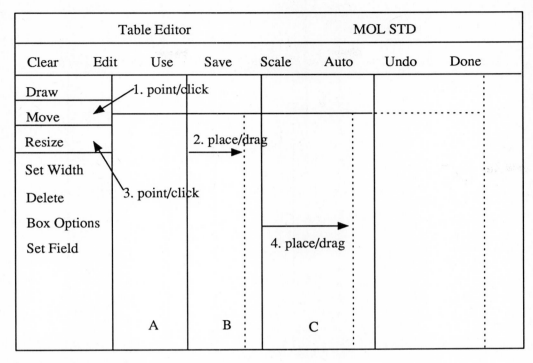

Figure 2/77. Resizing columns in Table

	columns cannot be edited in Table Editor, only in Main Manu.
SCALE	switch from BIG/STD size of Table, menubar shows the current scale. STD size shows the entire table on the screen, BIG part of it. Both can be scrolled (F8) by left and right arrow keys.
AUTO	create Table with columns for all defined fields.
UNDO	cancel all actions and return to the original table (or none).
DONE	switch to the Main Menu with the current table (table will be shown in BIG scale if we were in STD scale).

SAVING/RECALLING FORMS AND TABLES

When we create a Form or a Table we can:

Table Editor						MOL STD		
Clear	Edit	Use	Save	Scale	Auto	Undo	Done	
Draw								
Move								
Resize								
Set Width								
Delete								
Box Options								
Set Field								
			A		B		C	

Figure 2/78. Resized column in Table

1. Use them just for the current operations.
2. Save them in the Clipboard in files with the file name extension .FRM and .TBL, resp., and enter the file name in the SETTINGS as our new default Form and Table.
3. Recall the Form or Table Files as needed.

Using Form/Table for Current Operations

When we complete the new or finish modifying the default Form or Table in the Form and Table Editor, respectively, we exit the Editor by point/click DONE on the menubar. We will be returned to the Main Menu where we can execute further operations for which the Form and Table are the necessary prerequisites.

Saving Forms

In the Form Editor:

SAVE(Definition) - saves the outline of the Form in a .FRM file including labels, (if any), in the boxes but excluding data (if any) in the boxes.

CREATING FORMS AND TABLES

SAVE(Form) - saves the Form in a .FRM file including label and data (if any) in boxes.

In the Main Menu:

SAVE(Form) - saves the current Form in a .FRM file. If the Form contains data in boxes, they will be saved with the Form. If only the outline of the Form is current, it will be saved (equivalent to SAVE(Definition) in the Form Editor).

Recalling Forms

In the Form Editor:

USE - displays the Clipboard from which we select the .FRM File. Note that if the Form had been saved by SAVE(Form) in the Form Editor or in the Main Menu, the data will be displayed.

In the Main Menu:

USE(Form) - displays the Clipboard from which we select the .FRM file. Note that if the Form had been saved with data (by SAVE(Form) in Form Editor), the data will not be displayed.

Saving Tables

In the Table Editor:

SAVE - saves Table in a .TBL file.

In the Main Menu:

SAVE(Table) - saves Table in a .TBL file.

We can save only the Table definition by these options. Data in Tables are saved by SAVE(Data File) in the Main Menu.
Note that if we wish to store .FRM and .TBL files in a Clipboard other than \cpss\clip, we have to specify the new directory in our SETTINGS (or change the current Clipboard by F2 and CHANGE(Clipboard) prior to SAVE(Table).

Recalling Tables

In the Table Editor:

USE - displays the Clipboard from which we select the .TBL file.

In the Main Menu:

USE(Table) - displays the Clipboard from which we select the .TBL file.

Note that if we want to recall .FRM and .TBL files from a Clipboard other than \cpss\clip, the Clipboard has to be specified in the SETTINGS prior to USE(Table) or we have to change the current Clipboard by F2 and CHANGE(Clipboard).

LINKING FORMS AND TABLES TO DATA FIELDS

Each box in a Form and each column in a Table has to be tied with the data field in a database to input or output data.
In an example, let us assume that we have a database in which there are the following data fields: *structure, name, and *formula.
We have created a Form with boxes for these data.
Figure 2/79 shows the actions taking place when we search (in) or register (out) data through the Form linked to a database and the results of this linkage.

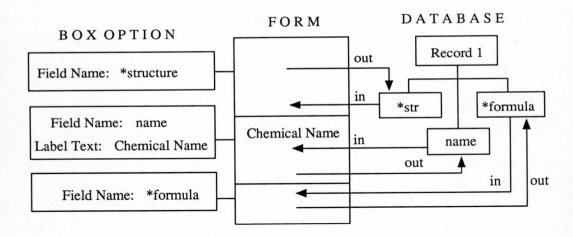

Figure 2/79. Linkage of boxes and data fields

An analogous process occurs with a Table in which more than one record of data can flow in both directions (Figure 2/80).
The linkage of the boxes in a Form and the columns in a Table is accomplished by using the side menu commands in the Form and Table Editors, BOX OPTIONS and SET FIELD.

CREATING FORMS AND TABLES

Figure 2/80. Linkage of columns and data fields

FORM BOX OPTIONS

When we point/click the BOX OPTIONS Command, a window will appear (Figure 2/81) showing the options for linking a box with a data field. To work with a specific option, highlight the option line by clicking the line and type the appropriate specifications:

Field Name: (enter the name of the data field defined in a database for the data to flow through the box).
        ChemBase 1.3 metafiles (see p. 368):
            a. enter the Field Name that has been defined
                for metafiles)
External Field Name: (see p. 368).
Default Data: (leave blank or click the line to select):
        a. today's date
        b. current time
        c. automatic ID.
        ChemBase 1.3 has additional options:
            d. formula
            e. molweight
            f. composition
            g. metafile (see p. 368)
Data Position: (click the line to select):
        a. top left(center)(right)

b. middle left(center)(right)
c. bottom left(center)(right).
Label Text: (enter optional permanent label for the box).
Label Position: (same as above for Data Position).
Box Frame: (click for yes(no)).
Box Highlight: (click for yes(no)).
Box Shadow: (click for yes(no)).
Box Reverse: (click for yes(no)).
Auto Fit: (click for yes - turns word wraparound on)
One-line Box: (click for yes - limits box to one line text).
Scientific Notation: (click for yes(no)).
Decimal Places: (click for 0-9).

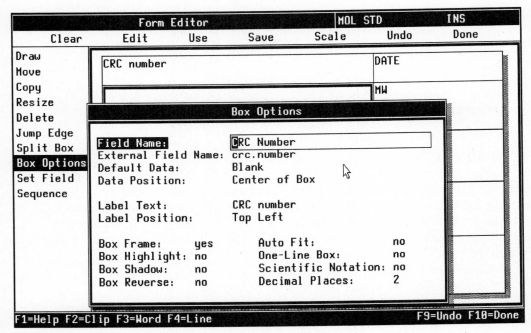

Figure 2/81. Linking box and data field

TABLE BOX OPTIONS

The window for Table Box Options is shown in Figure 2/82. Although it is called Box Options, the options relate to the column in a Table. Selecting the options is done similarly to the Form Box Options.
Column titles (which are analogous to the Box Labels in a Form) can be entered by activating EDIT on the Menubar and typing the title followed by F10. Jump from column to column by pressing TAB.

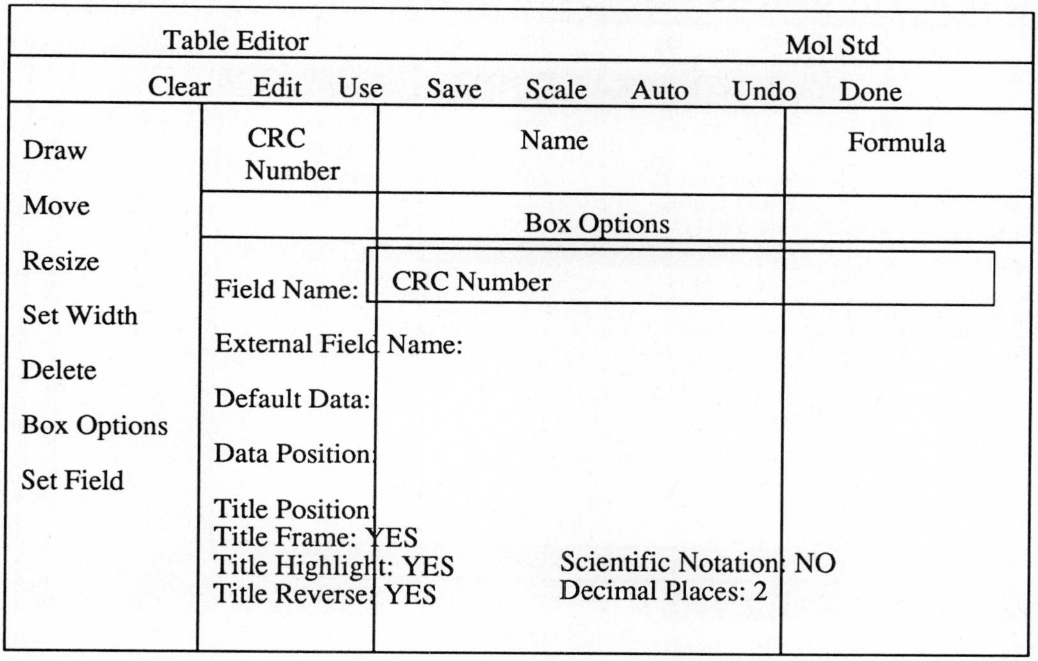

CREATING FORMS AND TABLES

Table Editor							Mol Std	
Clear	Edit	Use	Save	Scale	Auto	Undo	Done	
Draw	CRC Number		Name				Formula	
Move			Box Options					
Resize	Field Name:	CRC Number						
Set Width	External Field Name:							
Delete	Default Data:							
Box Options	Data Position:							
Set Field	Title Position: Title Frame: YES Title Highlight: YES Title Reverse: YES			Scientific Notation: NO Decimal Places: 2				

Figure 2/82. Linking column and data field

SET FIELD

Activating SET FIELD on the side menu followed by clicking a Box in a Form or a column in the Table will display a window showing all the data fields of our current database (Figure 2/83). We can enter the data field name from this window into the Field Name of the box or column we have clicked by highlighting the data field line and clicking the line. The window will disappear and the data field name will be displayed in the box/column.

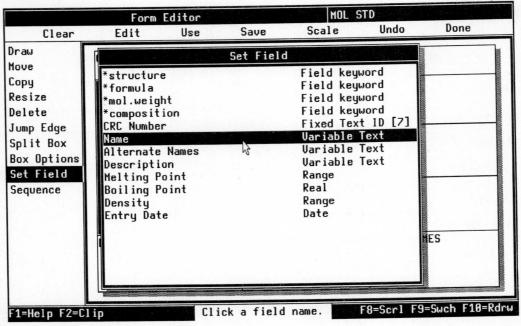

Figure 2/83. Setting data field

10   BUILDING DATABASES

Unless we wish to take an advantage of the databases developed for ChemBase from external sources (e.g. ISI), we are now ready to begin building our own database.
In this section, we discuss the techniques which will enable us to fully utilize ChemBase capabilities for building databases by registering our own structures and data or those transferred from MACCS-II or REACCS.
We have used the TUTORIAL database in section 8 - Setting Up Databases. We have said that the Molecule Database contains 100 structures/records. However, in Section 11 - Molecule Searching, we will see that we have added nine structures/records into the TUTORIAL database to illustrate the discussion.
We will now learn how these records were registered. The registration techniques for molecules can be equally applied to registration of reactions. The only difference is that to register molecules we have to be in the Molecule View, while to register reactions, we have to be in the Reaction View (see p. 276).

DATA REGISTRATION

The operations in data registration can be classified as follows:
1. registering new records
2. registering edited/updated existing records
3. deleting registered records

The input on the screen is carried out through the FORM since TABLE data entries cannot be edited (we can edit only the title boxes/data field names in TABLE). However, both TABLE and/or FORM can be used for registration if we transfer structures/data from files.

REGISTRATION OF NEW RECORDS

We will register the structure 13 shown in Figure 2/106 (p. 385):
1. Build the structure in the Molecule Editor and transfer by DONE (Figure 2/84) to the Main Menu.
On transfer to the Main Menu, the structure is displayed in the box that has been linked with the *structure field. The System calculates the values (i.e., those which are derived from the structure) and automatically displays the data in the boxes linked with *formula and *mol.weight fields. These boxes have been assigned by us the labels "Formula" and "MW," respectively, on creating the Form (see p. 356). We can now register this structure with the existing data. First, we have to supply the ID number. The registration cannot be performed without an ID number. We recall that the ID field is the first data field in the database structure (automatically created on setting up the database). However, in displaying the ID, we can link any box in the Form with the ID data field. It does not have to be the first box in the Form.

Example: *Registering new record*

Building structure

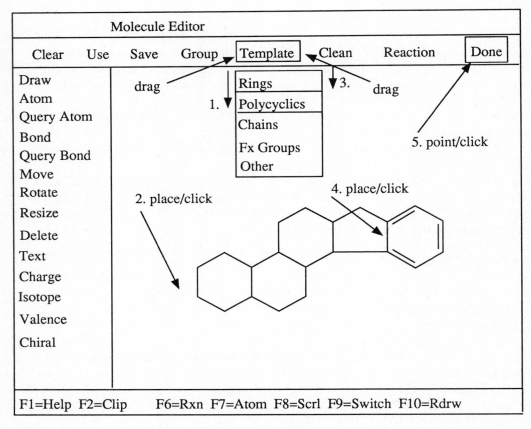

Figure 2/84. Building structure for registration

2. Specify the ID number (Figure 2/85): drag EDIT(Text).
The text cursor will appear in the first box (the label temporarily disappears).
Type the ID (we have selected Z-08888) and press TAB. The cursor jumps to the second box, the Name box. The first box becomes shaded or highlighted to remind us that we have entered new data in that box. The data and the box label is displayed. If we do not want to register additional data at this point, we are ready to register the record since the ID is the only data required for registration. But let us name the structure (we have selected "teststructure").
3. Type the name at the cursor in box 2.
The box becomes shaded (highlighted). The structure name and the box label are displayed in the box.

BUILDING DATABASES

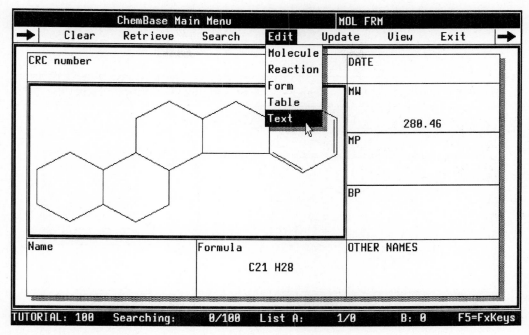

Figure 2/85. Inputing data (editing boxes)

4. Press F10.
Finish entry/editing of the boxes by F10 (Figure 2/86)
5. Drag UPDATE(New Entry)(Figure 2/87).
The System displays a meassage that the registration is being carried out. When the process is finished, the shaded (highlighted) boxes are returned to normal display to indicate that the data have been registered.

REGISTRATION OF UPDATED/EDITED RECORDS

If we want to change any data in a registered record or add new data to the data fields of a registered record, we retrieve the record, edit the data, and register it back under the same or different ID number.
We will add data to our record ID Z-08888.
1. Retrieve the record by RETRIEVE(ID) (see p. 403) or by RETRIEVE(Current) (see p. 374).
The record will be displayed (Figure 2/86).
2. Drag EDIT(Text) (Figure 2/85.).
The cursor will appear in the first box. If we want to register the edited record under a new ID, we would erase the ID and type the new ID in the box (the record Z-08888 will remain in the database unless we delete it (*vide infra*)). In this example, we will leave the ID.
3. Press TAB to move the cursor into the Name box.
We change the registered name "teststructure" to "mystructure" by clearing the box (F7

Inputing new data

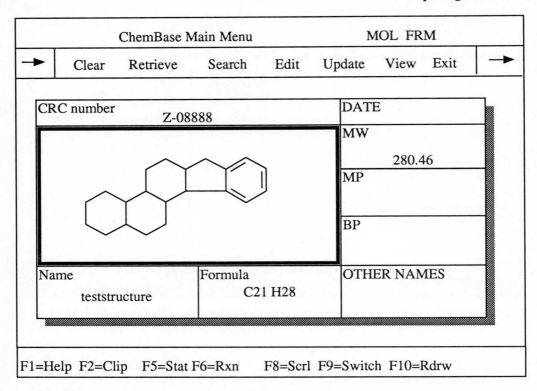

Figure 2/86. Editing boxes - inputing data

clears the entire box, F3 deletes a word, F4 deletes a line) and typing the new name "mystructure" in the box.

When we type a text in the box, the text will be displayed depending on the parameters for the box:

If we assigned Auto Fit (in Box Options, see Figure 2/81) to the box, the text will be automatically fit into the box (word wrap is on). If we did not use Auto Fit, the text will be displayed as typed. To fit the line into the box, press F6. If we did not use Auto Fit, we can still turn the word wrap on by F5. Note that if we type text which extends beyond the box border, the text will be saved on registration of the record but it cannot be displayed within the box. On recalling and displaying the record, the text can be scrolled in the box by ← → keys.

4. Press TAB to move the cursor into the box with the label Date.

Type a date which is to be associated with the record. If the box is linked with the Default Data field Today's Date (see p. 358, Figure 2/81), the current date from the computer system will be automatically filled in the box on registering the record. If the box is linked

Record registration

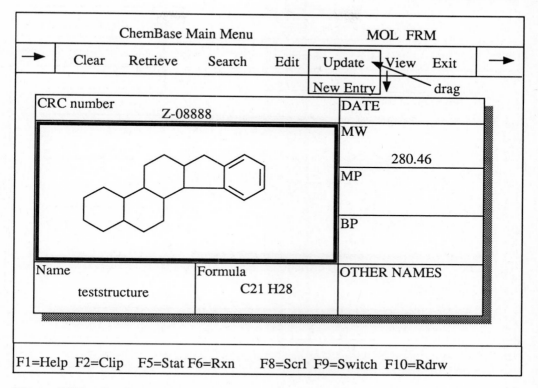

Figure 2/87. Registering new record

with a regular data field, type the date in the box (we used 4/12/88).
5. Press TAB to move the cursor into the MP box
Type the melting point value (in our case a fictional 350).
6. Press TAB to move into the remaining boxes and fill in the values (Figure 2/88).
7. Press F10.
The boxes with the edited data will be shaded (in our case the Name, Date, MP, BP, and Other Names boxes).
8. Drag UPDATE(Change) (Figure 2/88).
The changes in the record are registered.

DELETING REGISTERED RECORDS

To delete a registered record, we retrieve the record and drag UPDATE(Delete). The record is deleted after we confirm.

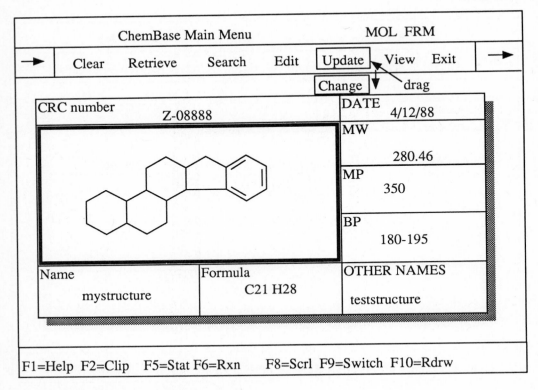

Figure 2/88. Registering updated record

REGISTRATION USING FILES

The above examples of data registration work well if we have a few new records to register or a few records to edit/update. For transferring structures from MACCS-II and REACCS we can use Molfiles and Rxnfiles. What do we do if we have a large quantity of data to register or to edit/update?

STRUCTURES

For editing/updating registered structures (molecules or reactions), we cannot escape from doing it one by one by retrieving the structure(s) from the database and editing the structures in Molecule/Reaction Editor and re-register them in the Main Menu.
For registering new structures:
1. Write SDfile or RDfile (see p. 367, 761) from one ChemBase database and read the files (*vide infra*) into another ChemBase database. Note that in ChemBase, we can write either of these files for molecules or reactions. However, on recalling the files we have to

BUILDING DATABASES

be in Molecule View if the files contain molecules and in Reaction View if the files contain reactions.
2. Write SDfiles in MACCS-II or RDfiles in REACCS and transfer the files and read them into a ChemBase database (see Part Five, p. 761)

TEXT AND NUMERIC DATA

If we have structures and wish to transfer them including the associated data, we can do it, of course, via the SDfiles and RDfiles which store multiple structures including data.
However, we can also transfer only alphanumeric data in any data field except *stext, *formula, *mol.weight, and *composition:
1. Write Data Files in Table View from a ChemBase database and read the Data Files unchanged into another ChemBase database for registration.
2. Edit data in Data Files in ChemText or in any system which reads/writes ASCII files, save the edited data in a ASCII file, and read the file in Table View by USE(Data File) to register the edited data in a ChemBase database.
Note that prior to registering the data, they should be previewed if the format fits the Table column width:
1. In Table View select USE(Data File Preview).
2. Select the .ASC file from the displayed Clipboard.
If needed, the column width can be adjusted. Then switch to the Main Menu and USE(Data File) and proceed with registration.

As an example, we want to develop a ChemBase database for the project STEROIDS. We have set up the database. We know that the TUTORIAL database contains structures of steroids. It would be a good start to transfer these structures to our STEROID database. We proceed as follows:
1. Perform retrieval or search (see p. 372) for the molecules.
2. Save the retrieved records in an SDfile by SAVE(SDfile) (we will name it steroids.sdf).
3. Exit the TUTORIAL database and open the STEROID database by USE(Database) in the Main Menu.
4. Drag VIEW(Table).
5. Drag USE(SDfile) and select steroids.sdf from the displayed Clipboard.
6. The TABLE will show the first highlighted entry. Use:
    down arrow - to register
    Del - to skip the record
    <cr> - to continuously register all records
    ESC - to abandon the registration process
The registration can be, of course, carried out in the Form View as well.

TRANSFERRING DATA

Transfer of structures and data between ChemBase/MACCS-II/REACCS requires a PC-to-mainframe communication software capable of transferring files, such as ChemTalk-ChemHost.

We will discuss ChemTalk transfer function in Part Five. Here, we will only outline in general terms the workings of the transfer process.

We know that data I/O in ChemBase is channeled through Forms and Tables (see p. 274, 346). The same I/O vehicle is used to transfer data between ChemBase and the mainframe systems.

In an ideal case, the ChemBase and the MACCS-II/REACCS databases will have all data fields of the same data field names. In that case, we can freely exchange data between these databases without any special measures.

In any case, we can do so for the automatic data generated from structures, i.e.:

in ChemBase:	in MACCS-II:
*structure	*structure
*formula	*regno
*mol.weight	*extreg
*composition	*comments

We can specify these keywords for any box or any column in the I/O formats to successfully transfer data.

However, if the regular data fields in MACCS-II or the datatypes in REACCS have different names than the ChemBase data field names, we have to establish a correspondence between these different data field names.

EXTERNAL FIELD NAME

In the discussion of Box Options for the Form or Table I/O formats (see p. 359), we have seen that there is one option called External Field Name. This is the specification to establish a correspondence between a ChemBase data field name and a different MACCS-II/REACCS name (Figure 2/89).

When we assign a box or a column an External Field Name, the ChemBase Field Name is translated into the MACCS-II or REACCS data field name/datatype on uploading data, and the MACCS-II or REACCS data field name/datatype is translated into the ChemBase data field name on downloading data.

The transfer process is illustrated in Figure 2/90.

USING METAFILES IN CHEMBASE 1.3

Metafiles contain graphics output generated by MDL mini/mainframe programs. We can also generate metafiles from the data captured in a session in the CAS Registry system by using the MDL program FROM4010 which is part of the CPSS system (see p. 780).

When we create a metafile, it can be stored in any directory, including the Clipboard, under the name we assign to the metafile. ChemBase will retrieve the file by the path we

BUILDING DATABASES

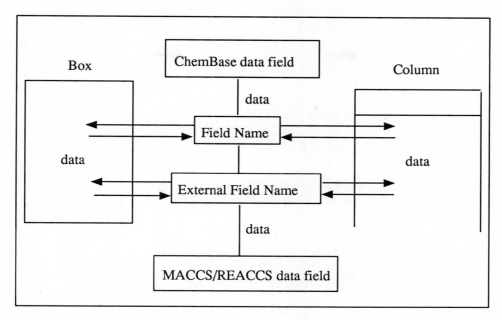

Figure 2/89.   Data field correspondence

specify or, if no path is specified, ChemBase first looks in the current directory, and then in the Clipboard. However, we can create a special directory for all metafiles and, using an Editor, enter the path in the CPSS.CFG file which has the entry:
Metafile Path:
Input: c:\cpss\cas\ (as an example of a directory for CAS-metafiles). The backslash must be included.

Defining Field Name for Metafiles in Database

We can "register" metafile structures in a ChemBase 1.3 database (*vide infra*). For this purpose, we have to define in the database the following:

Field Name: casstruc (or any other name)
Field Type: variable (or fixed) text

Registration of Metafile Structures

We have used quotation marks for "register" because it is not registration of structures *per se*. It is registration of the names under which we have stored the metafiles. Upon recalling a metafile name, the structure from the metafile is displayed in the Form in the specified box. We cannot perform structure or substructure searches for the structures in metafiles using structure queries.

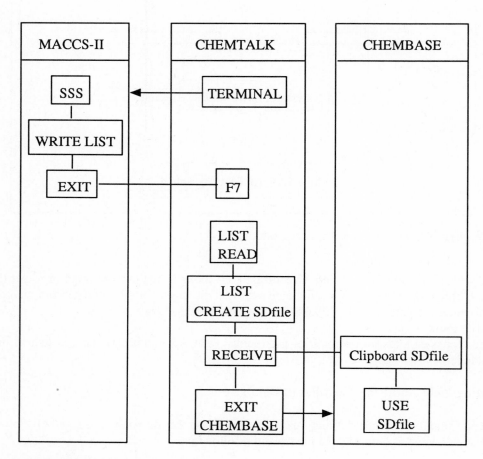

Figure 2/90. Transfer of data

BUILDING DATABASES

To "register" metafile structures we have to define a field for the metafiles when we create or modify a ChemBase 1.3 database (*vide supra*).
To "register" the metafile structure we use a Form with the specified box for the structure with the following Box Options:
Field Name: casstruc (or any other defined Field Name)
Default Data: metafile
See also Part Five, p. 737 for examples of capturing data, creating metafiles, and displaying the structures in Forms.

Displaying Metafile Structures

When we have a Form with the box that we have specified for the metafile structure, we select EDIT(Text) and enter the name of the metafile at the prompt, e.g., cas003.met. The structure in the cas003.met file will be displayed in the box. If we have not specified the path in the CPSS.CFG file (*vide supra*), we include the path in the metafile name.

11   MOLECULE SEARCHING

Molecule searching is the major function of ChemBase. We can search not only for molecules in databases which we have developed, but we can search databases created from data transferred to ChemBase from MACCS-II and REACCS, or we can purchase databases developed for ChemBase, e.g., ISI products.

ChemBase allows for searching molecules by using graphic structure queries to retrieve and display graphic molecular structures that can be printed to obtain hard copies.

Although we can retrieve structures by performing a Data Search (see p. 403), searching with structural queries will be our natural approach to finding information in a ChemBase database.

Structure searching in ChemBase can be classified as:
1. exact structure searching
2. substructure searching

The search operations are performed in the Main Menu by activating the RETRIEVE or SEARCH commands on the Menubar.

The search process produces statistical data that can be used to monitor the progress and effectiveness of the search and to carry out the search in selective subsets of a database. The statistics is shown on the Alternate Status Line (Figure 2/91).

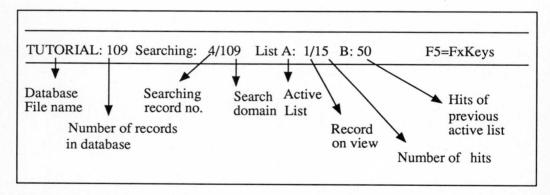

Figure 2/91.   Search statistics

SEARCH DOMAIN

The "Searching:" line in Figure 2/91 shows two numbers separated by a slash. The first number is a running counter showing, in the course of the search, which record is being examined as candidate for the hit list, List A.

The second number is the Search Domain which is the total of records in the database that are subjected to search. In Figure 2/91, the search domain was all the records in the database. This is the default search domain and it is the search domain for all RETRIEVE operations. As we will see, we can specify the search domain to be a subset of the entire database in SEARCH operations.

LISTS

A List in ChemBase is a collection of the ID numbers of the records retrieved by a search.
List A is the Active List showing two numbers separated by a slash. The first number is one of the records in the List A which is currently in view on the Form or Table. The second number is the total of records (hits), which is the result of the search satisfying the query criteria. The Active List is the List whose content of IDs can be displayed, manipulated to create another Active List, transferred to List B, or saved in a file.
List B is a temporary storage for List A. It is created by the System if we perform a second search after the first one which produced the List A, i.e. List A is transferred to List B. When the third search is performed, a new List A is created and the first one is transferred to List B (the previous List B is lost) (Figure 2/92).

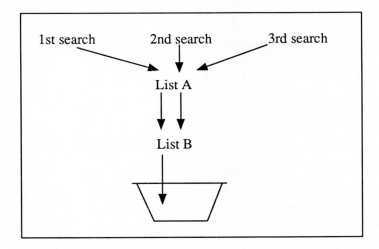

Figure 2/92. Search hits Lists

The contents of List B cannot be displayed or saved in a file. It can be manipulated to create a new List A. In these manipulations, the content of List B is preserved.
The numbers for List A in Figure 2/91 show that we are viewing the first retrieved record out of 15 hits. The previous search produced 50 hits (List B).

EXACT STRUCTURE SEARCHING

We will use the TUTORIAL database which comes with the ChemBase package and contains 100 molecules (9 structures have been added by us to demonstrate the searches in the following discussion) and 100 reactions.

To search for an exact structure, we have to construct the query structure identical to the structure we wish to retrieve. In building the query structure, we proceed as usual in the Molecule Editor. We cannot use the following specifications in building a query for exact search:
1. ATOM(Other(Special))
2. QUERY ATOM
3. QUERY BOND

Stereobonds can be indicated in the query structure; however, ChemBase 1.2 disregards the stereobonds in searching and, consequently, all stereoisomers will be retrieved. However, stereochemical structure queries can be used in ChemBase 1.3 or for transfer into MACCS-II or REACCS. Stereochemical structure queries can be built in all of these systems and stored in Molfiles for transfer between the systems.

Implicit hydrogens (which can be displayed by the F9 switches) (see p. 298) have the same effect in Exact Search as the explicit hydrogens in Substructure Search, i.e., no substitution is allowed.

When we complete a query structure in the Molecule Editor, we exit the Editor by point/click DONE. We are switched to the Main Menu with the query as the current structure displayed on the Molecule Form.

INITIATING EXACT STRUCTURE SEARCH

Exact Structure Search is initiated by dragging RETRIEVE(Current) on the menubar (Figure 2/93).

The System starts the search and when the first hit has been located, the structure is displayed automatically in the Form. The numeric and textual data associated with the structure of the data fields which have been linked to the boxes in the Form are displayed in those boxes (Figure 2/94).

To browse through the hits in List A, use the up/down arrow keys, Ctrl-Home, Ctrl-End. The second and third hit will be displayed (Figures 2/95, 2/96).

Note that all stereoisomers have been retrieved with the query structure in ChemBase 1.2 databases.

To see the hits in List A in Table View, drag VIEW(Table). The retrieved records are shown in Figure 2/97.

In ChemBase 1.3 databases, we can selectively search for the individual isomers. Thus, marking the asymmetric center in the query in Figure 2/93 by QUERY ATOM(Stereo) (or the stereobond by QUERY BOND(Stereo) and labeling the query CHIRAL in the Molecule Editor will result in retrieving only the isomer in Figure 2/96. Similarly, drawing the query with the stereobond DOWN, marking the asymmetric center, and labeling the

MOLECULE SEARCHING

Example: *Exact structure searching*
                                                Command: RETRIEVE(Current)

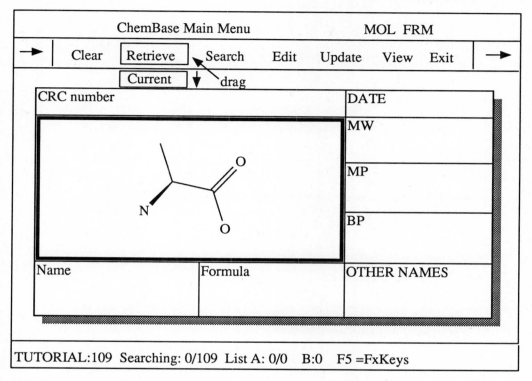

Figure 2/93. Exact Structure Search

query CHIRAL will selectively retrieve the isomer in Figure 2/94.
See the discussion on stereochemical searching in MACCS-II and REACCS for further details.

PRINTING RETRIEVED DATA

If we want to make a hard copy of the retrieved structures and data, we can print the Form with the entries by activating the PRINT option on the Menubar.

PRINT(Entry) - prints the currently displayed record.
PRINT(List) - prints continuously the entire List A.

Figure 2/98 shows the print of the first record.
In Table View, we can print one entry (that one which is highlighted) by PRINT(Entry):

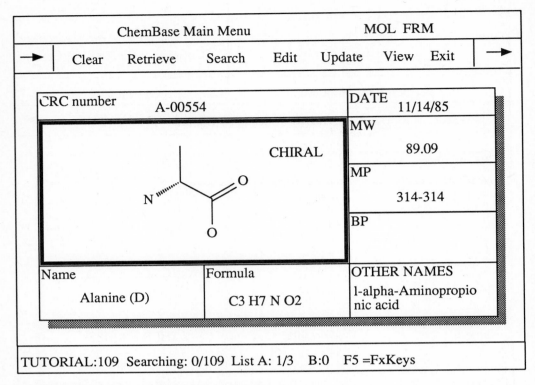

Figure 2/94. Display of Exact Search hit

3/17/88	TUTORIAL.DB Table Data	page 1
CRC Number	Name	Formula
A-00554	Alanine (D)C3	H7 N O2

or the entire Table by PRINT(List):

3/17/88	TUTORIAL.DB Table Data	page 1
CRC Number	Name	Formula
A-00554	Alanine (D)	C3 H7 N O2
A-00555	Alanine (DL)	C3 H7 N O2
A-00556	Alanine (L)	C3 H7 N O2

To print structures in Tables, we assign the *structure field to a column. The column will

MOLECULE SEARCHING

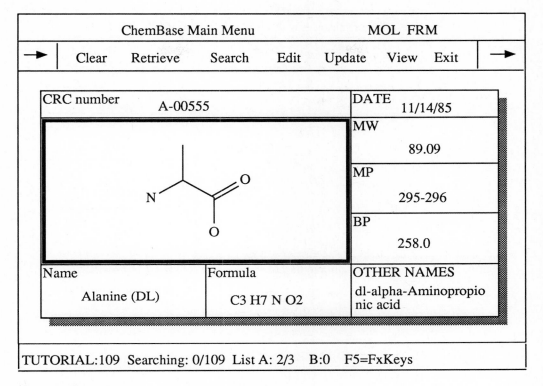

Figure 2/95. Display of Exact Search hit

show YES or NO to indicate whether the retrieved record contains a structure. The structure is not displayed on the screen; however, it is printed when we use one of the commands discussed above.

In ChemBase 1.3 we can set margin widths for printing Forms and Tables. It is done by editing the CPSS.CFG file which has entries for left, right, top, and bottom margins of Forms and Tables.

SAVING RETRIEVED DATA

We can save:
1. structure only in a .MOL file
2. data only in the column format in a .ASC file
3. structure and data with the Form in a .FRM file
4. List A in a .LST file
5. structures and data in a .SDF or .RDF file

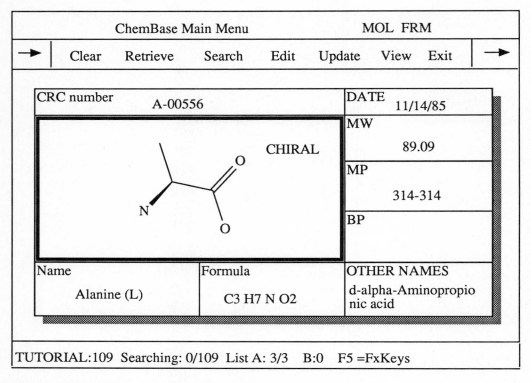

Figure 2/96. Display of Exact Search hit

SAVE(Molfile) - saves the current retrieved structure
SAVE(Form) - saves the current retrieved structure and data including the Form
SAVE(List) - saves the IDs of all hits in List A
SAVE(Data File) - saves data in column (Table) format excluding structures
SAVE(SDfile) - saves multiple structures and data
SAVE(RDfile) - saves multiple structures and data

RECALLING SAVED DATA

The saved data can be recalled by activating the USE option.

Molfile can be recalled in the Main Menu or in the Molecule Editor:
USE(Molfile) - calls up the Clipboard; click the file to recall; the structure will be displayed.

Form File should be recalled only in the Form Editor to display the data, including structure, as they were saved:
USE(Form) - calls up the Clipboard; click the file to recall; the Form including structure

MOLECULE SEARCHING

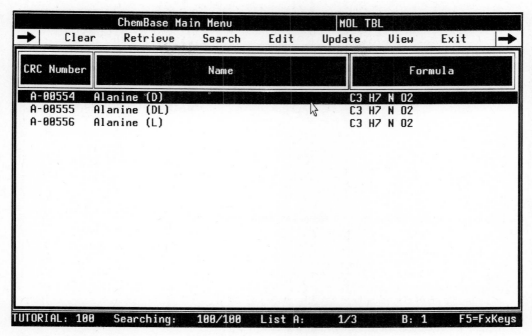

Figure 2/97. Table view of search hits

and data will be displayed

List file can be recalled in the Main Menu either in Form View or Table view:
USE(List) - calls up the Clipboard; click the file to recall; in Form View, the first record will be displayed. In Table view, all records will be displayed.
Note that to recall a List File, we have to be in the same database in which it was created since it essentially contains the hits from List A.

Data File can be typed in DOS or can be used in ChemText or other systems which utilize ASCII files.
In ChemBase, Data Files are used only for registration of new entries in a database.

SDfiles and RDfiles are recalled in the Main Menu by:
USE(SDfile)
USE(RDfile)

SDfiles and RDfiles are used to register entries into a ChemBase database.

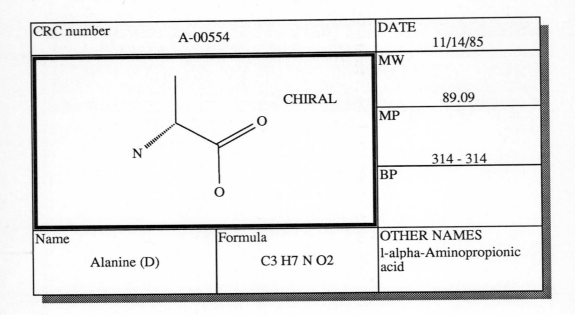

Figure 2/98. Printed record of Exact Search

SUBSTRUCTURE SEARCHING

Although the Exact Structure Search discussed above will in many cases meet our needs, the most potent capability of ChemBase is searching for substructures.

The concept of substructure searching can be well illustrated with a Markush structure.

Markush structures are widely used in chemical literature to structurally describe many chemical substances by one structure. An example of the Markush structure for steroids is shown in Figure 2/99.

The Markush structure is composed of a **structural fragment** and alphanumeric description to indicate which other **structural moieties** may be in the structural fragment. The Markush structure, in effect, represents all the substances that **must have** the structural fragment in their structures and **may have** the moieties at the positions, indicated by the boxes in Figure 2/99, described by the alphanumeric specification.

However, while a Markush structure limits the moieties to those specified by the alphanumeric description, a Substructure Search does not have any limits and will permit all moieties that can be attached or embedded in the structural fragment at the positions where the attachments can be (e.g., at the arrows in Figure 2/99), unless we prevent the substitution by explicit hydrogens.

Hence, the premise of Substructure Searching is:

the structural fragment must be, exactly as it is specified in the query, in the retrieved structures. The other moieties may be or may not be attached at any position where attachment is possible while the valences of the atoms involved in the attachments

MOLECULE SEARCHING

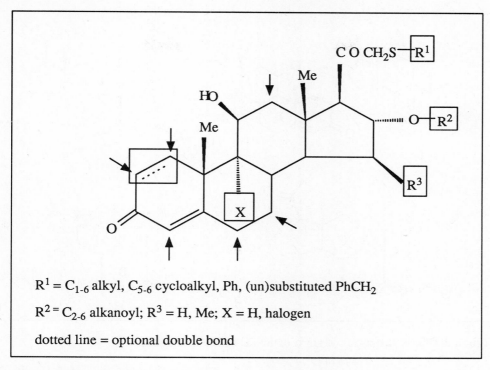

$R^1 = C_{1-6}$ alkyl, C_{5-6} cycloalkyl, Ph, (un)substituted $PhCH_2$

$R^2 = C_{2-6}$ alkanoyl; R^3 = H, Me; X = H, halogen

dotted line = optional double bond

Figure 2/99. Markush structure

are met.

To perform a Substructure Search, we have to build a generic structure query. We have learned how to do it in the Molecule Editor. When we have completed the query, we exit from the Molecule Editor by point/click DONE. We are returned to the Main Menu with the query displayed on the Form (it is now the current structure).

INITIATING SUBSTRUCTURE SEARCH

To search for substructures, we drag the Menubar option SEARCH(SSS/RSS). Since we are in the Molecule View, the SSS option is valid here (the RSS is valid for Reaction Substructure Search if we are in the Reaction View).
The System starts the search and produces the same statistics on the Alternate Status Line as in the Exact Structure Search (see p. 372).
As an example of substructure searching, let us search for the steroids shown in Figure 2/100. These are the compounds registered in the TUTORIAL database. We could retrieve them by the Exact Structure Search; however, we would have to know the exact structures as they are shown in Figure 2/100. We would have to draw each structure and individually search for it by the RETRIEVE(Current) Command.

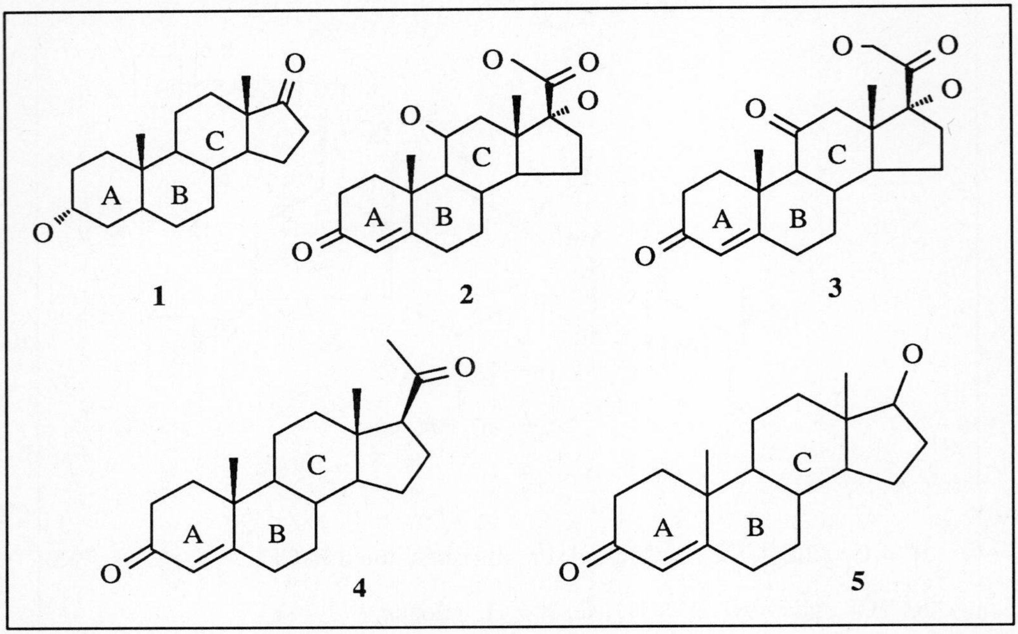

Figure 2/100. Structures of steroids in TUTORIAL database

But, suppose we do not know the exact structures or we do not know if any steroids are registered in the database. We perform a Substructure Search to avoid building the exact structures, one by one, and to find out whether there are any steroids in the database.

Query 1

The obvious structure query would be the tetracyclic steroidal system shown in Figure 2/101.

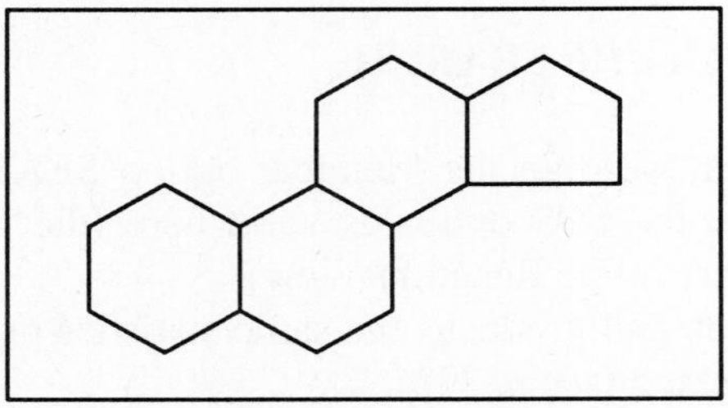

Figure 2/101. Query 1

The result of SEARCH(SSS) of Query 1 is shown in Figure 2/102.

MOLECULE SEARCHING

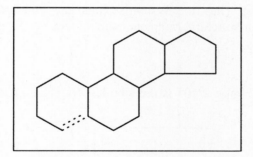

Figure 2/102. Hit for Query 1

We did retrieve a steroid which is the structure **1** shown in Figure 2/100, one of the registered substances. However, we have not retrieved the four other structures that are in the database. Why?
Because the structures **2-5** in Figure 2/100 have a double bond in the 4-5 position, and our Query 1 has a single bond in that position. As we have said above, the structure query must be an exact match with the structural fragment in the retrieved structures.

Query 2

This query shown in Figure 2/103 has been built with a specification of QUERY BOND(S/D) in the 4-5 position. We can build Query 2 from Query 1.

Figure 2/103. Query 2

The structure query of the last search can be redisplayed at any time by dragging RETRIEVE(Query) as shown in the following Example (Figure 2/104).
The structure is then transferred to the Molecule Editor by point/drag EDIT(Molecule) and modified to Query 2.

On searching Query 2 by the SEARCH(SSS) option, we retrieve the structures **1-5** shown in Figure 2/100. Why?

Example: *Query redisplay*

Command: RETRIEVE(Query)

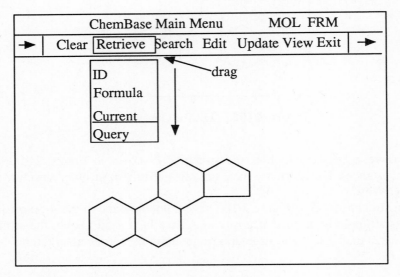

Figure 2/104. Redisplay of structure query

Because Query 2 permits the structural fragment to have **either a single bond or a double bond** in the 4-5 position. Thus, the System selected the structures that have both single and double bonds in their structural fragment matching the query.

Query 3

The query 3 shown in Figure 2/105 is the simplest fragment contained in steroids. If we

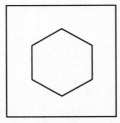

Figure 2/105. Query 3

submit Query 3 to SSS, we retrieve the structures **1-5** shown in Figure 2/100 in addition to the structures **6-14** shown in Figure 2/106.

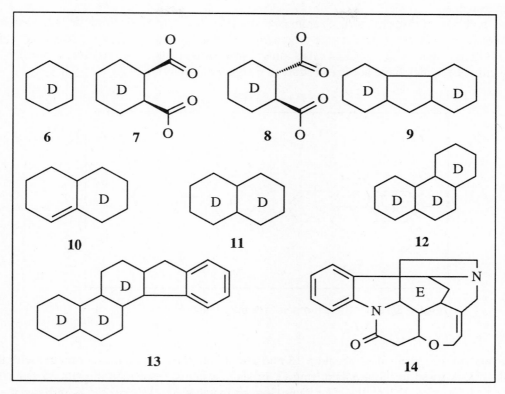

Figure 2/106. Structures retrieved with Query 3

(Note that structures 9-13 are not in the TUTORIAL database. They have been added to the database to illustrate our discussion.)
Query 3 retrieved the structure **1** in Figure 2/100 as we have expected since the query is the exact structural fragment, i.e., the ring A in structure **1**. However, we have retrieved also the structures **2-5** although the Query 3 does not allow for the 4-5 double bond. Why? Because the query structure is also the exact fragment, i.e., the ring B and/or C in the structures **2-5**.
The structure **6** has been retrieved because it is the fragment exactly matching the query and it happens to be, by itself, a full registered structure of a substance. The same structure, of course, would have been retrieved if we used Query 3 for Exact Structure Search by RETRIEVE(Current) (this search would not have retrieved the structures **1-5** and **7-14**).
The structures **7-13** have been retrieved because the query is the exact match with either of the fragments D in these structures. The substituents in the structures **7** and **8** comply

with the premise of the substructure search - any kind of substitution is allowed unless we prevent it by explicit hydrogens in the query and as long as it meets the valency requirements.

The potency of Substructure Search is shown by the retrieved structure **14**, a rather complex structure in which the fragment E matches the Query 3. However, the TUTORIAL database contains also other structures (Figure 2/107) very similar to structure **14**, which have not been retrieved by Query 3. Why?

Figure 2/107. Structures not retrieved with Query 3

Because the fragment F in structure **15** and **16**, although by itself an exact match with the Query 3, is fused to the aromatic ring G. In such systems, resonance imparts an aromatic bond in the ring F and the ring F no longer matches the Query 3. To retrieve these structures, and other similar to these, we have to modify the Query 3 to allow for the aromatic bond in fragment F.

This resonance effect also explains why we have not retrieved the structure **13** using the Query 1. The fusion of the benzene ring imparts an aromatic bond in the cyclopentane ring in Query 1. To retrieve structure **13**, we would have to modify Query 1 and specify the cyclopentane bond using QUERY BOND(Aromatic) or QUERY BOND(Any).

ISOMERISM IN SSS

Stereoisomers, as we have learned, are not recognized in ChemBase 1.2 searches; consequently, if we searched for structures with chiral centers and used a query that has stereobonds, we would have retrieved all stereoisomers as well as structures with no stereobonds as long as there was a fragment match with the query. ChemBase 1.3 allows for searching with stereochemical queries.

Geometric isomers are also not recognized in ChemBase 1.2 searches, so that if we searched for *cis*-cinnamic acid which is in the TUTORIAL database with Query 4 shown in Figure 2/108. we would retrieve both *cis*- and *trans*-cinnamic acid which are registered

MOLECULE SEARCHING

in the database (Figure 2/109). ChemBase 1.3 allows for selective stereoisomer searching.

Query 4

Figure 2/108. Query 4

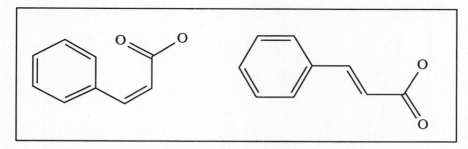

Figure 2/109. Structures retrieved with Query 4

However, with Query 4 we would have retrieved a large number of structures in which Query 4 is embedded in ring systems. To selectively retrieve only structures in which the fragment is in chains, we have to modify Query 4 to indicate that we want the bonds to be Chain Type Bonds by activating the QUERY BOND(Chain) option (Figure 2/45)(see also p. 138 for discussion on Bond Types).

Query 4 modified

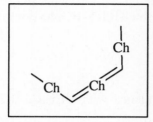

Figure 2/110. Query 4 modified

Tautomerism may affect the result of Exact Structure Search and Substructure Search. Consequently, we have to carefully consider whether structures we want to retrieve may be tautomeric and formulate our query accordingly. (See also p. 97 for discussion of Tautomerism.)

The TUTORIAL database contains the structures **17-20** shown in Figure 2/111. Structure **21** has been added to the database to illustrate our discussion.

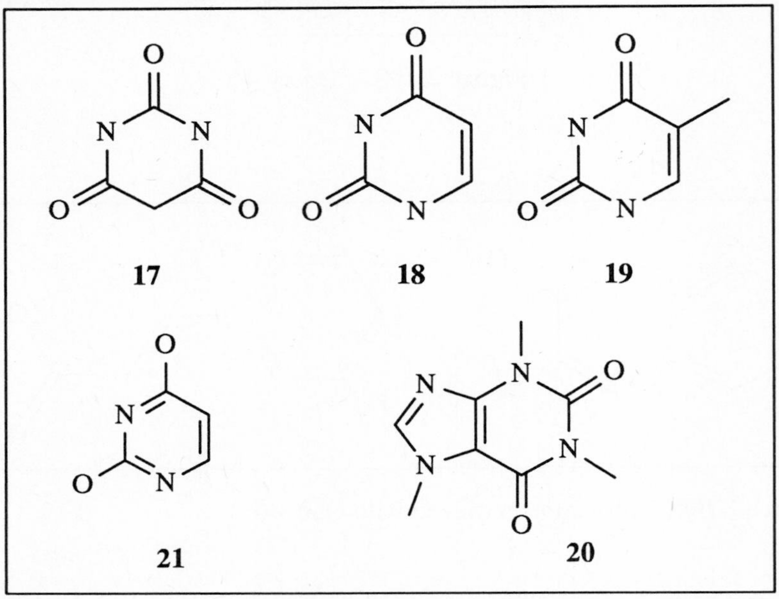

Figure 2/111. Tautomeric structures 18, 21

To retrieve the tautomers **18** and **21** we have to evaluate the structures by the bonds present in the structures. Structure **18** has an exact double bond in the ring. Structure **21** has also double bonds in the ring, but these are not exact double bonds but alternating single and double bonds in a 6-membered ring. ChemBase treats these as aromatic bonds. Therefore, we cannot use in our query the QUERY BOND(S/D) specification because we would retrieve only structure **18**. If we use QUERY BOND(Aromatic), we would retrieve structure **21** but not structure **18**.

Query 5

Query 5 shown in Figure 2/112 is formulated with QUERY BOND(Any) and retrieves structures **17-21**.

If we wanted greater selectivity of the SSS to eliminate structure **17** and **20**, we would prevent fusion and substitution in the 6 position by an explicit hydrogen to specify that there must be a hydrogen attachment at that position.

MOLECULE SEARCHING   389

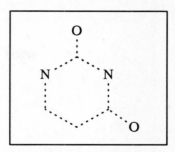

Figure 2/112. Query 5

Query 5 modified

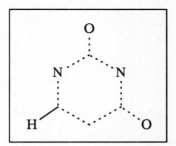

Figure 2/113. Query 5 modified

The query (Figure 2/113) retrieves only structures **18, 19,** and **21**.
Further selectivity could be achieved by an explicit hydrogen at position 5. Only the tautomers **18** and **21** would be retrieved.

VARIABLES IN SSS

ChemBase allows for searching generic structures which are, in fact, the Markush structures we have discussed. Generic structures are built as usual in the Molecule Editor using the options QUERY ATOM and QUERY BOND to specify the variables (see also Part One, p. 118; Part Three, p. 465, and Part Five, p. 773 for more comprehensive discussion of generic structures).
The TUTORIAL database contains the structures **22-28** shown in Figure 2/114. Structure **29** has been added to the database to illustrate our discussion.
To retrieve all structures that have these fragments we build Query 6 (Figure 2/115).

Query 6

To retrieve both 2,5 unsubstituted and substituted structures and to prevent fusion at the

Figure 2/114. Structures in TUTORIAL database

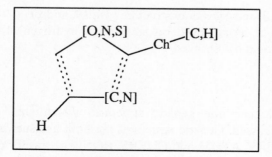

Figure 2/115. Query 6

2,3 and 4,5 positions, the query has been formulated with a chain bond and C,H variables in position 2 and with an explicit hydrogen in position 4. Thus, the position 2 can be substituted either by hydrogen or an alkyl/aryl attached by a chain type bond and position 4 is not available for fusion since there must be a hydrogen in the retrieved structures.

USING LISTS IN SSS

We have used the TUTORIAL database which now has 109 registered structures. It is not a large database and in our SSS we have retrieved up to 22 structures. We can easily browse through the answers in Form View or Table View in a short time. However, later in our use of ChemBase we may have databases with thousands of records. In these databases, the SSS would take considerable time to completion and we would retrieve a large number of answers. Again, browsing through the answers would be time consuming. In addition, there will be many searches when we successfully perform the first search and upon reviewing a few answers we want to further refine the query to achieve higher selectivity of the search. We do not have to carry out the new search over the entire database. We can create a subset of the database and perform the search in that subset. On another occasions, we can conduct a very approximate data search, e.g., a search for a certain elemental composition or certain biological properties, and use the retrieved records as a subset of the database for SSS or Exact Structure Search.
We, in fact, create a new search domain.

Example: *Setting search domain*
Command: LIST(Set Domain)

```
TUTORIAL: 109    Searching: 109/109  List A: 1/14  B: 0  F5=FxKeys
```

Figure 2/116

Figure 2/116 shows the result of our search over the entire database, which is the default search domain, with Query 3 that was broadly defined. List A has 14 hits and we want to refine the query and perform another, more restrictive and selective search to retrieve only the steroids.
We modify Query 3 to make it more specific. We create Query 2.
We make List A the new search domain. Point/drag LIST(Set Doamain) in the Main Menu. Figure 2/117 shows the changed status line.

```
TUTORIAL: 109    Searching: 0/14  List A: 1/14  B: 0  F5=FxKeys
```

Figure 2/117

Now, we have a search domain which is the 14 records of the first search. We have Query

2 as the current structure. We point/drag SEARCH(SSS/RSS) to initiate the new search. Figure 2/118 shows the changed status line.

```
TUTORIAL: 109    Searching: 14/14  List A: 1/5  B: 14  F5=FxKeys
```

Figure 2/118

The new search produced a new List A with 5 hits which are the structures **1-5**. List B now contains the original 14 hits.
If we wanted to use the 14 hits in List B to display all the structures we switch List B for List A. Figure 2/119 shows the new status line.

Example: *Switching lists A,B*
                                                        Command: LIST(Switch)

```
TUTORIAL: 109    Searching: 14/14  List A: 1/14  B: 5  F5=FxKeys
```

Figure 2/119

If we want to reset the List A search domain back to the default, i.e., the entire database search domain, we use CLEAR(Domain) option in the Main Menu (Figure 2/120)

Example: *Resetting search domain*
                                                        Command: CLEAR(Domain)

```
TUTORIAL: 109    Searching: 0/109  List A: 1/14  B: 5  F5=FxKeys
```

Figure 2/120

The alternate status line shows search domain 109. List A and B remain.

MORE ON USING LISTS

LIST(Merge) - creates new List A with all records that are either in List A or List B (removes duplicate records). List B remains unchanged.

LIST(Intersect) - creates new List A with these records present both in List A and List B. List B remains.

LIST(Subtract) - creates new List A with records that are in List B but not in List A (B-A). List B remains. To do A-B, switch List A for List B.

LIST(Sort Ascending) - sorts records in List A by selected data field.

LIST(Sort Descending) - sorts records in List A by selected data field.

CLEAR(List) - clears List A. The records are lost. Clears the current record on display in Form. It is not necessary to clear List A before starting a new search. The System clears the List A and places the records in List B.

SAVE(List) - saves records (IDs) in List A in a .LST file.

12 REACTION SEARCHING

Reaction Searching, like Molecule Searching discussed in Section 11, is the major function of ChemBase. Again, we can search reaction databases that we ourselves have created from internal or external data, by data transfer from REACCS or using data retrieved in CAS ONLINE, or we can purchase reaction databases designed for ChemBase and ready to search, e.g., ISI products.
Searching for reactions can be performed with graphic structure queries or with queries composed of numeric and textual data.
Alphanumeric queries are more important in reaction searching than in molecular structure searching since with reactions we can register a record of reaction conditions, e.g., reagents, solvents, catalysts, temperatures, etc., and yields. These data are, at times, of greater interest than the structures involved in the reactions.
In addition to searching for structures involved in reactions, ChemBase offers a unique capability to search for **reaction transforms**, i.e., for reacting centers of a molecule or molecules.
We can classify reaction searching as follows:
1. exact reaction searching
2. substructure reaction searching
3. reaction center searching

Both molecule searching and reaction searching is based on structural queries. However, there is a marked difference in how we treat structures in molecule and reaction searches.
In molecule searching, there is no chemical role of the structure we are searching for. In reaction searching, the structures have a distinct chemical role, i.e., they are reactants, intermediates, or products of the reaction we are searching for.
Graphic structure queries for searching reactions are built and edited in the Molecule Editor and the Reaction Editor (see p....).
The search operations are carried out in the Main Menu in Reaction View.
Reaction search produces similar statistics as the molecular structure search. We already know what is search domain and Lists and how to manipulate them.
We will use the TUTORIAL database which contains 100 reactions for reaction searching described below.

EXACT REACTION SEARCHING

To search for an exact reaction, we have to construct the reaction exactly as it has been registered in the database. The structures in the reaction must be exact structures, i.e., no variables. Stereobonds can be drawn; however, they are disregarded in ChemBase 1.2 searching and have meaning only if we want to use the query in REACCS or ChemBase 1.3. The structures must be in their proper role position in the scheme.

CHEMBASE

As an example, we will perform an exact search for the reaction shown in Figure 2/121.

Example: *Exact reaction search*

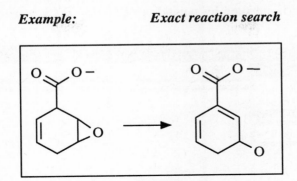

Figure 2/121.   Query 7

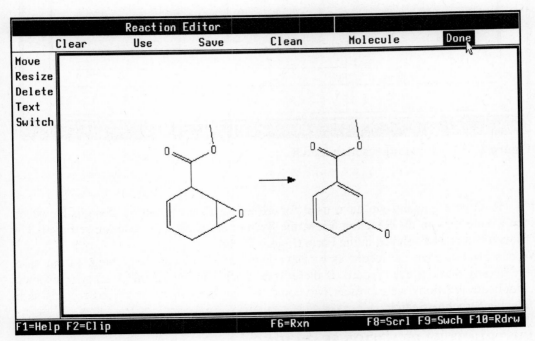

Figure 2/122.   Exiting with reaction query to Main Menu

When we complete the reaction Query 7, we review the query in the Reaction Editor and exit to the Main Menu by point/click DONE on the Menu (Figure 2/122). We enter the Main Menu in Reaction View with the current reaction on display in the Reaction Form.

INITIATING EXACT REACTION SEARCH

Exact reaction search is initiated by point/drag RETRIEVE(Current) on the Main Menu (Figure 2/123).

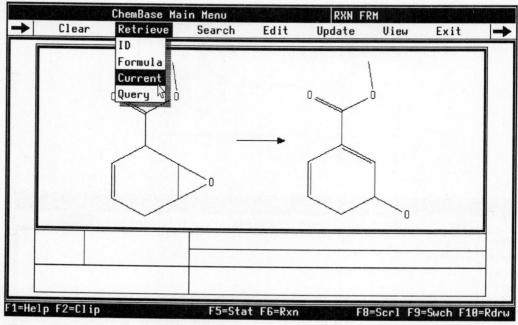

Figure 2/123. Initiating reaction search

It is the same command we have used for Molecule Exact Structure Search. However, since we are now in the Reaction View, the System proceeds with reaction retrieval. The retrieved record is displayed in the Form (Figure 2/124).
We can print or save the record as we learned in Molecule Searching. In our case, there is only one retrieved record. If there were more than one record, e.g., stereoisomers or duplicate reactions were present, we could browse through the retrieved records as in Molecule Structure Search.

SUBSTRUCTURE REACTION SEARCHING

Similarly to Molecule Substructure Search (SSS)(see p. 380), Substructure Reaction Search (RSS) retrieves the reactions in which the query structure matches exactly the structural fragment in the reaction structure participant to which the query structure has been linked by the reaction label, i.e., reactant, intermediate, or product. In constructing queries for RSS we can utilize all the specifications for generic structures (see p. 314).
RSS is employed mostly in finding whether a class of substances participates in a reaction,

REACTION SEARCHING

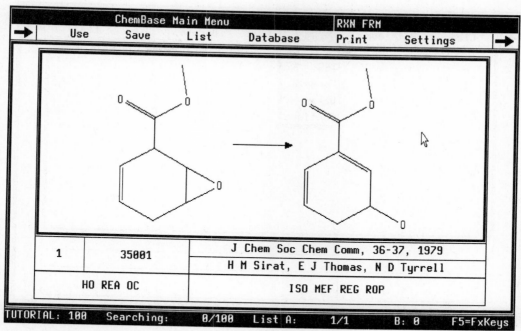

Figure 2/124. Retrieved reaction and data display

disregarding the type of reaction that is involved. (See "Reaction Transform Searching," p. 398 to search for specific reaction types.)

For example, to find out whether there are other epoxides than the one in Query 7 as reactants in any reaction, we formulate Query 8 shown in Figure 2/125.

INITIATING REACTION SUBSTRUCTURE SEARCH

The reaction substructure search is activated by point/drag SEARCH(RSS) in the Main Menu with the current query in the Reaction View (Figure 2/125).

Query 8 retrieves the same reaction as in Figure 2/124, in addition to three more reactions in which an epoxide is the reactant (Figure 2/126).

The reverse approach, i.e., using only a structure labeled as product in RSS could also produce fruitful results in certain cases.

For example, in our case of searching for epoxides, we might want to find all reactions that produce epoxides. We would formulate Query 9 shown in Figure 2/127.

Query 9

This query would retrieve reactions producing epoxides shown in Figure 2/128.

However, we have to be prepared that for common classes of compounds there may be a

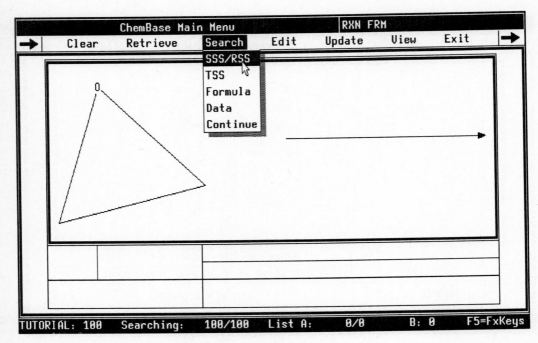

Figure 2/125. Initiating reaction substructure search

large number of hits if the database is large.
In our search with Query 8, we see that the reaction **2** in Figure 2/126 does not produce hydroxy compound. If we wanted to restrict the search for opening of epoxides to produce alcohols, we would formulate Query 10 (Figure 2/129).

Query 10

Query 10 would retrieve reaction **1**, **3** (Figure 2/126) and the same reaction (Figure 2/124) as Query 7.

REACTION TRANSFORM SEARCHING

Reaction transform concept is well described by using a functional group as an example of transformation.
The TUTORIAL database contains the reactions shown in Figure 2/130.
The structures in reaction **4-7** are very dissimilar except for the fragments in each of the structures which are highlighted in Figure 2/131.
We see that as far as the fragment/functional group is concerned, all the reactants belong to the compound class of enamines and all the products to the compound class of amines. Upon further examination of the structures, we also see that the only structural change

REACTION SEARCHING

Figure 2/126. Reactions retrieved with Query 8

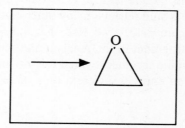

Figure 2/127. Query 9

caused by the reaction is the **transformation of the enamino group to the amino group**. The SEARCH(TSS) option in the Main Menu/Reaction View makes it possible to retrieve the reactions **4-7** by using Query 11 shown in Figure 2/132.
It is true that Query 11 is composed of functional groups that, at the same time, are

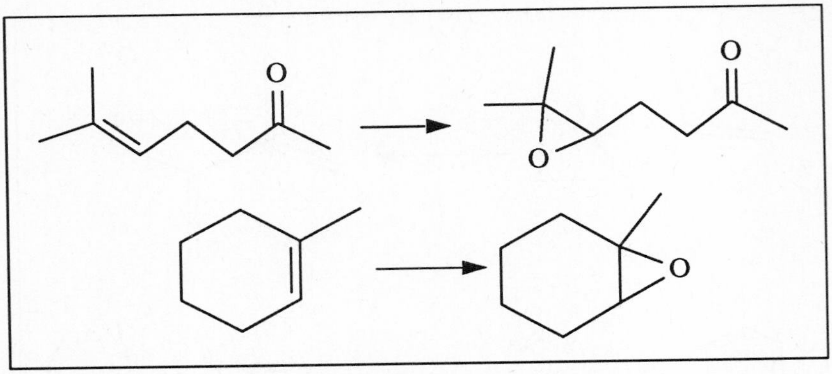

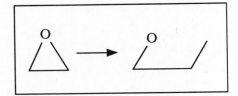

Figure 2/128. Reactions retrieved with Query 9

Figure 2/129. Query 10

structural fragments of the structures in reactions **4-7**. Then why don't we do SEARCH(RSS)?

We can use Query 11 in SEARCH(RSS). We do retrieve reactions **4-7**; however, we also retrieve the reaction shown in Figure 2/133 which is registered in the TUTORIAL database. In large databases we would retrieve many other reactions.

Examining the reaction in Figure 2/133, we find that although the structures have the enamine and amine structural fragment in the reactant and the product, the transformation has not taken place in the enamine fragment which remains unchanged. Rather the amine fragment has been formed by an entirely different type of reaction in which we have not been interested.

WHEN TO DO RSS

We do RSS when we want to find out whether:

1. a substance has been used as reactant in a reaction
2. a substance has been an intermediate in a reaction
3. a substance has been a product of a reaction and we are not concerned about the nature of the reaction in which the substances are involved.

REACTION SEARCHING

Figure 2/130. Reactions in TUTORIAL database

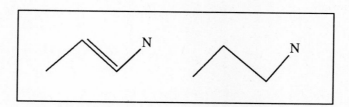

Figure 2/131. Fragments in structures

We are looking for a class of substances participating in a reaction.

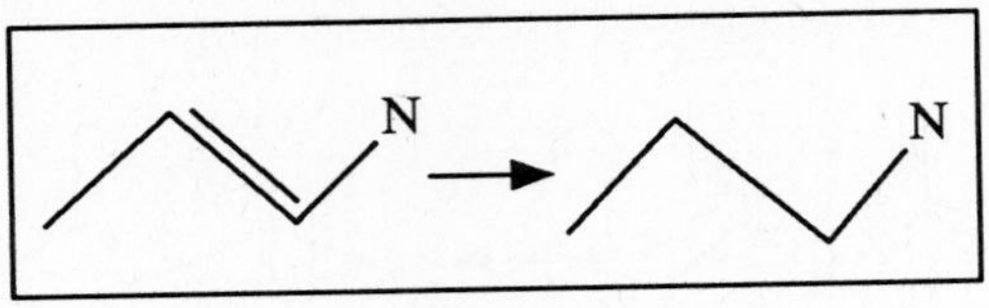

Figure 2/132. Query 11

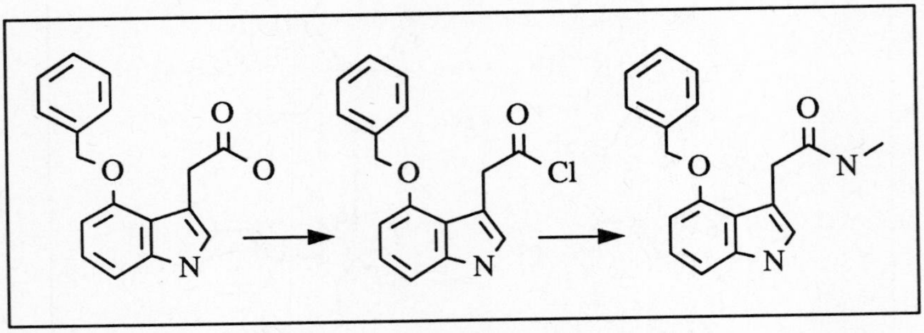

Figure 2/133. Reaction retrieved with Query 11

WHEN TO DO TSS

We do TSS when we want to find out:

whether a specific fragment in a reactant has been transformed into another specific fragment in an intermediate/product while the exact structure of the reactant and intermediate/product is immaterial

We are looking for a type of reaction while the class of compounds involved in the reaction is immaterial.

13 DATA SEARCHING

ChemBase is a truly comprehensive data management system because it handles not only the commonly used and required data by chemists, i.e., graphic chemical structures, but also alphanumeric data which may be any description related to the molecules or reactions in ChemBase databases be it chemical, physical, or biological. We can even define databases that do not use the *structure field but only alphanumeric data fields. We can store the data and search for the data.

In databases with both structures and alphanumeric data, they are tied with a common record ID number. If we design separate databases for structures and alphanumeric data, we can have the same record ID in many databases and link the data through the IDs, search for the records in one database, save the IDs in List Files, and transfer them to the related database to display the data records with the same IDs.

We have learned how to input alphanumeric data in a database (see p. 361). In this section, we discuss alphanumeric data searching.

Data Search is performed in the Main Menu and, as in molecule or reaction searching, we have two options of how to conduct a data search: RETRIEVE and SEARCH.

In **Molecule View**, the operations are performed over the current molecule database records; in **Reaction View**, the operations are performed over the reaction database records.

DATA RETRIEVAL

In RETRIEVE operation, we search for exact data.
We can retrieve:
1. ID number(s) of molecules or reactions
2. formulas of registered molecules in a database

We know that the ID data field is required as the first data field in any database we define. Any new entry in the database must have an ID to be registered in the database.

Formulas are automatically calculated by the System for the *formula special keyword data field. We can, of course, create our own data field and store any formulas we calculate and wish to register. These, however, would be searchable only by the SEARCH option (*vide infra*).

RETRIEVING ID

In retrieving ID, we have to specify the ID of the record we wish to retrieve exactly as it is in the database. If the ID number has embedded spaces, hyphens, slashes, etc. they have to be entered in the ID retrieve specification. If the ID is a text string, the whole string must be specified. We can retrieve in one RETRIEVE operation:

1. single ID
2. multiple IDs

Example:

Retrieving ID
Command: RETRIEVE(ID)

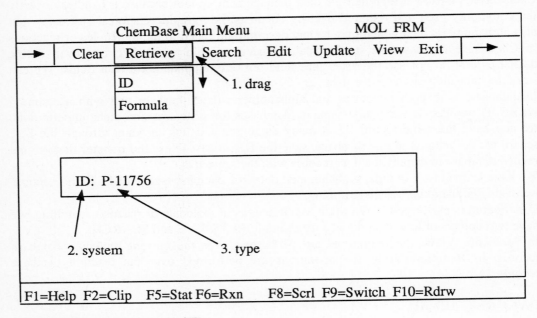

Figure 2/134. Retrieval of ID

Figure 2/134 shows retrieval of the structures in Figure 2/100 that we have retrieved with Query 2. They have the following IDs:
A-00753, C-05245, C-05246, P-11756
To retrieve structure **4**, enter the ID in the System prompt box.
The printed retrieved record is shown in Figure 2/135.
To retrieve multiple structures in Figure 2/100, we would enter:
A-00753, "C-05245"-"C-05246", P-11756
Note that the quotation marks are necessary in a range of IDs when there is a dash within the ID number.
Similarly, we would retrieve ID(s) of reactions in the Reaction View.
To retrieve all records in the database, we would input "all" in the prompt box.

RETRIEVING FORMULA

We recall that formulas are automatically calculated by the system for the special field *formula and that if we wish, we can define a regular data field for formulas we calculate.

DATA SEARCHING

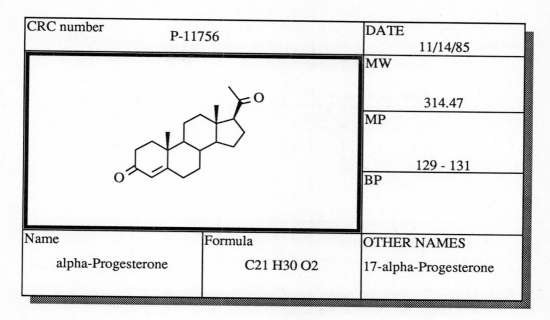

Figure 2/135. Printed record of ID retrieval

The formulas in the *formula field are retrieved by RETRIEVE(Formula) in Molecule View only. In Reaction View, the option is not valid. To retrieve the formulas we calculate, we have to search the defined data field.

By retrieving formulas, we actually mean to retrieve the structure and the associated data in the record. A record is retrieved by specifying a single exact formula of the structure we wish to retrieve. Ranges or lists of formulas are not allowed in RETRIEVE(Formula). Only the elemental formula composite format is allowed for RETRIEVE. Line structural formulas, e.g., C2H5OH are not allowed.

On retrieving the formula in the TUTORIAL database (Figure 2/136), we would retrieve the record shown in Figure 2/135.

However, in databases where multiple structures of the same formula are registered, we would retrieve all the structures.

This technique may be used to create a subset of a database (search domain) followed by a RETRIEVE(Current) or SEARCH (RSS/SSS).

DATA SEARCH

In contrast to DATA RETRIEVE, in DATA SEARCH operations we can search generically for formulas and for any data in the data fields that are in the current database.
In Data Search we can specify any numeric or text string in our data queries contrary to the RETRIEVE operation where we are limited to the exact format of the data query.

Example:

Retrieving formula
Command: RETRIEVE(Formula)

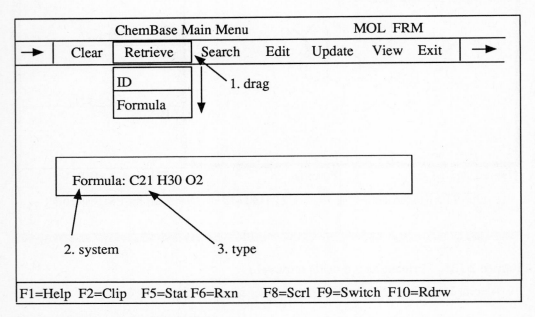

Figure 2/136. Retrieving formula

SEARCHING FORMULA

SEARCH(Formula) is valid, similarly to RETRIEVE(Formula), only in the Molecule View.
We can search for any element count that may be embedded in a formula. We can negate an element or elements in the searched formula.
The formula string can be specified as:
1. exact count of element)s)
2. a range of element(s)
3. zero element(s) (negation)
The query in Figure 2/137 would retrieve structure **2, 3, 4, 13,** and **14** in Figure 2/100 and Figure 2/106.
If we formulated the query:
C(19-21) H(28-30)
we would retrieve structures **1-5** in Figure 2/100, and **13** in Figure 2/106.
In negated queries, we exclude an element by 0(zero) suffix to the element to be negated:
C21 H28 N0
would retrieve all structures with embedded C21 H28 count. None of the structures would have nitrogen atom(s).

Example:

Searching formulas
Command: SEARCH(Formula)

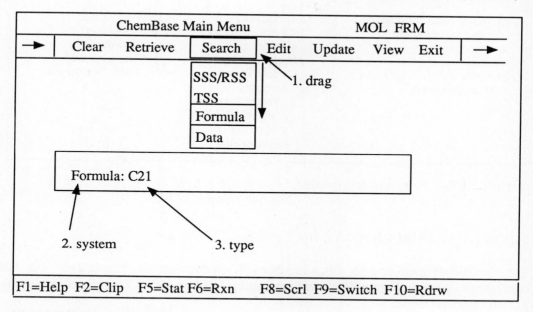

Figure 2/137. Search of formula

SEARCHING DATA

In searching for data, we formulate data queries consisting of:
1. data field name
2. relational operator symbols
3. numeric or text string of data

The queries have the format shown in Figure 2/138.

SEARCHING COMPLEX QUERIES

The queries for data searching can be simple queries or we can formulate complex queries with the use of the Boolean Operators AND, OR, NOT (see p. 247). For unambiguous meaning of the complex queries, use spaces between the Boolean operators and parentheses.

field	relation	value
can be any data field name in the database except: *structure *formula The fields *mol.weight and *stext are allowed	= equal to <> not equal to < less than > greater than <= less than or equal to > greater than or equal to : text string is embedded	can be any data in the database provided the data types in query and data field correspond

Figure 2/138. Format of data queries

SEARCHING NUMERIC DATA

The = relational operator used with a number will retrieve the value rounded to -+0.5. For example the query in Figure 2/139 will retrieve bp 149.5 to 150.5. This rounding also applies for decimal digits.

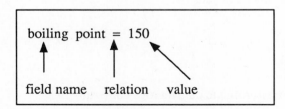

Figure 2/139. Data query

The <, >, <=, and >= operators will retrieve numeric data exactly of the specified minimum or maximum value.
Specifying a range by a hyphen between two numbers is not allowed. We have to use the AND operator to retrieve numeric data between minimum and maximum with a query of the format shown in Figure 2/140.
Molecular weights can be searched for by specifying a query shown in Figure 2/141.

SEARCHING TEXT DATA

Any text data field can be searched with a query composed of character strings. The strings to search can be:

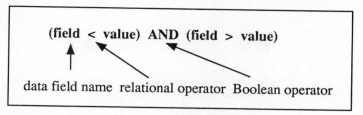

Figure 2/140. Data range query

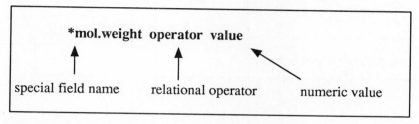

Figure 2/141. Molecular weight query

1. a full text
2. at the beginning of a text (right-hand truncation)
3. at the end of a text (left-hand truncation)
4. anywhere in the text (embedded strings)

The = operator is used for right-hand truncation.
The : operator is used for left-hand truncation or embedded strings.
Text data searches treat lower- and uppercase the same, i.e., strings in the queries can be typed in both cases to retrieve identical data.
Text strings are optionally enclosed in " " marks. However, the quotation marks must be used in strings where ambiguity may arise, i.e., if a string itself contains quotation marks, or right parentheses (frequently found in chemical names).
The query in Figure 2/142 would retrieve the structure **1** in Figure 2/100.
The query:
name =androst (right-hand truncation)
would retrieve the same structure.
The query:
name:sterone (left-hand truncation)
would retrieve structures **1, 2, 4,** and **5** in Figure 2/100.

The special field name *stext must be searched with a query of the format in Figure 2/143.

Example: *Searching text strings*

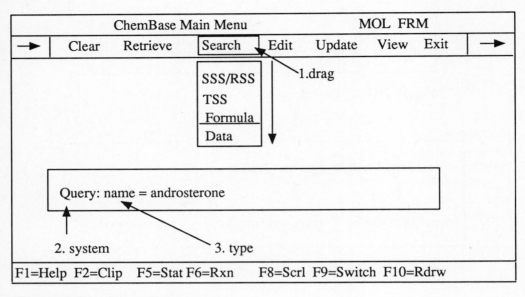

Figure 2/142. Text string search

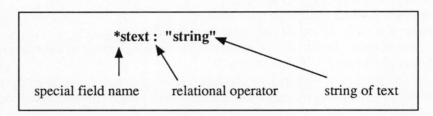

Figure 2/143. Text description search

SEARCHING REACTION SYMBOLS

In searching for reactions, the SEARCH(Data) proves to be a useful method to create subset of a database by searching the Theilheimer Reaction Codes.
The TUTORIAL database contains the data field "Rxn symbol."
The reaction in Figure 2/124 is classified by the code:
HO REA OC
If we formulate the SEARCH(Data) query:

rxn symbol = ho rea oc

we will retrieve the reaction in Figure 2/124 and reaction **1** in Figure 2/126.

Of course, since this is a general classification code for all rearrangements resulting in HO bond by breaking of the OC bond, we would retrieve, in a large database, many other reactions belonging to this type of rearrangement. This technique can be used to create a subset, search domain, of a database.

SEARCHING REACTION KEYWORDS

The TUTORIAL database also contains the data field "Keywords" which has been used to register special keywords describing each registered reaction. These keywords come from a list of Reaction Keywords that has been developed for the reaction databases. These keywords can be used in formulating a query to retrieve reactions.

The reaction in Figure 2/124 is coded by ISO MEF REG ROP keywords. If we formulate a query:

keywords:iso

we would retrieve the reaction in Figure 2/124. Other reactions coded by the same keywords would be also retrieved.

PART THREE

MACCS-II

1 SYSTEM BACKGROUND

MACCS-II (Molecular Access System) is a database management system (DBMS) for storing, searching, retrieving, and displaying/printing/plotting chemical structures and numeric/textual data associated with the structures.
In Part Two, we have discussed ChemBase, a similar system which is, in fact, an offspring of MACCS-II. While ChemBase is run as a standalone system on microcomputers, MACCS-II has been developed as a system for minicomputers. However, it is also available for mainframe computers.
The systems can exchange data through ChemTalk and ChemHost (Figure 3/1).

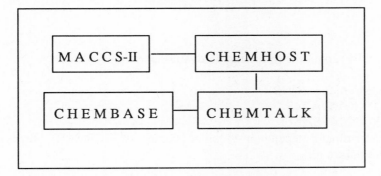

Figure 3/1. Systems interface

MACCS-II incorporates the report generator program DATACCS and other subsystems into one comprehensive system. MACCS-II Version 1.5 with the Power Search Module allows for searching with generic structure queries containing defined R groups. MACCS-II can be programmable and can be interfaced with other DBMS, like ORACLE.
MACCS-II has been designed for the end-user chemist and information scientist to develop corporate, departmental, project, and individual personal databases on site. MDL's Database Division offers services to develop databases for MACCS-II using the data supplied by the client. MDL also developed databases that can be purchased to use with MACCS-II, e.g., Fine Chemicals Directory, Aldrich Catalog of Chemicals, etc.
MACCS-II is installed under a license agreement with MDL and the system is supported by MDL throughout. One person at the client side functions as the System Supervisor who is responsible for maintaining the MACCS-II program after installation. The MACCS-II Supervisor is also responsible for the public databases, for the network of Users who can access these databases, and for the security and passwords of those public databases that may have a limited acces.

MDL annually calls meetings of the Users of MACCS-II to provide a forum for exchange of experience with the system.

MACCS-II databases can contain a few structures or hundreds of thousands of structures and related data. MACCS-II system is therefore suitable for large corporate databases from which subdatabases can be developed for use by departments, project groups, and individuals and possibly transferred to ChemBase.

Input of structures into a MACCS-II database can be facilitated and sped up by using the unique ATTACH technique of structure building. It is based on a repeated use of a parent structural fragment from which the desired structure is built followed by its registration. The parent is then regenerated and used again for building another structure. By this method, hundreds of related structures can be input in a database within a short time.

Searches for structures can be performed with graphic structure queries or with alphanumeric queries. The retrieved data can be manipulated to incorporate them into reports, manuscripts, and other hard copy material.

HARDWARE

MACCS-II can be run on the following hardware/operating systems:
VAX/VMS
Fujitsu FACOM/F4, X8, OVIS
IBM 43XX/30XX/VM/CMS
IBM 43XX/30XX/MVS/TSO

MACCS-II's terminal-independent graphics library, UNIGRAPH, allows for using a variety of terminals, e.g.:
IMLAC
TEKTRONIX 4105, 4107, 4114
HP 2623A
LUNDY 568X
ENVISION, ENVISION II
VT 240/241/340/640
DQ650M
and microcomputers that can emulate PLOT10 graphics, i.e. the TEKTRONIX 4010/4014 terminals, and are equipped with a communication software, e.g., ChemTalk, a product of MDL (see Part Five).

MACCS-II system allows for plotting of the structures we have drawn or those retrieved from a database. The plotting can be performed by a remote plotting device, e.g.:
Versatec V-80
Calcomp 1012
HP 7221T
HP 747xA,7550A
Apple LaserWriter
Auto-Imlac

Plotting can be also carried out on the screen of the terminals listed above, or by interfacing a terminal with a plotting device, e.g., IMLAC with Versatec V-80.

2 SYSTEM DESIGN

While ChemBase manages both molecular structures and chemical reactions and related data in the same database, MACCS-II handles only molecular structures (molecules) and associated data.
In MACCS-II we can have databases of hundreds of thousands of structures and associated data while in ChemBase we are limited in size of the databases to 8 megabytes.

STRUCTURES

MACCS-II recognizes stereochemistry due to a unique treatment of the connection tables of structures drawn and registered in MACCS-II. MACCS-II connection table (see p. 29 for discussion of Connection Table) stores the usual codes for atoms and bonds describing topologically the structure, the X and Y coordinates, to reassemble the structures for display. However, the structures are also coded for any stereocenter and stereobond if they are indicated in the structure diagram.

SEMA NAME

The stereochemical description of a structure is accomplished by encoding the structure using the Stereochemically Extended Morgan Algorithm (SEMA) and assigning to the structure a name (SEMA Name) which is arrived to by this algorithm. The encoded structure name by which the structure is identified in the system database describes the structure *in toto* by the connection table and including any stereocenters and stereobonds. The SEMA name is an integral part of the structure. The structures are stored in the system database with the two-dimensional coordinates which allow for regenerating the structures and display them on the terminal screen or plot them on paper with a plotting device.
Those readers who wish to read on this subject will find the following references as the basic ones on the topic:

Stereochemically Unique Naming Algorithm
W. Todd Wipke and Thomas M. Dyott
J. Amer. Chem. Soc. 1974(96), 4834-42

Simulation and Evaluation of Chemical Synthesis. Computer Representation and Manipulation of Stereochemistry
W. Todd Wipke and Thomas M. Dyott
J. Amer. Chem. Soc. 1974(96), 4825-34

The structures are also assigned structural fragment keys (see p. 424) that serve as screens in structure searching.

STRUCTURE SKELETONS

Structure skeletons that are drawn in MACCS-II are composed of lines that represent single bonds as the default value. Atoms connected by the bonds are, by default, carbon atoms. Carbons are implied at the junction of bonds or at the end of bonds. Any element value from the periodic table can be used to reassign the carbon value. Bond values can be reassigned to double and/or triple.

STEREOCHEMISTRY

MACCS-II recognizes stereochemistry at singly bonded carbon atoms and at carbon-carbon, carbon-nitrogen, and nitrogen-nitrogen bonded by a double bond.
MACCS-II perceives both absolute and relative configuration in stereochemical structures.
In structures with stereocenters, the stereobonds can be drawn conventionally as wedge bonds pointing to or from the stereocenters. The Cahn-Ingold-Prelog notation applies for these configurational isomers (Figure 3/2).

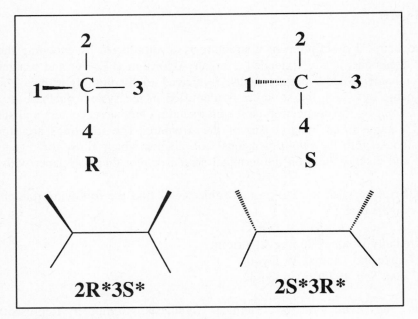

Figure 3/2. Stereoisomeric structures

MACCS-II has the capability to recognize incorrectly assigned chirality, e.g., if two or more of the attached groups to the asymmetric carbon atom are chemically equivalent,

SYSTEM DESIGN

MACCS-II removes the stereobond marking on registration of the structure.
In addition to the bonds indicating the configuration, labels are applied to structures which are to be registered as configurational isomers, or to structure queries to search for these isomers. The options we have in creating these structures are (Figure 3/3):
For structures with one stereocenter:
1. Structures with UP or DOWN bond are labeled CHIRAL (see p. 445, 455) to specify R or S absolute configuration.
2. Structures with one stereobond (UP or DOWN) with no CHIRAL label to indicate a racemic mixture.
3. Structures with a bond marked EITHER to specify a mixed or undefined configuration.
For structures with more than one stereocenter:
1. structures with UP or DOWN bonds and CHIRAL label to indicate absolute configuration
2. structures with UP or DOWN bond and no CHIRAL label to indicate relative configuration. For substructure search we also mark the stereocenters STEREO (see p. 525).
Figure 3/3 illustrates the choices for drawing and labeling configurational isomers.
In structures with relative configuration, MACCS-II will recognize the relative configuration if the bonds of at least two stereocenters are drawn as the wedge bonds UP or DOWN, and the structure has no CHIRAL label (Figure 3/3).
The specifications of configuration become part of the structures on registration and, consequently, we can selectively search for these structures by including the specifications in structure queries.
MACCS-II has certain rules for drawing stereochemical structures. The basic ones are discussed below. Those readers who will frequently register or search stereoisomers should acquire the rules from MDL and study the full set of these rules.
**Stereobonds should be never placed between two stereocenters.
**Drawing only one stereobond at an asymmetric carbon is preferred. If two stereobonds have to be drawn, like stereobonds have to be placed opposite each other and unlike stereobonds should be placed adjacent to each other (Figure 3/4).
It is recommended that an explicit hydrogen be drawn at a trisubstituted asymmetric carbon since MACCS-II may interpret the drawn structure incorrectly if the hydrogen is not included (Figure 3/5).

GEOMETRIC ISOMERISM

MACCS-II recognizes geometric isomerism. The configuration at C=C, C=N, and N=N is drawn by the conventional method. We can register structures with *E* and *Z* configuration, or with mixed (undefined) configuration (Figure 3/6) and search selectively for the isomers.
The convention for drawing structures with trisubstituted double bonds is illustrated in Figure 3/7.
MACCS-II does not recognize the stereochemistry of allenes and other hindered rotational isomers.

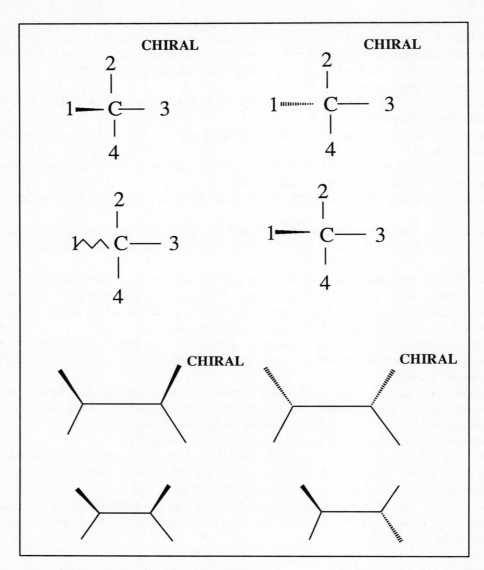

Figure 3/3. Stereoisomeric designations

TAUTOMERIC STRUCTURES

MACCS-II recognizes tautomerism in structures and on searching will retrieve the tautomers if they have the same molecular formula and if the only structural difference is in the value and position of single/double bonds and in the position of hydrogens, charges, and/or isotopes attached to the bonds (see also p. 98 for discussion of tautomerism).

SYSTEM DESIGN

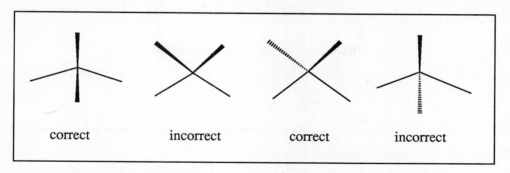

Figure 3/4. Drawing stereo bonds

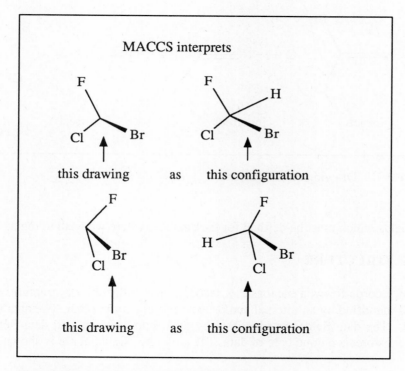

Figure 3/5. Drawing stereoisomers

NUMERIC AND TEXTUAL DATA

Structures registered in a MACCS-II database can be associated with any numeric and textual data we wish to register with the structures. These data are contained in categories

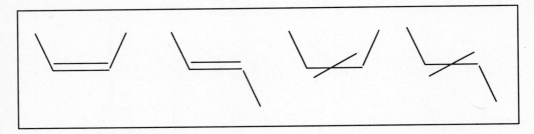

Figure 3/6. Geometric isomers

Figure 3/7. Drawing stereoisomers

called data fields which must be defined for the database where we wish to register data.

DATABASE STRUCTURE

Collection of records forms a database. A record is the collection of structures and data. The record is identified by an internal and/or external registry number. The data are stored in data fields. The data fields are identified by data field numbers and data field names. Each data field stores a unique type of data. The structure of a database is shown in Figure 3/8.

SPECIAL DATA FIELDS

MACCS-II automatically derives certain data from the registered structures. We have seen similar special data in ChemBase (see p. 271, 368) called special data fields.
The data automatically generated by the System and available for searching are:
1. internal registry number
2. keys
3. molecular formula

SYSTEM DESIGN

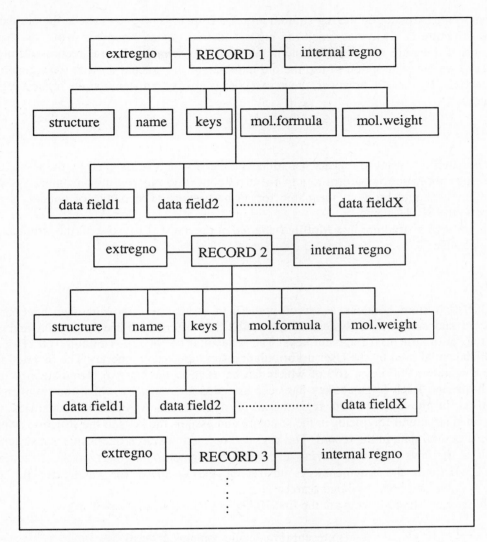

Figure 3/8. Database structure

The data fields "external registry number" and "name" are also automatically appended to a structure and are searchable; however, the name and the external registry number must be supplied by the User (MACCS-II can automatically generate even the extregno if the format is specified and automatic generation requested at the time when the database is set up).

INTERNAL REGISTRY NUMBER

The internal registry number is a sequential number, starting with 1, assigned by the

System to the structure in the registration process. If there is no gap between the numbers, a new structure is always assigned the highest internal registry number in the database, plus one. If there are gaps between the numbers due to deletion of some records from the database, we have an option to use the old numbers of the deleted records for registering new records. By default, however, the new record is always registered at the highest registry number.

NAME

Each structure we want to register has to have a name which is assigned by the User. The name does not have to conform to any chemical convention or any nomenclature system. However, it should be a unique name for each structure. The name, of course, is not related to the SEMA name.
When we save a structure in a Molfile (*vide infra*) the name is saved with the structure as the "molname."

KEYS

MACCS-II system contains a set of predefined structural fragments that have been collected and stored in a System File. MACCS-II offers both the sequential and the inverted set of structural fragment keys. The sequential file of keys contains 160 keys, 52 of which can be used by the User in formulating structure search queries. The inverted file of keys contains 900 keys, 165 of which can be used by the User to formulate structure search queries. Each fragment/key has been assigned a number by which the fragment is identified in the system. On registration of a structure, MACCS-II recognizes the predefined structural fragments in the structure and assigns the keys to the structure as part of the structural data in the record. The keys can be used in search queries for substructure search to retrieve structures having a particular fragment(s) as part of the structures. MACCS-II uses the fragment/keys to screen the database for candidates for the atom-by-atom and bond-by-bond search.
For illustration, the following are the first 10 keys of the sequential set of keys:

Number	Key	Key is present if molecule contains at least
1	N	1 nitrogen
2	N>1	2 nitrogens
3	N>2	3 nitrogens
4	N>3	4 nitrogens
5	O	1 oxygen
6	O>1	2 oxygens
7	O>2	3 oxygens
8	O>3	4 oxygens
9	F	1 fluorine
10	Cl	1 chlorine

SYSTEM DESIGN

The following are examples of the inverted set of keys:

Number	Key
14	S-S
19	7-membered ring
36	S-heterocycle
40	S-O

MOLECULAR FORMULA

Molecular formula is calculated by the system and is displayed with each retrieved structure unless we turn the display off (see p. 508). It is also searchable. We can define a data field for formulas we wish to calculate.

MOLECULAR WEIGHT

Molecular weight is calculated by MACCS-II for each registered structure. It is displayed with the retrieved structure unless we turn the display off. Molecular weight is not searchable. However, we can define a data field for molecular weight we wish to calculate. The data in the field will be searchable.

DATA FIELDS

In MACCS-II, the data we wish to register and which are associated with structures reside in data fields (compare ChemBase data fields, p. 271 and REACCS datatype treenames, p. 579). The data fields are of two kinds:
1. fixed data fields
2. flexible data fields

FIXED DATA FIELDS

Fixed data fields are defined when a database is set up and cannot be changed after the structure of the database has been saved in a file (however, minor alterations, e.g., of data field name, are permited).

Fixed data fields are destined to hold alphanumeric data of **short length and predefined format** in which the data will be registered and displayed on retrieval. The storage space for the fixed data fields is reserved even if no data are registered. If too many fixed data fields are defined and are not used, the disk capacity will be wasted.

The formatted data require the specification of **field width** on creating the data field.
The fixed data field can hold the following data:

1. a single numeric value, or
2. a pair of numeric values, or
3. alphanumeric string up to 20 characters long

For numeric data, we can specify a **units label** which is displayed on registering data to remind the User which units to use for the data being registered.

Fixed data fields are displayed, by default, with each retrieved structure unless we turn the display off (see p. 508).

Fixed data fields are fast to search. Thus they are suitable for data searching which will quickly retrieve the related record. Examples of these data are the external registry number, molecular weight, date of preparation of a substance, a date of registration of a substance, etc.

EXTERNAL REGISTRY NUMBERS

The internal registry numbers (see p. 423) are integers assigned by MACCS-II to the registered structures/records. MACCS-II gives the User an option to create a regular fixed data field as the first data field in the database for the external registry number. In the external regno data field, we can store a specially formatted registry number that can be composed of letters, numbers, and special characters. An example may be a company code and number for a substance, or CAS Registry Number.

FLEXIBLE DATA FIELDS

Flexible data fields do not have to be defined when we set up the database. They can be defined at any time and can be changed or deleted at any time.

Flexible data fields can hold up to **300 lines of alphanumeric text**, 120 characters per line, and the data can be of free format or can be formatted.

The data in the flexible data fields are displayed only when we specifically search for the data.

DATA FIELD DESCRIPTORS

The fixed and flexible data fields are assigned a number by the system and a name by the User. These are the data field descriptors we will need to describe the data fields in many of the operations we will be executing in MACCS-II.

The data field names may be up to 30 characters long but may not contain blanks, hyphens, or commas.

The data field numbers are assigned to the fixed data fields first starting with number 1 so that they have always the lowest numbers. The flexible data fields follow in sequential order.

DATA TYPE AND FORMAT

The fixed and flexible data fields may contain the folowing kind of data:
1. numeric
2. text

SYSTEM DESIGN

The data can be in the following formats:
1. free format with limitation in line length
2. formatted with limitation in line length

Figure 3/9 shows the data types and formats.

```
Text Data
─────────

Fixed data field    XXXXXXXXXXXXXXXXXXXX    One line only with
                                            a maximum of 20 characters

Flexible data field XXXXXXXXXXXXXXXXXX      Up to 300 lines with a maxi-
                                            mum of 120 characters per line

Numeric data
────────────

Fixed data field    0.0000 - 0.0000    One line with one or two values
                                       separated by space, comma, hyphen

Flexible data field 0.0000 - 0.0000 XXXXXX  Up to 300 lines with one or
                                            two values plus up to 112
                                            characters of text per line
                                            separated by space, comma,
                                            or slash

Formatted data
──────────────

Fixed data field    XXX-00-000-XX-X         One line with a single for-
                                            matted string of up to 20
                                            characters

Flexible data field XX-00000 XXXXXXX        Up to 300 lines with one
                                            formatted string plus text
                                            to a total of 120 charac-
                                            ters per line
```

Figure 3/9. Data fields and data types and format in MACCS-II

FORMAT SYMBOLS

A format is a series of short sections which are defined by using the following symbols:

0.........numeric section matched by any digit (0-9)
@.........nonnumeric section matched by any nonnumeric
          printing characcter except #, -, ^, or space
%.........character section matched by any print character,

space, punctuation, except #, and ^. A comma or leading blank can be included only if the entire data item is enclosed in single quotes

| ....optional section matched by any character except #, and ^, or none

— ......delimiter between sections. A delimiter between sections is essential between sections defined by identical format symbols but is not essential between sections defined by different format symbols

A constant formatted field may also be defined by any capital letter, blank, or a punctuation mark. The same character, then, must be in every field of data we wish to register.
For example, SC-0000 formatted data must always begin with the letters SC followed by hyphen and four digits.
The hyphen has a special meaning in formatting. It **may be used** between two data of any kind but it **must be used** between two data of the same kind. For example:
SC1234 will be stored and displayed without hyphen.
SC-1234 will be stored and displayed with the hyphen.
1234 will be stored as one number.
1-234 will represent two numbers with some relation between the two numbers.
If we assign a format to data, the format must be matched on the data registration.
MACCS-II fills the formatted fields as follows:
1. Numeric data
are filled from right to left into the format.
2. Non-numeric data
are filled from left to right into the format.
Figure 3/10 shows the formats with no hyphen, data input into these formats, and how MACCS-II interprets the data input.
Figure 3/11 shows the formats with hyphens, data input, and MACCS-II interpretation of the data.
For numeric data consisting of a single value followed by a numeric comment line, the format is as follows:
0000//0000 or 0000,,,0000

FILES IN MACCS-II

MACCS-II system uses two kinds of files:
1. System files
2. User files

SYSTEM DESIGN

Format	Data Input	Result
%%%%%	AB	AB
	123	123
	AB123	AB123
	A-B	A-B
	A,B1	no registration
	"A,B1"	A,B1
	A BC	A BC
	"A BC"	A BC
	spaceABCD	space ignoredABCD
@@@	AB	AB
	ABC	ABC
	A-B	no registration
	A1	no registration
	A B	no registration
	"A B"	no registration
0000	1	0001
	23	0023
	456	0456
	7890	7890
	12345	no registration

Figure 3/10. Formatted data without hyphen

SYSTEM FILES

System files are:
1. The program files that are installed in the computer and are the basis of MACCS-II functions. The program files are interfaced with the computer operating system from which they are invoked.
2. Database keys files that contain the structural fragments and their numeric codes.
3. File containing the periodic table.
4. File containing templates for structure building.
5. File containing the help system.

DATABASE FILES

Databases in MACCS-II may be public or private. The files containing the databases may be accessible by all the Users within a corporation, or they may be specialized files accessible only by selected Users. As we have noted in Part Two, the database files behave differently from the other files of the system. They are not written or read as the ordinary files. Database files are opened by the system at the start of the MACCS-II session and they are closed at the end of working with the database.
MACCS-II provides for a sophisticated Security System for databases. Passwords control

Format	Data Input	Result
@@-@@@	A-B	A -B
	ABC	AB-C
	AB-C	AB-C
	A-BC	A -BC
	ABCD	AB-CD
	A-BCD	A -BCD
	ABCDE	AB-CDE
	A-BCDE	no registration
00-000	123	12-003
	1-23	01-023
	-123	no registration
	0-123	00-123
	1234	12-034
	12345	12-345
	12-345	12-345

Figure 3/11. Formatted data with hyphen

access to a database. The registered structures can be assigned to classes and the data fields into groups. A controlled access to the classes and groups can be given to selected Users. The Security System is maintained by the System Supervisor for the public databases. Departmental and individual private databases can be similarly controlled.

USER FILES

User files contain data to be used in the MACCS-II databases. They are written by the system or by the User and can be read by the system or used for transfer of data.
The User files are the following:
1. Listfiles
2. Molfiles
3. Datfiles

LISTFILES

Listfiles store the internal registry numbers or the external registry numbers of selected records from a database. Listfiles can be created out of the list of records retrieved from a database, i.e., the hits (results) of our search, or they can be created in a Text Editor and transferred to MACCS-II.
There are three types of Listfiles in MACCS-II:

1. **Binary files containing the internal registry numbers.**

These can be written only by MACCS-II since the internal registry numbers are coded in the binary number system.
2. **Numeric files containing the internal registry numbers**.
These can be written either by MACCS-II or by the User in a Text Editor since the internal registry numbers are stored as ASCII characters.
3. **Extreg files containing the external registry numbers**.
These can be written either by MACCS-II or by the User in a Text Editor since the external registry numbers are stored as ASCII characters.

MOLFILES

Molfiles store one encoded structure.
The Molfile stores the connection table, three-dimensional coordinates, if any, or two-dimensional coordinates with the Z coordinate = 0 (the MACCS-II structures). In addition, the file also stores the molecule name (molname), the registry number if the structure has been retrieved from a database (however, the regno is not used on recalling the file), date, and other information for creating the file. No numeric or textual data associated with the structure can be stored in the Molfiles. However, we can input "comments" which is an optional short description (memo) of the structure.
The Molfiles can be used in other MDL programs, e.g., ChemBase, REACCS, and the molecular modeling programs.

DATFILES

Datfiles store registry number(s) and numeric and textual data of a single structure/record or multiple structures/records. Datfiles can be written by MACCS-II or by the User in a Text Editor. Datfiles can be used to transfer data between MACCS-II databases and other MDL programs that accept Datfiles. Note that ChemBase Data Files are not equivalent to Datfiles and both cannot be used to transfer data from MACCS-II to ChemBase or *vice versa*.

3 SYSTEM OPERATION

MACCS-II can be operated at two levels:
1. Command level (not covered in this book).
2. Menu level (the basis of this book).
At the menu level, MACCS-II offers the User over 150 commands, called options, to operate the system. Using these options we can perform the following major operations:
1. Create graphic chemical structures and register (enter) them into a database.
2. Create graphic structure queries for exact structure search or substructure search.
3. Display/print/plot graphic structures.
4. Define data fields and register data into a database.
5. Create alphanumeric queries to search for exact data or for generic data.
6. Display/print alphanumeric data associated with the structures.
7. Create new databases.

The User's actions in MACCS-II are carried out in operational Modes. There are seven Modes. Each Mode has a Menu. The Mode names are displayed in each Menu and we can move directly from one Mode to the other by selecting the Mode on the Menu. The options of MACCS-II are contained in the Menus. They can be selected and activated graphically or by keyboard input. In addition, when we are in the graphic selection mode, there are options which are selected/activated by keyboard input. Figure 3/12 illustrates the system of Modes and Menus in MACCS-II.

MAIN MENU

Main Menu is the hub of the MACCS-II system (Figure 3/12) from which we can run MACCS-II either at the command level or at the Menu level. The Main Menu is shown in Figure 3/13. We can select the MEditDraw option on the menu to enter the Draw Mode for creating structures or to enter the other Modes by selecting the Mode names.
In Main Menu we can perform many operations, e.g., to create a new database, we procedd by entering:
access create (see p. 550).

DRAW MODE

Draw Mode is the center of action for building structures in order to register them, to use them as queries for searching, to store them in Molfiles, and to print or plot them. Draw Mode has three menus:
1. Structure Menu
2. Query Menu
3. Rgroup Menu

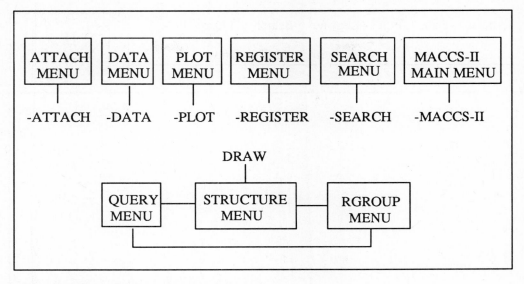

Figure 3/12. System of modes and Menus

Note that this menu is available only for those Users with MACCS-II Power Search Module.
From Draw Mode, we are always returned to the Mode from which we entered the Draw Mode.

STRUCTURE MENU

Structure Menu (Figure 3/14) allows for creating structures containing no variable features.

QUERY MENU

Query Menu (Figure 3/15) allows for building generic structures, i.e., structures with variables, to use as queries for generic structure and substructure searches.

The Users who have installed the MACCS-II Power Search Module will also be able to use the Rgroup Menu for building more complex generic structures with defined R groups (see p. 476). The discussion "Generic Structures" in CAS Registry (p. 118) will further illustrate the nature of generic structure building and use.

ATTACH MODE

In Attach Menu (Figure 3/16), we can build structures from a parent by attaching atoms and structural fragments to the parent. After the built structure is registered, the parent is

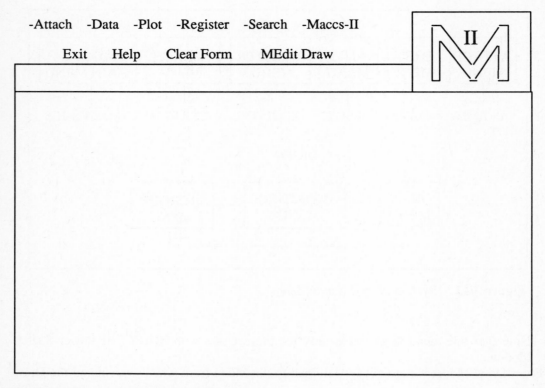

Figure 3/13. MACCS-II Main Menu

regenerated for building of another structure. This Mode makes it possible to speed up the building of structures and to develop databases with a large number of related structures.

REGISTER MODE

Register Menu (Figure 3/17) allows for retrieving single molecular structures and/or the associated data, updating databases, writing of Molfiles, and assigning the registered structures to classes according to accessibility by the Users.

SEARCH MODE

Search Menu (Figure 3/18) options allow for searching substructures, stereoisomers, tautomers, and keys, as well as numeric and textual data, viewing hit lists, and manipulating lists.
The Users who have installed MACCS-II Power Search module will utilize Search Menu (see Figure 3/12-PS, p. 521) which allows for Similarity Search and Substructure Search with generic structure queries containing R groups defined in Rgroup Menu (see p. 488).

SYSTEM OPERATION

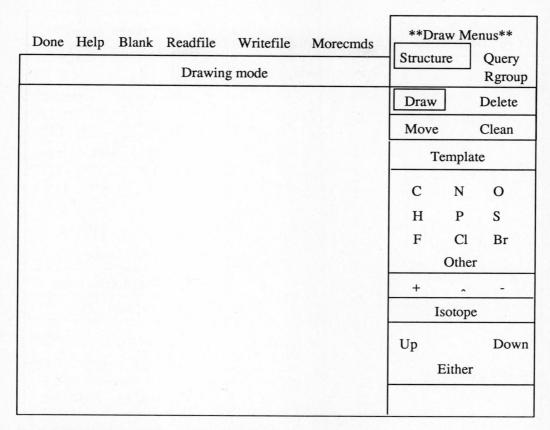

Figure 3/14. Draw mode - Structure Menu

DATA MODE

Data Menu (Figure 3/19) allows for searching and displaying numeric and text data only. Structures cannot be searched or displayed in this Mode. The Data Menu has options for searching data, updating data, defining flexible data fields, and transferring data within and between databases.

PLOT MODE

Plot Menu (Figure 3/20) makes it possible to plot structures locally or on remote plotters to obtain structure images on the screen or as a hard copy.

```
  Done   Help   Blank   Readfile   Writefile   Morecmds  | **Draw Menus**        |
                                                         | Structure  | Query    |
                      Drawing mode                       |            Rgroup     |
                                                         | Draw       | Delete   |
                                                         | Restore    | Exclude  |
                                                         | Hyd   Link   Stereo   |
                                                         |  C     N     O        |
                                                         |  H     P     S        |
                                                         |  F     Cl    Br       |
                                                         |  A    Other  Q        |
                                                         | Ring# Subst Unsat     |
                                                         | Chain      Ring       |
                                                         | Any        Arom       |
                                                         | S/A   D/A   S/D       |
                                                         | IsoRing    Nosubst    |
```

Figure 3/15. Draw mode - Query Menu

SELECTING OPTIONS

The options instructing MACCS-II to perform specific actions are selected by the following methods:
1. graphic selection
2. keyboard switch
3. keyboard selection

The methods 1 and 2 are used if we operate MACCS-II on graphics terminals. The method 3 is for terminals which do not have graphics capability. However, this method is also available on graphics terminals if we switch to keyboard operation (*vide infra*).
The options are displayed on the Menus. They are either a single-word, e.g., Name (Figure 3/18), or two-word option, e.g., Read List (Figure 3/18). The two-word options are shown on the Menus in two versions. One version is the full two-word option which is selected, e.g., Write List (Figure 3/18). The second version is composed of the first word

SYSTEM OPERATION

```
    -Attach  -Data  -Plot  -Register  -Search  -Maccs-II      | fcd:
                                                              | Curlist 64939
    EXit  Help  Blank  DRaw  SETtings  More Cmds              | Regno        0
                                                              | On File  64939
                                                              |----------------
                                                              |    ATTACH
                                                              |
                                                              |    Attach
                                                              | a g, p; a g, p
                                                              | [g=group name]
                                                              | [p=posn number]
                                                              |
                                                              |    Numbers
                                                              |    Restore
                                                              | Replace Name
                                                              | Stereo Bond
                                                              |
                                                              | View Regno
                                                              | Read File
                                                              |
                                                              | Register Current
                                                              | Show Templates
                                                              |----------------
                                                              |    MEDIT
                                                              |----------------
                                                              |    GRAPHICS
```

Figure 3/16. Attach Mode Menu

highlighted by **   ** followed by a list of second words, e.g., **Register** Current (Figure 3/17). These two-word options are activated by selecting a word from the list of the second unasterisked words.

Graphic Selection of Options

We will use the following terms to describe the procedures of selecting options and build structures. The procedure will depend on what type of graphic input device is used. Some options require additional keyboard input at the System prompt.

Point......move the cursor to and over an item on the Menu
     or put the light pen tip over an item
Select......click (press and release) the mouse button, or
     press and release the light pen tip, or press the

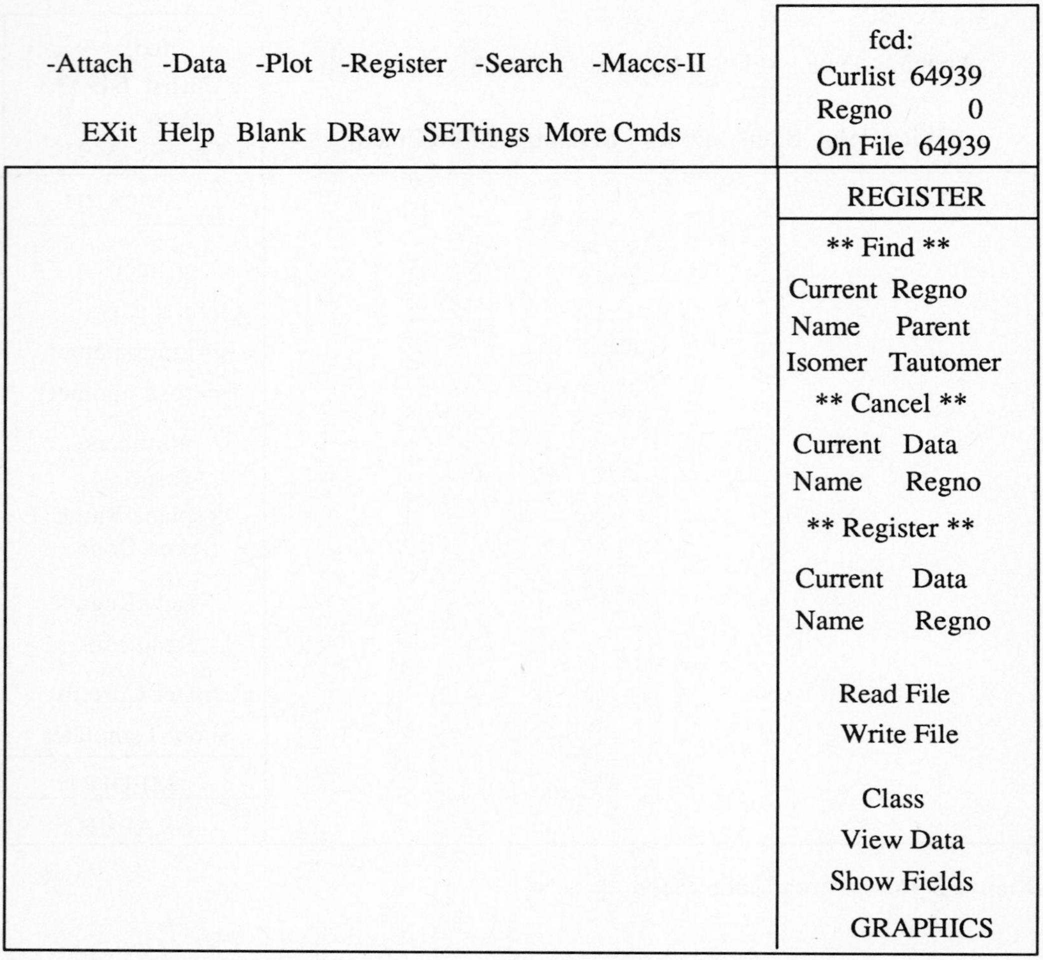

```
   -Attach  -Data  -Plot  -Register  -Search  -Maccs-II

      EXit  Help  Blank  DRaw  SETtings  More Cmds
```

	fcd: Curlist 64939 Regno 0 On File 64939
	REGISTER
	** Find ** Current Regno Name Parent Isomer Tautomer ** Cancel ** Current Data Name Regno ** Register ** Current Data Name Regno Read File Write File Class View Data Show Fields GRAPHICS

Figure 3/17.   Register Mode Menu

space bar (depending on the equipment used)
Place......move the cursor on a blank position in the display area of the Menu or on an atom or bond in the displayed structure
Type.......at the system prompt type on the keyboard and press carriage return(enter)

Keyboard Switch

The keyboard switch is activated by pressing the specific character on the keyboard without carriage return. This will give two results:

SYSTEM OPERATION

						fcd:
-Attach	-Data	-Plot	-Register	-Search	-Maccs-II	Curlist 64939
						Regno 0
EXit	Help	Blank	DRaw	SETtings	More Cmds	On File 64939

	SEARCH
	Name Formula
	Keys Query
	Data Isomer
	Sss Tautomer
	PArent CUrrent
	Read List
	Write List
	Show List
	Switch List
	All List
	Read File
	Write File
	View Query
	Show Fields
	** View **
	First Last Next
	Prev Regno Item
	Auto Time Data
	Print Continue
	Class GRAPHICS

Figure 3/18. Search Mode Menu

1. Typing the character(s) on the keyboard activates or deactivates an operation (on/off toggle)
2. Typing the character(s) on the keyboard executes a specific operation (push a button to execute)(switch)
The keyboard switches can be used only if we are in the graphic operation of MACCS-II. They are available in Register Mode and Draw Mode.

	fcd:
-Attach -Data -Plot -Register -Search -Maccs-II	Curlist 64939
	Regno 0
EXit Help Blank DRaw SETtings More Cmds	On File 64939
	DATA
	Cancel Data
	Register Data
	Find Data
	Search Data
	Transfer from/to tty, file, regno, database
	Read List
	Write List
	Switch List
	All List
	Show Fields
	Print
	Overwrite-append
	GRAPHICS

Figure 3/19. Data Mode Menu

Switching from Graphic to Keyboard Selection

From the graphic selection mode we enter the keyboard selection mode by pressing "k." The single and two-word options can be then activated via the keyboard while the graphics display is in effect. To return to the graphic selection mode we press "g" and <cr>. In Draw Mode, we can use keyboard selection by preceding the option name by a / (slash).
MACCS-II has the capability of creating structure queries by keyboard input; however, this method will not be discussed in this book which is oriented toward the graphics.
If we switch to the keyboard operation, MACCS-II issues the following prompts:
MACCS-II>
ATTACH>
REGISTER>
SEARCH>
DATA>

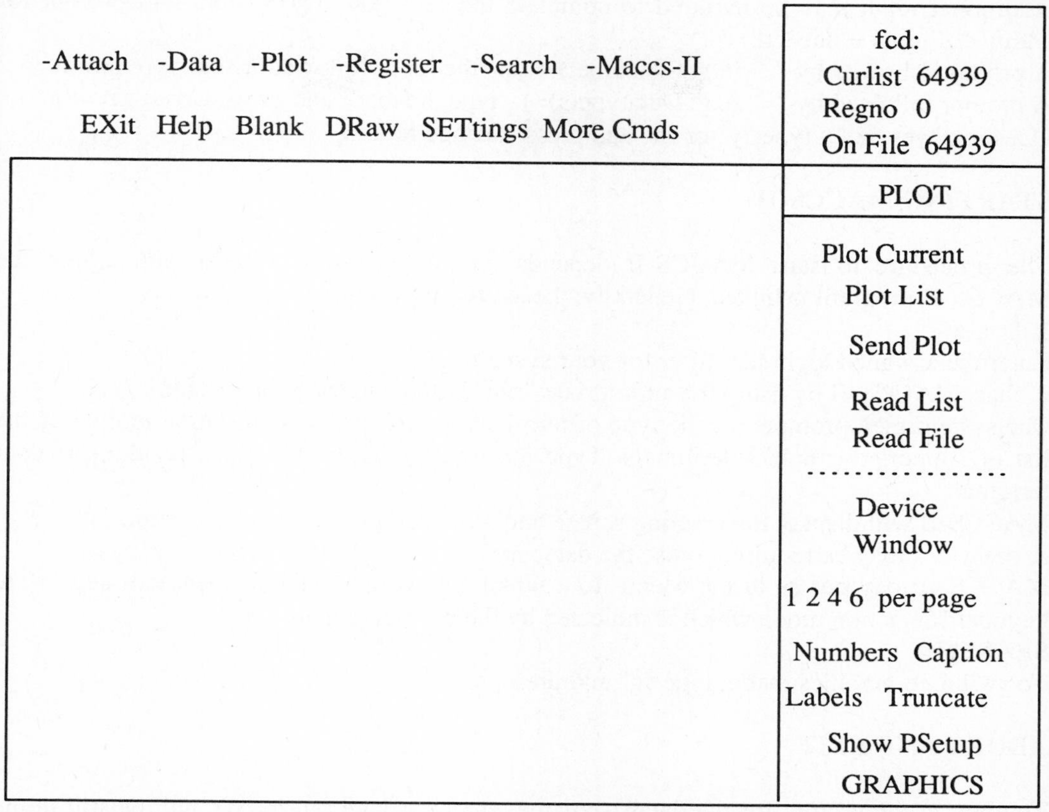

SYSTEM OPERATION

	fcd:
-Attach -Data -Plot -Register -Search -Maccs-II	Curlist 64939
EXit Help Blank DRaw SETtings More Cmds	Regno 0
	On File 64939
	PLOT
	Plot Current
	Plot List
	Send Plot
	Read List
	Read File
	....
	Device
	Window
	1 2 4 6 per page
	Numbers Caption
	Labels Truncate
	Show PSetup
	GRAPHICS

Figure 3/20. Plot Mode Menu

PLOT>
The keyboard options are entered at these prompts.

Keyboard Selection

We have said that some of the options have to be activated by keyboard selection even if we operate in the graphics mode. The keyboard-activated options are not shown on the Menus; however, we can display them in each of the Modes by activating the MORE CMDS option on the Menu. The appropriate option is then typed on the keyboard.

If we switch to the keyboard operation (*vide supra*), we type the option name after the prompts as follows:

One-word option: type as many letters to make the name unambigous for MACCS-II.

Two-word option: type the first letter of the first word, space, and the first letter of the second word. This type of selection includes the **xxx** marked two-word options.

Additional input may be required to complete the selection. The system will prompt for additional input if needed.
A prompt followed by ":" (e.g. Filename:) - type the input followed by carriage return.
A prompt followed by "=" (e.g. Datatype(s)=) - type the input and press carriage return.
Question "yes, no" - type "y" or "n" and press carriage return.

STARTING MACCS-II

The procedure to start MACCS-II depends on the operating system with which the MACCS-II program is linked. Generally, the steps are as follows:
1. Login
Enter the assigned login identifier for your system.
2. Start MACCS-II by using the appropriate "run" command for your system.
The system then prompts for the type of terminal we use or gives an option to display the list of supported graphics terminals. Type the number in the list corresponding to your terminal.
MACCS-II will display the greeting screen and will prompt for a database name.
A password may be required to use the database.
MACCS-II may put us in the Menu from which we exited in the last session, and in the keyboard selection mode which is indicated by the prompt, e.g.
SEARCH>
To switch to graphics mode, type "g" and press <cr>.

ABORT AND EXIT

To abort at the prompt for the database name, enter carriage return. We will be still in the MACCS-II Menu.
To abort an active option or operation, use the "!" or the break key (ctrl-C).
To exit from MACCS-II, select the EXIT option on the Menu or type "exit" at the MACCS-II> prompt in the keyboard mode.

SETTINGS

When we run MACCS-II, we want to be able to specify the conditions under which an operation will be executed. The initial conditions we encounter in using MACCS-II have been preset by MACCS-II and they are the default conditions which come into effect automatically when we start MACCS-II. For example, structures are displayed, by default, without hydrogens. However, we may wish to show the hydrogens in the structures. When we retrieve a structure by a search, the structure is displayed, by default, including the associated fixed data. However, we may wish to display structures without the data.
The Search, Register, Attach, Data, and Plot Menus contain an option, the SETTINGS, which enables the User to set and override the default conditions by executing one or more of a set of SETTINGS suboptions. Some of the suboptions are of general character; some of them are of specific use. We will list here the SETTINGS suboptions of the general

SYSTEM OPERATION

character. The specific ones will be discussed in the operations and Modes in which they are applicable.

To enter the SETTINGS suboptions, point/select SETTINGS to receive the prompt:
SETTINGS>
at which the suboptions can be typed in.

Some SETTINGS suboptions function as a toggle, i.e., when the function of the suboption is active it is inactivated by reentering the suboption at the SETTINGS> prompt.

General settings suboptions:

autoscale......sets on/off automatic scaling and centering of structures on screen display. If "off," the structures are displayed as they were when they were registered (default is "off").
database.......changes the current database to another specified database
done...........exits the SETTINGS process
help...........activates the help pages for the SETTINGS process
password.......allows for switching to another password for the current database. Note that there may be more than one password to the same database controlling the level of access to the database
set graphics...sets the graphics or nongraphics operation of MACCS-II. The switch is "on" when we specify at the start of MACCS-II that we are using a graphics terminal.
show settings..displays all of the current settings
set scroll.....determines the scrolling area for MACCS-II messages

4   CREATING STRUCTURES

We build structures in order to register them in a MACCS-II database, to create structure queries for searching, to save them in Molfiles for later recall, and to plot them to obtain hard copies.

As we have seen in Part One and Part Two, we can build structure skeletons and specify the values for atoms and bonds in the skeletons. The values may be single values or they may be a list of variables at specific locants in the structure.

In MACCS-II, we employ the following Menus to create these structures:
1. Structure Menu to create structures with single values.
2. Query Menu to create structures with variable values, i.e., simple generic structures.
3. Rgroup Menu to create more complex and/or very involved generic structures with defined R groups. Note that this Menu is available only if the MACCS-II Power Search Module had been installed.

STRUCTURE MENU

Structure Menu is entered from Main Menu by point/select MEditDraw (Figure 3/21). From the other Menus, we select the DRAW option (Figure 3/22).

In Draw Mode, we select the options graphically from the Menu or by keyboard selection (see p. 436, 441).

In addition, the following switches and toggles (see p. 438) are executable (in upper or lowercase) in Draw Mode:

Switch	Name	Result
*		redraws (refreshes) the entire terminal screen display
/		permits to activate any option by keyboard input
- or =	ATOMS	changes atom type, charge, isotope in structure (switch)
/B	BLANK	deletes the current structure (switch)
B		manual refresh (redraw) of structures (toggle)
C	CONTINUOUS DRAW	only one click needed for drawing a bond (toggle)
D	DIAGRAM UPDATE	cleans and redraws structure (switch)
E or >	ENLARGE	enlarges the current structure (switch)
F	FREE VALENCE	allows exceeding the atom valence in drawing bonds (toggle)

MACCS-II

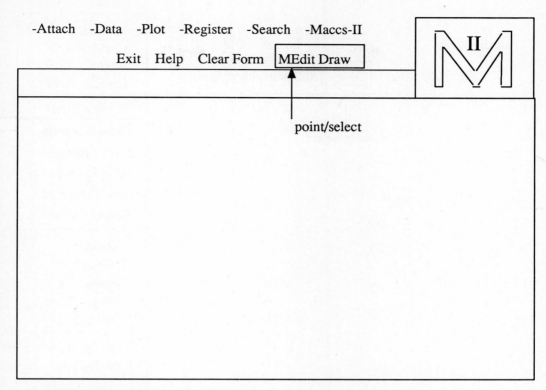

Figure 3/21. Switching to Draw Mode

G	GROUP OPTIONS	allows manipulation of fragments within a structure (switch)
H	HORIZONTAL	reorients structure so that the specified bond is horizontal (switch)
I	ISOMER	labels structure CHIRAL (toggle)
J	JOIN	forms a new ring incorporating the specified atom or bond, forms spiro or peri-fused ring (switch)
L	LABELS	displays hydrogens in structures (toggle) (see Settings "Label Level", p. 447)
M	FORMULA/WEIGHT	calculates and displays molecular formula and weight of the current structure (switch)
N	NUMBERS	displays locant numbers (toggle) (see Settings "Numbers", p. 447)
O	ORIGIN	centers and rescales a structure

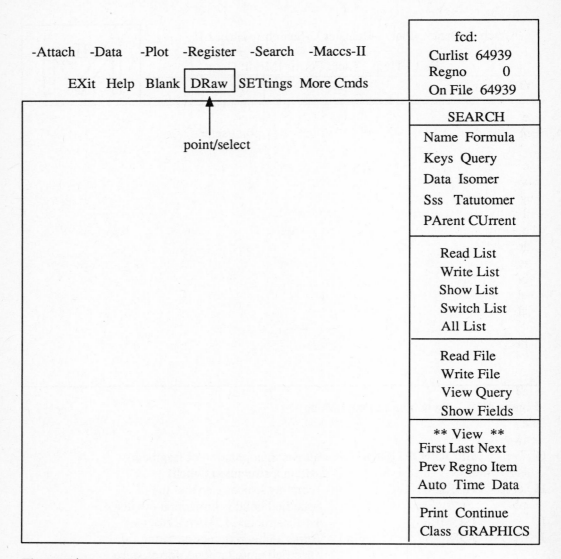

Figure 3/22. Switching to Draw Mode

Q	QUERY CHECK	checks the validity of queries
R or <	REDUCE	reduces structure size (switch)
S	SHIFT	moves structure (switch)
/T or +	TEMPLATE	attaches templates in structures (switch)
T		attaches last template (selected by +, TEMPLATE, or /+)(switch)
U	UNSPECIFIED STEREO CENTER	displays pointers to stereocenters

CREATING STRUCTURES

V	VERTICAL	that do not have stereobonds (toggle) reorients structure so that the specified bond is vertical (switch)

HYDROGENS IN STRUCTURES

The structures created in Draw and Attach Modes do not show hydrogens unless we draw bond(s) and assign the explicit hydrogen value at the end of these bond(s) by point/select the "H" option on the Menu and place/select at the bond(s) to have the explicit hydrogens. The explicit hydrogens will be retained in the query structure, if it is used for a search; however, if we register a structure with explicit hydrogens they will be stripped off unless they are used to designate configuration. Therefore, the structure retrieved from a database will not show hydrogens when displayed or plotted unless it is a stereoisomer.
The switch "L" (see p. 345) and the Settings suboption "label level" enable us to set the hydrogen display on or off.
When the "L" is on, hydrogens will be displayed. The Settings suboption (see p. 442) "label level" allows for specifying:
label level.....specifies which hydrogens will be displayed
    all.....hydrogens at heteroatoms and terminal carbons will
        be displayed when "L" is on (default)
    hetero..only hydrogens at heteroatoms will be displayed
        when "L" is on

DISPLAYING ATOM LOCANTS

The "N" switch (see p. 445) turns on/off display of atom locants in the structures in the Draw and Register Modes.
However, we may want to have automatically turned the display on every time when we enter the Draw or Attach Mode. We can use the Settings suboption (see p. 442) to specify the display.
numbers........turns on/off the automatic display of structure numbering (default is "off")

DRAWING STRUCTURE SKELETONS

To draw structure skeletons, the Draw option must be activated. When we enter Draw Mode, the Draw option on the Menus is automatically active. If we select another option from the Menu and wish to return to drawing, we point/select DRAW (Figure 3/23). We can use normal draw or continuous draw. There is no rubber band draw in MACCS-II (compare ChemBase, p. 295). To draw double or triple bond, we place/select the single bond or place/select the atoms at the end of the bond (compare drawing bonds in ChemBase, p. 292).

CREATING CHAINS

For continuous draw, we toggle "c" and click only once for each end of the bond. To start a bond at a new point, toggle "c" or click the end (but not the beginning) of the previous bond (Figure 3/23). In normal draw (i.e., the "c" toggle off), we click twice, once for each end of the bond.

Example: *Continuous draw*

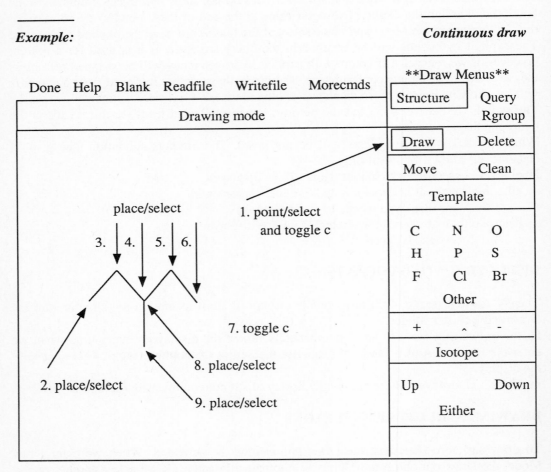

Figure 3/23. Continuous draw

CREATING RINGS

To create rings, we can draw chains of the appropriate length and shape, and close the chain at the terminal atoms. Alternatively, we can use the quick way by pressing the switch "j" and indicating the number of atoms in the desired ring (Figures 3/24, 3/25).

CREATING STRUCTURES

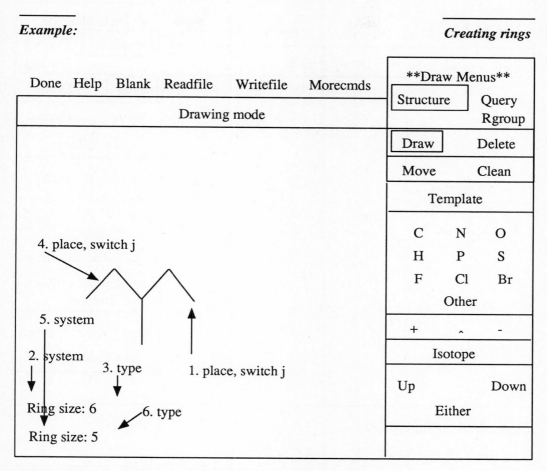

Figure 3/24. Creating rings with J toggle

Note that switching "j" on a ring atom with two connections gives a spiro ring system and a bridgehead atom gives a peri-fused ring system.

ASSIGNING ATOM VALUES

To specify atom values we, in fact, respecify the default implied carbon atoms in the created structure skeletons. Similarly, any other atom value can be respecified. We select the Menu items showing the atom values (Figure 3/26) and the atom(s) will be inserted into the structure (Figure 3/27; the structure was cleaned by CLEAN, see p. 495). The atom values which are not on the Menu are selected by activating OTHER and typing the element symbol at the prompt:

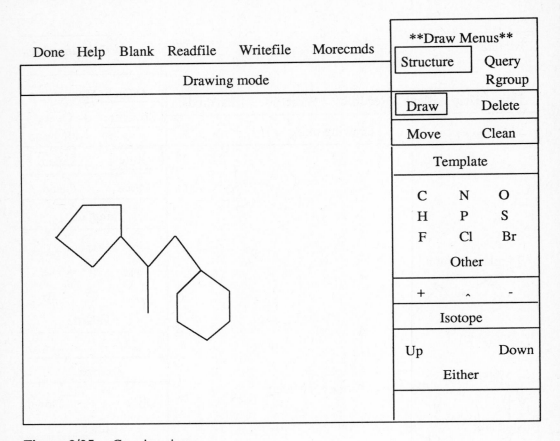

Figure 3/25. Creating rings

Atom symbol:
followed by place/select on the position to have the value.

KEYBOARD ASSIGNMENT OF ATOM VALUES

We can also use a shortcut method to assign atom values without having to select the atom symbols on the Menu and thus deactivate the Draw option (if we wanted to continue in drawing bonds after an atom selection, we would have again to select Draw).
If we place on the position where the atom value is to be assigned and depress "=" or "-," the System will prompt:
Atom type:
We can enter any of the elements in the Periodic Table or any of the special symbols (pseudoatoms). Note that the special symbols must have been predefined in the System or they will not have any chemical meaning in the structure. The element or the pseudoatom will be assigned at the position we placed on.

CREATING STRUCTURES

Example: *Specifying atom values*

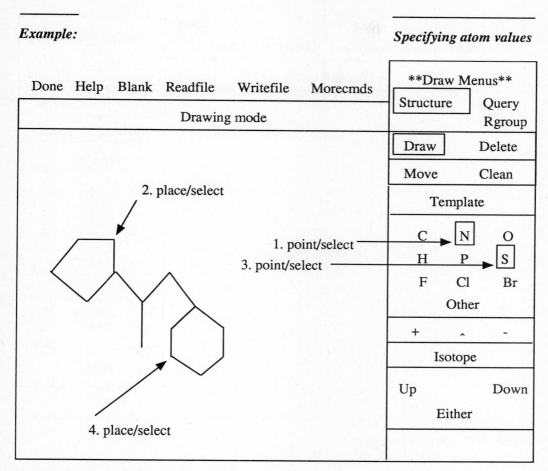

Figure 3/26. Specifying atom values

MODELS IN STRUCTURE BUILDING

To facilitate and speed up structure building we can use models to build structures. The models may be the System Models or the User models.

STRUCTURES AS MODELS

We can use any structure we retrieve from a MACCS-II database as the model for building a new structure. We simply use the FIND option in the Register Menu (see p. 506) to retrieve a structure using a Regno or Name query and exit to the Draw Mode as usual. The structure will be the current structure displayed in the Draw Menu and ready to

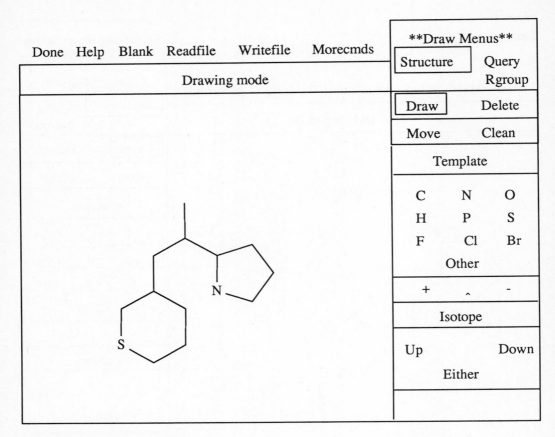

Figure 3/27. Clean structure

use as the model for structure building.

TEMPLATES

MACCS-II contains a set of predrawn structures that are stored in the TEMPL directory and can be recalled as models.

To use one of the templates, we activate TEMPLATE on the Menu (Figure 3/28) and select the desired template from the displayed Menu of the templates that are available.

For example, we select CO2H and, following the prompts, place/select in the structure where we want to attach the CO2H group. In addition to the system templates, we can use templates predrawn by the User. The fragment we wish to store as a template is drawn as usual in the Draw Mode. The first drawn atom in the template is the default point of attachment of the template. The User templates can be stored in the TEMPL directory or they can be stored in a Molfile in any directory. When we want to recall the User template we select FILE from the template list. At the System prompt we enter the name of the

CREATING STRUCTURES

Example: *Using templates*

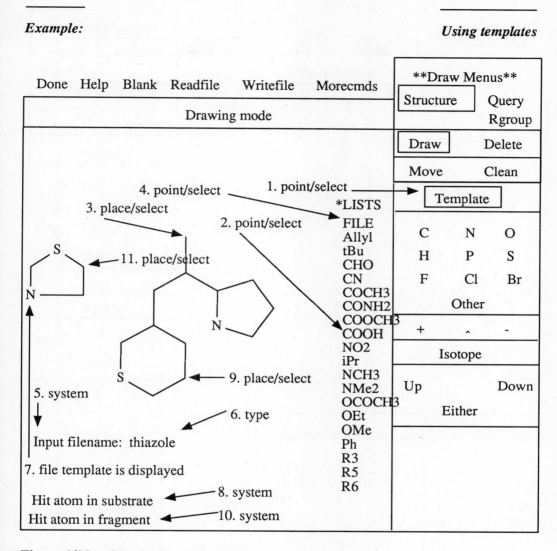

Figure 3/28. Substituting with templates

template under which we stored it in the TEMPL directory. If we want to use a fragment stored in a Molfile we enter the file (path)name. Following the System prompts, place/select in the structure to attach the template. The final structure drawn using templates is shown in Figure 3/29.

A selected template can be repeatedly used for attachment or for fusion using the "T" switch.

If we select *LISTS (see list of templates, Figure 3/28), MACCS-II displays categories of templates we can choose from:

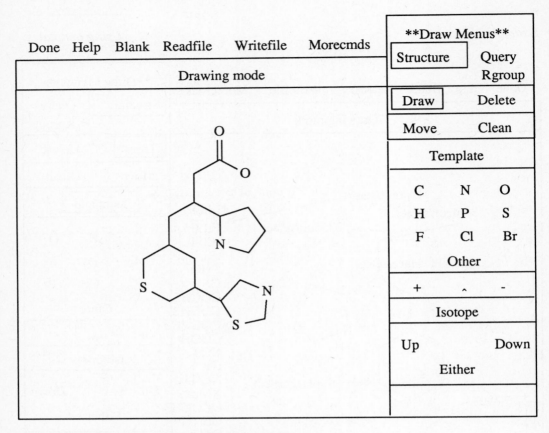

Figure 3/29. Final structure built with templates

```
*Chain
*Alkyl
*Rings
 :
 :
 :
*HeteroRings<-------selecting this item displays---------------------------->  *LISTS
 :                                                                             Ceph
 :                                                                             Indole
 :                                                                             Morph
                                                                               Morpho
                                                                               Pen
                                                                               Porph
                                                                               Py
                                                                               Pyrim
```

CREATING STRUCTURES

                                                              THF
                                                              THP
                    selecting--------------->   Thiazole

displays the thiazole ring as the template. Selecting *LISTS displays again the list of categories of templates.

We can also use the keyboard attachment of a template similarly to assigning atom values (*vide supra*). If we place on a position in the existing structure and press "+," the System will prompt:
template:
Entering a template name (for templates in the TEMPL directory) or the Molfile name (in any directory) will result in connecting the template, at the atom we placed on, by the default (1st drawn) atom of the template. We can also input an atom symbol and thus specify an atom value similarly to the = or - keyboard technique in ChemBase (see p. 304). If we do not place on a position in the structure, we will be prompted to "Hit atom of attachment" in the parent to which the template will be attached by its default atom.

CREATING STEREOISOMERIC STRUCTURES

See the discussion of Stereochemistry for more on creating and using stereochemical structures (p. 418).
The options UP and DOWN mark the bonds in stereochemical structures. The EITHER option is used for unspecified or unknown configuration at an asymmetric center. To register a stereoisomer with UP or DOWN marked bonds so that on searching only the specific isomer will be retrieved (see p. 511), we label the structure CHIRAL using the toggle "i."

CREATING *R* AND *S* ISOMERS

Specify one bond at the stereocenter UP or DOWN as required and label the structure CHIRAL by the "i" switch (Figure 3/30).

CREATING STRUCTURES OF RACEMATES

Specify one bond at the asymmetric center UP or DOWN (either one can be used) and do not label the structure CHIRAL.

CREATING STRUCTURES OF UNDEFINED CONFIGURATION

Mark any bond EITHER and do not label the structure CHIRAL, or draw unmarked single bonds.

Example: *Creating chiral structure*

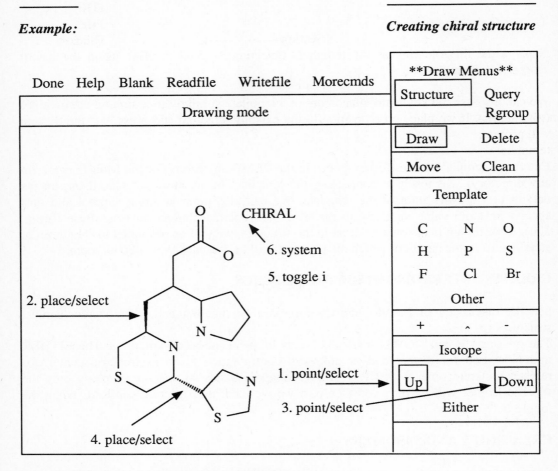

Figure 3/30. Creating configurational isomer

HYDROGENS IN STEREOISOMERS

It is a good practice to indicate hydrogens along with the other substituents at the asymmetric center. MACCS-II may perceive the configuration incorrectly if the explicit hydrogens are not indicated in the stereoisomeric structures (see p. 421). MACCS-II retains the explicit stereo hygrogens in the structure on registration (explicit hydrogens which do not indicate configuration are removed by MACCS-II on registration).

E AND *Z* ISOMERS

Draw the bonds connecting the substituents at the double bond terminal atoms in the position defined by the rules of the highest substituent priority to obtain *E* or *Z*

CREATING STRUCTURES

configuration and register the structure or submit the query to search. For structures of undefined configuration at the double bond or for mixtures of *E,Z* isomers, mark the double bond EITHER.

MORE ON DRAWING STRUCTURES

The remaining options on the Menu are used to specify special features in structures. Similar options have been discussed in Part Two.

+ - ˆ.........creates charged atoms or radicals.
      The charge can be -3 to +3 and is assigned by place/select on the atom. Plus and minus cancel each other. The first caret removes the coordination shell, subsequent caret adds one electron.
ISOTOPE.......assigns abnormal mass to an atom from -3 to +4.
      Point/select ISOTOPE will result in a prompt:
      mass difference:
      Enter any integer in the above range followed by place/select on the atom to be isotope. To label H as D or T, the latter must be defined similarly to pseudoatoms (*vide supra*).

If we for any reason wish to abandon structure building and start from scratch:

BLANK or /B.........clears screen (deletes structure(s))
DONE..........finish structure drawing and return to the
      Mode from which the Draw Mode was entered
      The structure will be the current structure.
READFILE......reads a Molfile and displays the stored
      structure which can be used as a model (see p. 464)
WRITEFILE.....stores a structure in a Molfile (see p. 463)
MORECMDS......displays options available in this Mode

See also "Correcting and Modifying Structures" (p. 495) for additional options to manipulate structures.

CONTINUOUS BUILDING OF STRUCTURES

In the above discussion of using templates, we have seen that a structure can be quickly created by attaching templates.
A similar technique, i.e., using keyboard selections while the graphic structure is on display, is used in the Attach Mode.
If we are confronted with a task to set up and build a new large database, or if we have to

register many structures into an existing database and the structures can be divided into classes of structures derived from the same parent (structural fragment) that does not change, the ATTACH Mode is very advantageous to use.

Although the options in the Attach Mode can be selected graphically from the Menu, we will find that the keyboard selection will be faster. The following example will use the keyboard selection.

The requirement for using the Attach Mode is that the structural fragment we wish to use as the parent must be the current structure with which we enter the Attach Mode or we have to make a structure current while in the Attach Mode by:

1. recalling a structure from a Molfile by the READ FILE
   option (see p. 464)
2. retrieving a structure from the database by the VIEW REGNO
   option (see p. 513)

The fragment skeleton must have all bond values we desire to have in the fragment. We cannot specify bond values for the fragment in the Attach Mode; however, we can specify or remove stereobonds in the fragment. Of course, we can enter Draw Mode, make any changes as desired and return.

ATTACH MODE

Attach Mode is entered from any of the Modes by point/select -ATTACH (Figure 3/31).

When we enter Attach Mode, the Menu is displayed (Figure 3/32) with the current structure.

The options on the Menu can be activated, after we press "k," by keyboard input at the prompt:

ATTACH>

The steps to build a structure in the Attach Mode are shown in Figures 3/33 and 3/34.

MORE ON ATTACH MODE

blank............clears the screen and the memory of
        the current structure
counters.........displays the number of molecules in the
        current list
exit.............terminates MACCS-II session and returns to
        the operating system
number...........functions as a toggle, turns on/off
        the locants display
show templates...displays a screen of templates in the TEMPL
        directory

CREATING STRUCTURES

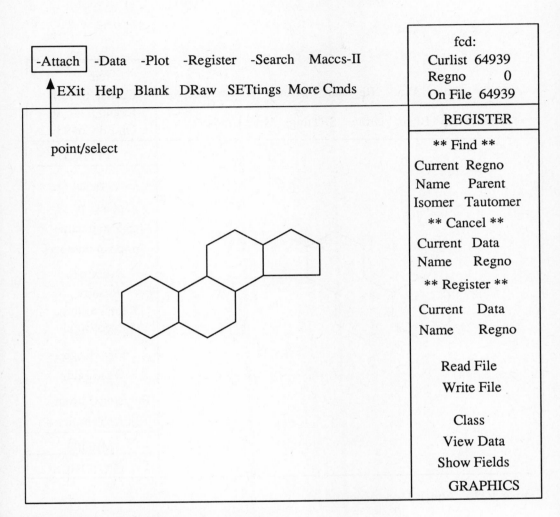

Figure 3/31. Switching to Attach Mode

	fcd:
-Attach -Data -Plot -Register -Search Maccs-II	Curlist 64939
	Regno 0
EXit Help Blank DRaw SETtings More Cmds	On File 64939

	ATTACH
	Attach a g, p; a g, p [g=group name] [p=posn number] Numbers Restore Replace Name Stereo Bond View Regno Read File Register Current Show Templates
	MEDIT
	GRAPHICS

Figure 3/32. Attach Mode Menu

CREATING STRUCTURES

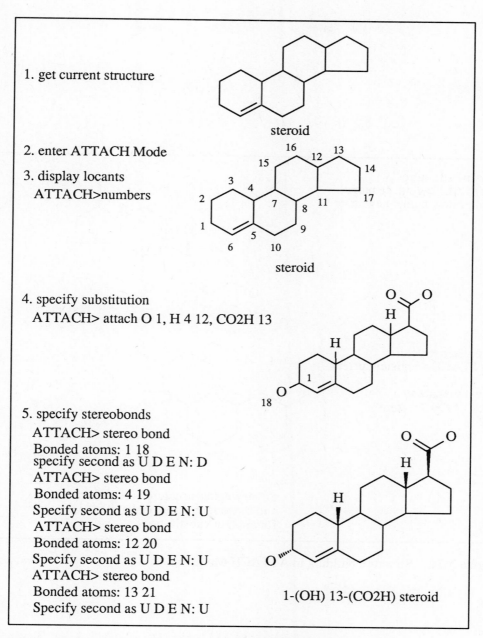

1. get current structure

2. enter ATTACH Mode

3. display locants
   ATTACH>numbers

4. specify substitution
   ATTACH> attach O 1, H 4 12, CO2H 13

5. specify stereobonds
   ATTACH> stereo bond
   Bonded atoms: 1 18
   specify second as U D E N: D
   ATTACH> stereo bond
   Bonded atoms: 4 19
   Specify second as U D E N: U
   ATTACH> stereo bond
   Bonded atoms: 12 20
   Specify second as U D E N: U
   ATTACH> stereo bond
   Bonded atoms: 13 21
   Specify second as U D E N: U

Figure 3/33. Structure building in ATTACH Mode

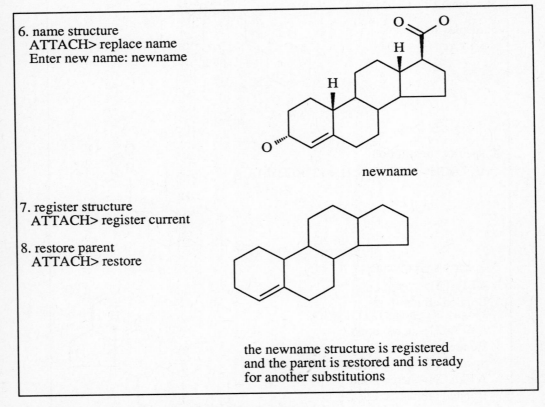

Figure 3/34. Structure building in ATTACH Mode

CREATING STRUCTURES

SAVING STRUCTURES IN MOLFILES

When we create a structure, we might wish to store the structure in a Molfile and recall it when needed either as a building block for another structure or as a query for searching. To use the structure as a template, we already know, the first drawn atom will be the default point of attachment of the fragment. We can also save structures we retrive in searching a database. The writing of the file is carried out by activating the WRITEFILE option (Figure 3/35).

Note that if we want to use the file name extension ".mol" we have to type it in; MACCS-II does not append the ".mol" automatically. WRITEFILE option is available also in the other Modes.

To recall the structure in a Molfile, we use the READFILE option in Draw, Register (Figure 3/36), Search, Attach, and/or Plot Mode.

Example: *Writing Molfiles*

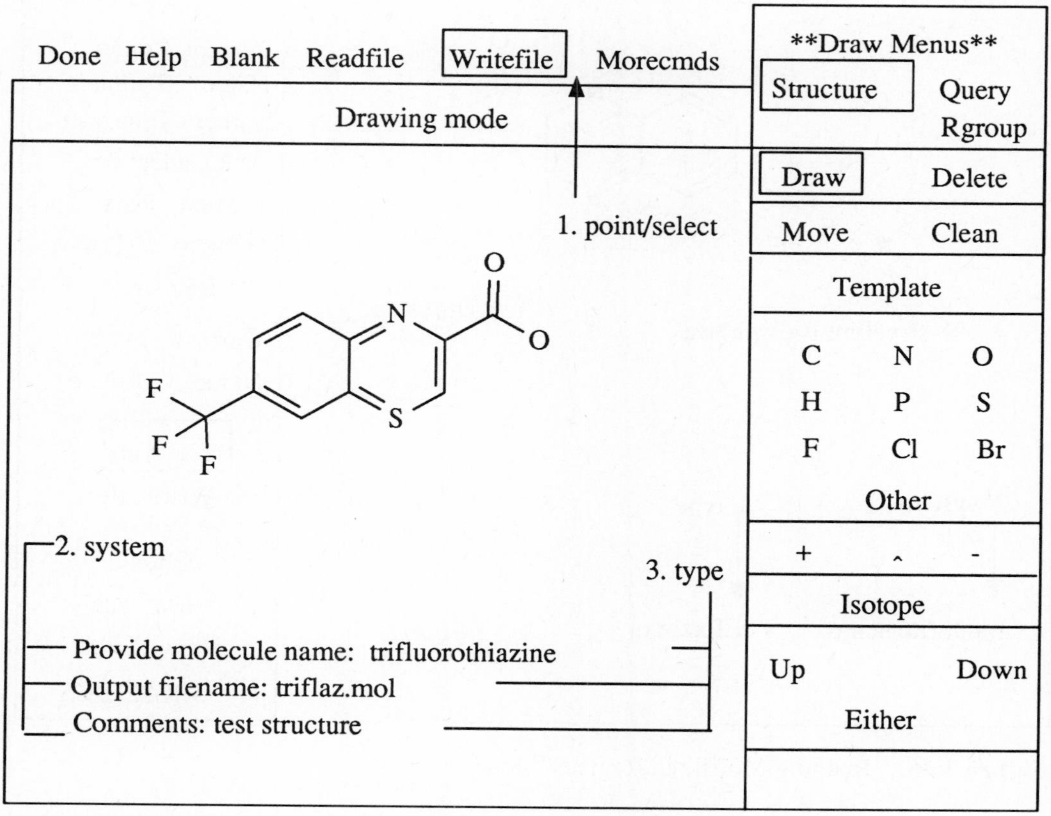

Figure 3/35. Writing Molfile

Example: *Reading Molfiles*

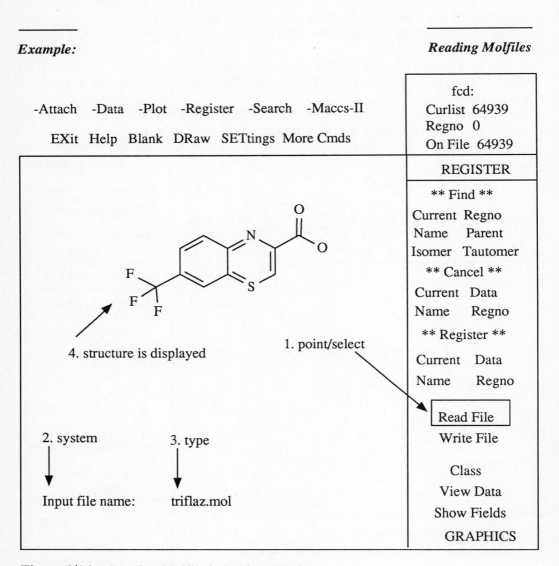

Figure 3/36. Reading Molfile in Register Mode

5   CREATING GENERIC STRUCTURES

Similarly to the other systems that we have discussed in Part One and Part Two, the MACCS-II System includes methods for building generic structures. These generic structures cannot be registered in a MACCS-II database although they can be saved in Molfiles. They are used as queries for substructure searching.
The Query Menu which will be discussed here allows for creating simple generic structures. The Users who have installed the MACCS-II Power Search Module will be able to construct complex generic structures by defining Rgroups attached to a parent structure (see p. 476).

The options for building simple generic structures are available in Draw Mode-Query Menu which is entered from the Structure Menu (Figure 3/37).
The Query Menu is displayed (Figure 3/38).
Most of the options in the Query Menu have the same function as those in the Structure Menu. We can build the structure skeleton in both Modes. Note that there are no options in the Query Menu for specifying UP, DOWN, or EITHER bonds which have to be drawn in the Structure Menu. The STEREO option on the Menu is used to mark the stereo centers in queries for searching relative configuration in substructures (see p. 525).

SYSTEM VARIABLES

The A and Q symbols on the Menu are used to specify the following variables in a structure (Figures 3/39, 3/40):
A........any atom except hydrogen
Q........any atom except carbon and hydrogen

USER VARIABLES

In addition to the System variables, we can create any combination of variables out of the elements in the periodic table. Those atoms are selected from the Menu as usual and place/selected on the locant where the list of variables is desired (Figures 3/40, 3/41).

ATOM NEGATION

If we want to create a query for searching structures in which specified atoms are to be excluded (atom negation), we activate the EXCLUDE option (Figures 3/42, 3/43).

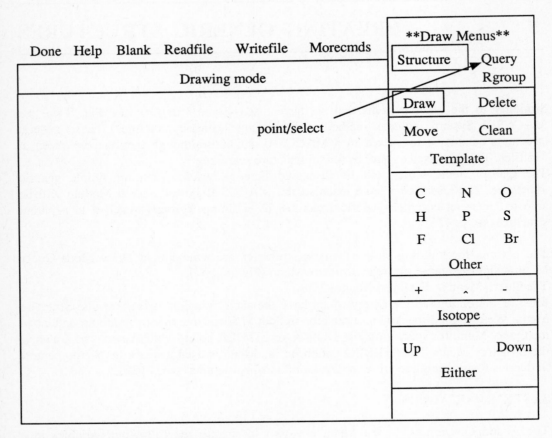

Figure 3/37. Entering Query Menu

HYDROGEN COUNT

If we create a structure query in which an atom has unfilled valences and submit the query to the Substructure Search (SSS), we permit any substitution at the atom by hydrogen(s), any element(s) and/or any structural fragment(s) in the retrieved structures to satisfy the atom valences.

To prevent substitution and/or to specify the degree of substitution at an atom, we either draw explicit hydrogens at the particular atom or use the HYD option to specify the minimum number of hydrogens required at the particular atom in the retrieved structures. The System will prompt for the number of hydrogens. We can enter a value 0-4 (Figures 3/44, 3/45). Note that the switch "L" (see p. 445) which turns on the hydrogens in structures does not specify hydrogen count and does not create explicit hydrogens for the purpose of searching.

In the example in Figure 3/45 the hydrogen count is a minimum of 2 hydrogens, consequently, no substitution is allowed at the locant. If we specify H1, there must be one

CREATING GENERIC STRUCTURES

```
Done   Help   Blank   Readfile   Writefile   Morecmds    **Draw Menus**
                                                         Structure    Query
                       Drawing mode                                   Rgroup

                                                         Draw      Delete
                                                         Restore   Exclude
                                                         Hyd    Link   Stereo
                                                          C      N      O
                                                          H      P      S
                                                          F      Cl     Br
                                                          A    Other    Q
                                                         Ring# Subst Unsat
                                                         Chain      Ring
                                                         Any         Arom
                                                         S/A   D/A   S/D
                                                         IsoRing   Nosubst
```

Figure 3/38. Query Menu

hydrogen and any substituent including a hydrogen in the retrieved structures. If we specify H0, there must be 2 nonhydrogen attachments.

SUBSTITUTION AND NO SUBSTITUTION

In the example in Figure 3/45, we control the substitution degree by the number of hydrogens that are required at the specified position.
The Query Menu also enables us to specify the number of nonhydrogen attachments at an atom (Figure 3/45-A).
Note that the degree of substitution is the sum of the nonhydrogen attachments already present at the atom and those we wish to specify as nonhydrogen attachments (see also CONNECT, p. 152 for further illustration of substitution degree).
In Figure 3/45-A we specified the substitution s2. There are already two nonhydrogen attachments at that atom so that only two hydrogens will be present at that atom in the retrieved structures. If we specify s3, we require one nonhydrogen substituent and one

Example:

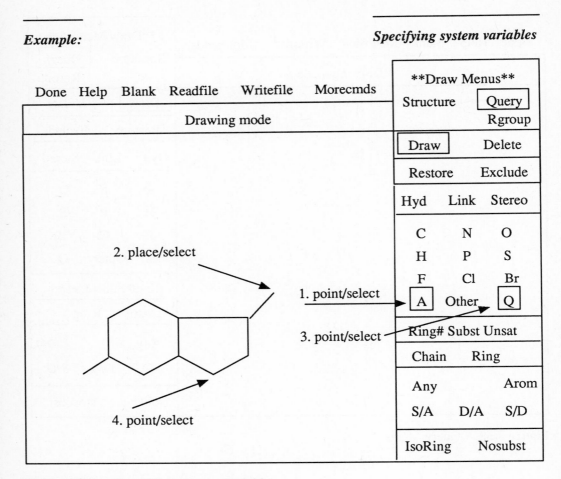

Figure 3/39. Specifying system variables

hydrogen in the retrieved structures. If we specify s4, there will be no hydrogens, only two nonhydrogen substituents. MACCS-II offers the following choice in specifying the substitution:
off.........removes the existing substitution specification
0...........no substitution permitted
1,2,3,4,or 5...substituent(s)
6..............6 or more substituents
*..............substitution as drawn in the query
To cancel the substitution label we use RESTORE, or point/select Subst and enter "off."

If we wish to prevent nonhydrogen attachments at all atoms in the structure query, we point/select Nosubst (Figure 3/45-B). All atoms will be marked by s*; however, the spe-

Example: *Specifying variable atoms*

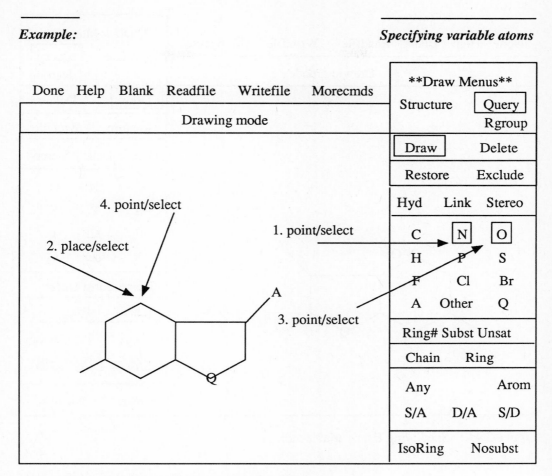

Figure 3/40. Specifying system variable atoms

cified substitution (s2) will be retained.

To remove all s*, point/select NOSUBST. The (s2) will be retained. Any s* labeled atom can be relabeled with a specific substitution degree using the SUBST option.

RELATIVE CONFIGURATION

In creating queries for substructure searching, we can specify relative configuration in the query fragment which must be contained in the hit structure. We draw the stereobonds in the Structure Menu.
If no relative configuration was specified, or if only one stereocenter is marked, all stereoisomers would be retrieved including structures with no stereochemistry.
To mark the stereocenters, we activate the STEREO option and select the desired

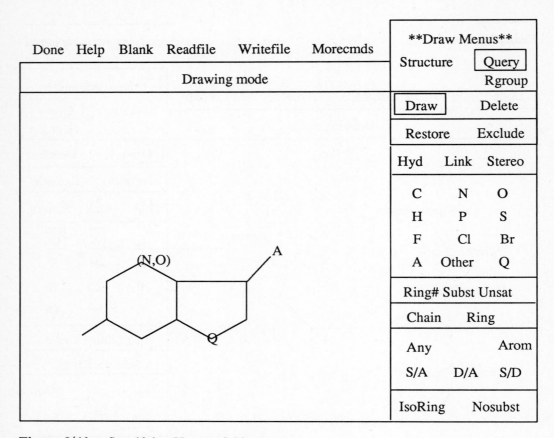

Figure 3/41. Specifying User variable atoms

stereocenters (Figure 3/46). The stereocenters will be marked by a square (Figure 3/47).

VARIABLE BONDS

Variable bond values are specified using the following options on the Menu:

ANY..........the bond can have any value
AROM.........the bond must be aromatic
CHAIN........the bond must be in a chain
RING.........the bond must be in a ring
S/A..........the bond may be single or aromatic
S/D..........the bond may be single or double
D/A..........the bond may be double or aromatic

The variable bonds are graphically represented as shown in Figure 3/48.

CREATING GENERIC STRUCTURES

Example: *Atom negation*

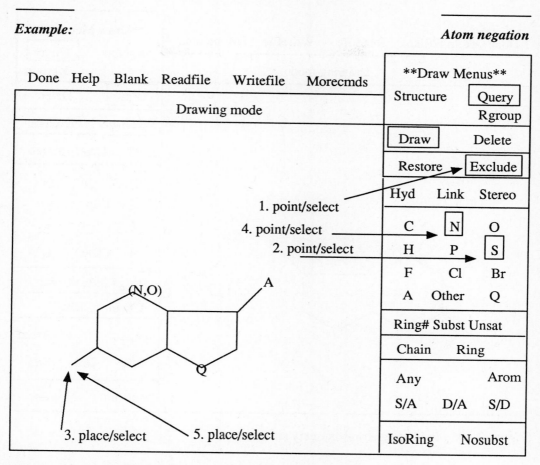

Figure 3/42. Excluding atoms in query structure

For discussion of variable bond values, see also p. 142.
Figure 3/49 and Figure 3/50 show the assignment of single/ double value to the bond.

UNSATURATION AT ATOMS

We can specify that an atom or atoms must have at least one multiple bond - double, triple, or aromatic (Figure 3/50-A).
The atom will be labeled with "u." To remove the label, we repeat the step or point/select RESTORE.

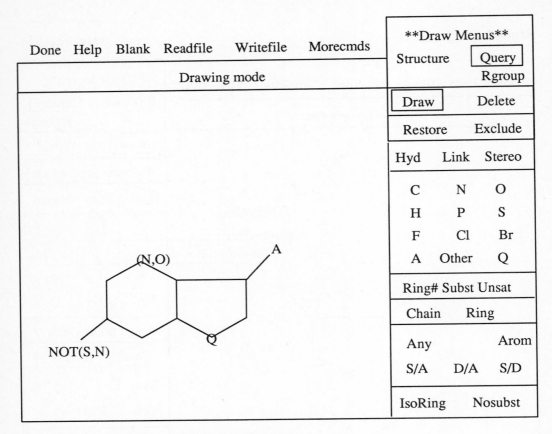

Figure 3/43. Excluding atoms in query structure

RING ATOMS

The atom(s) in query structures can be specified to have a definite number of ring connections. Therefore, we are able to require that an atom is in a cyclic or acyclic environment (compare NODE SPECIFIC, p. 144, 147). The Query Menu has the option Ring# which is activated to input the following choices:

0........no ring connections at the atom (the atom is acyclic)
2........2 ring connections at the atom (a simple ring atom)
3........3 ring connections at the atom (atom in fused ring
           or a bridgehead atom)
4........4 or more ring connections at the atom (spiro ring)
*........ring connections as drawn in the query structure.

In the example in Figure 3/50-B, the atoms labeled "r" will have the specified ring connections. Since there are already two ring connections at those atoms, fusion at this

CREATING GENERIC STRUCTURES

Example: *Specifying hydrogen count*

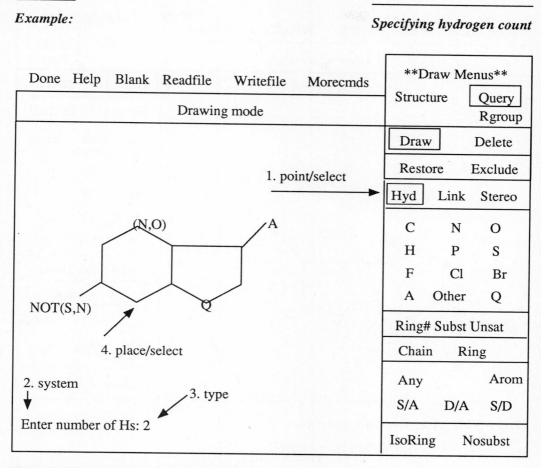

Figure 3/44. Specifying hydrogen count

side of the ring will be prevented.
To remove the "r" label, use RESTORE or again the option RING#.

ISOLATED RINGS

In the example in Figure 3/50-B, we prevented fusion at one side of the ring. To specify that the entire ring is not to be embedded in larger ring systems, we use the option ISORING (Figure 3/50-C).

Note that the ring systems that share ring bonds are isolated as a unit (structure **1** in Figure 3/50-C) while in the ring systems that share acyclic bonds (stucture **2** in Figure 3/50-C), we can isolate either the 5-membered ring (by place/select on an atom in the ring) or the

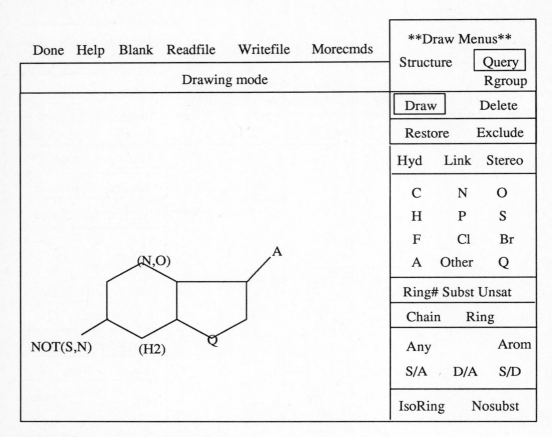

Figure 3/45. Specified hydrogen count

6-membered ring (by place/select on an atom in that ring), or both by place/select on an atom in both rings.
To remove the isolated ring labels, point/select IsoRing and place/select any atom with the r* label. Those atoms labeled with "rn" (i.e., the specific ring bond count) are retained. Compare RING SPECIFIC, p. 147.

REPEATING ATOMS

The atom(s) in a query structure can be designated to represent a linear chain of atoms. The Link option (Figure 3/50-D) specifies the link.

The linked atom in Figure 3/50-D is labeled with L and the range of the chain, i.e., 1-3 and with arrows embracing the linked atom. If the link atom has two bonds, MACCS-II marks the bonds with the arrows automatically. If there are three or more bonds, we have to mark the atom by place/selecting the bonds which connect the linked atom to the other atoms

CREATING GENERIC STRUCTURES

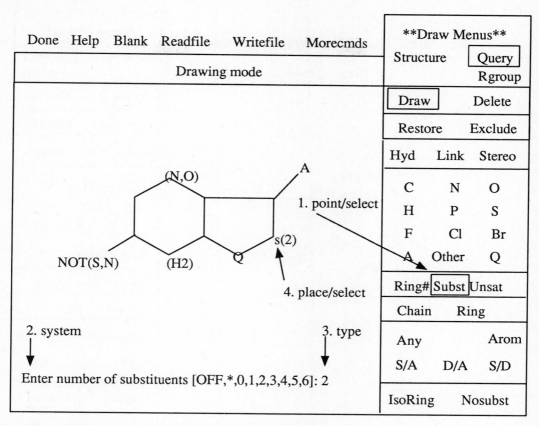

Figure 3/45-A. Specified degree of substitution

with which it is connected to make the query fragment.

MORE ON QUERY MENU OPTIONS

RESTORE........removes labels from atoms and/or restores heteroatoms
　　　　　　　to neutral carbons, or bonds to single bonds
READFILE......reads a Molfile
WRITEFILE.....writes a Molfile (we can store generic queries;
　　　　　　　however, they cannot be registered in a database)
MORECMDS......displays available options
HELP..........calls up the Help System
DONE..........exits to the Mode we came from into the Draw
　　　　　　　Mode, with the created current structure

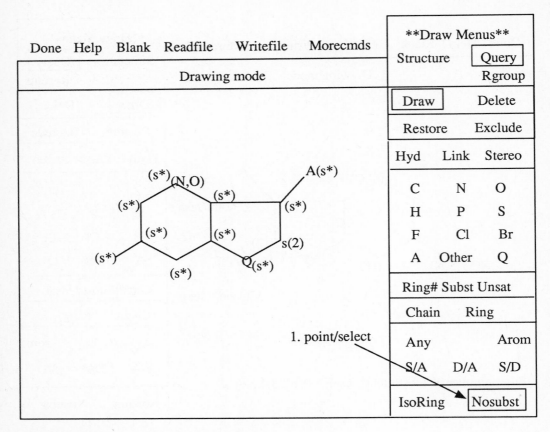

Figure 3/45-B. Specified no substitution

GENERIC STRUCTURES WITH DEFINED R GROUPS

The Users who have access to MACCS-II system with the add-on Power Search module will be able to construct generic structures with defined R groups. These are the structures which we have discussed in Part One (see p. 119) and in describing the Markush structure to illustrate substructure searching in ChemBase (see p. 380).

Generic structures consist of a parent structure in which R group(s) are attached at specific locants. The R group(s) are defined to represent a set of elements or structural fragments. A generic structure, when submitted as a query to the Substructure Search in the Power Search module, will retrieve structures containing the parent structural fragment and at least one unit of the set defined in the R group(s).

In MACCS-II, the parent structure is called the **root**. The units in a set of an R group are called **members**.

Since this MACCS-II capability to create generic structure queries is available only if the Power Search (PS) module has been installed, the examples and figures in the following

CREATING GENERIC STRUCTURES 477

Example: *Designating relative configuration*

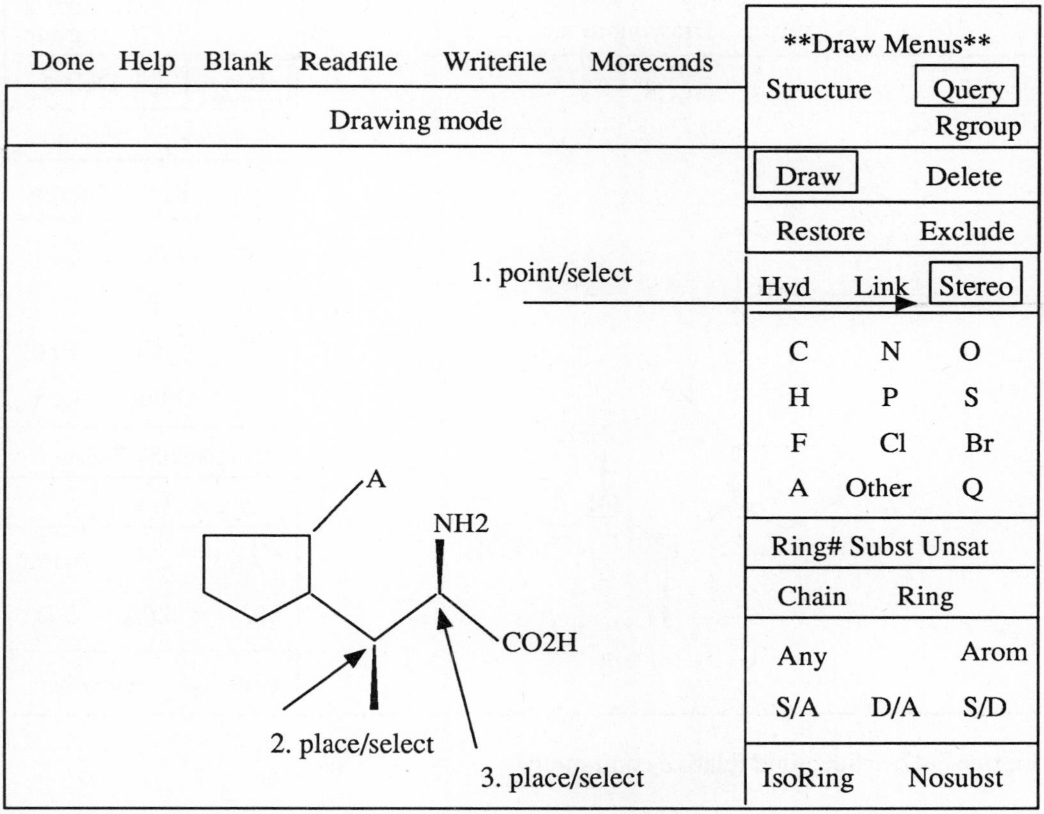

Figure 3/46. Specifying relative configuration

discussion are numbered with the suffix PS independently of the other figures of Part Three. An example of a generic structure with R groups is shown in Figure 3/1-PS.
The root is created in the Structure Menu or the Query Menu by the usual techniques that we have already described.

ATTACHING R GROUPS TO ROOT

When we have completed the root, we enter the Rgroup Menu from Structure or Query Menu. The root is displayed on the screen which is called the **root screen**. The R groups are attached as though they were atoms (Figure 3/2-PS).
There can be 1-32 R groups in a root. R1-R9 can be selected from the menu; higher Rs are attached by selecting >R9 and by inputting the number "n" at the system prompt. More

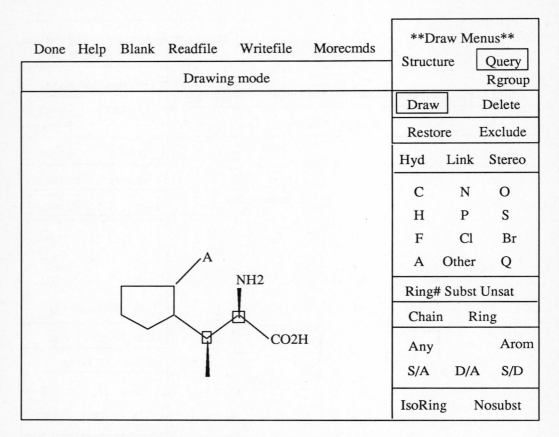

Figure 3/47. Specified relative configuration

than one R group can be attached at a locant by the same or different bond type (e.g., R4R5 in Figure 3/1-PS). An R group, and consequently the members of the R group, can be attached to the root by one or two attachments (e.g., R1 and R6 in Figure 3/1-PS). R groups can be nested to the second level, e.g., R3 in Figure 3/1-PS; however, the nested R group can be attached to the root by only one attachment. Nesting to the third level is not permitted (e.g., in this example members of R3 cannot have R groups).

DEFINING R GROUPS

The members of R groups are defined by point/select DEFINE (Figure 3/3-PS).
The Menu is divided into the Definition Screen and the Drawing Screen (Figure 3/3-PS). The Definition Screen will show the members of the selected R group which is being defined. The Drawing Screen will show the members being created. Upon completion, each member is moved to the Definition Screen.
In defining the members, we use all the techniques in Structure and Query Menu.

CREATING GENERIC STRUCTURES

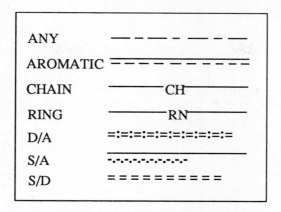

Figure 3/48. Graphic variable bonds

To define the members H and Cl for R1, we enter the Query Menu and create an atom list (Figure 3/4-PS).
Note that when we switch to the Query Menu, the Definition Screen is retained. The steps 5 and 6 in Figure 3/4-PS can be performed in the Query or Structure Menu. They accomplish the following:
Step 5.: toggle "a" specifies the attachment point for the
   atoms in the list
Step 6.: point/selecting R1 makes the atom list the member
   which is moved to the definition screen. The atom
   list is removed from the drawing screen. The point
   of attachment is marked by the arrow and *.
Note that the atoms H, Cl could have been defined in Structure Menu as separate members by the same, but two steps. Placing any number of atoms in a list in Query Menu saves steps for defining the separate members.
To define COCH3, we enter Structure Menu and select TEMPLATE and from the template list select COCH3 (Figure 3/5-PS).
The other members of R1 would be defined similarly.
The definition screen can show up to four members. If there are more than four, we can browse through them when MACCS-II displays More ▼ or More ▲ by clicking the appropriate arrow.
To define R2, we can create the fragment in the Structure Menu and enter the Rgroup Menu or draw the pyrrolidinylmethyl member in the Rgroup Menu and then define the R2 member (Figure 3/6-PS).
In Figure 3/6-PS, we have selected in step 6 the option AttachPt to specify the attachment point of the member. This is the same function we accomplished with the toggle "a" in the Structure or Query Menu.
To define the nested R3 group, we proceed in the same steps (Figure 3/7-PS).

Example: *Specifying variable bonds*

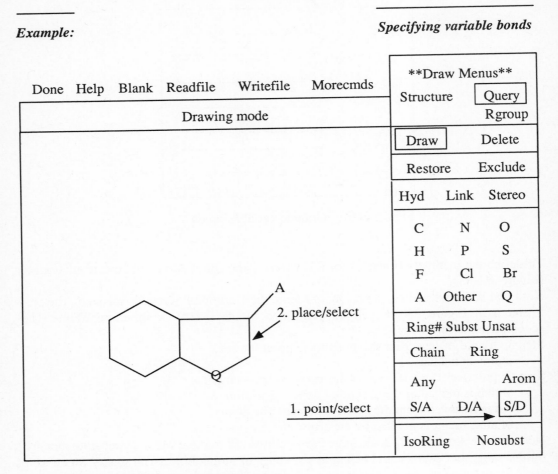

Figure 3/49. Specifying variable bond values

Step 5 in Figure 3/7-PS, selecting the NewMem option, accomplishes the same function as point/select R3 in the definition screen, i.e., the fragment becomes the member and is moved to the definition screen. The ethyl fragment for the R3 group member is defined similarly.

The members that we have defined for R1, R2, and R3 have one point of attachment which is indicated by the ---->* in the definition screen. A member can have also two points of attachment. For example, if we had R7 group in the pyrrolidinyl fragment (Figure 3/6-PS) and specified the R7 connection to the root S/D and defined a member for this R7 as O (oxygen) with one point of attachment, there would be a hydroxyl in the retrieved structures. If we marked O with two points of attachment, there would be a carbonyl in the retrieved structures. The R groups in the root have one attachment. However, if we have a root in which an R group has two attachments, the members of the

CREATING GENERIC STRUCTURES

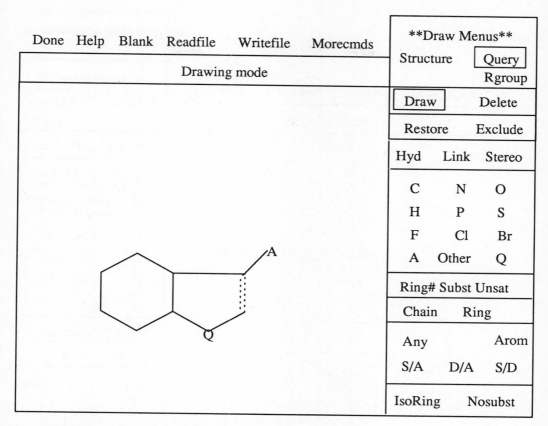

Figure 3/50. Specified variable bond value

group will have also two points of attachment. Depending which atom in the member is marked as the points of attachment, we retrieve different structures. When we insert an R group with two attachments, MACCS-II marks the second attachment by the " label (e.g., R6 in Figure 3/8-PS).

The members with the nitro group in the query (Figure 3/8-PS) were marked at different points of attachment, consequently, two different structures would be retrieved (Figure 3/9-PS).

When we are finished with defining R groups, we point/select Display to display the root and the members of R groups in the display screen (Figure 3/10-PS).

OCCURRENCE OF R GROUPS

The display of the root and R group members (Figure 3/10-PS) shows values of R1>0, R2>0, and R3>0 for the R groups. These are the default values indicating the number of times an R group can be present in the root. The >0 specifies that at least one or more of

Example: *Specifying atom unsaturation*

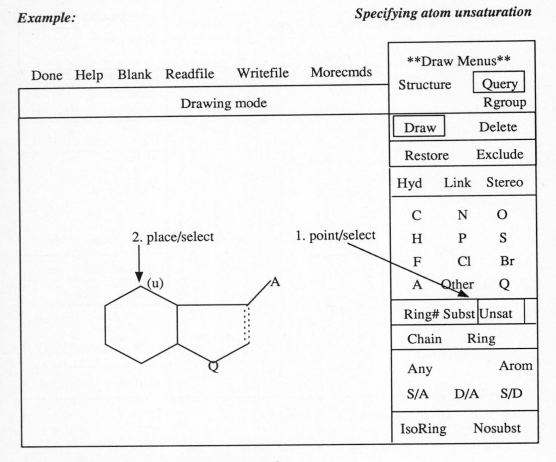

Figure 3/50 -A. Specifying atom unsaturation

the R groups must be present. We can redefine the default by point/select the Occur R option in the Rgroup Menu. On MACCS-II prompt, we can enter the following values:
0.....R group is not permitted to occur
n.....exactly n (e.g., 3) occurrences of the R group
n-m...a range of n-m occurrences (e.g., 1-4)
>n....greater than n (e.g., >4, more than 4)
<n....less than n (e.g., <4, less than 4)
Any combination of the above values (except 0) separated by commas can be used.
The specification Occur R is useful when more than one of the same R group is in a root. Figure 3/11-PS shows an example of a structure with identical R groups.
The occurrence in Figure 3/11-PS was set for R1 = 2. The search is thus forced to retrieve *o-*, *m-*, and *p*-isomers. The occurrence for R2 = 1 which results in one R2 group in either

Example: *Specifying atom ring bonds*

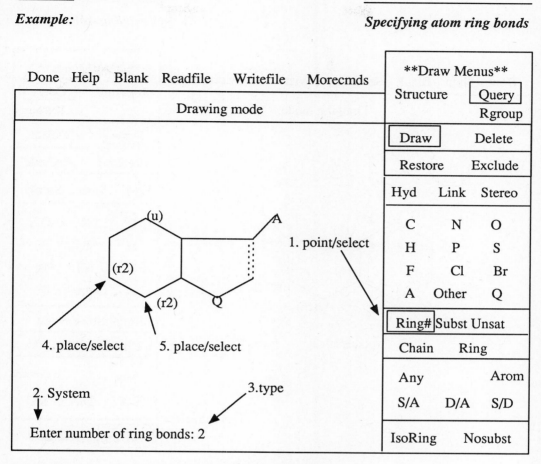

Figure 3/50 -B. Specifying number of bonds at atom

of their position in the *o-, m-,* and *p-*isomers.
If we set R1=4 and R1=2, the fully substituted benzene with R1 and R2 groups would be retrieved (but see Rest H, *vide infra*).

IF NOT R GROUP THEN HYDROGEN

In Figure 3/11-PS we have set the occurrence for R1 = 2 and claimed that the positional isomers would be retrieved. However, that would be the case only if all the positional isomers were present in the database. If some of them are not in the database, then according to the rules of Substructure Search, any substitution is allowed, i.e., by hydrogen or any other element or fragment. We may wish to specify that in the case that some of the isomers are not found, only hydrogens be present at the positions unfilled with

Example: *Specifying isolated rings*

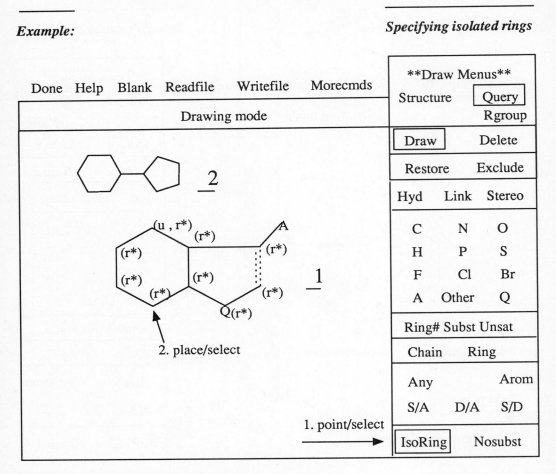

Figure 3/50 -C. Specifying isolated ring

R1 groups. We can do so by toggling the option Rest H.

The default for Rest H is off. If we toggle Rest H on, all unfilled positions of R1 or R2 will be filled with hydrogens in the retrieved structures.

The Rest H toggle can be used when we have the definition screen for R group on display. For an R group not currently in the definition screen, we point/select Rest H followed by point/select the appropriate Rn from the Menu.

MORE ON RGROUP MENU

Delete..deletes atoms and/or bonds
Root....displays the root in the root screen

CREATING GENERIC STRUCTURES

Example: *Atom linking*

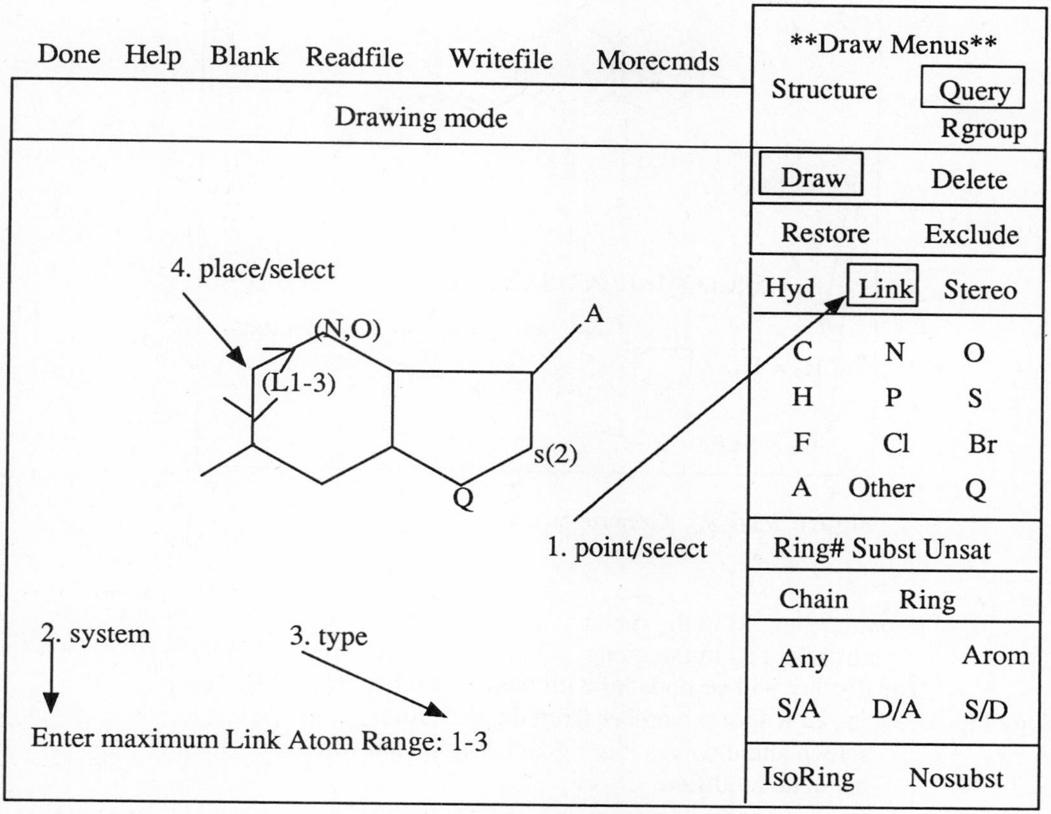

Figure 3/50-D. Specified atom linkage

Rnew....selects the next available Rn (the next
  higher, or lower Rn due to previous deletion)
Copy R..copies the members defining one R group (Ri) into
  a second R group (Rj) (Occur R and Rest H values are
  not copied in this process)
  1. Copying in the definition screen:
     a. point/select Copy R
     b. point/select Ri in the menu
     The members of Ri will be displayed in the definition
     screen.
  2. Copying in the root screen:
     a. point/select Copy R

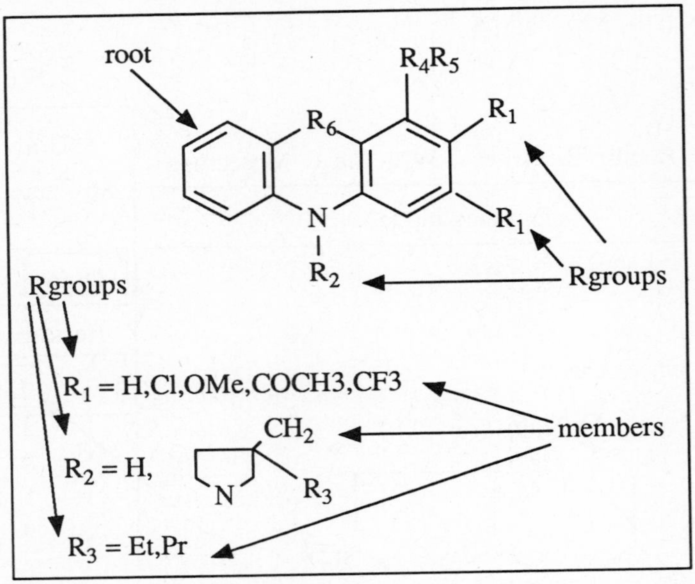

Figure 3/1-PS. Generic structure

       b. point/select Ri in the menu
       c. point/select Rj in the menu
      The display will be updated with the Ri members for the Rj group.
Copy Mem...copies an R group member from the definition
        screen and displays it in the drawing screen where
        it can be modified
    1. Copying a member from current definition screen:
       a. point/select Copy Mem
       b. point/select the member in the definition screen
      The member is displayed in the drawing screen.
    2. Copying a member from noncurrent definition screen:
       a. point/select Define
       b. point/select the desired Rn in the Menu
       c. the members are displayed in the definition screen
       d. point/select the desired member
      The member is displayed in the drawing screen.
      If the drawing screen contains a fragment, the member
      behaves like a template.
Del R....deletes all members defining an R group. The R group
        label remains in the root and can be used for
        another definition.
       a. point/select Del R

Example: *Attaching R groups to root*

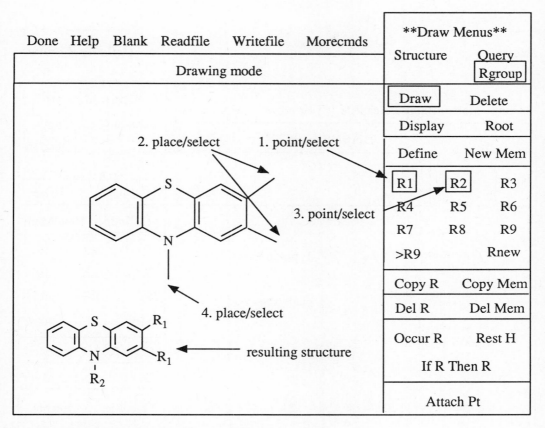

Figure 3/2-PS. Attaching R groups to root

        b. point/select the desired Rn in the menu
        If the R group label is not deleted from the root,
        MACCS-II informs: Rn: >0 *Undefined
Del Mem.....deletes a member of an R group
    The deletion can be performed both in the definition
    or the display screen.
        a. point/select Del Mem
        b. point/select the member to be deleted
If a member is both in the definition area and the
drawing area, the member must be deleted by point/selecting
it in the definition area.

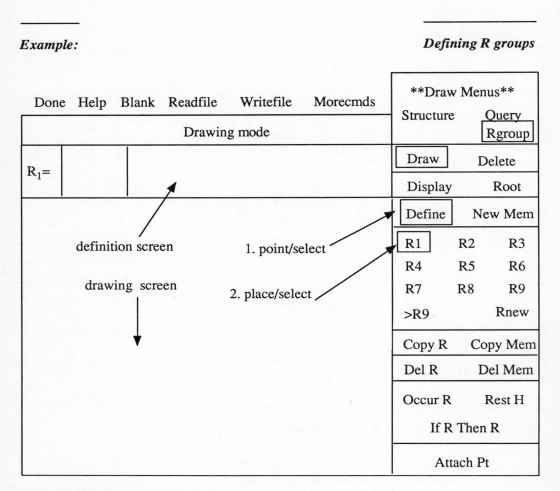

Figure 3/3-PS. Defining R groups

CREATING GENERIC STRUCTURES

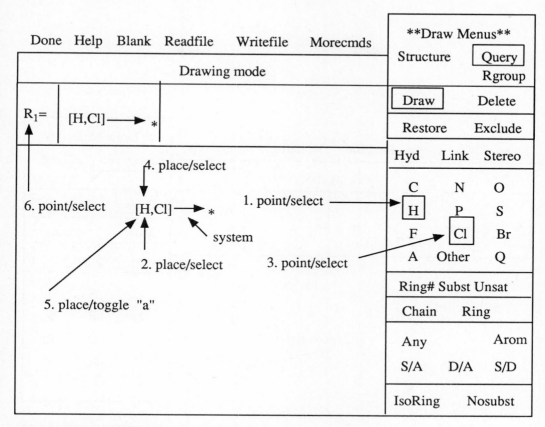

Figure 3/4-PS. Defining members of R1

If R Then R.....sets conditional logic for Ri and Rj groups
        If Ri is found in the search with the query
        then Rj must be also found to produce a hit.
        If Ri is not found, Rj is not searched for.
        The hit will be produced regardless of the
        presence or absence of Rj.
  a. point/select If R Then R
  b. point/select Ri in the Menu
  c. point/select Rj in the Menu
        i => j will be displayed with the query
To remove the condition after it has been set:
repeat the a, b, c. MACCS-II will ask if the If
condition is to be deleted. i => j is removed.

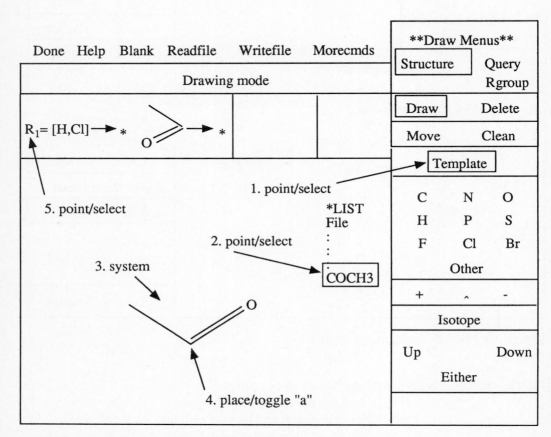

Figure 3/5-PS. Defining member of R1

CREATING GENERIC STRUCTURES

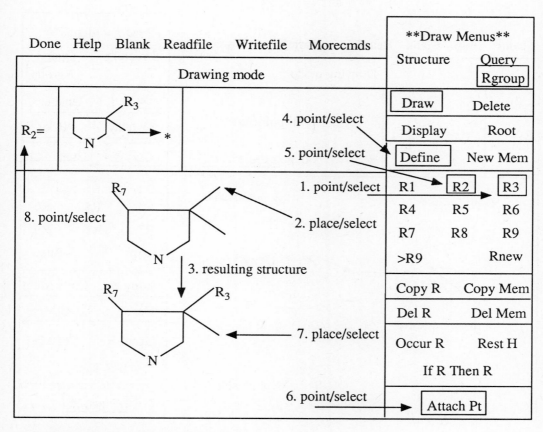

Figure 3/6-PS. Defining R2 member

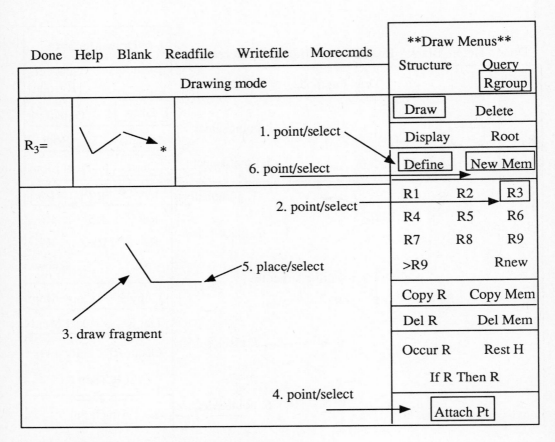

Figure 3/7-PS. Defining R3 group member

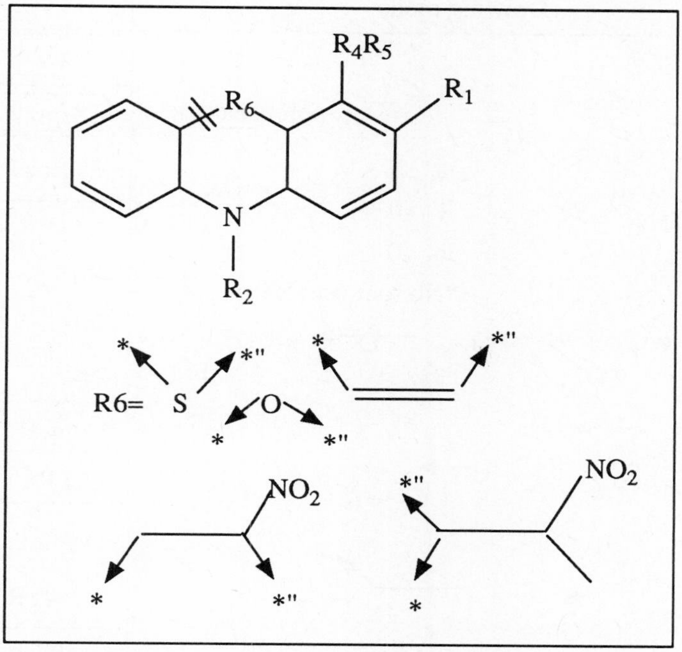

CREATING GENERIC STRUCTURES

Figure 3/8-PS. Members with two attachment points

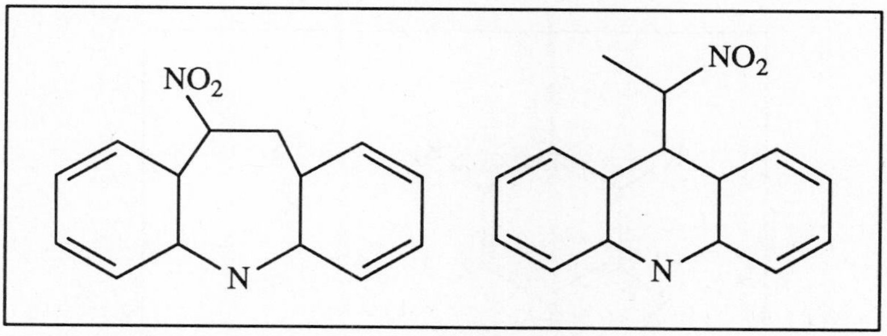

Figure 3/9-PS. Retrieved structures

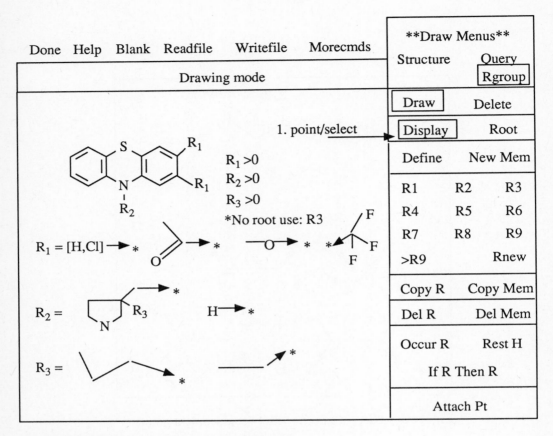

Figure 3/10-PS. Displaying root and members

Figure 3/11-PS. R group occurrence

6 CORRECTING/MODIFYING STRUCTURES

The structures we have created, retrieved from a database, or recalled from a Molfile can be corrected, modified, and manipulated on the screen. In addition to activating the Menu options, we use the switches and toggles to work with structures.

DELETING STRUCTURES, ATOMS, AND BONDS

To delete all current structures:
point/select BLANK or use the switch /B.
The screen will clear and we can start drawing a new structure.
To delete atoms and bonds, we activate the DELETE option on the Menu Figures 3/51, 3/52).

CLEANING STRUCTURES

At times, when we draw structures and specify atom and bond values and/or respecify the values, we may end up with a structure that shows the inserted symbols in inexact positions. We may also get some strange characters due to the telephone line noise, etc. To clean the structures from these minor blemishes, we use the switch "d."
To reshape structures with major distortion we use the CLEAN option (Figures 3/53, 3/54).

RESIZING STRUCTURES

The size of structures can be changed by using the switches:
E or >......makes the structure larger (enlarge)
R or <......makes the structure smaller (reduce)

REORIENTING STRUCTURES ON THE SCREEN

The structures can be reoriented on the screen so that a bond within the structure becomes vertical or horizontal. The entire structure is redrawn by MACCS-II to match the indicated bond.
To make a bond horizontal or vertical, place on the bond and use the switches:
h...........the structure is reoriented so that the selected
          bond becomes horizontal
v...........the structure is reoriented so that the selected
          bond becomes vertical

Example: *Deleting atoms and bonds*

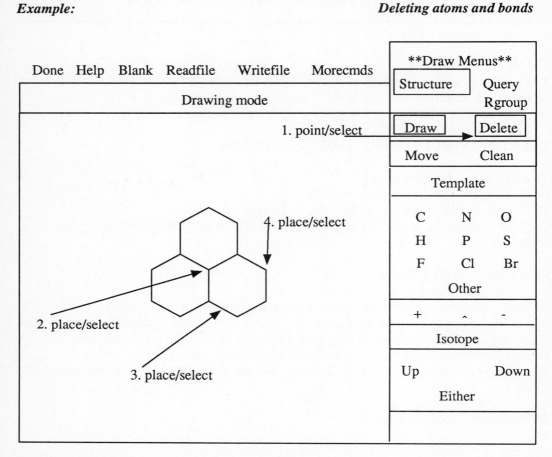

Figure 3/51. Deleting atoms and bonds

MOVING STRUCTURES ON THE SCREEN

To move the entire structures about the screen to create a page layout which we may want to print or plot, or to make room on the screen for incoming structures from Molfiles or TEMPL file, or for starting to draw another structure on the same screen, we place/select on an atom in the structure to be moved and activate the switch:
s...........shift
followed by place/select on the new position where we want to shift the structure. The structure is moved with no changes in the structure shape.
To simply move the structure to the center of the screen, activate the switch:
O(zero).......moves structure and centers it on the screen

CORRECTING/MODIFYING STRUCTURES

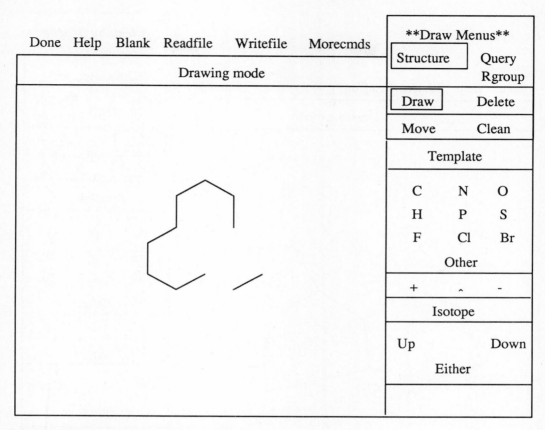

Figure 3/52. Deleted atoms and bonds

MOVING ATOMS IN STRUCTURES

Moving atoms within a structure to a new position results in a new shape of the structure. We activate the MOVE option (Figures 3/55, 3/56) to move atoms. The length of bonds may be altered by moving atoms (this is like the Draw Rubber Band technique, see p. 295). Moving atoms does not change connectivities, i.e., if we move an unconnected atom to an existing bond, the atom is still unconnected even if we place it at the end of the bond. Also, a double or triple bond cannot be made single by moving and superimposing the bonds.

MANIPULATING GROUPS IN STRUCTURES

Groups in structures are assemblies of the atoms which represent a structural fragment connected to the main structural skeleton by an acyclic bond (see also p. 332, 335). The group has to be defined before it can be manipulated.

Example: *Cleaning structures*

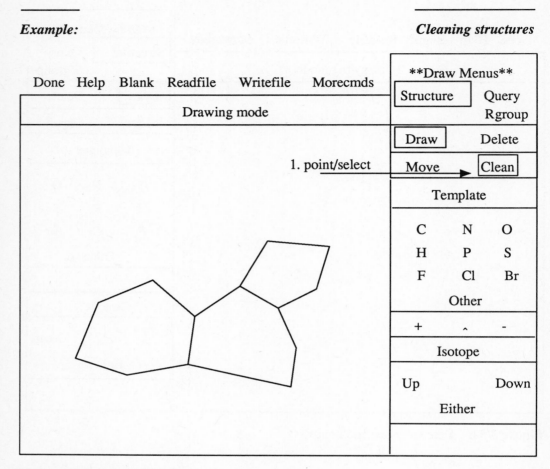

Figure 3/53. Cleaning structure

GROUP MODE

The groups are manipulated in the Group Mode which can be invoked in the Draw Mode and/or in the Register Mode by the switch:
g...........switches to Group Mode which is driven by the
       prompt GMODE:
The following options are available in the GMODE:
d.......deletes the defined group
l.......makes the group larger
s.......makes the group smaller
f.......flips the group 180 degrees around the acyclic bond

CORRECTING/MODIFYING STRUCTURES

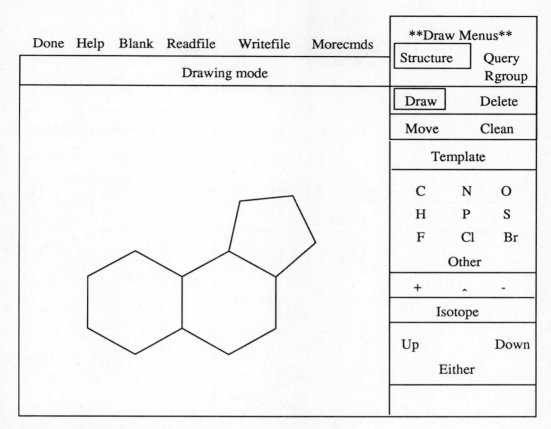

Figure 3/54. Cleaned structure

r.......rotates the entire group to a new position
t.......translates group in the screen plane
The following steps are involved in manipulating groups in the GMODE:
1. Depress the key to activate an option.
2. Define the group by place/select on the acyclic bond
   followed by place/select on an atom in the group.
3. For the options d, l, s, f: place/select on an atom in the group.
4. For the options r, t: place/select on an atom in the group
followed by place/select on the new position on the screen
where we wish to rotate or translate the group.
To abort GMODE:
press <cr> at the prompt GMODE:
Figure 3/57 shows the group manipulations.

Example: *Moving atoms in structures*

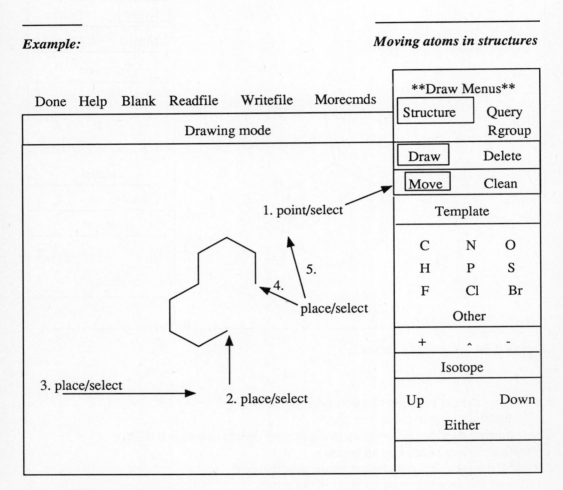

Figure 3/55. Moving atoms

CORRECTING/MODIFYING STRUCTURES

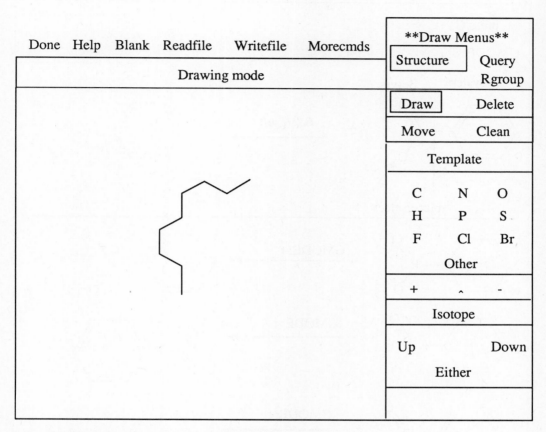

Figure 3/56. Structure after moving atoms

Example: *Manipulating groups*

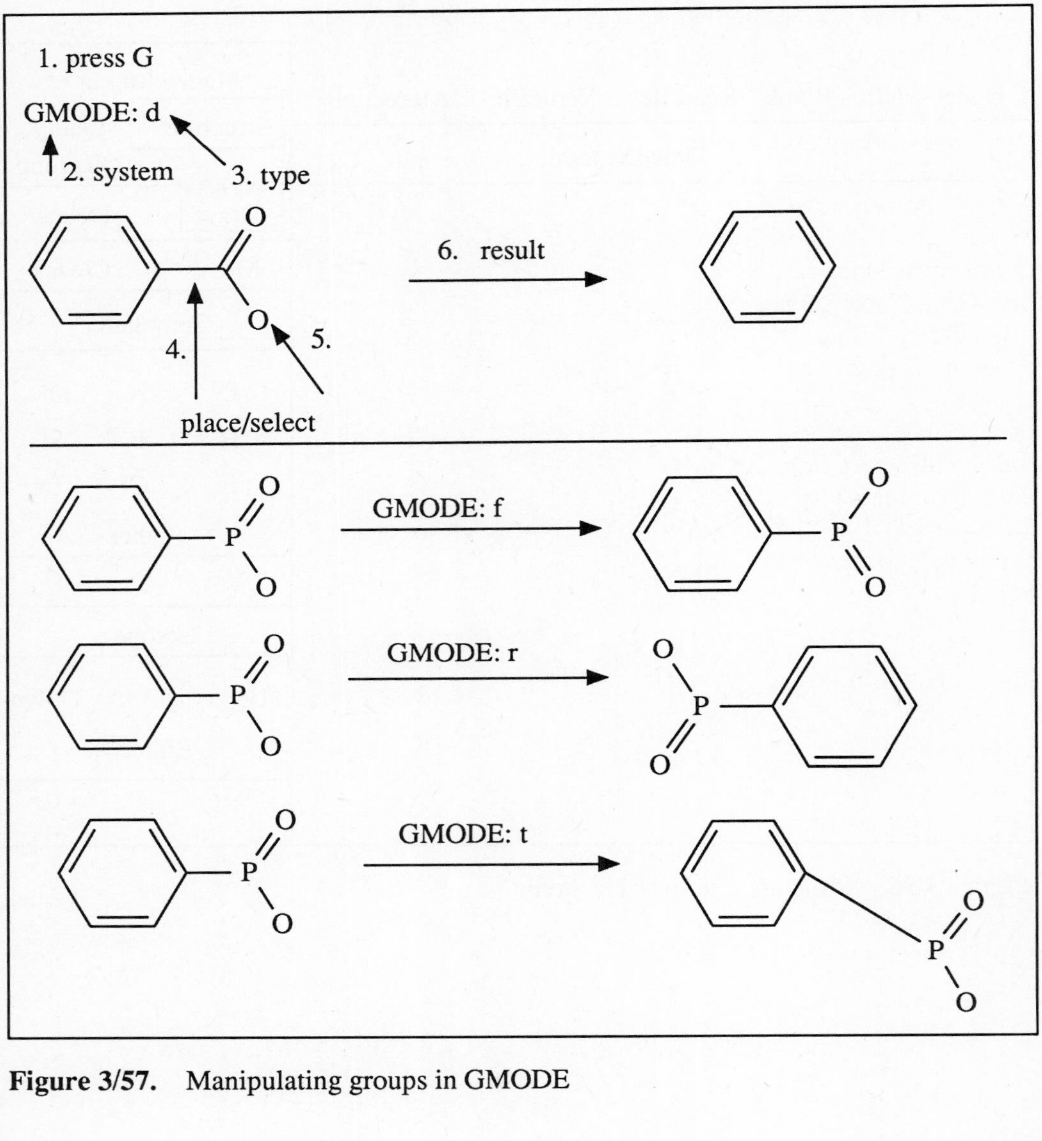

1. press G

GMODE: d

2. system   3. type

4. place/select   5.

6. result

GMODE: f

GMODE: r

GMODE: t

Figure 3/57. Manipulating groups in GMODE

7 STRUCTURE SEARCHING

MACCS-II like the other systems we have discussed in Part One and Part Two, allows for structure searching using graphic structure queries. In MACCS-II we can formulate queries specifying configuration at stereocenters so that we can selectively search for:
** absolute configuration
** relative configuration
** enantiomers
** diastereomers
** racemates
** geometric isomers
MACCS-II also offers options to search for:
** tautomers
** parent of salts

As usual, we can divide structure searching into two classes:
1. Exact Structure Search
2. Substructure Search
The search processes which MACCS-II goes through for these classes of structure searching are shown in Figure 3/58 and Figure 3/59.

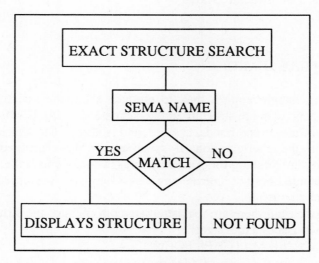

Figure 3/58. Exact structure search

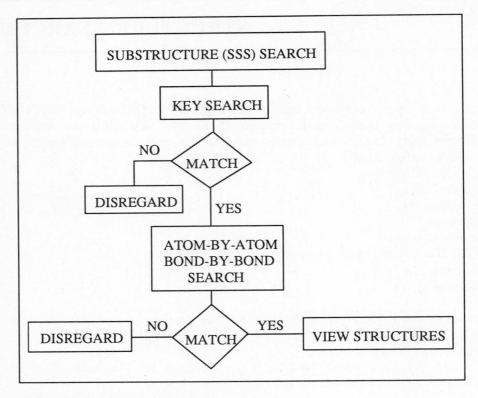

Figure 3/59. Substructure search

EXACT STRUCTURE SEARCHING

The exact structure search requires that, to produce a hit, the structure query must be identical in all aspects with the structure in the database. In MACCS-II, the identity is not only in the nature of atoms and bonds, charges, and isotopes. The structure query must be also identical with the configuration, if specified, of the registered structure in the database. Therefore, in searching for exact structures, we may obtain the following results:

1. If the query structure does not indicate any stereochemistry, we retrieve an exact match, if any, with the query structure.

2. If there is no match but there are registered stereoisomers of the structure query in the database, MACCS-II issues a message:

(SIOF) STEREOISOMER MATCHES FOUND

and performs a search for the isomer (*vide infra*).

3. If the structure query indicates stereochemistry, the hit, if any, will be the configurational isomer matching the query. If there is no match but there are isomers registered in the database, we receive the same message:

(SIOF) STEROISOMER MATCHES FOUND
and the isomer search is automatically performed (*vide infra*).
4. If there is no match and there are no stereoisomers on
file, the System gives a message:
NOT FOUND
MACCS-II then performs tautomer search depending on our Settings (see p. 516, 517).

SEARCH IN REGISTER OR SEARCH MODE?

In MACCS-II, the Exact Structure Search can be executed in the Register Mode or in the Search Mode. Both Modes contain identical options for exact structure searching. What, then, is the difference between searching for exact structures in the Register and Search Mode?
MACCS-II is an enhanced version of the previous version of MACCS in which the Exact Structure Search was performed in one of the Modes which was called the Executive Mode. This Mode has been redesigned and has become the Register Mode in MACCS-II. The search options have been retained; however, they have been also made available in the Search Mode, a Mode which was already in the previous version of MACCS but did not contain the options for Exact Structure Search.
The differences in the Register and Search Modes are:
1. In the Register Mode the search is always automatically carried out in the entire database while in the Search Mode we have to pay attention what is the current search domain (*vide infra*).
2. In the Register Mode we cannot manipulate lists whereas we can in the Search Mode (*vide infra*).
3. In the Registry Mode, only a single structure is displayed even if the hit list contains multiple items (and if the Settings "multiple find hits" is off), while in the Search Mode we can view all items, one by one.
4. In the Register Mode we can retrieve multiple structures with a query of multiple regnos, while in the Search Mode we can search (View Regno) with one-regno query only.
5. In the Register Mode we cannot recall the last query structure whereas we can in the Search Mode.
6. In the Register Mode, we can use the following switches (see p. 444) (they are not available in the Search Mode):
*, /B, E or >, G, H, L, M, N, R or <, S, V, 0(zero) and
  the switch:
    t.......truncate, replaces common substituents in
        structures by text abbreviations, e.g., NO2, Ph,
        Pr, etc.
  Note that we can make the truncate feature the default for
  all structures if we turn the Settings suboption (see
  p. 442) on:
    truncate.....turns on/off the abbreviation of common frag-
        ments in structures (default is "off")

On entering the Register Mode with the current structure, the counters show (Figure 3/60) that the current list (search domain) is the entire database. The "On File" total (at this writing) equals the Curlist if the search domain is the entire database. No Regno is shown since there is no retrieved structure on display (the current structure in Figure 3/60 is the query structure).

The exact search is initiated by FIND(Current) (Figure 3/60).

Example: *Exact structure search*

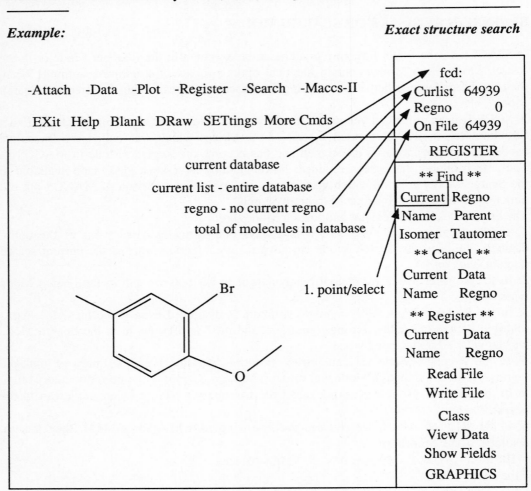

Figure 3/60. Counters in Registry Mode Menu before search

DISPLAYING SINGLE HIT STRUCTURE

If the exact structure match is located in the database, the structure is displayed and the counters change as shown in Figure 3/61.

STRUCTURE SEARCHING

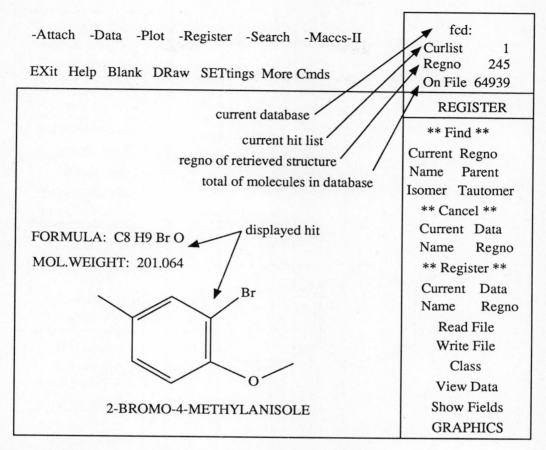

Figure 3/61. Counters in Registry Mode Menu after search

The Curlist in Figure 3/61 now shows the number of hits of the search, i.e., 1. The Regno is the regno of the retrieved structure. If we perform another search, MACCS-II will run it again in the entire database. The Curlist will change to 64939 (at this writing).
However, if we perform this search in the Search Mode (Figure 3/62), the Curlist becomes the search domain (Figure 3/63) and if we run a new search in Search Mode it will be run only over the Curlist of 1 which is the current hit list (Figure 3/63) unless we switch the Curlist for the entire database (*vide infra*).

The structure query for the Exact Search must be the current structure which may be the result of:
1. creating it in the Draw Mode (see p. 444)
2. recalling a Molfile (Figure 3/64)

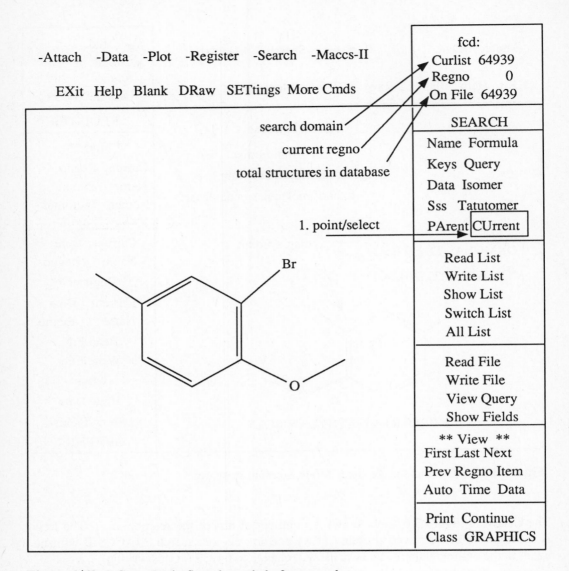

Figure 3/62. Counters in Search mode before search

DISPLAYING FIXED DATA FIELDS

Figure 3/61 and Figure 3/63 show the retrieved structures with the fixed data, i.e., FORMULA and MOL.WEIGHT. We can turn on/off this fixed data display by the Settings suboption (see p. 442):
fixed data display......turns on/off the fixed data display
       (default is "on")

STRUCTURE SEARCHING

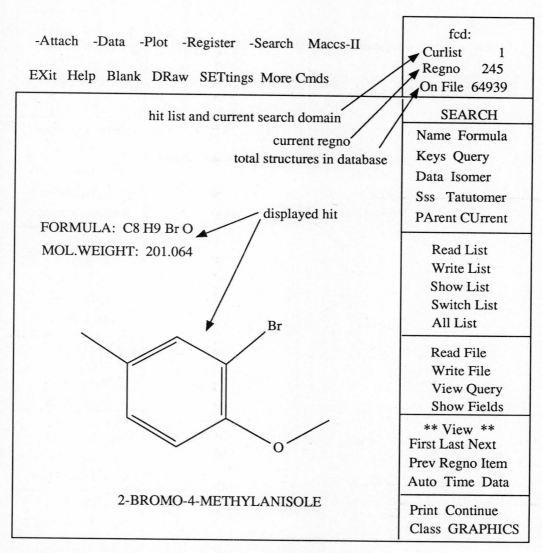

Figure 3/63. Counters in Search mode after search

DISPLAYING FLEXIBLE DATA FIELDS

If we want to display the data associated with the retrieved structure (the data in the flexible data fields), we activate the VIEW DATA option while the current structure is on display. MACCS-II will prompt:
View data field(s):
We can enter:

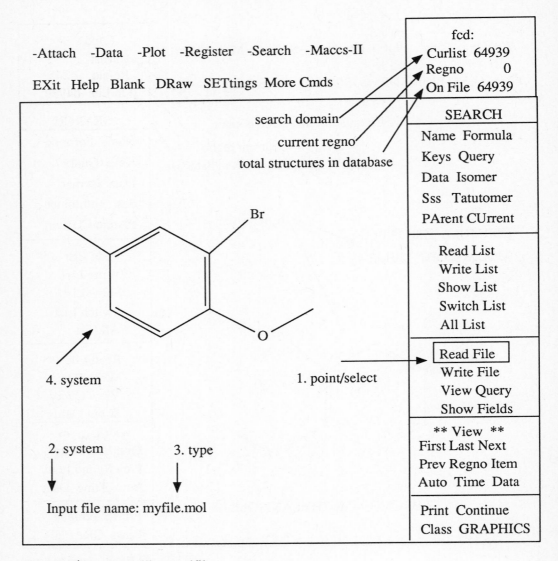

Figure 3/64. Recalling molfile

1. data field name(s)(use of the wild card "@" is permitted)
2. data field numbers
3. a combination of 1 and 2
4. all (for all data fields)

To find out which data fields have been defined for the current database, we activate the SHOW FIELDS option.

For example, if we wanted to find the CAS Registry Number for the retrieved structure in

STRUCTURE SEARCHING

Figure 3/63, we would select VIEW DATA and at the prompt:
Data field(s): we would input:
2 (or CAS.number)
(See also "Data Searching," p. 535 and the discussion on data fields, p. 421).

DISPLAY OF MULTIPLE HITS

If we search in the Register Mode and retrieve multiple hits, the items on the list are displayed one hit per line consisting of:

internal regno   (external regno)   name of substance

We can turn on/off the multiple hit display by the Settings suboption (see p. 442):
multiple find hits....turns on/off display of multiple lines
             of items on the hit list (default is "on")
Note that the display is turned off but the Curlist will show the total number of hits.
To view the structures in the multiple hit list, we have to transfer to the Search Mode and use the VIEW option (see p. 513).

SEARCHING FOR PARENT OF SALTS

In MACCS-II, the parent of a salt is defined as the molecular structure stripped of its counterion. The counterions that will be stripped from a structure have to be defined in a System File (the System Supervisor maintains it). They cannot be defined by the User in the course of Parent Search.
The purpose of the Parent Search is to obtain hits even if the counterion of the registered structure is different from that one in the structure query.
For example, the query in Figure 3/65 would not yield any hit in the CURRENT search if the registered structure was the ammonium salt. However, if we select PARENT, MACCS-II performs the search for the parent which is retrieved and displayed (Figure 3/66).

SEARCHING CONFIGURATIONAL ISOMERS

In Exact Structure Search we can search for specified absolute or relative configuration.
To perform the search, we enter the Search Mode with the stereochemical structure query as the current structure and select the CURRENT option (Figure 3/67).
The search can be also performed in the Register Mode by FIND (Current).
MACCS-II performs the search and issues one of the following messages:

IDENTICAL MATCH....hit is stereochemically identical with the query
ENANTIOMER......hit and query are enantiomers (both hit and query must have CHIRAL label)(see p....)
RACEMATE........hit is racemic mixture (the query is a chiral component and is not

Example: *Parent search*

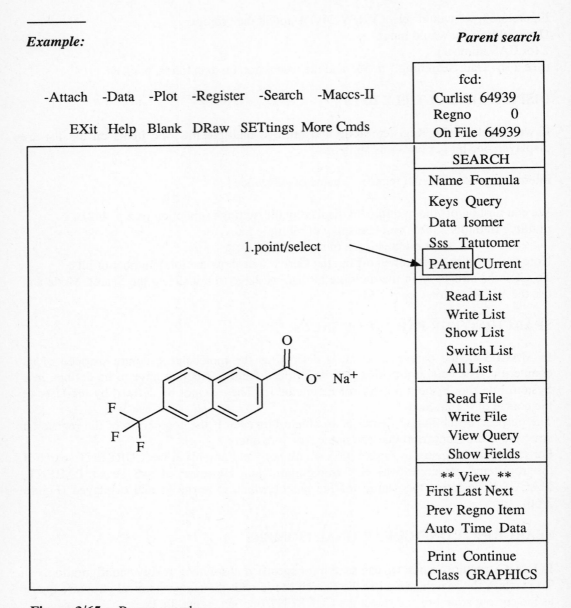

Figure 3/65. Parent search

labeled CHIRAL)
CHIRAL ANALOG...hit is one chiral component of racemic query
DIASTEREOMER....hit and query are diastereomers
GEOMETRIC ISOMER...hit and query have opposite configuration at one or more double bonds

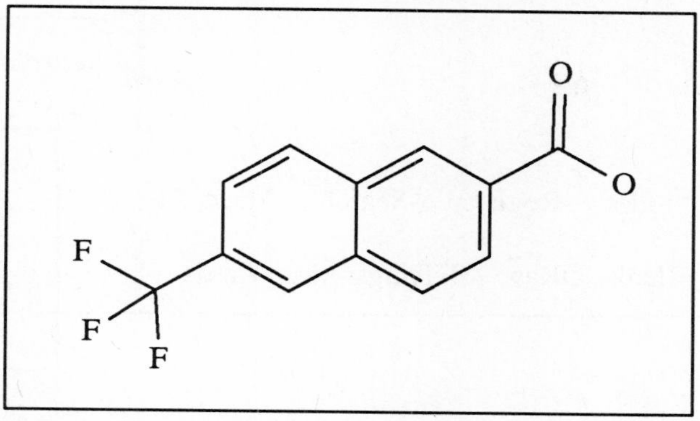

Figure 3/66. Parent of salt

We can also perform a search for stereoisomers of the structure query by selecting ISOMER in the Search Mode or FIND(Isomer) in the Register Mode.
If the search yields multiple hits, they are displayed as text entries as shown in Figure 3/68. The structures can be viewed by selecting the VIEW options (*vide infra*) (Figure 3/68).

DISPLAYING MULTIPLE RETRIEVED STRUCTURES

If we perform a search in the Search Mode and the result is a single hit, the structure is automatically displayed. However, if the result of the search is a hit list with multiple items, the hits are displayed as in the Register Mode, i.e., one hit per line consisting of (Figure 3/68):

regno   (external regno)   name of substance

We can turn on/off this display by activating the PRINT option on the Menu.
To make the display of the multiple hits in the Search Mode the default, we turn on the Settings suboption (see p. 442):
search print......turns on/off display of multiple hits in
          Search Mode (default is "on")

To display the structures in a multiple hit list, we activate the VIEW options Figure 3/68):
First.....displays item No. 1 in the Curlist
Item......displays specified item in the Curlist
Last......displays the last item in the Curlist
Next......displays next item sequentially
Prev......displays the preceding item
Auto......automatically displays all hits in the Curlist
Time......sets the display time of each structure in the

Example: *Searching stereoisomers*

```
-Attach  -Data  -Plot  -Register  -Search  -Maccs-II
EXit  Help  Blank  DRaw  SETtings  More Cmds
```

	example
	Curlist
	Regno
	On File
	SEARCH
	Name Formula
	Keys Query
	Data Isomer
	Sss Tatutomer
1. point/select →	PArent **CUrrent**
CHIRAL (cyclohexane-1,2-dicarboxylate structure)	Read List / Write List / Show List / Switch List / All List
	Read File / Write File / View Query / Show Fields
	** View ** / First Last Next / Prev Regno Item / Auto Time Data
	Print Continue / Class GRAPHICS

Figure 3/67. Search for absolute configuration

Auto display, e.g.:time 10 (sets the view time to
10 seconds for each structure)

The view time can be set as the default by the Settings suboption (see p. 442):

STRUCTURE SEARCHING

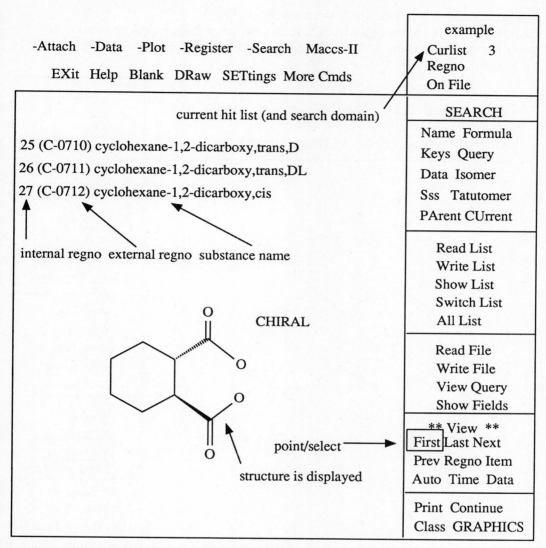

Figure 3/68. Result of search for stereoisomers

search view time.....sets the default for view time for
           View Auto

Note that if we display a structure from a hit list (Curlist) the structure becomes the current structure (if we leave the Mode, the structure will be carried over to the new Mode, except Data Mode, as the current structure). If we activate a search option, the search will be performed with the current structure as the query. Therefore, if after a search we wish to return to the original query structure and possibly modify it for another

search, we have to recall the query structure (Figure 3/69).
Only one query is retained by MACCS-II. A new query replaces the previous query. The query is erased on exiting MACCS-II unless saved in a Molfile by activating the WRITE FILE option (see p. 463).

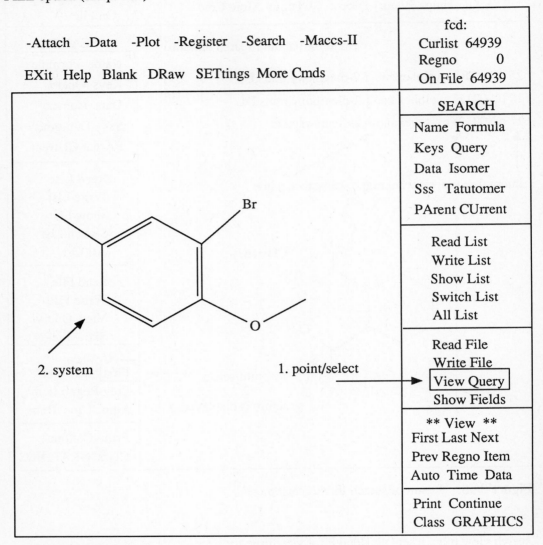

Figure 3/69. Recalling query

SEARCHING TAUTOMERS

We have learned that if no match is found in the CURRENT search, MACCS-II performs

the isomer search automatically. If no stereoisomer is located, MACCS-II performs tautomer search but only if the Settings suboption (see p. 442) is turned on:
tautomer search......sets the default for tautomer search
        if no match is found in the CURRENT
        searches (default is "off")

The tautomer search retrieves structures that have the same atom types and connectivities as the query structure but differing in the position of hydrogens, isotopes, and/or charges at double/single bonds, and/or in the stereochemical designations (see p. 97 for discussion of tautomerism).

Of course, we can perform a search for tautomers directly with a query by activating the TAUTOMER option (Figure 3/70).

The search in Figure 3/70 would retrieve the structures shown in Figure 3/71.

SUBSTRUCTURE (SSS) SEARCHING

Substructure searching can be performed only in Search Mode.

The concept of substructure searching has been discussed in Part Two (p. 380). It is valid in MACCS-II as well.

SSS WITH GRAPHIC STRUCTURE QUERIES

To perform SSS, the query structure must be the current structure in the Search Mode. The current structure is obtained by any of the following:
1. drawing in the Draw Mode and/or Draw/SSS Mode
2. recalling the structure query from a Molfile
   (remember, the generic structure query can be saved in a
   Molfile, however, it cannot be registered in a database)
   or from the database by FIND REGNO or VIEW REGNO

INITIATING SUBSTRUCTURE SEARCH

When we have the query structure ready to search, i.e. the query structure is on display as the current structure, we activate the SSS option (Figure 3/72).

We have illustrated in Figure 3/59 (p. 504) the substructure search steps MACCS-II goes through. The messages shown in Figure 3/72 reflect the process and inform us about the progress of the search and give us the opportunity to intervene and specify the conditions MACCS-II offers in the prompts.

The conditions of the search can be set by the Settings suboption (see p. 442):
search level.....determines whether MACCS-II stops SSS after
        keys search and asks if to continue
        There are two levels we can set:
        1.....kyes search is automatically followed
            by atom-by-atom/bond-by-bond search
            without interruption

Example: *Searching tautomers*

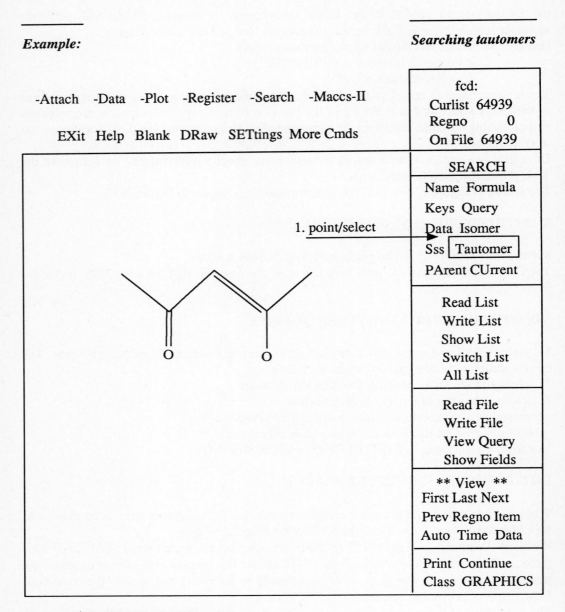

Figure 3/70. Search for tautomers

     2.....MACCS-II completes keys search and
         asks if to continue

The first stage, i.e., the key search is fast and can be interrupted by the User by "Q"(see

STRUCTURE SEARCHING

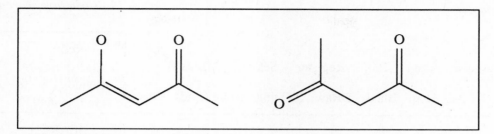

Figure 3/71. Tautomeric structures

the prompts in Figure 3/72).
Q......aborts the key search and, if not resumed (*vide infra*)
     skips atom by atom/bond by bond search
The result of the first stage is a list of candidate structures that contain the basic structural fragments, the keys, which will be ultimately subjected to the atom by atom/bond by bond search. This is reflected in the search statistics. If we interrrupt the search by "Q," the Curlist will show the hits of the interrupted search, i.e., the number of candidates generated by the key search up to the point of entering "Q." At this point we can decide to continue with the search or reformulate the query, if the number of hits is too large, and finish the search with the Curlist as the search domain.
We are also given the opportunity to narrow the number of hits and make the search more selective by answering "M" or "A":
M.....instructs MACCS-II to generate 10 more keys
A.....instructs MACCS-II to generate all remaining keys

To proceed automatically with the keys search followed by the atom-by-atom and bond-by-bond search, we enter carriage return <cr>.
<cr>.
SSS can be performed with any query structure created in the Structure or Query Menu. SSS with query structures containing defined R groups can be performed only with the add-on Power Search Module.

SSS WITH GENERIC RGROUP QUERY

The Users who have access to the MACCS-II Power Search (PS) will be able to search with queries comprising defined R groups which are created in Rgroup Menu (see p. 476). The Search Menu of the PS is the same as the regular Search Menu except for an added option, i.e., SIMILAR (*vide infra*) which replaces the QUERY option in the menu. The QUERY option is still available, however, by keyboard selection.
To perform a SSS with R group query, we enter Search Menu as usual with the current query containing R groups and point/select the SSS option. The search proceeds similarly

Example: *Substructure search*

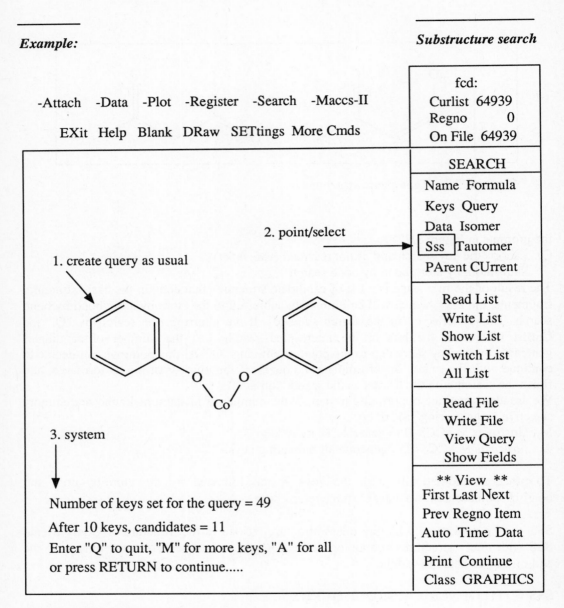

Figure 3/72. Substructure search

to the SSS we have discussed above.
The query we have created (see p. 486) has been searched in the FCD database (Figure 3/12-PS).
The result of the search was 12 substances whose regnos and names were displayed

STRUCTURE SEARCHING

Example: *Substructure search*

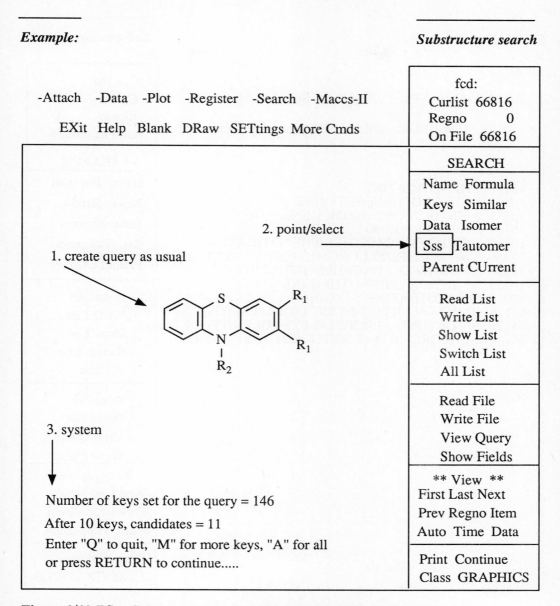

Figure 3/12-PS. Substructure search with generic query with R groups

(Figure 3/13-PS).
The eleventh item was viewed graphically (Figure 3/14-PS).

Example: *Substructure search*

fcd:
Curlist   12
Regno   0
On File 66816

-Attach -Data -Plot -Register -Search -Maccs-II

EXit Help Blank DRaw SETtings More Cmds

16351 PHENOTHIAZINE 16352 2-CHLOROPHENOTHIAZINE 16353 2-METHOXYPHENOTHIAZINE 16354 2-ACETYLPHENOTHIAZINE 16355 2-(TRIFLUOROMETHYL)PHENOTHIAZINE 16356 7-NITRO-2-TRIFLUOROMETHYLPHENOTHIAZIN 16357 3,7-DINITRO-5-OXOPHENOTHIAZINE 16358 3-METHOXYPHENOTHIAZINE 37014 PHENOTHIAZINE-5-OXIDE 61128 10-(1,3-DIETHYL-3-PYRROLIDINYLMETHYL)-PH 61129 10-(1-ETHYL-3-PROPYL-3-PYRROLIDINYLMETH 61140 2-CHLORO-10-(3-(2-METHOXYETHYL)-1-ME-3-	SEARCH
	Name Formula Keys Similar Data Isomer Sss Tautomer PArent CUrrent
	Read List Write List Show List Switch List All List
	Read File Write File View Query Show Fields
	** View ** First Last Next Prev Regno Item Auto Time Data
	Print Continue Class GRAPHICS

Figure 3/13-PS. Result of SSS with R group query

SIMILARITY SEARCH

In a Substructure Search, the features of the structure query which we assign either in a broad or in precise specifications, are the criteria by which the search will proceed and

STRUCTURE SEARCHING

	fcd:	
	Curlist	12
	Regno	61129
	On File	66816

-Attach -Data -Plot -Register -Search -Maccs-II

EXit Help Blank DRaw SETtings More Cmds

FORMULA: C22 H29 CL N2 S
MOL. WEIGHT 389.007

SEARCH
Name Formula
Keys Similar
Data Isomer
Sss Tautomer
PArent CUrrent
Read List
Write List
Show List
Switch List
All List
Read File
Write File
View Query
Show Fields
** View **
First Last Next
Prev Regno Item
Auto Time Data
Print Continue
Class GRAPHICS

Figure 3/14-PS.   Viewing the eleventh item

which the retrieved structures must meet.

The MACCS-II Power Search enables us to perform yet another type of search in which there is only a loose relationship between the structure query and the retrieved structures.

Similarity Search utilizes the 933 keys (see p. 424) that MACCS-II System sets for a database.

A query is statistically analyzed by MACCS-II to arrive at a similarity rating against a structure in the database based on the correlation of the weighted keys:

Rating = keys matching both query and structure against keys
     set in either query or structure.

The 100% similarity represents an exact match of the query with the structure in the database. The 0% similarity indicates that no keys are common for the query and the structure.

We can specify minimum percent of similarity to be used by MACCS-II in the search, i.e., we specify the percentage at the MACCS-II prompt, e.g.:

1. point/select SIMILAR
MACCS-II: Enter ......
Type: 80

Similarity Search can be used only in inverted-keys databases with key weights.

A query for the Similarity Search must be a structure with no generic features, i.e., no variables or R groups. It must be an exact structure.

In comparing the types of searches in MACCS-II, we arrive at the following precision of the hits:

Exact Search....exact match of the query with retrieved
        structure
Substructure Search...retrieved structures must contain the
        structural fragment of the query and
        may or may not contain the variables
        or R groups specified in the query
Similarity Search.....retrieved structures have a predefined
        percentage of keys in common with the
        query while structural feature may
        differ with the query

See also the discussion of Similarity Search in REACCS and examples of the search (see p. 692)

SUBSIMILARITY SEARCHES

Subsimilarity Search compares the number of common key settings to the keys set in the query. The rating is arrived at:

Rating = keys matching both query and structure against keys
     set in the query

The hits of a Subsimilarity Search will contain a substructure similar to the query and will be larger than the query.

The 100% subsimilarity will be met by any structure that passes the Key Search stage of a Substructure Search (see p. 504).

To run a Subsimilarity Search, point/select SIMILAR, and specify the percentage at the prompt:

Enter........sub 80

STRUCTURE SEARCHING

SUPERSIMILARITY SEARCH

Supersimilarity Search compares the number of common key settings to the keys set in the structure.
Rating = keys matching both query and structure against
    keys set in structure
The hits of a Supersimilarity Search will be smaller structures than the query and will contain features similar to the query.
To run a Supersimilarity Search, point/select SIMILAR and enter at the prompt for percentage e.g.: SUPER 80.

STEREOCHEMISTRY IN SSS

In SSS, we can search only for relative configuration at stereocenters (in Exact Search, we can search for both absolute and/or relative configuration).
The query structure is marked by the STEREO option (see p. 469) at the stereocenters. MACCS-II will search for structures which contain the query fragment with the specified relative configuration. If we do not mark the stereocenters, or if only one of the stereocenters bears a stereobond, SSS retrieves all stereoisomers and also structures with no stereochemistry in the fragment of the query. Figure 3/73 shows the type of stereochemical queries for SSS.

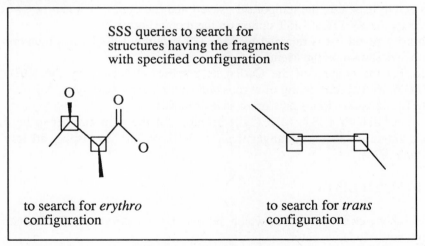

Figure 3/73. Stereochemistry in SSS

INTERRUPTING AND RESUMING SEARCH

If we use the response "Q" or if we interrupt the search by the "!," we can still resume the search by the option CONTINUE on the Menu provided we do not destroy the Curlist. The following options, if activated, do not interfere with continuing the search:

BLANK, HELP, PRINT, READ FILE, VIEW, SHOW, and WRITE
If we use the response "A" for the search in Figure 3/72, the result will be 3 hits which will be displayed first as the text entries (Figure 3/74) which we can follow by displaying the structures by the VIEW option (Figure 3/74)

Note that the option PRINT serves the same purpose as the toggle "search print" (see p. 513); however, the display is turned on/off only for the current display. Thus, if we anticipate a large number of hits, we can shorten and speed up the search by not diplaying the hits.

USING LISTS IN SEARCH MODE

The Curlist contains the records (hits) that are retrieved by the search from the database. The Curlist is transferred if we leave the Search Mode and go into Register Mode and *vice versa*. There are always two lists in a MACCS-II session in which we perform more than one search; one is the Curlist and the other is the temporarily saved list from the previous search. For example, if we perform three searches, the list of the first search will be ultimately erased while the list of the second search will be preserved and the list of the third search will be the current list (see also Part Two, p. 373):

Search No. 1.---->Curlist---->Temporarily saved ---->Erased
Search No. 2.---->Curlist---->Temporarily saved
Search No. 3.---->Curlist

These two active lists, i.e., the temporarily saved and the current list can be switched one for the other by the SWITCH LIST option on the menu (Figure 3/76).

The temporarily saved list is retained by MACCS-II without the User's intervention. The Curlist is always shown on the Menu.

We can display the content of the Curlist as graphics structures by the VIEW options (Figure 3/74) or as text (consisting of regnos and substance names) by:

SHOW LIST......displays items on Curlist as text entries

When we select SHOW LIST, MACCS-II prompts for the item number to be displayed. We can use single numbers, or ranges, e.g., 1, 3-10, or "all" to display all items on the Curlist (Figure 3/75).

WORKING WITH LISTS

If after a search we exit the Mode in which the search was performed, the Curlist is carried over into the new Mode we enter.

The Curlist can be manipulated, i.e., items can be deleted from the hit list. This is useful when we view the items on the list and do not want to include some of the hits in our final print of the search hits. To delete items on Curlist, we switch to the keyboard selection:
User: k
System: SEARCH>
User: delete item
System: delete item(s):
User: (enter the item number(s) to be deleted)

STRUCTURE SEARCHING

Example: *Substructure search*

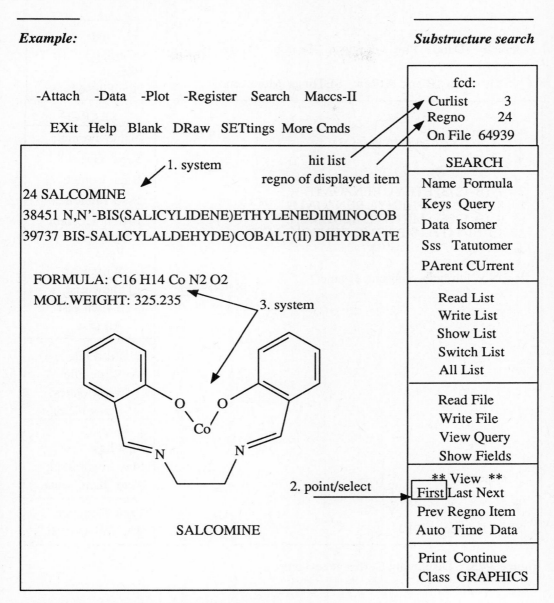

Figure 3/74. Result of substructure search

The Curlist can be manipulated by options SWITCH LIST and ALL LIST (Figure 3/76).

WRITING LISTFILE

The Curlist can be saved permanently in a file by WRITE LIST option (Figure 3/76)

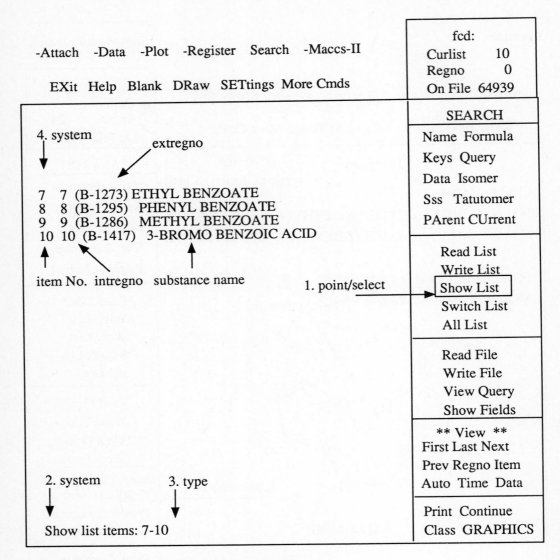

Figure 3/75. Displaying Curlist as text entries

which elicits the prompt:
Listfile name (H for help):
The input will depend on which type of file we want to write:
1. If we wish to save the binary regnos, we enter the file name.
2. If we wish to save the editable (ASCII) regnos, we enter "R."
   MACCS-II prompts:
   Enter NUMERIC file name:

STRUCTURE SEARCHING

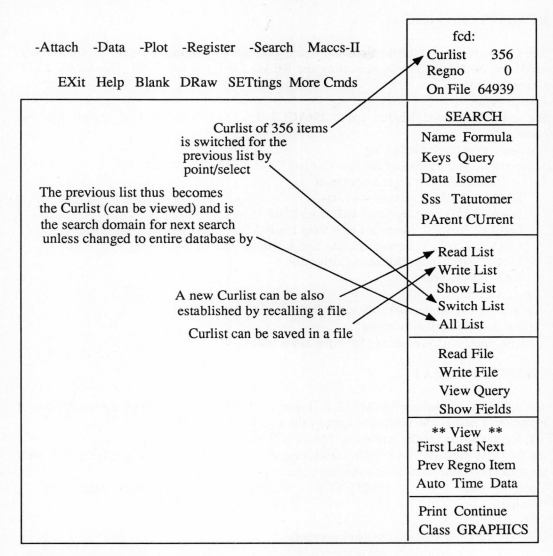

Figure 3/76. Manipulating Curlist

3. If we wish to save the extregnos (ASCII), we enter "E."
   MACCS-II prompts:
   Enter EXTREG file name:
(See also the discussion on p. 430 of Listfiles.)
Note that the temporarily saved list of a previous search cannot be saved unless we switch it for the Curlist.

READING LISTFILES

The Curlist can be established without running a search if we recall the hit list (previously saved by WRITE LIST) by activating the READ LIST (Figure 3/76). The READ LIST elicits the prompt:
Listfile name:
We enter the saved Listfile name (MACCS-II knows what type of regnos the file contains).
We can also enter the following:
1. prev (to read the previous list into Curlist)
2. Lists combined with logical operators:
    ListA + ListB.......merges the Lists
    ListA - ListB.......subtracts ListB from ListA
    ListB - ListA.......subtracts ListA from ListB
    ListA & ListB.......intersects ListA and ListB
    ListA/ListB.........performs exclusive OR
where List may be:
cur......Curlist
prev......previous, temporarily saved list
filename..binary listfile name
3. tty (to enter regnos or extreg from the keyboard)

LISTFILE FORMAT

The Listfiles are written by MACCS-II when we activate the options discussed above. However, Listfiles can be also written by the User in the Text Editor.
The format of the Listfile written by MACCS-II is as follows:
the first line contains the Listfile identification and is up to 80 characters long with each identification code starting in a specific column position:

Code:	C	L	MM/DD/YY	HH:m	I	S	database name
Column:	1	6	15	24	30	37	46

C = *R* for binary internal registry numbers
    N for numeric internal registry numbers
    E for external registry numbers
L = number of regno entries in the Listfile
MM/DD/YY = date the file was written
HH:m = time in hour and minutes when the file was written
I = User ID under which the file was written
S = size of database from which the file was written
database name = from which the file was written;

the second line is blank or can contain User comments (entered by the User when the file is written);

STRUCTURE SEARCHING

the following lines contain registry numbers.

The User written file (using a Text Editor) does not have to include all the identification in the first line.
The C is the only required identification.
The following would be a valid Listfile written by the User:

N
This is a test Listfile containing ASCII internal regnos
12345
23456
34567
45678

8   PLOTTING STRUCTURES

When we perform an Exact Structure Search or a Substructure Search, we display the retrieved structures on the screen. However, we may wish to create a hard copy of the structures for inclusion in reports, manuscripts, etc.
We could dump the screen to a printer but then we would print everything that is on the screen, including the Menu.
MACCS-II offers a technique by which only images of the structures can be displayed or printed. This is done by plotting the structures either on the screen and dumping the screen to a printer, plotting on a plotting device connected to the terminal/PC, or plotting on a plotting device connected to the mini/mainframe computer. We can plot not only structures retrieved by a search, but also structures recalled from a Molfile or a Listfile or structures drawn in Draw Mode.
Plot Mode is the Mode for plotting structures. The options on Plot Mode Menu (Figure 3/77) enable us to:
1. Format the page, i.e. to specify how many structures and at which positions on the page the structures will be.
2. Format the structures, i.e. to specify certain features we wish to display with the structures.
3. Specify which structures to plot.
4. Specify plotting on the screen or on a plotting device.

The conditions that we set by activating the Menu options are valid for our session and can be displayed by point/ select:
PSetup.......displays current setup for plotting

SELECTING PLOTTING DEVICE

MACCS-II supports several plotting devices which are of two kinds:
1. local or remote plotters for direct plotting:
    Versatec, Calcomp, HP 7221, Apple LaserWriter
2. terminals/PCs for plotting on screen and dumping the screen to a printer

To specify a device we will use, we activate the option DEVICE on the Menu. MACCS-II prompts:
Device: <L or device number>
If we enter "L," MACCS-II displays the list of supported devices.
Note that each time we activate the DEVICE option, the plot position (see WINDOW and PER PAGE options) is set to 1.

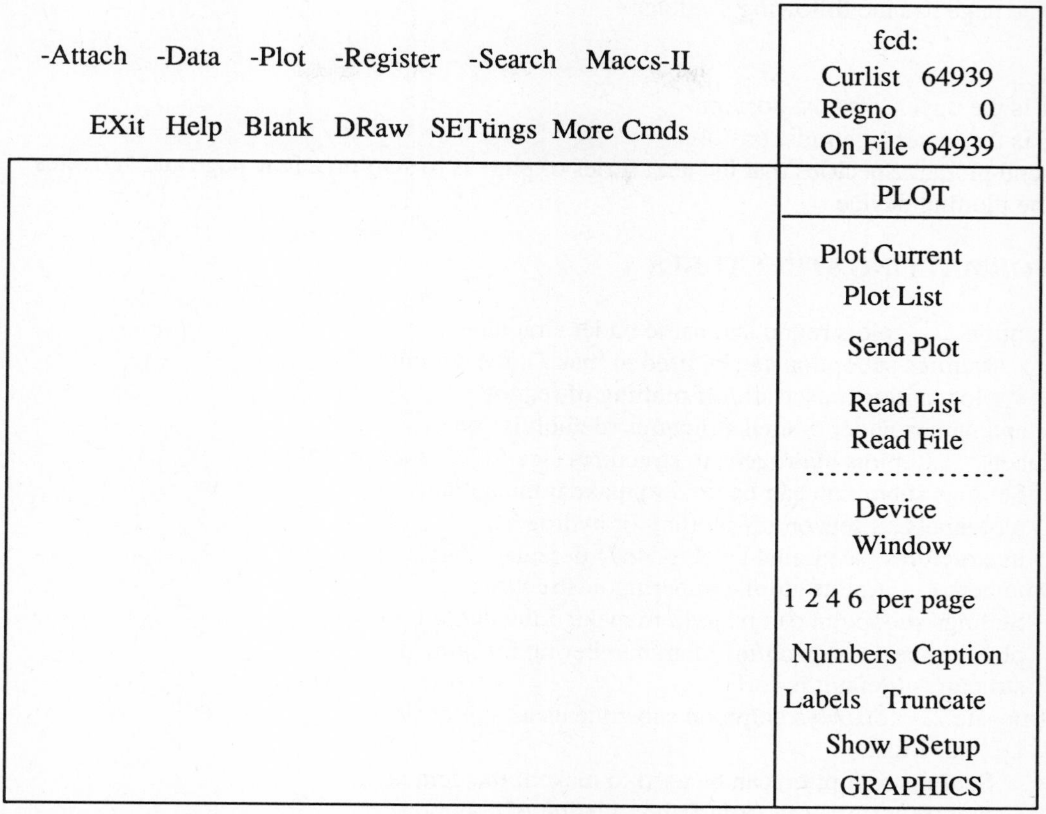

MACCS-II

-Attach -Data -Plot -Register -Search Maccs-II EXit Help Blank DRaw SETtings More Cmds	fcd: Curlist 64939 Regno 0 On File 64939
	PLOT Plot Current Plot List Send Plot Read List Read File · Device Window 1 2 4 6 per page Numbers Caption Labels Truncate Show PSetup GRAPHICS

Figure 3/77.   Plot menu

FORMATTING THE PAGE

1 2 4 6 per page......specifies to plot 1, 2, 4, or 6 structures per page
 1.....full page/one structure per page
 2.....half page/two structures per page
 4.....quarter page/four structures per page
 6.....sixth page/six structures per page
 Note that the setup applies for all subsequent plots until a new setup is executed. The underscored number on the Menu indicates the current perpage setup
 We can also set perpage by the Settings suboption:
 plot perpage........sets the default perpage values
window..........specifies the section of the page in which the structure will be plotted

1   2   3
4   5   6
1 is the upper leftmost position
6 is the lower rightmost position
send plot........specifies that the next series of plots is to start on a new page and activates the plotting device

FORMATTING STRUCTURES

caption..........plots regno and name under structures
    Settings suboption can be used to make it the default:
      plot caption......sets on/off plotting of regnos
  and names under plotted structures (default is "on")
labels...........plots hydrogens in structures (see Label level, p. 447)
    Settings suboption can be used to make it the default:
      plot labels......sets on/off plotting of hydrogens
    in structures (see Label level, p. 447)(default is "off")
numbers...........plots atom numbering in structures
    Settings suboption can be used to make it the default:
      plot numbers....sets on/off atom numbering for plotted
    structures (default is "off")
truncate.........displays common substituents as abbreviation
    i.e., Ph, Bu....
      Settings suboption can be used to make it the default:
      plot truncate.....sets on/off abbreviation of common
      substituents in plotted structures (default is "off")

SPECIFYING STRUCTURES TO PLOT

plot current.....plots the current structure
plot list........plots structures in the Curlist
read list........establishes a new Curlist
read file........recalls structure from a Molfile

9   DATA SEARCHING

In this section, we will discuss how to retrieve and/or search the data which are associated with the molecules registered in a database and, similarly, to retrieve/search data if we have a no-structure database.

Data retrieval/search is not limited to the Data Mode. However, this Mode offers other capabilities, e.g., data transfer, which are not available in the other Modes.

Data searching/retrieval in MACCS-II can be categorized as follows:

1. Searching/retrieving data which have been derived from the registered structures.

This data search allows for retrieving structures similarly to the structure search; however, we formulate alphanumeric queries, not graphic structure queries.

The data queries may consist of:
   a. registry numbers
   b. formulas
   c. names
   d. keys

2. Searching/retrieving data which are related to the structures but which have not been derived from the structures. These are the data that are stored in the flexible data fields and may be a variety of chemical, physical, or biological properties of the substances represented by the structures, i.e. data generated and input by the User.

Data can be searched in the Register, Search, and Data Mode.

Figure 3/78 illustrates the similarities and differences in these Modes.

RETRIEVAL WITH REGNO QUERY

Retrieval of registry numbers is executed by the FIND REGNO option in the Register Mode and by the VIEW REGNO option in the Search Mode. In the Search Mode, we can retrieve only one regno, in the Register Mode we can search for multiple regnos (Figure 3/79). The search results in retrieval of the exact structures registered under the registry numbers.

The registry numbers are entered at the prompt:

Registry number(s):

We can enter:
1. specific numbers, i.e.,
   1234,5678,24567
   1234-23456,34567
   SK-1234 to SK-1245

Both the internal and the external registry numbers can be used for the retrieval. The external registry numbers must be entered in the exact format. However, if the number we enter at the prompt starts with a numeric character, MACCS-II searches for the internal registry numbers in this operation. External registry numbers which start with a numeric

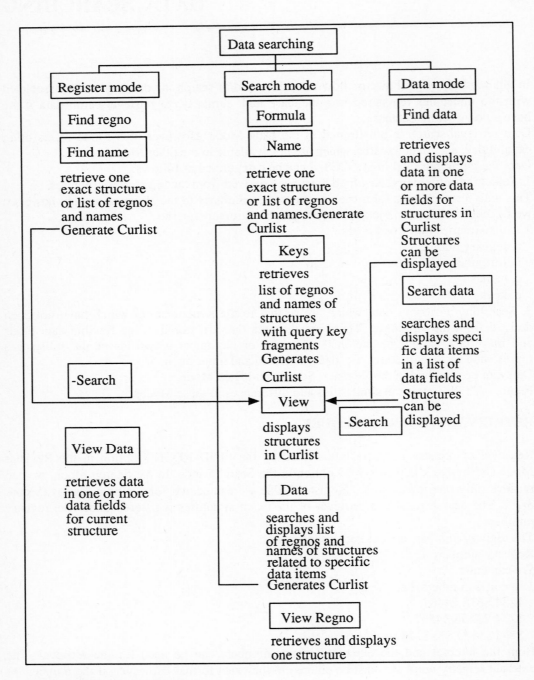

Figure 3/78. Data searching

character but which may also have non-numeric characters, must be preceded with the "@" symbol.
2. the registry numbers stored in a Listfile or
   Listfiles combined with the logical operators (see p. 247).
   System: Registry number(s):
   User: file
   System: Listfile name:
   User: (enter a filename to use a binary, numeric, or
              extregno listfile)
3. the regnos in the Curlist or the Previous list:
   System: Registry number(s):
   User: cur (or prev)
4. To retrieve all structures in the database:
   System: Registry number(s):
   User: all

SEARCHING WITH NAME QUERY

Name queries can be formulated in the Register and the Search Modes and searched by FIND(Name) and SEARCH(Name). The search technique and the result are equivalent in both Modes; however, we have to watch the Curlist in the Search Mode to determine what is the search domain (see p. 529 for switching lists).
Each of the structures registered in the database has to have a name which we provide on registration of the structure (see p. 555). The names can be used for exact structure or substructure searching by formulating queries that contain an exact name or segments of the names of the registered structures.
The segments of the names are specified using:
left-hand truncation
right-hand truncation
left-hand/right-hand truncation
The symbol of the truncation is "@."
The name segmentation must be at some of the punctuation character that breaks the name into segments. The "@" cannot be embedded within a character string.
For example, we have a database in which we have structures named as follows:
1. 3-benzoylbenzo(f)coumarin
2. 3-benzoyl-7-methoxy-coumarin
3. 4-bromomethyl-6,7-dimethoxycoumarin
4. 3,3'-carbonylbis(7-diethylaminocoumarin)
5. 3,3'-carbonylbis(7-methoxycoumarin)

Search 1
We are searching for all the coumarins:
SEARCH> name = @coumarin
will retrieve the structures 1-5.

Example: *Retrieving regnos*

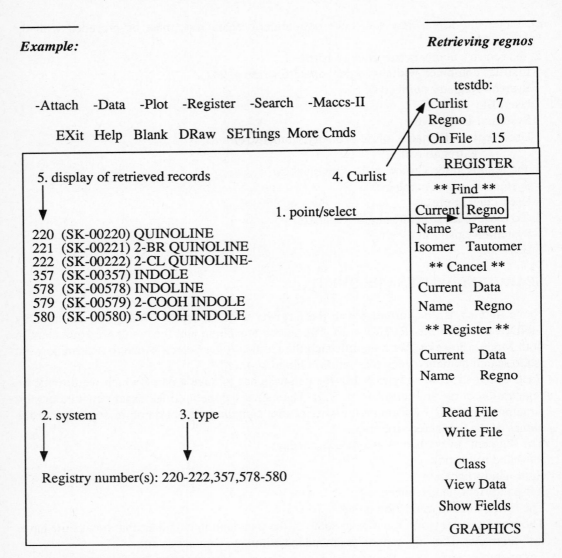

Figure 3/79. Retrieval of regnos

Search 2
We are searching for carbonylbiscoumarins
SEARCH> name = @carbonylbis@
will retrieve structures 4 and 5.
Search 3
We are searching for benzoylcoumarins:
SEARCH> name = 3-benzoyl@
will retrieve structures 1 and 2.

DATA SEARCHING

However, embedding the "@" symbol in names is not allowed, e.g.,:
bro@methyl is not valid.

An exact structure can be also retrieved with a query consisting of the complete name as it has been registered with the structure, e.g.:
SEARCH> name = 3-benzoylbenzo(f)coumarin

SEARCHING WITH KEYS QUERIES

We have learned (see p. 424) that each structure is registered with a set of keys that are part of the record in the database. The keys can be used to formulate queries to retrieve structures with specific structural fragments. The search can be performed only in the Search Mode by the SEARCH(Keys) option. In effect, this is a substructure search using nongraphic query.
Complex queries can be formulated using the Boolean Operators (see p. 247) which in MACCS-II are specified with the following bracket symbols (A and B are the keys):

(A,B)............AND - both A and B must be present in the
                                retrieved structures
[A,B]............OR - A or B or both may be in the retrieved
                                structures
<A,B>............NOT - NOT B excludes B, NOT A excludes A
                                from the retrieved structures
{A,B}...exclusiveOR - either A or B but not both may be
                                present in the retrieved structures

The format of the Keys Query is as follows:

$$\begin{array}{c}(key,key,key......)\\ [key,key,key......]\\ <key,key,key......>\\ \{key,key,key......\}\end{array}$$

To list the keys that have been defined for the current opened database, use the option:

show keys........displays the keys for current database
   SEARCH> show keys
    Show keys: (enter all, keyno, or a range keyno-keyno)
For example, the query with the sequential keys:
SEARCH> keys [(1,26),(1,29)]
would hit structures with 5-membered ring with nitrogen or a 6-membered ring with nitrogen

SEARCHING WITH FORMULA QUERIES

Formulas can be searched only in the Search Mode.
We can perform a broad search by the SEARCH(Formula) option for structures using element counts as the query. In large databases, of course, this may yield a fairly long list of structures. However, this technique can be used to create a subset of a database which is then used as the search domain for a more selective search using a graphic structure query. The element count format for the query is as follows:

$$atom(count) \ atom(count)......atom(count)$$

Zero(0) count specifies that the element must not be in the retrieved structures. If no count is specified, the count is equal to 1.
To retrieve no-structure records, the "H0" query can be used.

The spaces between atom(count)s are important to resolve ambiguous formulas. The following examples show the format for entering formulas:
CCl3............C Cl3   or   C1Cl3
CLi.............C Li   or   C1Li
Cl I............Cl I   or   Cl1I
C7H5O2Na........C7 H5 O2.Na
C(CH3)3Cl.......C (CH3)3 Cl

To search for exact formulas and thus for exact structures, hydrogens are included in the formula, e.g.:
SEARCH> formula C2 H6 O
retrieves ethyl alcohol and all isomers having the element counts.
To search for substructures, hydrogens are not included in the formula, e.g.:
SEARCH> formula C2 O
retrieves all structures having two carbons and one oxygen.

RETRIEVING DATA IN DATA FIELDS

We have seen in the discussion of structure searching that we can display the data associated with the retrieved current structure by the VIEW DATA option in the Register and Search Mode (see p. 509). However, if we retrieve multiple structures, we have to view each structure on the list and repeat the VIEW DATA operation for each current structure. We might want to display the data in specified data fields for the entire list of structures. We can do so by switching to Data Mode (the Curlist will be carried over) and selecting FIND DATA. The data will be displayed (Figure 3/80.)
In the example in Figure 3/80 we had 3 structures in the Curlist (the hits of our search in Register or Search Mode) and we wanted to find mol.weight and CAS.Regno for the compounds.
The data header for each of the data field consists of:

DATA SEARCHING

Example: *Data retrieval in Data mode*

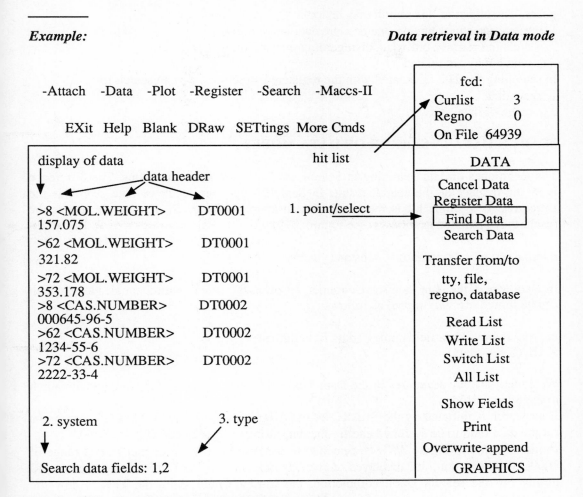

Figure 3/80. Result of data retrieval in Data mode

intregno   data field name   data field number

We can turn off the display of the data headers by the option PRINT on the menu to display only the data. Note that this option has a different function in Data Mode than in Search Mode (see p. 513).

This operation is simply a retrieval of the exact data related to the structure by specifying the data fields we wish to see. The subject of the search was a structure query and the primary object was to find structures having either the exact match or having the structural fragment in common.

We can reverse the process and perform searches in which the search subject is a data

query. The object of this search may be to:
1. retrieve structures having one or more data in common
2. search and retrieve only data disregarding the structures
   to which they are related

In searching for data associated with the registered structures, we can search for:
1. exact data
2. a range of data

SEARCH DATA IN SEARCH OR DATA MODE?

The techniques of data searching in Search and Data Mode are analogous. The difference is in the display of the search result. In Search Mode, the result of data search is the structures. If a single hit is the result, the structure is displayed. If multiple hits is the result, they are listed as follows (see Figure 3/79):

internal regno   (extregno)   substance name

In Data Mode, the data themselves are displayed on the screen with Data Headers (which can be turned off, *vide supra*) as follows:

internal regno   data field name   data field number
XXXXXXX (data)

We cannot display structures in the Data Mode. To view the structures, we have to switch to the Search Mode.

Data search is initiated by the SEARCH DATA option (Figure 3/81). MACCS-II prompts for the data field to be searched and for the data string to be searched (Figure 3/81).

In the example in Figure 3/81 we searched in the entire database for mol.weight range of 1000-1100. The data are displayed, screen by screen, and the Curlist is updated by 10 throughout the search. Upon completion, the Curlist would show 64. This is now the search domain. If we transfer to Search Mode, the Curlist is transferred.

If we perform the same data search shown in Figure 3/81 in Search Mode by selecting SEARCH(Data), the result will be as follows:

69 TRIS(6,6,7,7,8,8,8-HEPTAFLUORO-2,2-DIMETHYL-3,5-OCTANEDI
93 TRIS96,6,7,7,8,8,8-HEPTAFLUORO-2,2-DIMETHYL-3,5-OCTANEDI
1283 PERFLUOROEICOSANE
17073 TOMATINE
38126 DIPENTAERYTHRITOL HEXAPELARGONATE
:
:

DATA SEARCHING

Example: *Data search in Data mode*

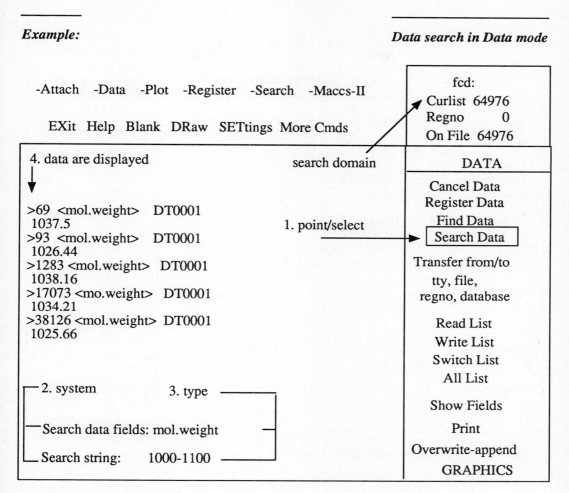

Figure 3/81.   Search of data in Data mode

DATA FIELDS INDEX

To display the data fields that have been defined for the currently opened database, their kind, and type of data they contain so that we know the exact format, we activate the option SHOW FIELDS on the Menu (Figure 3/82).
Note that one of the specified fields in Figure 3/82 has not been displayed. If access to a particular data field has not been granted, the field does not display. The same option is available in Search Mode and Register Mode.

If we specify "all," all data fields will be displayed (excluding those with limited access).

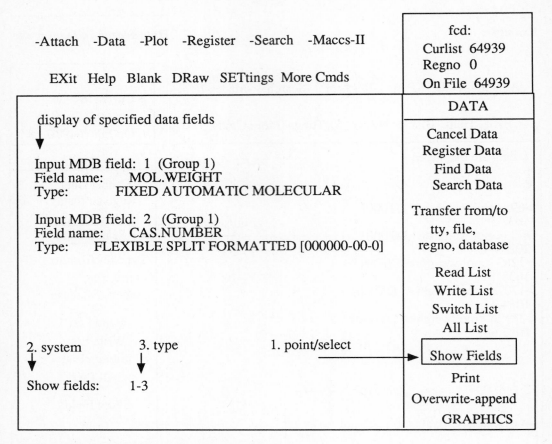

Figure 3/82. Displaying database data fields

DEFINING REFERENCE LIST IN DATA SEARCH

We have already learned that the Curlist, if any, is carried over from Mode to Mode. We have discussed how to manipulate lists and how to establish lists (see p. 526). The same options for these operations are available in the Data Mode.

DATA SEARCHING TECHNIQUES

MACCS-II offers various techniques to search for:
1. text strings in text data
2. numeric strings within text data
3. numeric data ranges within text data
4. numeric data
5. formatted data

SEARCHING TEXT DATA

A text is composed of lines of alphanumeric characters.
The fixed data fields can have up to 20 characters per line.
The flexible data fields can have up to 120 characters per line.
Each character in the line of text data is in a column position (Figure 3/83-A). The first character is in column position 1, the last character is in column position 120:

```
ABC ABC DE F GH I JK LMN O P Q 123-4....Z
1 2  34 5 6 7 8 9 10.....................            .120
```

Figure 3/83-A. Column positions of text data

Column Range Searching

Any string of characters in the line of text data can be searched for by formulating a data search query.
The query has the following format:
<C1-C2> STRING1 <C3-C4> STRING2...........
To search for a string in any position in the line:
no column range required
To search for a string in a specific position in the line:
column range required
To search for strings with leading or trailing spaces:
use the symbol @
To search for numbers within text data:
use the symbol @

Example: *Data searching*

The following examples refer to the strings in Figure 3/83-A.
(In our examples, the data field to be searched is specified by the data field number = 2.)

Query 1
To search for the GH and LMN string, i.e., for a string in any position in the line:
Search data field(s):2
Search string:gh lmn

Query 2
To search for the second ABC string:

Search data field)s):2
Search string:<5-9> abc

Query 3
To search for I and O in their exact positions in the line:
Search data field(s):2
Search string:<17> I <26> O

Query 4
To search for DE anywhere in the line and GH in the exact position:
Search data field(s):2
Search string:DE <14-15> gh

Query 5
To search for Q with the space:
Search data field(s):2
Search string:<28-30> @ Q

Query 6
To search for GH with the space:
Search data field(s):2
Search string:<14-16> GH @

Query 7
To search for the number in the exact position:
Search data field(s):2
Search string:<32-36> @123-4@

Alphanumeric Range Searching

We can perform alphanumeric range searching in a combination with the column range searching. The query has the following format:
TEXTA TO TEXTB

```
ABCBCDEFGH              846                    Z
1 2 3 4 5 6 7 8 9 10......    20 21 22......   120
```

Figure 3/83-B. Column positions of text data

DATA SEARCHING

Example: *Data range searching*

The following queries refer to Figure 3/83-B:
Query 8
To search for any string between text C and F:
Search data field(s):2
Search string:C to F
retrieves DE

Query 9
To search for a string in the exact column position:
Search data field(s):
Search string:<5> B to D
retrieves the second C in the line

Query 10
To search for a digit in an exact position:
Search data field(s):2
Search string:<21> 0 to 9
retrieves 4

SEARCHING NUMERIC DATA

Numeric data are one or two numbers per line. Multiple lines of number(s) form one or two columns of numbers.
The numbers can be separated by a space, comma, or a hyphen. Flexible numeric data fields may have text of comment, up to 112 characters per line.

Example: *Searching numeric data*

We can search for the numbers in either of the columns.
The formats of numeric data queries are as follows:

Format 1: single value - V
The hits will be any line of numbers in which the value V is in either of the columns, or any line with two numbers A, B in which the V is in the range of A - B.
Query 11
Search data field(s):
Search string:50
Result of Query 11:

50.0
45.0-50.0
45.0-55.0

Format 2: two values - U,V separated by a comma, space, or hyphen
The hits will be:
a value A between the range U-V
a range of values A-B in which A or B is in the U-V range or U or V is in the A-B range

Query 12
Search data field(s):
Search string:30-35
Result of Query 12:
30.0
32.0
25-30.0
25-38.0

Format 3 - U,V/W,X
The hits will be any pair of numbers A-B in which A is in the U-V range and B is in the W-X range. Either of the column of numbers can be searched for exact values or for a range of values

Query 13
Search data field(s):
Search string:50-60/1000-1200
Result of Query 13:
50.0-1200.0
58.0-1145.0
60.0-1145.0

Query 14
Search data field(s):
Search string:58/1000-1200
Result of Query 13:
58.0-1145.0
58.0-1341.0

Query 15
Search data field(s):
Search string:50-60/1145
Result of Query 15:
58.0-1145.0
60.0-1145.0

DATA SEARCHING

Query 16
Search data field(s):
Search string:58/1145
Result of Query 16
58.0-1145.0

Query 17
Search data field(s):
Search string:/1000-1200
Result of Query 17:
38.0-1002.0
50.0-1200.0
58.0-1145.0
60.0-1145.0
60.0-1200.0

Query 18
Search data field(s):
Search string:50-60/
Result of Query 18:
50.0-1200.0
58.0-1145.0
58.0-1341.0

Query 19
Search data field(s):
Search string:58/
Result of Query 19:
58.0-1145.0
58.0-1341.0

Query 20
Search data field(s):
Search string:/1145
Result of Query 20:
58.0-1145.0
60.0-1145.0

10   SETTING UP DATABASES

In MACCS-II, the databases are classified as follows:
1. private databases
2. public databases

PRIVATE DATABASES

Private databases are those which the User creates and maintains in a directory in the operating system. Access to private databases is controlled by the User. Other registered MACCS-II Users may use the private databases with the consent of the creator. The private databases can be converted into the public databases.

PUBLIC DATABASES

Public databases are those which may have been part of the MACCS-II package (MDL databases that have been developed for MACCS-II), or have been created within the organization by an individual or a group, or are private databases converted to public ones.
The public databases are controlled and maintained by the System Supervisor who arranges access to the databases for qualified Users.
The public databases are subject to the database security system. The security system can be tailored to each environment of MACCS-II.
The structures/records registered in the public databases can be divided into classes and access can be selectively granted for these classes to Users.
The data fields can be divided into groups and access can be selectively granted for these groups to Users.
Passwords are assigned to Users to be able to access, view, register, change, or delete specific data categories in the public databases.
Multiple passwords can be assigned to Users to selectively grant them access to these data.

SETTING UP A DATABASE

Set up of a database in MACCS-II is simplicity itself. MACCS-II will initiate a question-answer process which will drive the database creation.
Before we start the process, we should determine whether:
1. We wish to be able to register duplicate structures.
   If we elect this feature, on registering a duplicate
   structure MACCS-II will issue the following message:
   MOLECULE CURRENTLY REGISTERED
   and will give us a chance to cancel or go ahead with the registration.

2. How many and what type of fixed data fields we want in the database. We recall that the fixed data fields are always displayed (unless turned off by Settings) with the retrieved structures and if too many of them are present they may obscure the structure image. The space for fixed data fields is reserved in the database even if we do not use the fixed data fields for registering data. If too many are defined and not used, the disk storage capacity will be wasted.

INITIATING DATABASE SETUP

1. Create a directory in your system where the database will reside.
2. Enter MACCS-II Main Menu and initiate the setup process by switching to the keyboard selection and activating the ACCESS CREATE option followed by entering the database name (Figure 3/84).

The database name can be up to 60 characters and must begin with a letter. Blanks, commas, <>, and special characters except period, comma, and hyphen, are not allowed. The pathname must conform to the operating system in use.

After entering the database name MACCS-II will continue in the dialog:

Enter estimated database size (number of compounds):
(A carriage return results in size of 500 molecules)

Define fixed field number 1 [N]? Y(es)
Answering yes will initiate definition of fixed data field.

DEFINING FIXED DATA FIELDS

MACCS-II: place external registry number in this field [N]?:
User : y (for YES), <cr> for NO
  Must be the first data field in the database.
  YES will result in MACCS-II prompt:

MACCS-II: enter name for extreg field:
User : (enter a name for the data field)
  Up to 30 characters with no blanks, commas, hyphens. The data field names should be descriptive of the data since they are always displayed when working with data.
MACCS-II: provide format pattern:
User : (enter the desired format)

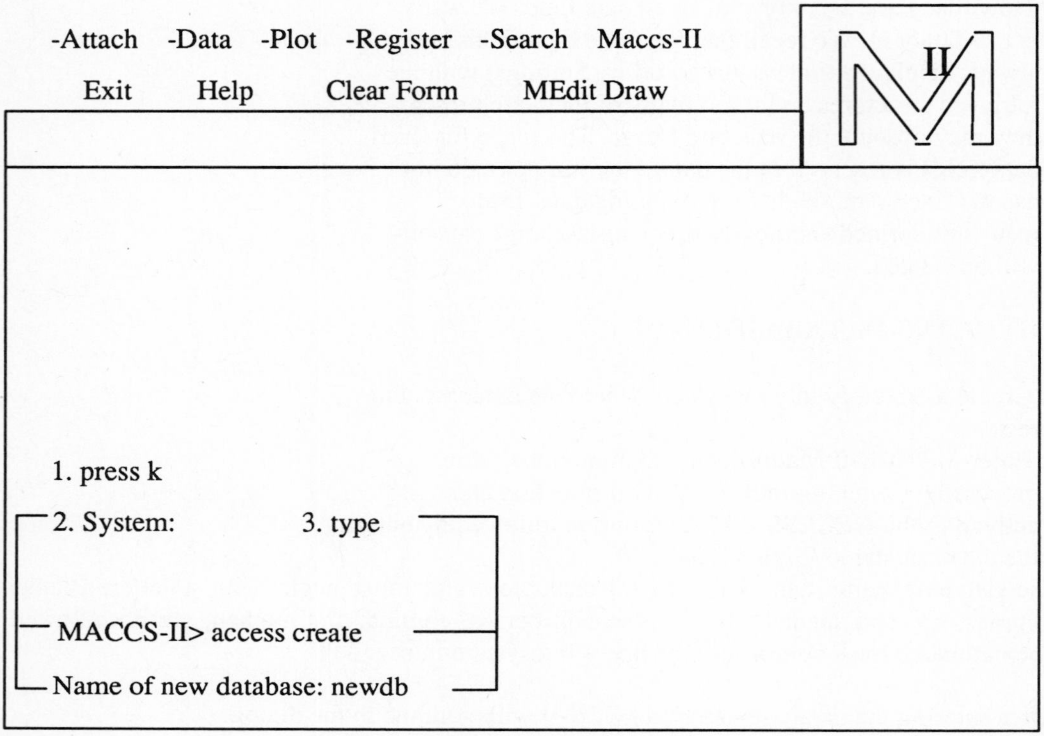

```
   -Attach  -Data  -Plot  -Register  -Search  Maccs-II
    Exit     Help        Clear Form        MEdit Draw




     1. press k
     2. System:          3. type


      MACCS-II> access create
      Name of new database: newdb
```

Figure 3/84. Maccs-II Main menu to set up a database

The format pattern is assigned using the symbols we have discussed on p. 427.

The first fixed data field has been defined and MACCS-II displays the definition for approval, e.g.:
  Data field number 1
  Name: SK.NUMBER
  Data storage: fixed
  Field type: external registry number, unique_chk
  Database size estimate for Ext_Hash = 500
  Format: SK-0000
  Security - data group: 1
  Is this correct (enter No to erase the definition) [N]?
Y(es) will confirm the definition and continue the dialog:

MACCS-II: define fixed field number <n> [N]?
  n = is the number of the next fixed data field we wish

SETTING UP DATABASES

to create
To finish fixed data field definition process, enter <cr>
Y(es) will continue the dialog:

Place date in this field [N]? Y
American (M-D-Y) or European (D-M-Y) [A]: <cr>
Automatic system date [Y]?: <cr>
(If we enter "N", we can name the field):
Enter name for date field: prepn.date

Define fixed field number <n> [N]? Y
Place initials in this field [N]? Y
Enter number of characters for initials (1-4): 2
Automatic initials [Y]? N
Enter name for initials field: grp.leader

Define fixed field number <n> [N]? Y
Place molecular weight in this field [N]? Y
Automatic calculation [Y]? <cr>
(The field is named by MACCS-II "MOL.WEIGHT")
If we answer "N" MACCS-II will ask:
Field type (Num/Form/Text): n
("n" stands for numeric data type)
Enter label for units:
(Label may be up to 20 characters with blanks, commas, hyphens.)
If we specify "f" (for formatted), MACCS-II asks:
Provide format pattern:
(we use the symbols, see p. 427)
If we specify "t" (for text), MACCS-II asks:
Data width in characters (1-20):

To finish definition of fixed data fields, enter <cr> at the MACCS-II prompt:
Define fixed field number <n> [N]?
Do you want to define another fixed field first [Y]:
(Note that if we answer no, we are not able to return back to the fixed field data definition.)

At this point, the database structure has been essentially created and the database is ready to accept structures and data for the defined fixed data fields.
However, we can continue in the process and define flexible data fields.

DEFINING FLEXIBLE DATA FIELDS

Define flexible field number <n> [N]? Y

Enter name for this field:
  Names can be up to 30 characters long, may not contain
  blanks, commas, or hyphens
Enter descriptive comments:

  A comment/memo up to 65 characters can be entered to identify the nature of the data field. The comment is displayed
  when the option SHOW FIELDS (see p. 543) is activated.
Field type (Num/For/Text):
Enter label for units:
  If data field is numeric, we can indicate any unit up to
  20 characters long. However, this is a label only, i.e., if
  incorrect unit is registered, MACCS-II does not detect the
  error.
Provide format pattern:
  If the data field is formatted, we enter any of the format
  symbols (see p. 427) to pattern the data

When we are finished with creating flexible data fields, we enter <cr> at the prompt:

Define flexible field number <n> [N]?

Do you want to define another flexible field first [Y]? N

The database data fields have been defined and we can register data.
To create additional data fields after the database structure has been created or to change/modify existing data fields, help of the System Supervisor should be sought.

11 UPDATING DATABASES

To update existing databases, we perform operations by which structures and the associated data are registered (entered) in the database, delete structures and/or data, or modify/change structures and/or data.
The same operations are performed to fill a new database whose structure has been defined and set up with structures and data.
Note that if the security system applies to the database in which we wish to perform the update operations, specific authorization is required to perform specific update operations. If the authorization is not in effect, MACCS-II will not proceed with the update.
We will divide the operations of registration (entry) in a database into operations for registering:
1. structures
2. alphanumeric data
These classes of data can be registered independently of each other although they may have the same registration number, i.e., they are part of one record. We can first register into a database all structures we have and later append the alphanumeric data to the structures. We can first register the alphanumeric data and later enter the structures that are related to the alphanumeric data.

REGISTERING STRUCTURES

Structures are registered in Register Mode or in Attach Mode.
In both Register and Attach Mode, the structure we wish to register has to be the current structure.
To get a current structure, we can:
1. draw the structure in Draw Mode
2. assemble the structure in Attach Mode
3. recall a structure from a Molfile
When the current structure is established, the REGISTER CURRENT option initiates the registration process (Figure 3/85).
The registration involves a dialog between MACCS-II and the User:
Register into class <default class> [Y]?
: y (for YES), n (for NO)
  YES registers the structure into the class 1
  NO elicits the prompt:
  Enter class(es) for registration:
  (enter the optional class)(the System Supervisor
  should be contacted for authority to class registration)

555

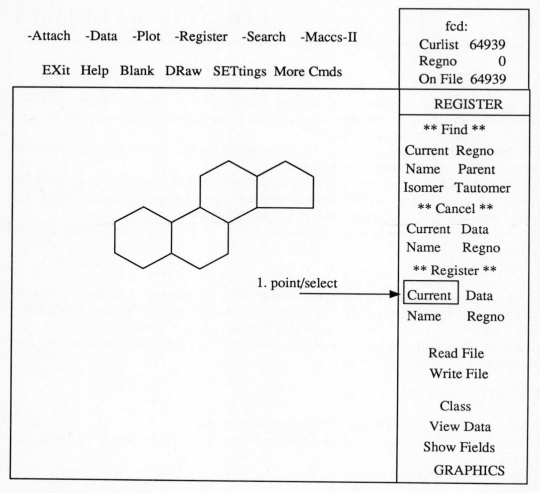

Figure 3/85. Structure registration

Provide molecule name:

The name can have up to 80 characters (note that if we recall structure from a Molfile, the structure is already named).

Provide SK.NUMBER (SK-0000): SK-1234
    <DUPE> Duplicate external regno SK-1234 on file
    OK to register duplicate [N]? <cr>
    SK-1235 OK [Y]? <cr>
If a duplicate extregno is located, MACCS-II offers the choice of registering the duplicate, or, if the automatic regno feature has been selected (see Setting up databases p. 550), the

alternative extregno is displayed for approval.
If an exact duplicate structure exists in the database and the database has been set up for duplicate structures, MACCS-II prompts:
MOLECULE CURRENTLY REGISTERED
Wish to register duplicate [N]?
   Y proceeds with the registration, N (or <cr>) aborts the
   registration.
If the database definition does not allow for duplicate structures, MACCS-II will prompt:
MOLECULE CURRENTLY REGISTERED
and aborts the registration process.

If stereoisomers of the current structure are registered in the database, MACCS-II will issue a warning:
--WARNING--
STEREOISOMERS ON FILE
Proceed with registration [N]?
   N (or <cr>) aborts the registration, Y results in continuing registration process

If there are tautomers of the current structure in the database and the Settings suboption "Tautomer" is on MACCS-II checks on the tautomers and issues a warning:
--WARNING--
MAY BE TAUTOMERS ON FILE
Proceed with registration [N]?
N (or <cr>) aborts the registration, Y registers tautomer.

GENERATING SEMA NAME
   The system informs us that the SEMA name for the current structure is being generated.

SETTING KEYS
   The system sets automatically the keys of the structure.

REGISTERING NEW RECORD
   The system completes the process. The structure is now registered.

MACCS-II assigns automatically an internal registry number to the structure. The internal regno is sequentially generated from the highest intregno in the database plus 1. If we have a database from which we have deleted some records, the internal registry numbers of the deleted records are retained in the database. On registration of new structures/records, we have an option to use these existing regno for the new structures. The Settings suboption (see p. 442) sets the reuse of intregnos:
reuse deleted regnos.......turns on/off reuse of existing
               free intregnos (default is "off")
If a fixed data field for external registry numbers has been defined in the database, it is the

first fixed data field in the database structure. We receive prompts consisting of the data field name and format to fill in the data.
If additional fixed data fields have been defined, they can be entered now or at any time later.

Note that it is possible to create a record with no structure and generate a regno for it. Data for this record can be registered into the data fields defined for the database. The structure can be registered later. To carry out registration with no structure, we simply have the screen with no current structure and initiate the registration process and answer N at the prompts (*vide supra*) when appropriate.

REGISTERING ALPHANUMERIC DATA

Registration of alphanumeric data into the data fields can be carried out in the Register Mode or in the Data Mode. The difference is:
1. In the Register Mode, we can register data for only one
   current structure (Figure 3/86).
2. In the Data Mode we can register data for a list of
   structures in the Curlist (Figure 3/87).

When we activate the REGISTER DATA option we receive the prompts for data registration.

The defined data fields in the database can be displayed by activating the SHOW FIELDS option.

At the prompt:
register data field(s):
we can enter:
all - to register data for all data fields in the database
data field names - one or more names separated by commas or
              spaces. The @ symbol can be used in names.
data field numbers - single, multiple numbers separated by
              commas or spaces, or a range
combination of numbers and names, or
" (quotation marks) to specify the preceding data field(s)

If the specified data field(s) are fixed data field(s), MACCS-II displays the data field name and, for formatted data fields, the format:
Provide CAS.NO (0000-00-0):
<cr> following the data entry results in registration of the data. The new data overwrite any old data that may have been previously registered in this data field for this structure. If data are already registered in this data field, it will be displayed. To leave the data field empty or the existing data unchanged, enter <cr>.

UPDATING DATABASES

Example: *Registering data in Register mode*

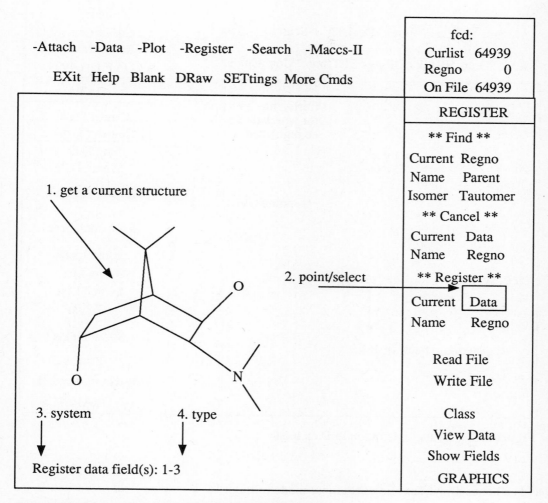

Figure 3/86. Registration of data in Register mode

If the specified data field(s) is flexible data field(s), MACCS-II displays the header:

>internal regno   (ext regno)   data field name   data field no

and a prompt:
(To register data, enter SAVE or QUIT to abort.)

Example: *Registering data in Data mode*

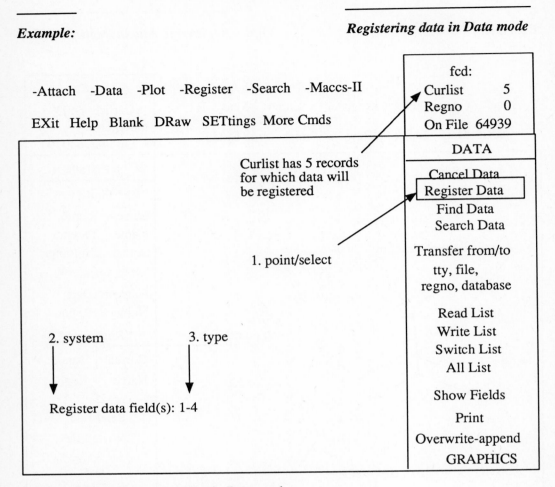

Figure 3/87. Data registration in Data mode

and switches to the Text Editor in input mode (prompt I>) if no data field is present. In the Text Editor, we can enter new data, correct or modify the entries of new data, or change, correct, or modify the existing data.
The Text Editor prompt is:
E>
The Text Editor is a line editor with a set of commands. We will list the major ones:
help.......displays help for edit commands
append.....inserts text at the end of the line
bottom.....sets the current line to the bottom of the buffer
delete.....deletes lines of text from the buffer
get file...places text in a file in the buffer

UPDATING DATABASES

input......switches to input mode and places the pointer at
       the bottom line
locate.....forward search of specified text
next.......moves the pointer X line up (X) or down (-X)
position...moves the pointer to specified line
put file...saves the buffer in a file
quit.......clears the buffer and exits the Text Editor
replace....replaces a line with specified text
rlocate....backward search of specified text
set ccase..sets uppercase for Change
set lcase..sets lowercase for Change
set numbers...displays line numbers
suspend....exits Text Editor, does not clear the buffer
top........moves the pointer to the top of buffer
type.......displays the content of the buffer
where......reports the status of the buffer
+ ....inserts text in the current line
- ....deletes a character from the current line
= ....replaces text in the current line
* ....joins the current line with the one that follows
/ ....splits the current line

Example:

> 36 (SK-1234) <LIT.REFERENCE>   DT0015
J. Org. Chem. 36, 134-200(1980)
(To register data, enter SAVE or QUIT to abort)
<LIT.REFERENCE>
E> quit
This operation results in leaving the existing data unchanged.
<LIT.REFERENCE>
E>top
E>type
J. Org. Chem. 36, 134-200(1980)
E>delete
E>input
E>J. Amer. Chem Soc. 45, 134-200(1980)
J. Amer. Chem Soc. 45, 134-200(1980)
E>save
This operation results in deletion of the existing data and typing and registering the new data.

RE-REGISTERING RECORDS IN DATABASE

Changes in the registered data can be done by operations in the Register Mode which offers the following options:

REGISTER NAME - stores the current structure with a new name
          (Figure 3/88)
This operation is performed if we want to change the name of a registered structure without changing the structure itself.
The structure must be the current structure which must be established as a single hit by:
1. retrieval in the Register Mode
2. search in the Search mode
REGISTER REGNO - stores the current structure under the
          specified internal registry nymber
          (Figure 3/89)
This operation is performed to change/modify/replace a registered structure under the same or different intregno, or to register a structure for a no-structure record.
If we respond to the prompt in Figure 3/89 by "N," the re-registration is aborted and the original structure/record remains unchanged. Note that we can use:
1. the same registry number of the original structure
2. a new registry number (if we enter a regno higher than the highest one+1 in the database, MACCS-II registers a number of the highest one+1)
3. a vacant regno number which remained after deletion of a record

This operation allows for changing a structure while the associated data are unchanged.
MACCS-II will prompt:
Overwrite existing data [N]?
and then prompts for registration of new fixed data.

DELETING STRUCTURES

The structures that are registered in a database can be deleted by operations carried out in the Register Mode. Note that if we delete a structure, all associated data are automatically deleted. On the other hand, we can selectively remove only data while the structures remain (*vide infra*). If we delete a structure or structures, the registry numbers are available for registration of new structures (*vide supra*).
The operations for deleting registered structures from a database are, in fact, registration in reverse. We use options that manipulate the same objects as in the registration (*vide supra*), i.e., CANCEL CURRENT, CANCEL NAME, and CANCEL REGNO on the Register Mode Menu.
The structure to be deleted must be the single current structure (Figure 3/90).
If we answer Y at the prompt (Figure 3/90) the structure, its name, and all data in the data fields will be erased. However, the structure will remain on display and we can use it

UPDATING DATABASES

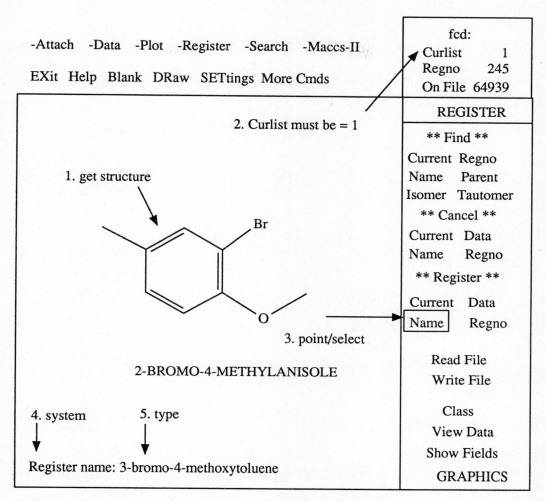

Figure 3/88. Registering existing structure with a new name

either to re-register it or modify it for another structure.
Answering N will abort the process.
Note that if we perform FIND CURRENT which results in multiple hits and the Settings Multiple Find Hits is on (see p. 511), only the structure with the lowest intregno will be deleted.

The options CANCEL NAME and CANCEL REGNO allow for deletion of a single structure or multiple structure.
When we point/select CANCEL NAME, MACCS-II prompts as follows:
Cancel name:
If we input an exact name or a segment of a name which identifies a single structure, the

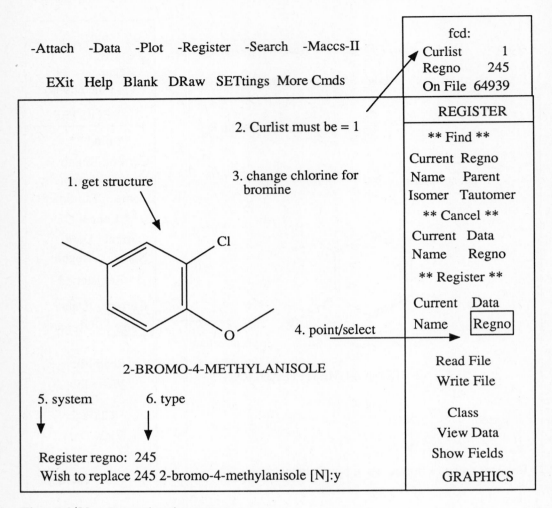

Figure 3/89. Re-registering structure

structure and associated data will be deleted from the database. If we input a name or a segment of a name (using the "@" symbol) which identifies multiple structures and the Settings Multiple Find Hits is on (see p. 511), all the structures will be deleted. MACCS-II will display the regno of each of the structures on the list and ask:
Delete structure <regno> and all data [N]?
The structures will not be displayed.
If the Settings Multiple Find Hits is off, we cannot perform this operation on the entire list (*vide supra*).

CANCEL REGNO allows for input of a list of regnos.
MACCS-II prompts:

UPDATING DATABASES

Example: *Deleting single structure*

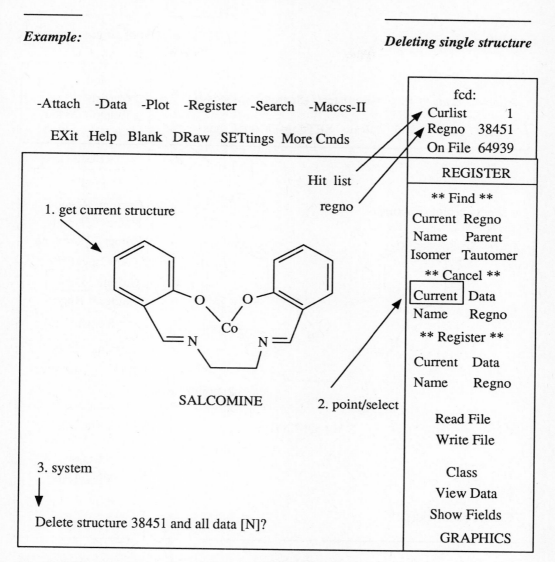

Figure 3/90. Deletion of structure from database

Cancel regno:
The input can be:
1. a single or multiple intregnos (via the keyboard)
    e.g.:   14
               1,2,7-12
               134-280

Example: *Deleting data of single structure*

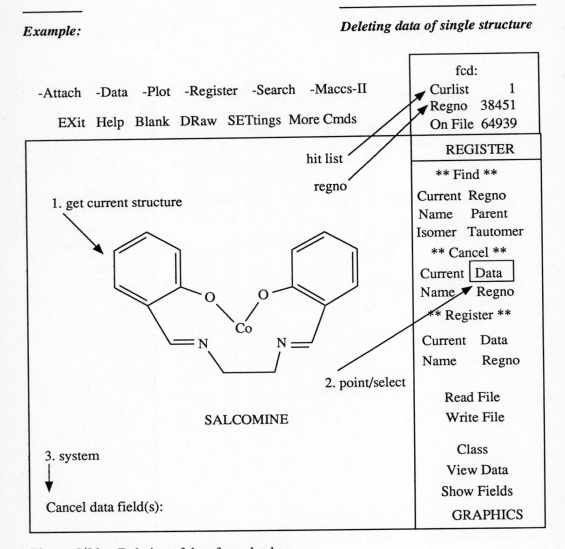

Figure 3/91. Deletion of data from database

2. a single or multiple extregnos (via the keyboard)
   e.g.   SK-0023
                SK-0024, SK-0028
                SK-0024 to SK-0045
3. file
   Listfile name(s):   myfile.lst
                             curr
                             prev

UPDATING DATABASES

The files can be binary, regno, or extregno files (see p. 423, 426, 430 for discussion of regnos and listfiles).

If multiple regnos are to be deleted, MACCS-II asks for approval for each of the structures on the list. The structures are not displayed.

DELETING DATA

In the Register Mode, we can delete selectively data for a single current structure and preserve the structures in the database.
The CANCEL DATA option deletes the specified data fields (Figure 3/91).
At the prompt:
Delete data field(s):
we can enter:
1. name(s) of the data field(s)
2. number(s) of the data fields
3. "all" to delete all data associated with the structure
4. " (quotation mark) to delete data in the fields specified
   in the last CANCEL DATA operation.

The data field header and the data are displayed for each data field specified for the CANCEL DATA option. MACCS-II prompts for confirmation to delete, e.g.:
> 5 (SK-0245) <EXT.REGNO>   DT0001
   SK-0245
Delete external regno for regno 5 [N]?
(Y deletes, N or <cr> aborts the CANCEL process)

> 5 (SK-0245) <BOIL.POINT>   DT0003
245.0
Delete data field 3 from regno 5 [N]?

Note that these operations delete data, not the data fields which have been defined for the database, and can be again filled with new data.

While in Register Mode we delete data of a single structure, in Data Mode we can perform the same operation for a list of structures, i.e., for the Curlist.

12   TRANSFERRING DATA

The Data Mode comprises a sophisticated TRANSFER option for transfer of data in the data fields in or out of the database and within the database.

To transfer data, we have to have the source (location) of the data and the destination (location) for the data being transferred.

There are four locations where the transfer of data can take place:
1. database
2. file (datfile)
3. registry number (a record in a database)
4. tty (terminal screen or keyboard)

Combination of these data locations in the data transfer results in:
1. data retrieval from the database
2. registration of data in the database
3. writing data in a datfile
4. retrieval and display of data on the terminal screen

Figure 3/92 illustrates the allowed combinations of locations in data transfer.

The combinations which are invalid are shown in Figure 3/93.

The Datfiles for data transfer are written by MACCS-II or can be written by the User in a Text Editor. The format of the Datfile is shown in Figure 3/94.

The **Compound Identifier** may be the internal registry number or the external registry number. The external registry number must be enclosed in parentheses. If more than one record in the database have the same external registry number, MACCS-II uses the first record of this number in the database.

The **Data Field Identifier** may be the data field name or the data field number. The data field name must be enclosed in <> brackets and may be either the full name or truncated name with the @ symbol. The data field number is preceded by DT.

If the Compound Identifier consists of both the internal and external registry number, MACCS-II uses the external registry number. If the Data Field Identifier consists of both the data field name and number, MACCS-II uses the data field name.

The following is an example of the Datfile named "mydataf" prepared in the Text Editor:

105 (XX-1234) <Literature.ref> DT 12
J. Org. Chem. 100, 234-235(1988)

25 (XX-5678) <Literature.ref> DT12
J. Amer. Chem. Soc. 90, 45-68(1988)

(XX-8912) DT12
Synthesis 23, 23-56(1970)

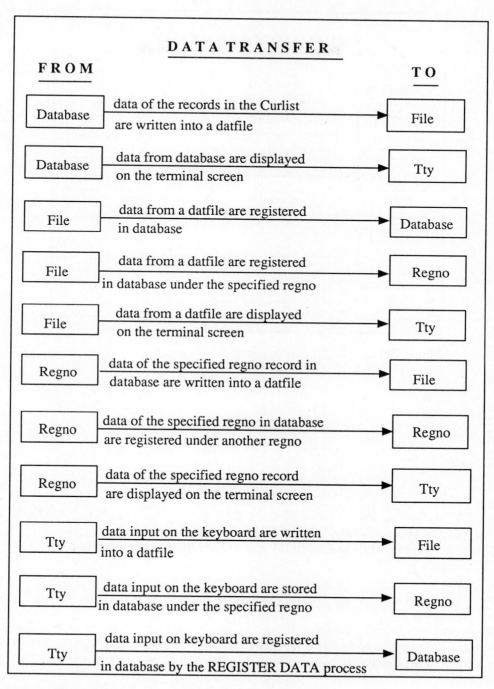

Figure 3/92. Transfer of data in Data mode

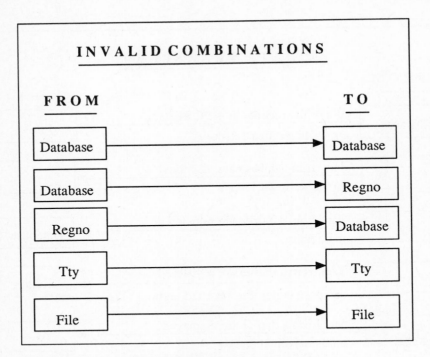

Figure 3/93. Invalid combinations in data transfer

25 <Mol.weight>
235

105 <Mol.weight>
467

If we are transferring data into or within the database, we, in fact, are registering the data in the database (Figure 3/95).
The source and destination at the FROM: and TO: prompts can be specified by:
d...........database
f...........file
r...........regno
tty.........keyboard
Entering <cr> will abort the operation.
The transfer in Figure 3/95 would register data in the data fields DT 0003 to DT0006 for the records in the Curlist (10 records) by typing the data on the keyboard at the MACCS-II prompts.
If data have been previously registered in the data fields into which we want to transfer data, MACCS-II will prompt if the old data should be overwritten, appended, or the regis-

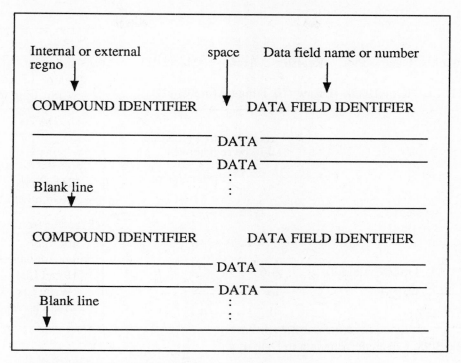

Figure 3/94. Format of Datfile

tration skipped, e.g.:

> 105 (XX-1234) <Literature.ref> DT 0003
J. Org. Chem. 100, 234-235(1988)
Overwrite, append, or skip (O/A/S): o

We can answer this prompt for each of the data field involved in the transfer. Or, we can set up the conditions by the OVERWRITE-APPEND option on the menu (Figure 3/96). The choices in Figure 3/96 are outlined below:
1......automatically overwrites old data with the new data
    (the old data are lost)
2......appends the new data at the end of the existing
    data (for flexible datatypes). For fixed data fields,
    this option performs the skip
3......automatically skips the registration if there are data
    already registered in the data field
4......prompts what to do (it is the default)

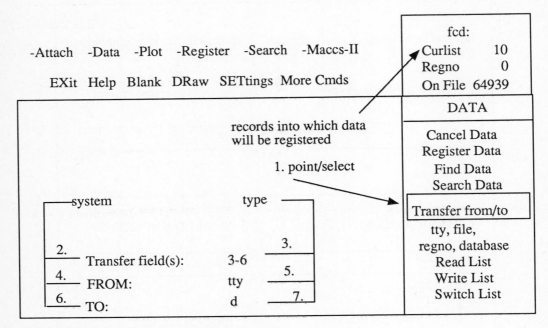

Figure 3/95. Transferring data in Data mode

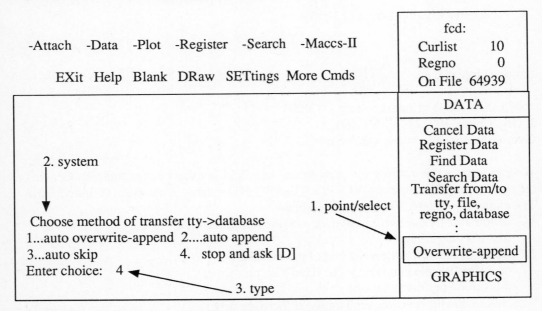

Figure 3/96. Setting up transfer conditions in Data mode

For transfer of files from/to ChemBase or Reaccs, see Part Five, p. 761.

PART FOUR

REACCS

1   SYSTEM BACKGROUND

REACCS (Reaction Access System) is a database management system (DBMS) for storing, searching, and retrieving molecular structures (molecules) and chemical reactions represented by molecular structures in the conventional chemical reaction schemes (reactions). The structures are displayed as graphics on the terminal screen or they can be printed or plotted to obtain hard copies.
Molecules and reactions can be appended with numeric and textual data. These data can be searched to retrieve exact numeric values or range of values or exact text strings or range of text strings, and/or to retrieve the molecules or reactions associated with the data.
REACCS has been developed by Molecular Design Ltd. (MDL) (see p. 265) for the end-user chemist and information scientist. The development of REACCS followed the well-accepted MACCS system (see p. 413) and has many features that are the same or similar to MACCS-II. MACCS-II and REACCS have been later emulated in the microcomputer system ChemBase. Those readers who studied Part Two and Part Three will find REACCS very easy to master.
REACCS, similarly to MACCS-II, has been developed as a minicomputer based system, however, it can be run on the mainframes as well. With the introduction of the CPSS system (see p. 265), the utility of REACCS is further expanded to the microcomputer realm since REACCS can exchange data with ChemBase, the microcomputer system for molecules and chemical reactions (see p. 265). Structures and data can be also exchanged between REACCS and MACCS-II. The systems interface is illustrated in Figure 4/1.

REACCS has been acquired by many pharmaceutical and chemical companies and has become the standard of the systems for managing chemical reaction data and information by chemists.
There are many databases that have been developed by MDL for REACCS that can be purchased in a package with REACCS or separately following the installation of REACCS.
The databases contain structures and data covering retrospectively over 50 years of published literature of synthetic organic chemistry and/or synthetic reactions reported in the current literature.
A new database, CHIRAS, comprising reactions for asymmetric syntheses was established. The database is regularly updated with new reactions.
The current literature of synthetic reactions is also covered by the Current Literature File database.
All these databases can be searched with graphic queries or text queries or by a combination of both to achieve highly selective retrieval.

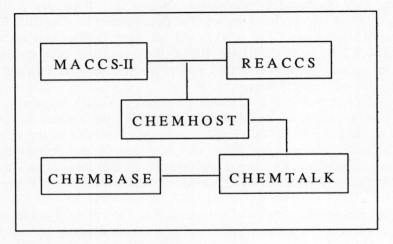

Figure 4/1. Systems interface

SYSTEM BACKGROUND

Some of the databases are listed below:
- **THEIL (Theilheimer's Synthetic Methods of Organic Chemistry)
- **REACCS-JSM (Journal of Synthetic Methods)
- **ORGSYN (Organic Syntheses)
- **CLF (Current Literature File)
- **CHIRAS (Asymmetric Syntheses)
- **FCDRX (Fine Chemicals Directory)

When the REACCS system is installed, a person in the organization functions as the System Supervisor who maintains the system files and manages the access to the system by the Users and is responsible for the security of the public databases.

The System Supervisor also assists the Users in setting up databases. Setting up databases is a somewhat involved procedure in REACCS, compared to the simple method in MACCS or ChemBase, and it is best done in collaboration with the System Supervisor. Therefore, Part Four does not cover setting up databases in REACCS.

MDL organizes courses for learning REACCS and periodically calls meetings of the Users to exchange ideas and experience with the system.

REACCS can be installed on the hardware systems we have discussed in Part Three (p. 415). The terminals and/or microcomputers emulating a terminal that we use with MACCS or ChemBase can be used with REACCS as well. For a list of terminals supported by REACCS, see p. 597.

2   SYSTEM DESIGN

REACCS has been designed to manage databases containing both molecules and reactions and associated data while MACCS-II manages only molecules and associated data. Thus, REACCS is analogous to ChemBase which can also manage databases with both molecules and reactions. However, REACCS is different from ChemBase in how the reactions are treated by the programs. In ChemBase, the individual structures in a reaction are not stored as single structures in the molecule part of the database, while in REACCS the structures participating in a reaction are stored automatically as single structure entries in the molecule part of the database. A reaction is then composed of the internal registry numbers of the molecules and registered in the reaction part of the database. Therefore, a registered molecule in REACCS may participate in more than one reaction and can have more than one role in the reactions.

MOLECULES

Graphic structures in REACCS are subjected to the same treatment as in MACCS and the other systems we have discussed.
The SEMA name (see p. 417) is generated for each structure on registration. The SEMA name is the encoded form of the structures. It identifies the structure as to the atoms, connectivities, charges, and/or isotopes. Similarly to MACCS-II, REACCS recognizes stereochemistry. We can register structures with absolute or relative configuration and, in turn, search for the isomers. REACCS will also recognize tautomeric structures and on searching will retrieve the tautomers if the query structures and the registered structures have the tautomeric differences. Structural keys (see p. 424) are assigned to each structure on registration. Each registered structure is assigned automatically an internal registry number. We can define a datatype for the external registry number.
The graphics of the structures are composed of the same elements as we have discussed in the other Parts. The discussion on p. 26, 268, 417 will illustrate these points.

REACTIONS

A reaction is stored in REACCS as a reaction scheme of the molecules participating in the reaction and labeled as reactants or products (in REACCS we cannot label structures in reactions as intermediates as in ChemBase, see p. 319). Catalysts (note that, by REACCS definition, catalyst includes both catalyst and reagent) and solvents are registered in special datatypes and can be also searched with structure queries. The molecule that we create and label as reactant or product is stored as a single entity. The reaction itself is then stored as the list of the internal registry numbers of the molecules in the reaction (Figure 4/2) and the reaction as a whole is assigned a registry number.

Catalysts/reagents/solvents, if involved in the reaction, are assigned registry numbers as molecular entities and are placed in the customary location, i.e., above or under the reaction arrow.

$$[24] + [12] \longrightarrow [36] + [150]$$

Figure 4/2. Registered reaction

If a reaction contained the same internal registry numbers of molecules as the other reaction but the molecules were assigned different roles in the reaction, the reactions would not be the same. We will see later that because of the availability of the registered molecules to participate in more than one reaction, we are able to assemble reactions by simply specifying the registry numbers of the molecules we want to have in the reactions.
In searching for reactions, REACCS searches for the single structures matching the reaction query structures, identifies them by their registry numbers, and then searches the lists of reaction internal registry numbers to find a match with the query as to the position of the structures (their role) in the reaction. The reaction is then displayed with the reactants before and the products after the reaction arrow (Figure 4/3).

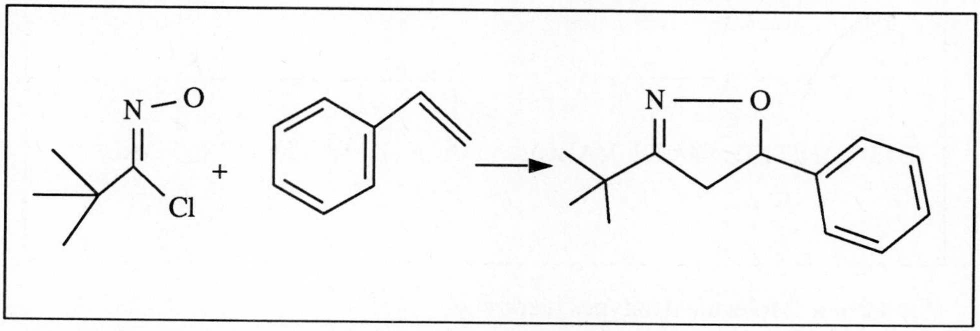

Figure 4/3. Graphics display of reaction

REACCS has also the capability to show reaction conditions, e.g., catalyst, solvent, temperature, pressure, yields, etc. in the customary format above or under the reaction arrow and/or under the reactant or product.

NUMERIC AND TEXTUAL DATA

It is the category of numeric and textual data which we will find at a substantial difference with MACCS-II or ChemBase as to their organization in the system.

In REACCS, the numeric and textual data are organized in a hierarchical system of datatypes. The datatypes are defined at the time when the database is created. The database has two branches of the hierarchy. One branch is for molecules, the other branch is for reactions. The datatypes in the hierarchy are arranged in a cascaded system of:

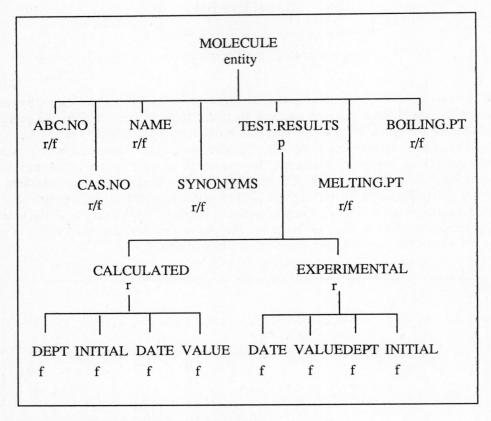

Figure 4/4. Molecule datatypes hierarchy

1. entity
   The entity is the root of the hierarchy, i.e. it is the highest item in the hierarchy. Molecules and reactions are the entities.
2. Parent
   The parent follows the entity and it is the second highest point in the hierarchy pointing further down in the hierarchy. The parent datatype contains no data; it is the branch point down the hierarchy pointing to records.

3. record
   The record is the datatype that is the branch point down the hierarchy to the data fields. The record, however, can contain data and in this case it is both the record and the data field.
4. field
   The field is the lowest point in the hierarchy. It contains only the actual data.

Datatypes are further classified as follows:
1. single datatypes
   The single datatype occurs only once in a particular molecule or reaction entry in the database.
2. multiple datatypes
   Multiple datatypes of the same name can occur more than once for the same molecule or reaction entry in the database. They are differentiated by line numbers, e.g., VARIATION(1), VARIATION(2).

We are familiar with the hierarchy of subject terms in an index to a book or to *Chemical Abstracts*. If we show the hierarchy of REACCS datatypes in this familiar form we would have the following entries (Figure 4/4 and Figure 4/5 show this hierarchy in a graphic network (p=parent, r=record, r/f=record/field, f=field), e.g.:

```
MOLECULE                         REACTION
   ABC.NO                           VARIATION(1)
   CAS.NO                              REACTANT(1)
   NAME                                   AMOUNT
   SYNONYMS                               WEIGHT
   BOILING.PT                             VOL
   MELTING.PT                          REACTANT(2)
   TEST.RESULTS                           AMOUNT
      CALCULATED                          WEIGHT
         DEPT                             VOL
         INITIAL                       VARIATION(2)
         DATE                             REACTANT(1)
         VALUE                               AMOUNT
      EXPERIMENTAL                           WEIGHT
         DEPT                                VOL
         INITIAL                          REACTANT(2)
         DATE                                AMOUNT
         VALUE                               WEIGHT
                                            VOL
```

While in ChemBase or in MACCS-II the specific fields holding data are fully characterized by their data field names, in REACCS the fields containing data are

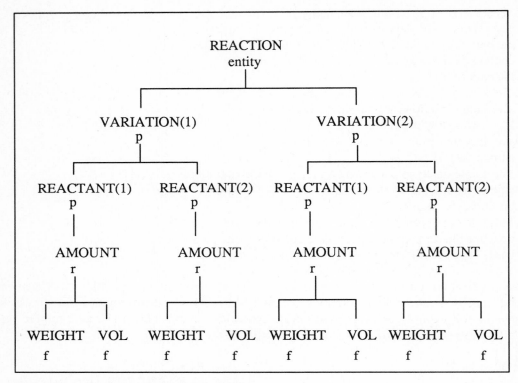

Figure 4/5. Reaction datatypes hierarchy

described by their datatype treenames (Figures 4/6, 4/7).

The type of data that can be stored in the datatypes are defined at the time when the database is set up. We can define the following types of data:
1. numeric data (integer or real)
2. fixed-length text
3. variable-length text
4. date

We have discussed the data types in Part Two, p. 271 and Part Three p. 426.

DATABASES

The collection of molecules, reactions, and their associated numeric and textual data forms a database. The structure of REACCS databases is shown in Figure 4/8.

The databases are stored as files in the system under the designated database name. At the start of the REACCS session, a database name must be specified to be able to run REACCS. The database file is opened by the system at the start of the session and it is closed at the end of the session. We do not have to save the database files like the ordinary

```
MOL:ABC.NO
MOL:CAS.NO
MOL:NAME
MOL:SYNONYMS
MOL:BOILING.PT
MOL:MELTING.PT
MOL:TEST.RESULTS:CALCULATED:DATE
MOL:TEST.RESULTS:CALCULATED:VALUE
MOL:TEST.RESULTS:CALCULATED:INITIAL
MOL:TEST.RESULTS:CALCULATED:DEPT
MOL:TEST.RESULTS:EXPERIMENTAL:DATE
MOL:TEST.RESULTS:EXPERIMENTA:VALUE
MOL:TEST.RESULTS:EXPERIMENTAL:INITIAL
MOL:TEST.RESULTS:EXPERIMENTAL:DEPT
```

Figure 4/6. Molecule datatype treenames

```
RXN:VARIATION(1):REACTANT(1):AMOUNT:WEIGHT
RXN:VARIATION(1):REACTANT(1):AMOUNT:VOL
RXN:VARIATION(1):REACTANT(2):AMOUNT:WEIGHT
RXN:VARIATION(1):REACTANT(2):AMOUNT:VOL
RXN:VARIATION(2):REACTANT(1):AMOUNT:WEIGHT
RXN:VARIATION(2):REACTANT(1):AMOUNT:VOL
RXN:VARIATION(2):REACTANT(2):AMOUNT:WEIGHT
RXN:VARIATION(2):REACTANT(2):AMOUNT:VOL
```

Figure 4/7. Reaction datatype treenames

files if we make changes or enter new data in the existing database. They are automatically registered and saved by REACCS.
The databases can be searched with either or both of:
1. graphic queries
2. alphanumeric queries
A variety of search techniques are available for searching:
1. simple search
2. conversion search
3. combination search
We will discuss these techniques in the section "Search Techniques", p. 637.

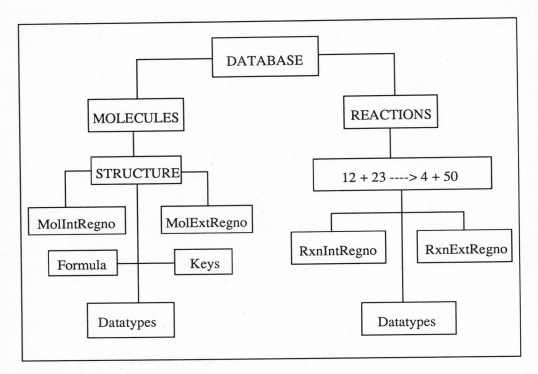

Figure 4/8. REACCS database

REACCS FILES

The REACCS system uses the following files:
1. program files
2. temporary files
3. permanent files

PROGRAM FILES

The program files are those which reside on the computer disk system and were installed when REACCS was loaded. The files contain the REACCS system program and associated files which may be used by the User when invoked, e.g., Help system, templates for drawing structures, the periodic table, etc., or by REACCS, i.e., the User Profile, Form Files, etc., that are automatically loaded when we start the REACCS session.

TEMPORARY FILES

The temporary files are those which are created during the session in REACCS and are

SYSTEM DESIGN

deleted when we exit from the current session. However, they can be saved in permanent files before we exit the current session. The temporary files are:
1. temporary List files
2. temporary Query files
3. temporary Omit list files

Temporary List File

A temporary List file is the result of a search.
Temporary List file contains the internal registry numbers of the molecules or reactions that are retrieved from the REACCS database. Temporary List file is created by REACCS after a search operation is completed. It is also called the List or Hit List. The temporary List file is named automatically by REACCS by the letter "L" and an integer "n," i.e., "Ln" (e.g., L3) (see also p. 42, 210 for Ln designations). The integer "n" can be 1 to 99 (0 is reserved for the current Reference list (see p. 658). The Ln designation enables the User to work with the Hit List e.g. in combination searches (see p. 645) or in establishing the Current list or Reference list (see p. 658). When we recall a permanent Listfile (*vide infra*) (note the terminology of "List file" and "Listfile"), the content of the file is designated "Ln" to be able to work with the list in the current session. Temporary List files of the current session can be displayed (listed) by INDEXLIST (see p. 663).

Temporary Query File

When we create a query structure, be it a molecule or a reaction, we can save the query in a temporary file. The temporary file is named by the System by the letter "Q" and an integer "n," i.e., Qn. The "n" can be any integer in the range 1-99, e.g., a structure query may be named Q15. The Qn designation is used to recall the temporary Query file to work with the query structure in the current REACCS session. The content of the temporary Query file is the same as the Molfile or Rxnfile (*vide infra*); however, the file is deleted after we exit from the current session, unless saved in the permanent files.

Temporary Omit List File

When we perform a search, REACCS creates a temporary list file (*vide supra*). We can display the items (i.e., structures and/or data) in this list (see ViewList, p. 652). If we so desire, we can omit (remove) an item from the list. The omitted item(s) are temporarily saved in the Omit list files. We can direct REACCS to transfer the omitted items into another temporary List file Ln (see p. 653). Omit list files can be displayed (listed) by the INDEXLIST option (see p. 663).

PERMANENT FILES

The permanent files are the files the User creates in the current REACCS session for later recall in the same session or in the subsequent REACCS sessions.

The external files are:
1. Listfiles
2. Molfiles
3. Rxnfiles
4. Rdfiles
5. Form Files

Listfile

Listfile is the temporary List file (Ln) (note the terminology of "Listfile" and "List file") that has been saved as a permanent file. It contains the internal registry numbers of the molecule(s) or reaction(s) we have retrieved by a search from a REACCS database. A saved Listfile can be used only in the database in which it was created because the internal registry numbers correspond to the records in the database in which the Listfile was created. On recalling a Listfile in the current session, it is again designated as the temporary List file "Ln" to be able to work with the list.
Listfiles are also used to transfer data between REACCS and MACCS-II or ChemBase by converting them into RDfiles (see section Updating Databases).

Molfile

The Molfile stores a single molecule. The file contains the following information:
1. the connection table of the structure
2. the internal registry number of the structure if it was retrieved from a REACCS database
3. the User initials, time, date, where the structure was created
4. a comment line (optional)
5. three-dimensional coordinates if the structure was created in one of the programs handling 3-dimensional graphics (structures created in REACCS have the z-coordinate = 0)

REACCS Molfiles can be read by MACCS-II and ChemBase.

Rxnfile

The Rxnfile stores a single reaction.
The Rxnfile contains all the information as the Molfile and also contains all information for the structures in the reaction scheme as to their appropriate role and identification as reactants or products.
We can create two types of Rxnfiles:
1. REACCS Rxnfiles for use in REACCS system
2. PC-Rxnfiles for transfer of reactions to ChemBase

RDfile

Reaction Data file (RDfile) stores single molecule with associated data or single reaction with associated data or multiple molecules with or without associated data or multiple reactions with or without associated data (in contrast to Molfile or Rxnfile which can contain only single molecule or reaction but no associated data).
The RDfiles are used to transfer records between REACCS databases, and from REACCS to ChemBase (see p. 727, 761).
We can create the following RDfiles:
1. REACCS RDfiles for use in REACCS system
2. PC-RDfiles for transfer of data to ChemBase

Form File

The Form File stores a set of molecule Form/Table and reaction Form/Table.

3   SYSTEM OPERATION

REACCS is operated by a system of Modes. There are six Modes in REACCS. Each of the Modes has a menu with the available options in that particular Mode. Some of the Modes have a submode with a menu. There are over 150 options in REACCS. The options are activated by graphic selection or keyboard selection. The graphic selection is, of course, available only on the terminals with graphics capability. Some of the options are not available for graphic selection and must be selected by the keyboard selection even if we operate in graphics mode.

We describe the graphic selection of the options by the same terms we have used in MACCS-II (see p. 437).

1. Main Mode (Figure 4/9)
   The Main Mode has options by which we can enter the other
   Modes, perform retrieval operations for single molecule or
   reaction, update the database by registering or deleting
   structures and data, create Molfiles and Rxnfiles, and select a database other than the current one.
2. Build Mode (Figure 4/10)
   The Build Mode enables us to create molecules and
   reactions by drawing the structure skeletons and
   specifying atoms and other attributes characteris-
   tic for the structures. The Build Mode has two submodes,
   Query Options, and HighlightRxn, in which we can
   create generic structures with variables and specify
   certain features for the query structures.
3. Search Mode (Figure 4/11)
   The Search Mode allows for searching multiple molecules,
   reactions, or data, with graphic queries or alphanumeric
   queries. The search results can be saved in permanent
   files or viewed by switching to the ViewList Mode.
4. Viewlist Mode (Figure 4/12)
   The ViewList Mode is the Mode for displaying molecules
   or reactions and/or data retrieved in the Search Mode. We
   can also perform other operations in this Mode, e.g.,
   define a new List to view, or redefine the search domain
   over which we wish to perform the search.
5. Plot Mode (Figure 4/13)
   The Plot Mode allows for plotting molecules and reactions
   with or without data on the terminal screen to be printed
   or, through a plotting device, creating hard copies of the
   structures and data.

Help Set Exit	Main Build Search ViewL Plot Forms	THEIL:	
REACCS	Clear Db ReadFile WriteFile IndexData	Rxn	46818
	Find Register	Mol	60748
	Delete　　Data　　Current	List	0M
	SELECT OPTION		

Figure 4/9. REACCS Main Mode Menu

6. Forms Mode (Figure 4/14)
   The Forms Mode is the Mode in which we create Forms
   for displaying structures and data in a prearranged
   format. From the Forms Menu, we enter the Tables Menu
   to create Tables.

GLOBAL OPTIONS

Each of the Modes (*vide supra*) has a set of options, selected either graphically or by keyboard selection, which are active in that particular Mode. Some of the options are available in more than one Mode and have the same function. In any case, we have to be in a Mode to be able to activate the options functioning in that Mode.
However, REACCS also has a set of options that can be activated in any of the Modes to invoke their function. These are called the global options. We have already encountered similar options in Part Three.

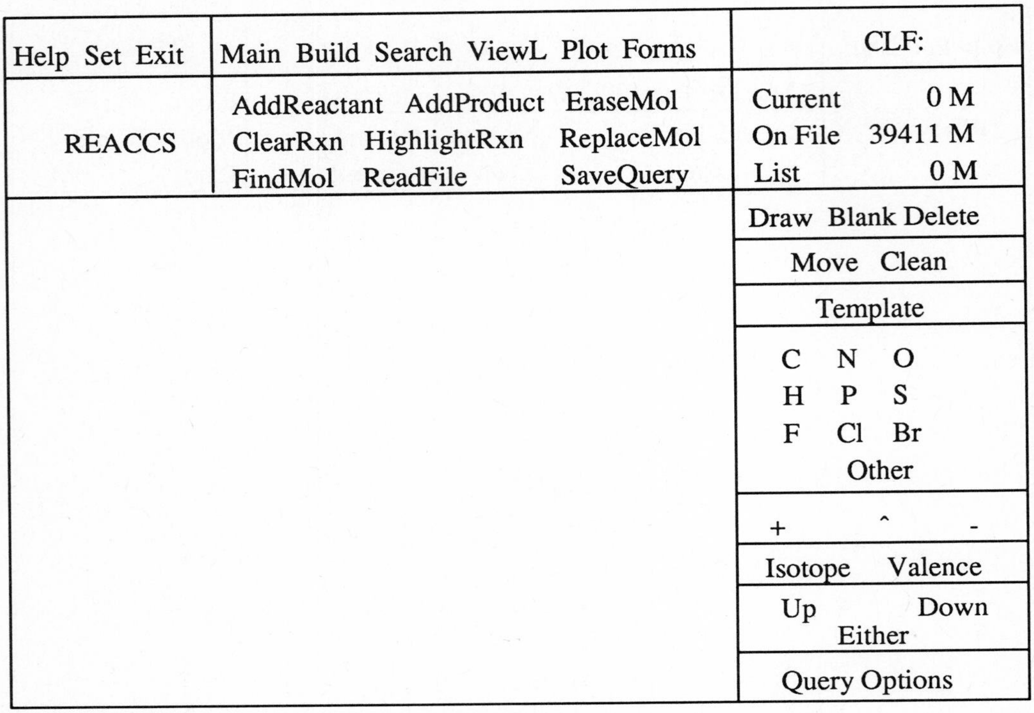

Help Set Exit	Main Build Search ViewL Plot Forms	CLF:
REACCS	AddReactant AddProduct EraseMol ClearRxn HighlightRxn ReplaceMol FindMol ReadFile SaveQuery	Current 0 M On File 39411 M List 0 M
		Draw Blank Delete
		Move Clean
		Template
		C N O H P S F Cl Br Other
		+ ˆ -
		Isotope Valence
		Up Down Either
		Query Options

Figure 4/10. Build Mode Menu

Keyboard Global Switches

Most of the global options are keyboard switches which are activated by depressing the appropriate key on the keyboard:

!.............interrupts options or process before completion (soft interrupt)
*.............refreshes the screen and redraws the menu and structures on display
#.............turns on/off highlighting of bonds in reaction centers by the option Center
d (isplay)........redraws the current structure
e (xpand) or >....enlarges the current structure
g (roup mode).....switches to Group mode in which group of atoms in current structure can be manipulated by these subswitches:
   d (elete)......deletes entire defined group
   l (arger)......enlarges defined group in size
   s (maller).....reduces defined group in size
   f (lip)........rotates defined group 180 degrees
   r (otate)......swings group in the plane of the screen
   t (ranslate)...moves group relative to the entire structure
h (orizontal).....rotates the current structure so that the specified bond becomes horizontal

SYSTEM OPERATION

Help Set Exit	Main Build Search ViewL Plot Forms	CLF:
REACCS	MolData Fmla RxnData Ref=List Db SSS Similar RSS ReadList WriteList CurAsAgt CurAsPrd DeleteL IndexList	Rxn 25110 Mol 39411 List 0 M

Figure 4/11. Search Mode Menu

k.................switches to keyboard operation from graphics operation
l (abels).........turns on/off display of hydrogens at hetero atoms and at terminal carbons
m (olecular formula/weight).....calculates and displays molweight/formula for current structure
n (umbers)........turns on/off display of atom locants in structures
o (rigin).........moves current structure to the center of the screen
r (educe) or <....reduces current structure in size
s (hift)..........move entire structure to another position on the screen
t (runcate).......replaces common substituents with symbols
u (nspecified stereo centers).......turns on/off identification of potential asymmetric centers
v (ertical).......rotates current structure so that the specified bond becomes vertical

Keyboard Global Options

Keyboard global options differ from the switches in that that they are typed in full (or as many characters to be unambiguous) followed by carriage return <cr>:

edit.............switches to the system Editor to create/edit datatypes and alphanumeric files

Help Set Exit	Main Build Search	ViewL Plot Forms	CLF:	
REACCS	First Next Prev Item List Table Data Omit Query	Ref=List Db ReadList WriteList DeleteL IndexList	Current On File List	0 M 39411 M 0 M

Figure 4/12. ViewList Mode Menu

graphics..........switches from keyboard operation to graphics
help..............invokes the help system
system............switches to the system level and permits execution of the following system commands:
  delete file
  directory
  help
  type file
  quit..........returns to REACCS

SETTINGS

Each of the REACCS Menus has the SET option which allows for specifying the Settings under which our REACCS session will run. We have encountered similar Settings in Part Two and Part Three, p. 284, 442).
Point/selecting SET in the Main Menu displays the Main Settings Window (Figure 4/15)
Selecting SET in other than Main Menu displays General Settings Window (Figure 4/16).
The difference in the Main and General Settings Windows is in the bottom half of the

SYSTEM OPERATION

Help Set Exit	Main Build Search	ViewL Plot Forms	CLF:
REACCS	List Current ReadFile FindMol FindRxn	Device Labels 1 2 4 Numbers NewPage Truncate	Rxn 25110 Mol 39411 List 0

Figure 4/13. Plot Mode Menu

windows. The Main Settings Window contains the settings for Registration operations. The General Settings Window contains the settings for Search and Plot operations.

The Settings are set by point/select the particular Setting line (on/off toggle) or point/select the Setting and typing in the prompt box which REACCS displays. To leave the Settings Window, click outside the Window or point/select QUIT (the Window option).

The Settings and their function will be discussed later in the description of the operations in which they are applicable. Here, we will describe those which are of general use:

Address............sets the address that will be printed on plots
ErrorFile..........if on, copies the messages issued by REACCS into an external file
FormFile...........sets the Form File with Forms and Tables to display structures and data in a preformatted outlay
Initials...........User initials which appear in Molfile and Rxnfiles
LengthUnits........
MassUnits
PressureUnits   the units settings specify
TempUnits   that data will be input and

Help Set Exit	Main	Build	Search	ViewL	Plot	Forms	CLF:	
Forms Tables	Clear	DrawBox		Delete		Field	Rxn	25110
Reactions	Read	Split		HideEdge		TextSize	Mol	39411
Molecules	Write	Move/Resize		Copy		TextPos	List	0

Figure 4/14. Forms Mode Menu

TimeUnits   output in the selected units
VolumeUnits..........
MolInfo.............sets the datatype to represent molecules in nonstructural display, e.g., catalysts and solvents are often not displayed as structures. Instead, they may be displayed as symbols or formulas, which is determined by this Setting.
Name................sets User name to appear on hard copy plots
Password............sets a password for a database
Scroll..............sets the size of scrolling buffer at the screen bottom

As we perform the operations for creating molecules and reactions, searching, registering, and plotting, we find that we need to set or reset some of the Settings. For the experienced User, it will be faster to do it via the keyboard than calling up the Settings Window. To set a Setting by keyboard selection:
1. User: press "k"
2. System: Search: (or other prompt, depending on which Mode we are in)
3. User: set (type the Setting name to toggle, and/or answer the prompt which REACCS issues)

SYSTEM OPERATION

Help Set Exit	Main Build Search ViewL Plot Forms	THEIL:	
REACCS	Clear Db ReadFile WriteFile IndexData	Rxn	46818
	Find Register	Mol	60748
	Delete Molecule Reaction Data Current	List	0 M

```
                        SELECT SETTING
┌─────────────────┬──────────────────────────────────────────────┐
│ Set Help Quit   │ MDL:[MFILES]REACCS.UPR   ReadUPR   WriteUPR  │
├─────────────────┼──────────────────────────────────────────────┤
│ AAMaps   OFF    │ DataDisplay:  ON     DBaseList:*             │
│ Labels:  OFF    │ ErrorFile:                                    │
│ Numbers: OFF    │ FormFile:   THEIL:THEIL2:FRM                 │
│ Password:       │ GlobalSearch: OFF    FindData: OFF           │
│ RxnCenter #     │ MolExtReg:                                    │
│ Scroll:  5      │ MolInfo:    SYMBOL                            │
│ Truncate OFF    │ RxnExtReg:                                    │
├─────────────────┼──────────────────────────────────────────────┤
│ Editor:    OFF  │ LengthUnits:                                  │
│ Overwrite: ON   │ MassUnits:                                    │
│                 │ PressureUnits:                                │
│                 │ TempUnits:                                    │
│                 │ TimeUnits:                                    │
│                 │ VolumeUnits:                                  │
└─────────────────┴──────────────────────────────────────────────┘
```

Figure 4/15. Main Settings Window

or:
show.........displays all current settings
all..........sets all Settings
REACCS will show the Settings with values, if any.
<cr>........leaves the unchanged Setting
new value...replaces the old one
! interrupts the process

USER PROFILE

The parameters that are set by the Settings (*vide supra*) can be saved in a permanent file, the User Profile (.UPR). The file is read by the system every time we start a REACCS session, thus making the Settings specified in the file valid for the session. The Settings we specify in the current session by SET in fact override the Settings in the User Profile (which are the default) if they are of different values. We can also set new Settings for the current session, if those are not in the User Profile, by SET. On terminating the current session, the changed Settings will return to the default values unless we save the new

Help Set Exit	Main Build Search ViewL Plot Forms	CLF:
REACCS	MolData Fmla RxnData Ref=List Db SSS Similar RSS ReadList WriteList CurAsAgt CurAsPrd DeleteL IndexList	Rxn 25110 Mol 39411 List 0 M

SELECT SETTING

Set Help Quit	MDL:[MFILES]REACCS.UPR ReadUPR WriteUPR
AAMaps OFF Labels: OFF Numbers: OFF Password: RxnCenter # Scroll: 5 Truncate OFF	DataDisplay: ON DBaseList:* ErrorFile: FormFile: THEIL:THEIL2:FRM GlobalSearch: OFF MolExtReg: MolInfo: SYMBOL RxnExtReg:
MolSim: 80 RxnSim: 80,20 StereoSearch: ON ViewHits: OFF	Address: 2132 Farallon Dr., San Leandro Initials: MDL Name: Molecular Design Ltd. PlotDisplay: OFF

Figure 4/16. General Settings Window

setup in the User Profile file. The .UPR file in Figure 4/15 and Figure 4/16 is MDL:[MFILES]REACCS.UPR and contains the default Settings which, unless we change them or create our own .UPR file, will be in effect.

To save the changed and/or new Settings, in both Windows, point/select WriteUPR and enter the file name at the prompt. Conversely, to read a .UPR file and thus set new Settings replacing the current ones, point/select ReadUPR.

STARTING REACCS

The login procedure requires that the User has an ID to access the system (and a password(s) and access rights, if required, for the database(s) to be used for searching and/or registering/updating data).

The login procedure and the messages we receive from the system will depend on the type of the system in which REACCS is installed. The following example has been run on the VAX system. After login the system, we receive the prompts:

User name:
Password:

                       Welcome to VAX/VMS version V4.6

SYSTEM OPERATION

User: reaccs

```
****************************************
                REACCS
****************************************
             Revision 7.1
```

Database List:

1. CLF:
2. THEIL:
3. JSM:
4. FCDRX:
5. ORGSYN:
6. CHIRAS:

Enter REACCS database number (or name if not on list): 1
Please choose terminal type
or L for list, RETURN for nongraphics: L

NONGRAPHIC	Alphanumeric (Nongraphic)
1	Imlac
2	Tektronix 4114
3	VT-640
4	Tektronix 4105
5	Lundy 568x
6	DQ650M
7	HP 2623A
8	GT40
9	Envision
10	Tektronix 4107
11	Envision II
12M	ChemTalk Terminal/Monochrome
12C	ChemTalk Terminal/Color
13	FACOM TERMINAL
15	VT240/241
16	Tektronix 420x
17	Pericom MG400/600
18	GraphOn 250

After choosing the terminal, the Main Mode Menu is displayed.

CHANGING DATABASE

If we select a database at the start of REACCS and wish to switch to another database in the same session, we activate the DB option which is in the Menus except the Build, Forms, and Plot Modes.

EXITING REACCS

To exit REACCS, activate the option Exit from any of the Modes.

4   CREATING MOLECULES

The purpose of creating graphic molecular structures in REACCS is to:
1. register molecules in the database
2. build molecules to assemble reactions for registration in the database
3. formulate graphic structure queries for exact structure, substructure, or similarity searches for molecules or reactions
4. save molecules or queries in Molfiles
5. transfer structures via Molfiles between the MDL systems
6. plot/print molecules

Building molecules in REACCS is very analogous to the methods that we have discussed in Part Two and Part Three. The reader is referred to these sections for a detailed description of structure building and correcting/modifying structures.

In REACCS, the structure building is carried out in the Build Mode.
As in ChemBase and MACCS-II, we can build:
1. exact structures
2. generic structures

BUILDING EXACT MOLECULES

We enter the Build Mode from the Main Mode (or any other Mode) by activating the Build option (Figure 4/17).
The MACCS-II Draw Menus and the REACCS Build Mode Menus are analogous.
To start drawing a structure skeleton, we activate DRAW (Figure 4/18). The following keyboard switches are available in the Build Mode:

c.........continuous draw (on/off)
   to turn on, press c, to turn off at the end of the bond
   (to start at a new position), press c or click the end atom
   twice
- or =....to specify a heteroatom
i.........to label the structure CHIRAL
j.........to build a single ring or fused or spiro rings
+.........to attach a template
b.........suppress refreshing of structures until CLEAN is activated

Atom values are specified in structures by selecting the elements on the Menu or the OTHER option for those elements not on the Menu. Any existing atom in a structure can be respecified.
Note the specific use of "H":
In building reaction queries, we might wish to draw explicit hydrogens to specify reaction

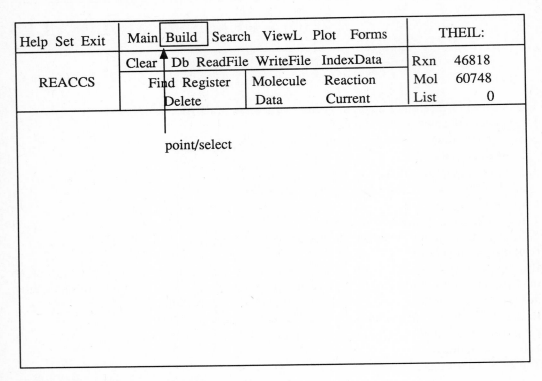

Figure 4/17. Entering Build Mode

centers. If this is intended, the query structure must be atom/atom mapped (see p. 616), otherwise explicit hydrogens have the same effect as those specified by HYDROGEN (see p. 604, 620, 623, 680).

The keyboard shorthand assignment for atom values (- or =) can be used as we have shown in Part Two (see p. 304).

Charges, isotopes, and abnormal valence are assigned by the +, -, and ∧ symbols on the Menu and by the ISOTOPE, and/or VALENCE option.

Stereobonds are assigned by the UP, DOWN, or EITHER options and configurational isomers are labeled "CHIRAL" by the "i" switch (see Part Three, p. 51, 418 for discussion of stereochemistry and drawing of stereochemical structures).

Unspecified stereocenters can be marked by the Global Option "u" (unspecified).

Templates can be attached to the current structural fragment by activating the TEMPLATE option on the Menu or by using the keyboard "+" shortcut attachment technique.

Models for structure building can be recalled from a REACCS database by the FINDMOL option (Figure 4/19) or from a Molfile by the READFILE option (Figure 4/20).

FINDMOL in Figure 4/19 will display the structure of intregno 245 which can be used as the model.

CREATING MOLECULES

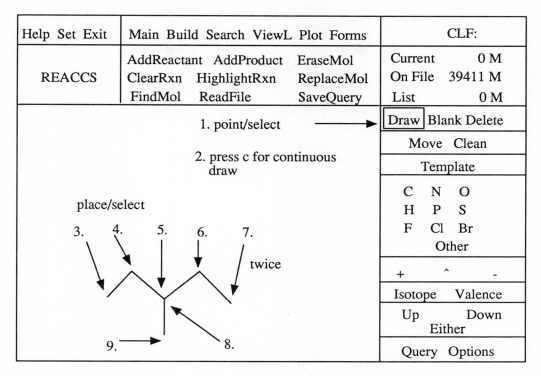

Figure 4/18. Draw menu

The Molecule ID in Figure 4/19 may be any of the following:
(see p. 637 for discussion of query descriptors):
internal registry number
external registry number
temporary list
omit temporary list
listfile
formula search query
data search query
READFILE in Figure 4/20 will display the structure in the myfile.mol and can be used as the model.
The filename in Figure 4/20 may be any of the following:
Molfile
Query File
Correct/modify structures by deleting atoms and/or bonds by activating the DELETE option on the Menu.
Clean structures by the CLEAN option on the Menu or, for minor refreshing of structures, use the Global switch "d."

Example: ***Recalling structures as models***

Help Set Exit	Main Build Search ViewL Plot Forms	CLF:	
REACCS	AddReactant AddProduct EraseMol ClearRxn HighlightRxn ReplaceMol [FindMol] ReadFile SaveQuery	Current On File List	0 M 39411 M 0 M
		Draw Blank Delete	
		Move Clean	
	↑	Template	
	1. point/select	C N O H P S F Cl Br Other	
2. system ↓ Molecule ID: 245	3. type	+ ^ -	
		Isotope Valence	
		Up Down Either	
		Query Options	

Figure 4/19. Getting structure model

The global switch "*" redraws/refreshes the entire screen.
Move atoms in structures about the screen by the MOVE option on the Menu.
Move entire structures about the screen by the "s"(shift) or the "o" (origin) Global switch to position the structure in the center of the screen.
Resize structures by the Global switch:
e or >........enlarges structure size
r or <........reduces structure size
Reorient structures by making a bond horizontal or vertical by the Global switch:
h.......reorients the bond horizontally
v.......reorients the bond vertically
Manipulate groups in structures by the Global switch:
g (roup mode)......switches to Group Mode with the options:
  d, l, s, f, r, t (see p 498)
Hydrogens and numbers display in structures by the switches:
l........displays hydrogens at terminal carbons and hetero-

CREATING MOLECULES

Example: *Recalling structure from Molfile*

Help Set Exit	Main Build Search ViewL Plot Forms	CLF:	
REACCS	AddReactant AddProduct EraseMol ClearRxn HighlightRxn ReplaceMol FindMol ReadFile SaveQuery	Current 0 M On File 39411 M List 0 M	
		Draw Blank Delete	
		Move Clean	
	1. point/select	Template	
		C N O H P S F Cl Br Other	
	3. type	+ ^ -	
2. system		Isotope Valence	
Input filename:	myfile.mol	Up Down Either	
		Query Options	

Figure 4/20. Recalling a Molfile

       atoms
n........displays atom numbering in structures
The Settings (see p. 592):
Labels......turns on/off hydrogens in structures
Numbers.....turns on/off atom numbering in structures

In other than Build Mode, in displayed structures:
Truncate common substituents by the switch:
t (runcate)......abbreviates common substituents by alpha-
        numeric descriptor (e.g., CH3CH2CH2CH2 to Bu)
The Setting (see p. 592):
Truncate.....sets display of common fragments as text ab-
      breviations (e.g., Ph, Bz).

CREATING GENERIC MOLECULES

Structures with variable atoms or bonds are built in the Query Options Menu of the Build Mode. The Menu is entered by activating the QUERY OPTIONS on the Build Mode Menu (Figure 4/21).

Help Set Exit	Main Build Search ViewL Plot Forms	CLF:	
REACCS	AddReactant AddProduct EraseMol	Current	0 M
	ClearRxn HighlightRxn ReplaceMol	On File	39411 M
	FindMol ReadFile SaveQuery	List	0 M
		Draw Blank Delete	
		Move Clean	
		Template	
		C N O	
		H P S	
		F Cl Br	
		Other	
		+ ^ -	
		Isotope Valence	
		Up Down	
		Either	
	point/select ⟶	Query Options	

Figure 4/21. Entering Query options menu

The Query Options are displayed on the right side of the Menu (Figure 4/22). Building generic structures with variables is the same process as in MACCS-II (see p. 465).

Variable atom values are assigned by activating the system symbols A and Q or by selecting the element symbols, or the OTHER option for elements that are not on the Menu.
Atom negation is specified by the EXCLUDE option.
Degree of substitution is assigned using the HYDROGEN option to specify the minimum number of hydrogens required at a locant in the structure.
Relative configuration is assigned by the STEREO option.
Variable bonds are specified by the CHAIN and RING options for bonds that are required to be in chains and rings. ANY specifies any bonds, AROMATIC specifies the

CREATING MOLECULES

Help Set Exit	Main Build Search ViewL Plot Forms	CLF:
REACCS	AddReactant AddProduct EraseMol ClearRxn HighlightRxn ReplaceMol FindMol ReadFile SaveQuery	Current 0 M On File 39411 M List 0 M
		Draw Blank Delete
		Restore Exclude
		Hydrogen Stereo
		C N O H P S F Cl Br A Other Q
		Center NotCenter
		Chain Ring
		Any Arom S/A D/A S/D
		Draw Options

Figure 4/22. Query options menu

aromatic bonds, and S/A, D/A, and S/D stand for single/aromatic, double/aromatic, and single/double bond values.

SAVING QUERY TEMPORARILY

The created structure queries can be saved temporarily for the current session of REACCS. The option SAVEQUERY on the Menu in both Build Modes saves the current query (Figure 4/23). On exiting REACCS, the structure query is erased (however, it can be saved permanently in a Molfile).
To recall the temporarily saved query, we use the same READFILE option as for recalling a Molfile. However, we specify the Qn number in response to the prompt, e.g.:
Input filename: Q4

SAVING QUERY PERMANENTLY

To save an exact or a generic structure in a Molfile, we exit the Build Mode into the Main Mode. The created structure will be transferred to the Main Mode and will be the current

Example: *Saving query temporarily*

Help Set Exit	Main Build Search ViewL Plot Forms	CLF:
REACCS	AddReactant AddProduct EraseMol ClearRxn HighlightRxn ReplaceMol FindMol ReadFile SaveQuery	Current 0 M On File 39411 M List 0 M
	1. point/select N—⌬—S 2. system ↓ Query is number Qn	Draw Blank Delete
		Move Clean
		Template
		C N O H P S F Cl Br Other
		+ ^ -
		Isotope Valence
		Up Down Either
		Query Options

Figure 4/23. Saving query temporarily

structure. The WRITEFILE option in the Main Mode creates the Molfile (Figure 4/24). The Molfile structure is appended automatically with the initials of the creator of the structure if we set the Setting "initials" on (see p. 592).

BUILDING MOLECULES - THE QUICK WAY

REACCS offers a method to build molecules by connecting templates to a structural fragment. The method is similar to the ATTACH technique in MACCS-II (see p. 457).
The steps of the methods are as follows:
1. Establish the current structure as the parent.
   The structure can be newly drawn or recalled from a Molfile or retrieved from the database by formulating a
   search/retrieve query.
2. Display the parent structure numbering.
3. Switch to the keyboard selection by typing "k."

Example: *Saving structure in Molfile*

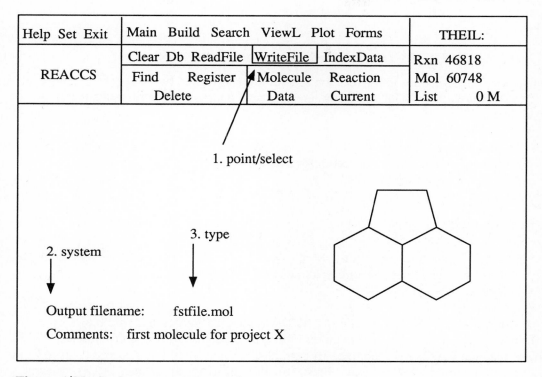

Figure 4/24. Saving query permanently

4. System: Build:
5. User: assemble
6. System: Assemble:
7. User: 7-co2h, 5-oh <cr>
8. System: assemble:
9. User: <cr>
10. System: Build:
The current parent structure has been substituted at position 7 and 5.
The same structure can be, of course, created by the "+" template attachment technique.

5   CREATING REACTIONS

Reactions in REACCS are composed of structures of the molecules taking part in the reaction. The structures in the reaction scheme are interrelated by + and ⟶ signs.
The structures in front of the arrow are reactants and the structures after the arrow are products.
We create reactions to:
1. register them in the database
2. formulate graphic queries to search for reactions
3. save them in Rxnfiles for recall
4. print/plot them

BUILDING REACTIONS

Reactions are built in the Build Mode from molecules that we draw as usual. The structures are then labeled according to their role in the reaction as:
1. reactants
2. products
by the options ADDREACTANT and ADDPRODUCT (Figures 4/25, 4/26).
In REACCS, we cannot label structures as intermediates in contrast to ChemBase.
The reaction building is monitored by the "reaction indicator" line that is displayed in the upper left corner of the Menu. Each structure is assigned a letter (A-Z) when it is designated as the reactant or product in the reaction. The System enters automatically the structures in the proper place in the scheme (Figure 4/27). The reaction indicator line behaves like a Global option. It can be activated in any Mode by point/select one of the letters in the reaction indicator line to display the molecule (Figure 4/28) or point/select the arrow to display the entire current reaction (Figure 4/27).

EDITING REACTIONS

If we have created a reaction and wish to make changes in the reactant or product structure, we use the option REPLACEMOL, and to erase a molecule from the scheme, we use the option ERASEMOL.
The REPLACEMOL option replaces a structure in the scheme by a current structure. The current structure may be the result of:
1. drawing it in the Build Mode as usual
2. retrieving a structure from the database by FINDMOL
3. recalling the structure from a Molfile by READFILE
4. recalling the structure from a Query File by READFILE
5. modifying a structure from the current reaction.

Example: **Building reactions**

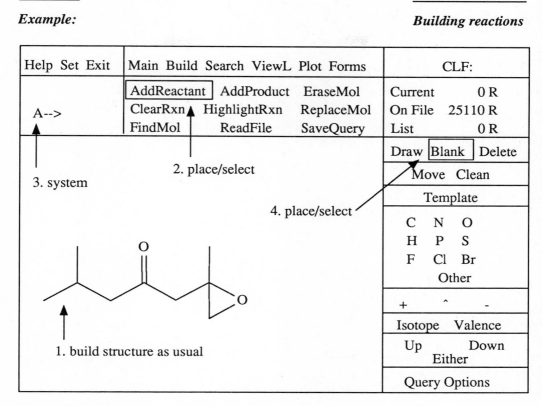

Figure 4/25. Building reaction - reactant

The structure is taken out from the reaction by point/select a letter in the reaction indicator line. The structure is displayed and can be modified as usual by the Build options.
The example in Figure 4/29 shows the product structure taken out from the reaction in Figure 4/27; it is modified and placed back as product by specifying the letter B at the System prompt, i.e., the structure is again the product of the reaction. If we wanted to interchange this structure for the reactant of the reaction, we would respond to the prompt: Enter molecule letter: A
Similarly, if we take out a structure and erase it by BLANK and make a new structure current, the structure will replace the specified structure in the reaction by the same steps (the original structure we are replacing is lost).
If we simply want to erase a structure from the reaction, we activate the ERASEMOL option. The structure will be removed (it is lost) and the reaction indicator line will be automatically adjusted to reflect the new reaction minus the removed structure. The letters

Example: *Building reactions*

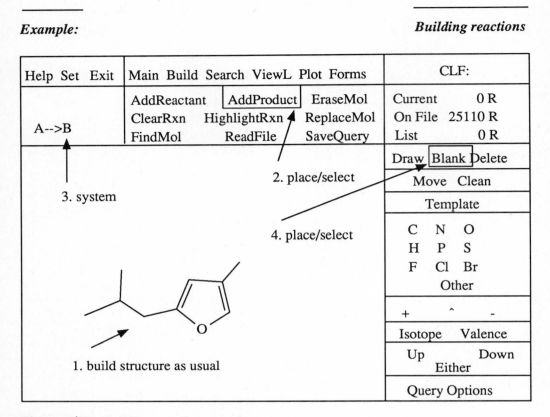

Figure 4/26. Building reaction - product

in the reaction indicator will be reassigned to the remaining structures, e.g., if the original reaction was A + B + C --> D + E and we removed the structure B, the C structure is assigned the letter B and D and E are reassigned to C and D, respectively. We can, of course, insert any new structure in the reaction after we erase one of the reaction participants.

SPECIFYING REACTION CENTERS

We will see later in the discussion of reaction searching that we can employ a search technique that allows for retrieving reactions by a query structure with indicated reaction center(s) of the molecules, i.e., indicating the specific structural fragment(s) in the reactant which are transformed in the reaction into another specific fragment(s) in the product (see also the discussion on "Reaction Transforms" in Part Two, p. 398).
To formulate the query structure for reaction center search, we build the structure as usual and then mark the bond(s) in the fragment(s) by activating the option CENTER followed by place/select the bond(s) that are changed in the reaction. Figure 4/30 shows

CREATING REACTIONS

Example: *Displaying reactions*

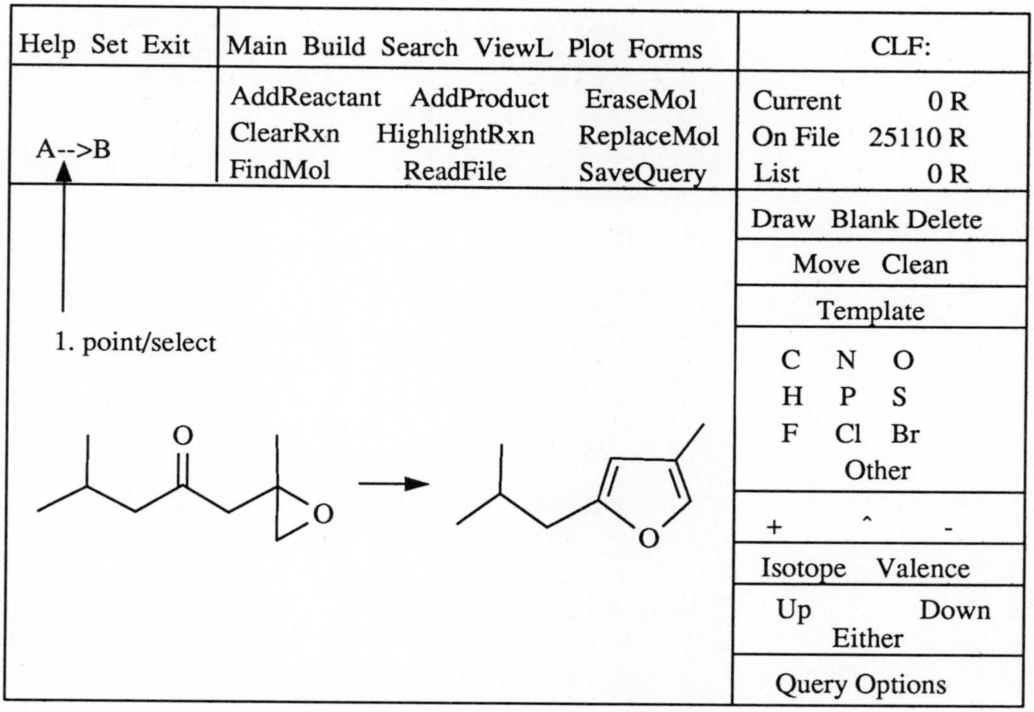

Help Set Exit	Main Build Search ViewL Plot Forms	CLF:
A-->B 1. point/select	AddReactant AddProduct EraseMol ClearRxn HighlightRxn ReplaceMol FindMol ReadFile SaveQuery	Current 0 R On File 25110 R List 0 R
		Draw Blank Delete
		Move Clean
		Template
		C N O H P S F Cl Br Other
		+ ^ -
		Isotope Valence
		Up Down Either
		Query Options

Figure 4/27. Displaying reaction

Example: *Displaying reaction product*

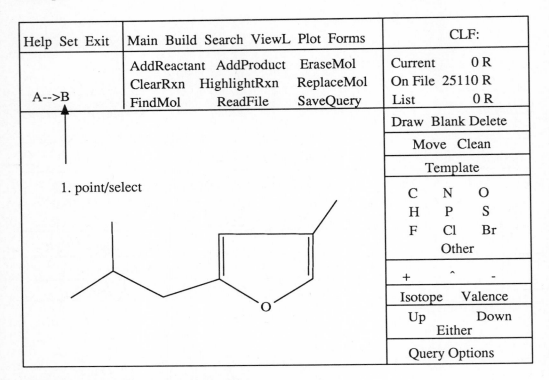

Figure 4/28. Displaying reaction product

the reactant in Figure 4/27 being marked for bond change. The bonds marked by the option CENTER will be highlighted or marked by "#" (depending on the terminal in use) (Figure 4/31). The structure is then labeled reactant and thus placed in the reaction as usual.

We can also indicate that specific bonds are not allowed to be changed (they are not in the reaction center) in the reaction we search for. We use the option NOTCENTER and place/select the bonds that are not to be changed (Figure 4/32).

The bond not allowed to be changed in the reaction is marked with "X" (Figure 4/33).

In the example in Figure 4/33, if we label the molecule as Reactant, the reactions would be retrieved in which the double bond is reduced, (this was the specified CENTER), while the keto group is preserved (this was the specified NOTCENTER).

The CENTER option followed by place/select the marked bond erases the NOTCENTER specification(s). The NOTCENTER option erases the CENTER specification(s).

Note that the Global switch "#" (see p. 589) must be on to highlight the reaction center bonds. The "#" is automatically turned on when we enter the Build Mode and cannot be

CREATING REACTIONS

Example: *Replacing reaction structure*

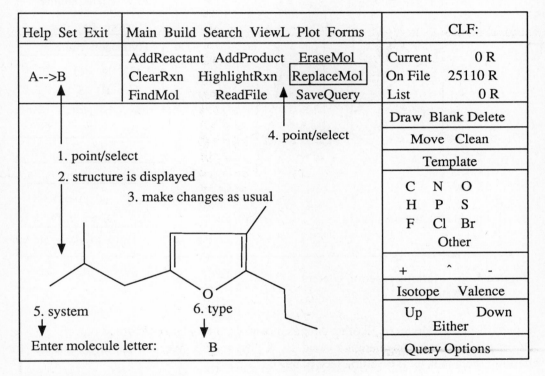

Figure 4/29. Editing reaction

Example:		Specifying reaction center in query
Help Set Exit	Main Build Search ViewL Plot Forms	CLF:

REACCS	AddReactant AddProduct EraseMol ClearRxn HighlightRxn ReplaceMol FindMol ReadFile SaveQuery	Current 0 M On File 39411 M List 0 M
(structure diagram: place select, with numbered arrows 3., 4., 5., 6. pointing to atoms/bonds of a ketone with an epoxide; 1. create structure as usual; 2. point/select)		Draw Blank Delete Restore Exclude Hydrogen Stereo C N O H P S F Cl Br A Other Q ⎡Center⎤ NotCenter Chain Ring Any Arom S/A D/A S/D Draw Options

Figure 4/30. Marking reaction centers

turned off in that Mode. In all other Modes, however, the "#" can be turned on/off.
The mark character for changing bond is "#" by default. We can set any character for marking the bonds by the Setting (see p. 592):
RxnCenter......sets the character used to mark reaction
       center bonds on monochrome terminals
The NOTCENTER option specifies that the the corresponding bond cannot be changed in the reaction. The CENTER option specifies that the reaction must produce the corresponding bond which may be the same or in which the bond order has changed.
Note that in HighlightRxn Mode (p. 616) we can request REACCS to mark automatically reaction centers which in many cases will be preferable to manual marking.
We are able to extend this kind of correspondence to the atoms in the structure fragments by atom/atom mapping and thus create queries that are even more specific than assigning the correspondence to the bonds only.

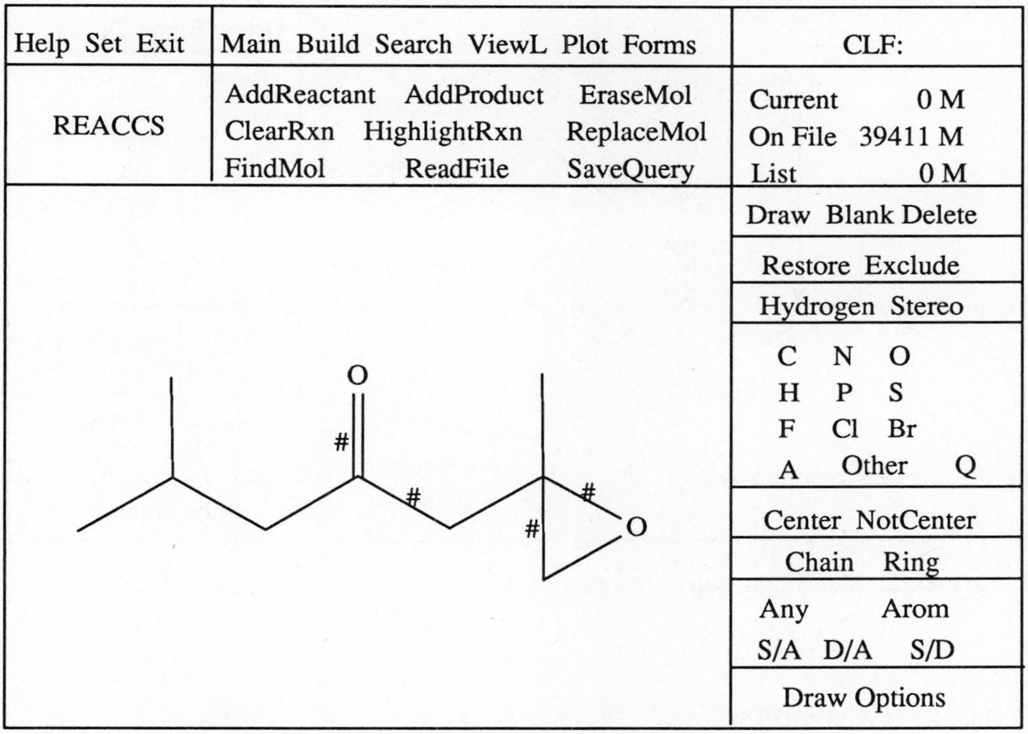

CREATING REACTIONS

Help Set Exit	Main Build Search ViewL Plot Forms	CLF:
REACCS	AddReactant AddProduct EraseMol ClearRxn HighlightRxn ReplaceMol FindMol ReadFile SaveQuery	Current 0 M On File 39411 M List 0 M
		Draw Blank Delete
		Restore Exclude
		Hydrogen Stereo
		C N O H P S F Cl Br A Other Q
		Center NotCenter
		Chain Ring
		Any Arom S/A D/A S/D
		Draw Options

Figure 4/31. Structure with marked reaction centers

Example: *Negative reaction center in query*

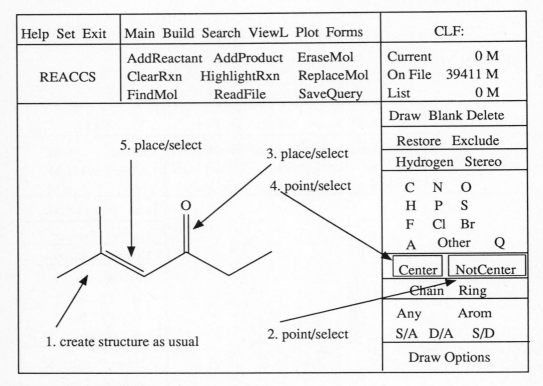

Figure 4/32. Marking negative reaction center

ATOM/ATOM MAPPING

We will see in the later discussion of Reaction Substructure Searching (RSS) (see p. 678) that we can search for reactions in which the atom(s) at specific locants in the molecule/reactant must correlate with the atom(s) in the molecule/product of the reaction. Correlation of the atoms is accomplished by atom/atom mapping of the query structures. To be able to use this technique of atom/atom mapping for reaction searching, all the registered reactions in the database must be atom/atom mapped. The databases supplied by MDL are atom/atom mapped. If we create a new reaction and register it in the database, the atom/atom maps, if applied, is stored with the structure(s).
The atom/atom mapping is carried out in the HighlightRxn Menu of the Build Mode. The Menu is entered by activating the HighlightRxn option (Figure 4/34). The reaction must be on display when we activate the HighlightRxn. It is transferred to the HighlightRxn submode (Figure 4/35).

CREATING REACTIONS

Help Set Exit	Main Build Search ViewL Plot Forms	CLF:	
REACCS	AddReactant AddProduct EraseMol ClearRxn HighlightRxn ReplaceMol FindMol ReadFile SaveQuery	Current On File List	0 M 39411 M 0 M
		Draw Blank Delete	
		Restore Exclude	
		Hydrogen Stereo	
		C N O H P S F Cl Br A Other Q	
		Center NotCenter	
		Chain Ring	
		Any Arom	
		S/A D/A S/D	
		Draw Options	

Figure 4/33. Structure marked with reacting and not reacting centers

The mapping is done either automatically by REACCS or manually by the User or by both. If the mapping is done by the system, every atom in the structures is usually mapped and the bonds in reaction centers are also marked. In User mapping, only selected specific atoms can be mapped and bonds can be marked by the User (*vide infra*). For automatic mapping, we point/select the AutoMaps option on the Menu (Figure 4/35) and REACCS numbers the atoms in the structures as shown in Figure 4/35.

Each atom in the reactant has a corresponding atom in the product of the same number. Note that the numbering can be in any order; it does not comply with any of the chemical numbering systems. If a reactant produces more than one product, the corresponding atoms can be in all of the products (Figure 4/36).

For manual mapping, we activate the option MARKMAP and selectively map the atoms in the structures by place/select on the corresponding atoms in the reactant and product(s) (Figures 4/37, Figure 4/38). The Global switch:

"a"....A mapping on/off

must be on to display mappings in the other than Build Mode (when we enter the Build Mode, the "a" is automatically turned on).

The Setting (see p. 592):

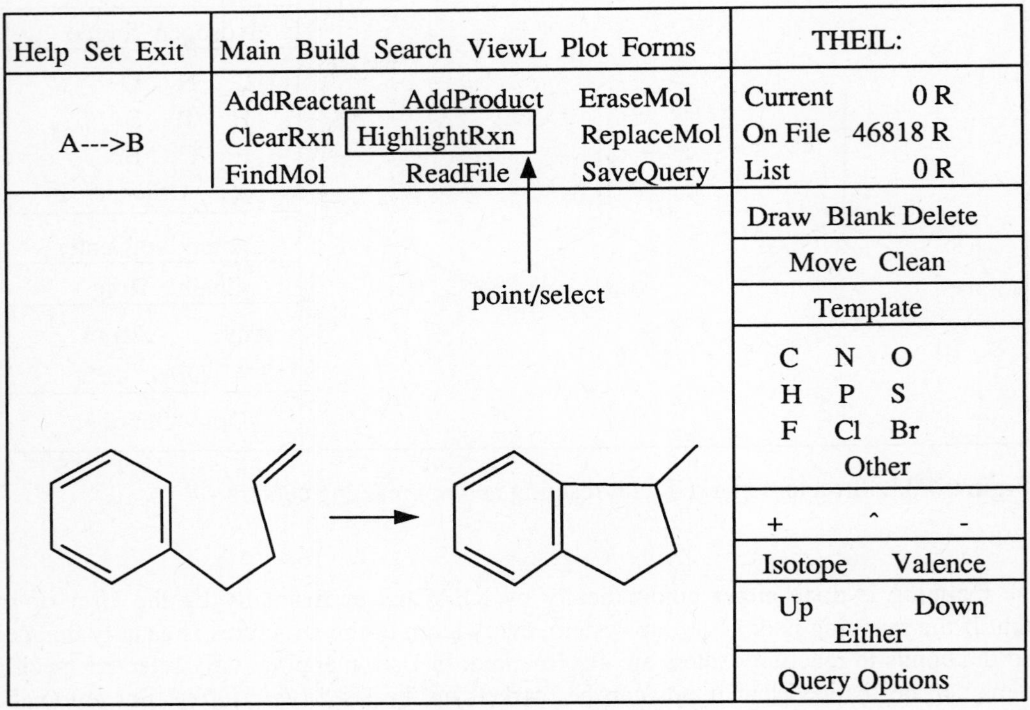

Help Set Exit	Main Build Search ViewL Plot Forms	THEIL:
A--->B	AddReactant AddProduct EraseMol ClearRxn HighlightRxn ReplaceMol FindMol ReadFile SaveQuery	Current 0 R On File 46818 R List 0 R
		Draw Blank Delete
		Move Clean
		Template
		C N O H P S F Cl Br Other
		+ ^ -
		Isotope Valence
		Up Down Either
		Query Options

Figure 4/34. Entering HighlightRxn Menu

CREATING REACTIONS

Example: *Auto atom/atom mapping*

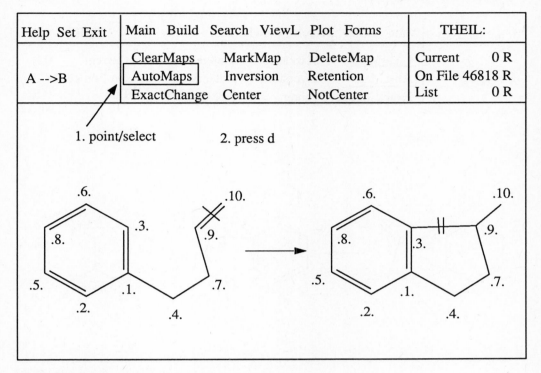

Figure 4/35. Atom/atom mapping

AAmaps........sets on/off display of atom/atom mapping
   in displayed reactions
When AAmaps is on, an .A. will appear in the left upper corner of the Menu.

MORE ON ATOM/ATOM MAPPING

The other options on the HighlightRxn Menu have the following functions:
center..........specifies that particular bond(s) must be
   in the reaction center(s)
The Center option in the HighlightRxn Menu allows for more specific reaction center (changing bonds) marking than that in the Build/Query Options Menu (see p. 623). When we point/select Center, the pull-down menu appears (Figure 4/39).
To mark the bond(s) that will be formed or broken in the reaction, select Make/Break. To mark bond(s) that will change in value, e.g., double to single, select Change (Figure 4/39). The bonds will be highlighted or marked as follows:

Example:			Auto atom/atom mapping
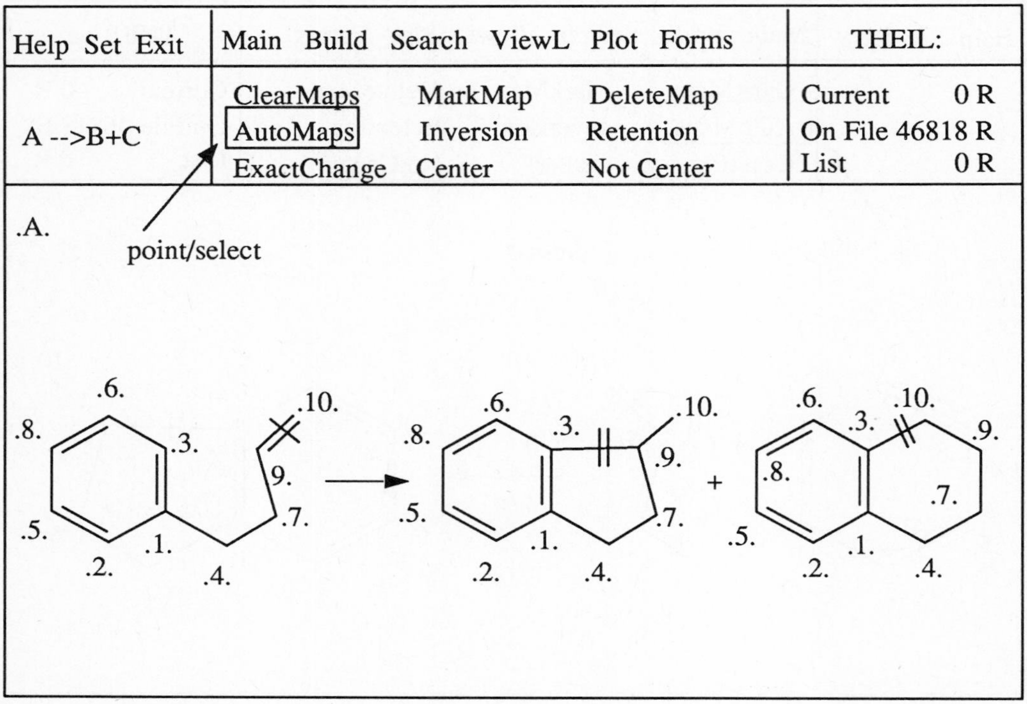 Help Set Exit	Main Build Search ViewL Plot Forms		THEIL:
A -->B+C	ClearMaps MarkMap DeleteMap AutoMaps Inversion Retention ExactChange Center Not Center		Current 0 R On File 46818 R List 0 R

.A.   point/select

Figure 4/36. AutoMapped reaction

1. Selecting CENTER marks the bond(s) with # or any character set with the RxnCenter Setting (see p. 592).
2. Selecting MAKE/BREAK marks the bond(s) with a double slash.
3. Selecting CHANGE marks the bond(s) with a single slash.

Figure 4/40 shows the marking. This Figure 4/40 also shows the reaction in which explicit hydrogens have been drawn to specify changes in the reaction centers (see p. 600 for note on hydrogens).

Note that if we use AutoMaps, the bonds in reaction center(s) will be automatically marked (Figures 4/36, 4/40).

notcenter......specifies that particular bonds are not
          changed in reaction
clearmaps......erases all atom/atom mappings and reacting
          center specification in the current reaction
deletemap......erases a specified atom/atom mapping from the
          reaction
  1. point/select deletemap

CREATING REACTIONS

Example: *Manual atom/atom mapping*

Help Set Exit	Main Build Search ViewL Plot Forms	THEIL:
A -->B+C	ClearMaps MarkMap DeleteMap AutoMaps Inversion Retention ExactChange Center NotCenter	Current 0 R On File 46818 R List 0 R

.A.

1. point/select
2. place/select
3. place/select
4. place/select
5. place/select
6. place/select
7. place/select
8. place/select
9. place/select
10. place/select

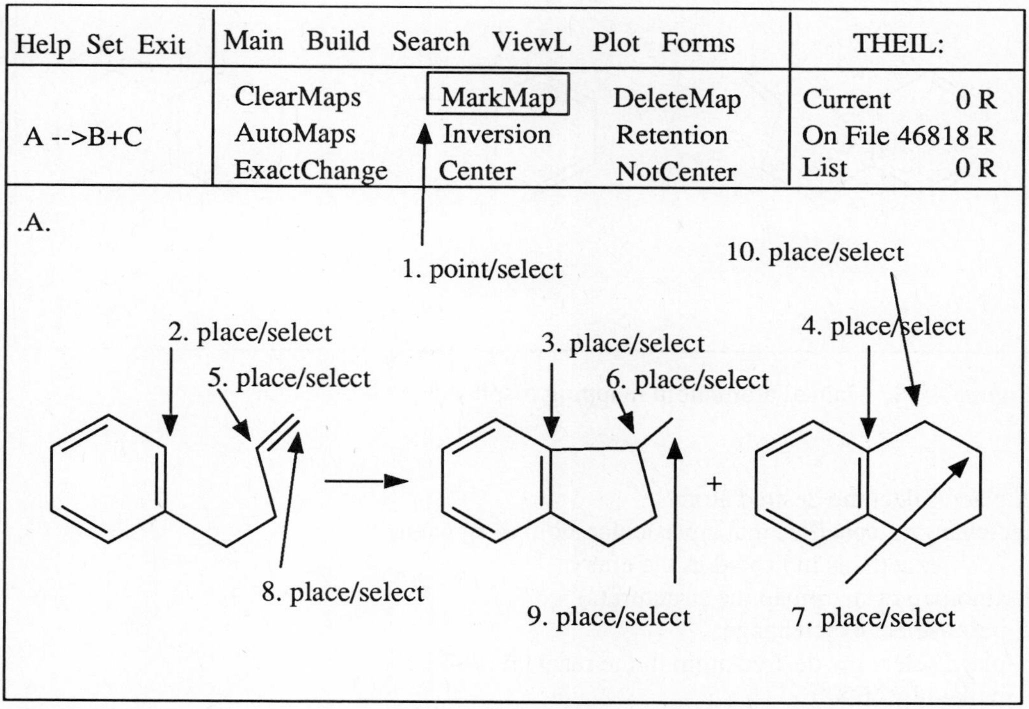

Figure 4/37. Manual atom/atom mapping

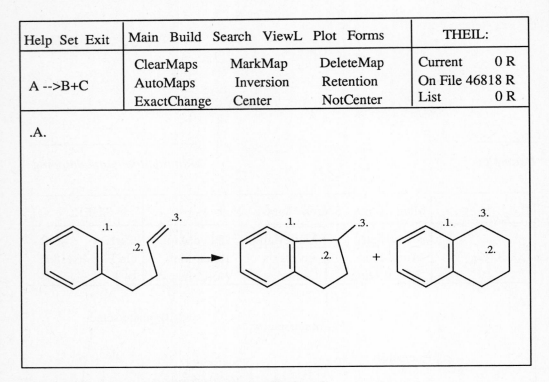

Figure 4/38. Manual atom/atom mapping displayed

   2. place/select the desired atom
exactchange....specifies that a particular atom must change
        exactly as indicated in the query
   1. automap or markmap the structure(s)
   2. point/select exactchange
   3. place/select the desired atom in the reactant (will be
     marked by "EX")
   4. place/select the desired atom in the product (will be
     marked by "EX")
inversion......specifies that the configuration at the parti-
        cular atom must be inverted in the reaction
retention......specifies that the configuration at the parti-
        cular atom must be retained in the reaction

BUILDING REACTIONS - THE QUICK WAY

REACCS offers an alternative method to build reactions by activating the ASSEMBLE option and entering the registry numbers of the molecules we wish to use as the

CREATING REACTIONS

Example: *Manual atom/atom mapping*

Help Set Exit	Main Build Search ViewL Plot Forms	THEIL:
A -->B+C	ClearMaps MarkMap DeleteMap AutoMaps Inversion Retention ExactChange Center NotCenter	Current 0 R On File 46818 R List 0 R

Figure 4/39. Marking bond change

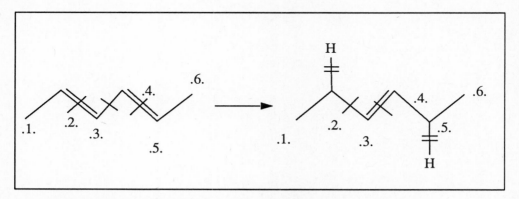

Figure 4/40. Atom/atom mapping and bond marking

participants in the reaction. The structures must be already registered in the database.

The following are the steps to build a reaction:
1. press k
2. System: Build:
3. User: assemble
4. System: Assemble:
3. User: 34 + 60 -> 128 +24
3. System: Assemble:
4. User: <cr>

The reaction has been created in which the reactants are the structure registered under the internal registry numbers 34 and 60, and the products are the structures with registry numbers 128 and 24.

6   CREATING FORMS AND TABLES

The readers who studied Part Two will find this Section very analogous to the discussion on creating display formats in ChemBase.
In REACCS, the retrieved molecules or reactions can be displayed as follows:
1. excluding data
2. including data
When we perform a search in the Main Mode (see p. 649), the retrieved structures are automatically displayed on the screen. The structures retrieved as a result of the search in the Search Mode (see p. 652) can be displayed in the ViewList Mode (see p. 658). In both Modes, the structures appear with no data that have been registered in the datatypes if the Setting DataDisplay (see p. 592) is off. To display the data associated with the retrieved molecules or reaction, the Setting must be on and we must have a display format in which specific data will be displayed.
The display format can be:
1. a Form
2. a Table

ABOUT DISPLAY FORMATS

The databases developed by MDL contain a standard set of molecule and reaction Forms and Tables. The set is contained in a file the name of which has been entered into the User Profile so that it is automatically loaded when we start the REACCS session. If we search the MDL databases, we can display structures with the associated data in these formats, molecules in the molecule Form/Table and reactions in the reaction Form/Table. REACCS switches automatically into the Molecule Form if we display molecules and into the Reaction Form if we display reactions.
However, we may wish to create our own Forms/Tables to be able to display data in a variety of formats. Although a database can be linked at one time only to one set of the Forms/Tables, we can have as many sets saved in the Form Files (.frm) as we desire and change one set for the other in the course of searching the particular database to which the Forms/Tables have been linked. The Forms/Tables are not interchangable between databases unless the databases have the same datatypes. The set of REACCS Forms/Tables cannot be used in the other MDL systems, e.g. ChemBase or MACCS-II.
We will now discuss how to create Forms/Tables and how to specify which data will be displayed in them.
The Forms/Tables in REACSS are created in the Forms Mode which is entered from any Mode by point/select Forms on the Menu (Figure 4/41).
The Forms Menu is displayed with the default Molecule Form of the current database. To create Forms/Tables for a particular database, we must be in that database. If the database

Help Set Exit	Main Build Search ViewL Plot Forms	CLF:
	AddReactant AddProduct EraseMol	Rxn 25110
REACCS	ClearRxn HighlightRxn ReplaceMol	Mol 39411
	FindMol ReadFile SaveQuery	List 0
		Draw Blank Delete
	point/select	Restore Exclude
		Hydrogen Stereo
		C N O H P S F Cl Br A Other Q
		Center NotCenter
		Chain Ring
		Any Arom
		S/A D/A S/D
		Draw Options

Figure 4/41. Entering Forms Mode

for which we wish to create the forms is different from that one we started REACCS with, we use the DB option (see p. 598) to change the database.

The current displayed form can be modified by any of the techniques we will describe later. However, we will start our form from scratch. To remove the current Molecule Form from display, point/select CLEAR (Figure 4/42).

To start creating a Reaction Form, switch to Reactions (Figure 4/43).

CREATING FORMS

The Form consists of boxes. The boxes indicated in Figure 4/43 by the broken lines are the "floating boxes." We may assign data to them or leave them blank. However, we never create these boxes; they are permanently fixed to these positions and the data will be always displayed in the proper position in the structure display (*vide infra*). The data assigned to these floating boxes will appear above or under the arrow and below the reactant and product structures.

The other boxes in the form can be positioned anywhere in the Form.

CREATING FORMS AND TABLES

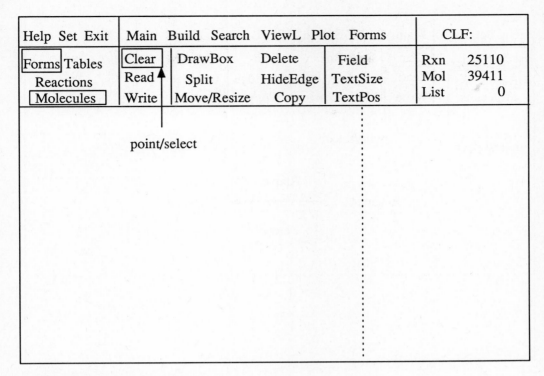

Figure 4/42. Clearing default form

STRUCTURE BOX

The most prominent of the boxes on the Form will be the structure box. If we wish, we can take advantage of the standard structure box that, by default, is assigned by REACCS. On the Molecule Form, the default structure box will be in the area left of the broken line (Figure 4/42), on the Reaction Form the structure box will be above the broken line (Figure 4/43). If we elect the standard structure box, the data boxes (*vide infra*) should be drawn only in the area to the right of the broken line in the Molecule Form, and below the broken line in the Reaction Form.

DRAWING BOXES

To draw boxes, point/select DrawBox followed by place/select on the diagonal corners of the box we wish to draw (Figure 4/44).
The other options on the Menu enable us to manipulate the drawn boxes:
COPY.........copies a box to another position on the Form
Select the box to be copied and place/select the new position where the box should be copied.

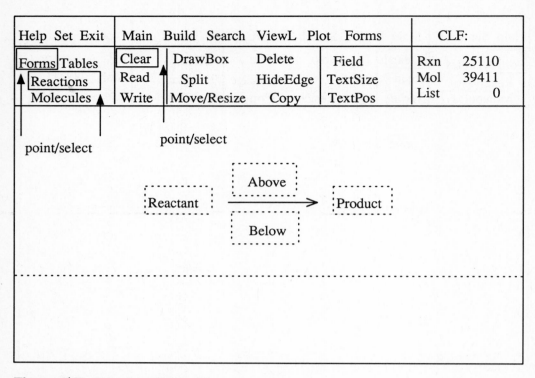

Figure 4/43. Reactions Forms Menu

DELETE......deletes a box
Select the box to be deleted. The box along with the Data Field assignment (*vide infra*) will be deleted.
HIDEEDGE....makes one edge of the box invisible
Select the edge to be hidden. Another HideEdge selection will redisplay the edge.
MOVE/RESIZE...moves a box to another position or resizes the box
To move a box, select inside the box followed by place/select the new position. To resize a box, select the edge to be moved followed by place/select the new position, or select a corner of the box followed by place/select the new position. Two boxes will share one edge if we place/select the edge to be moved followed by place/select the edge of the box which will share the moving edge.
SPLIT.....splits a box in half vertically or horizontally
To split vertically, place/select a horizontal edge, to split horizontally, place/select a vertical edge. To split

CREATING FORMS AND TABLES

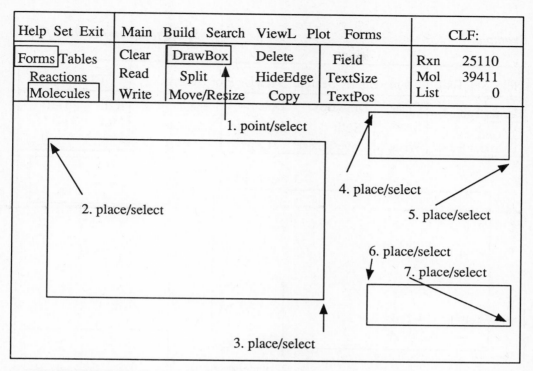

Figure 4/44. Drawing boxes in Form Menu

to multiple boxes, place/select on a vertical or horizontal edge and type a number from 3 to 9.

ASSIGNING STRUCTURE FIELD TO BOX

To display a molecule or reaction structures in the default structure box, we do not have to make any field assignment.
To assign to a created box the *structure field, we proceed as shown in Figure 4/45.

ASSIGNING DATATYPES TO BOX

To display the data that are associated with the molecule or reaction we retrieve from the database in a search, we have to assign the specific datatype to the box (Figure 4/46). The datatype must have been, of course, defined when the database was created.
In the example in Figure 4/46 we have assigned to the data box the datatype "symbol" (which in the CLF database contains molecule names) and the label "Name."
(Note that there must be at least one space between the datatype and label entries at the prompt.) The datatypes are assigned using the treenames (see p. 579, 583). The treename

Example: *Specifying box structure field*

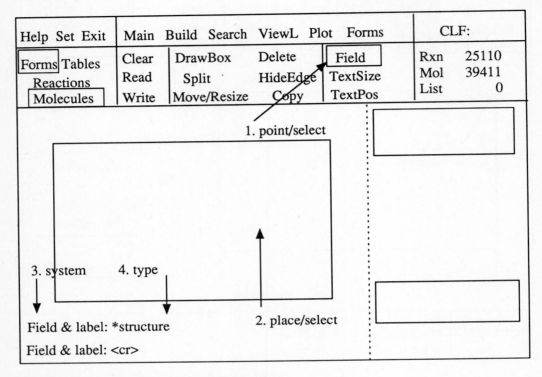

Figure 4/45. Assigning structure field to box

must be unique for the field we wish to assign. Thus in Figure 4/46 we used "symbol" since it unambiguously defines the field in the treename "mol:symbol." On the other hand, if we were assigning a data box in the reaction form the field "grade" we have to distinguish whether we want the reactant, catalyst, or solvent grade since the database has been set up with the following datatypes:

rxn:variation:reactant:grade
rxn:variation:catalyst:grade
rxn:variation:solvent:grade

We would use the following:
reactant:grade (for the reactant structure)

The label, which is optional, is a line of text which serves as a heading for the data in the box. It is displayed every time the box receives data for display. We can input more than one datatype and label for a box by entering the additional fields and labels at the prompt which REACCS repeats until we exit the prompt by <cr>. This concludes the assignment for the box. The remaining data boxes in the form are assigned similarly other datatypes

Example: *Specifying box datatype field*

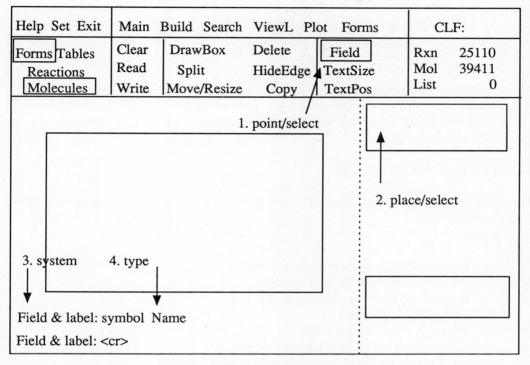

Figure 4/46. Assigning datatype field to box

and labels.
Note the special datatypes:
*formula......displays molecule formula
*regno........displays internal regno
*structure....displays structure
*text.........displays text which we enter
  To use them, we input, e.g.:
   *text (type text to be displayed in box)
The data in a box can be aligned by quotation marks as follows:
Field & label: journal:vol.   "Volume:"
Field & label: journal:year   "Year: "
The aligned data will appear in the box, e.g.:
Volume: 9
Year: 1988

MULTIPLE DATA DISPLAY

If the datatype treename we specify for a box consists of more than one multiple datatype (see p. 581), REACCS displays all lines of data for the first multiple datatype in the treename going from the right to left, and the current line (i.e., the first multiple datatype) for the other multiple datatypes in the treename. This is the default which can be overridden by "line numbers" "A" for all, and "C" for current.
Thus the datatype:
variation(a):catalyst(c):regno
will display all data for "variation" and current line (1) for "catalyst" (without the A and C, the result would be the opposite). To display additional lines of data for "catalyst," we use the NEXT option (see p....).

EDITING BOX DATA FIELDS

The assigned datatypes and labels in a box can be edited by the following suboptions typed at the prompt:
Field & label:

clear..........removes all datatypes and labels from box
delete.........removes a selected field and label
print..........displays the datatypes and labels that have
             been assigned to the box
index..........displays all parent and field datatypes in the
             database (type <cr>), or selected datatypes
             (type name or number to select from display)

MORE ON FORM MENU OPTIONS

TEXTPOS...determines position of text in a box
            positions: 1   2   3
                            4   5   6
                            7   8   9
TEXTSIZE...determines text size (1-9)

SAVING FORMS/TABLES IN FILES

When the Molecule and/or Reaction Form is complete, we either leave the default Tables unchanged or modify them, or create new ones (*vide infra*). Then we save the set in a .frm file (Figure 4/47). Note that WRITE saves a set of:
Molecule Form/Table and Reaction Form/Table.
If we do not change one of the default members of this set, the default will be saved in the set with any of the other modified or new members of the set.

CREATING FORMS AND TABLES

RECALLING FORMS/TABLES

To recall a .frm file containing a set of Forms/Tables, we point/select READ.
(See also the discussion of SETTINGS(FormFile) and User Profile (p. 592, 595).

Example: *Saving form in a file*

Help Set Exit	Main Build Search ViewL Plot Forms	CLF:
Forms Tables	Clear DrawBox Delete Field	Rxn 25110
Reactions	Read Split HideEdge TextSize	Mol 39411
Molecules	Write Move/Resize Copy TextPos	List 0

1. point/select

2. system 3. type

Output file name: myform

Figure 4/47. Saving form in file

The default Molecule and Reaction Form with assigned fields are shown in Figures 4/113, 4/114, p. 709.

CREATING TABLES

A Table displays data in columns. Each column contains one datatype of a record, each row of the Table contains one full or partial record. The top of the column contains a label, i.e., a name for the datatype displayed in the column. A Table can have up to 10 columns with a maximum width of 120 characters/column. Total Table width cannot exceed 800 characters. The Table can be wider than the display screen. To scan horizontally, use < and >.

To enter the Table Menu, point/select TABLES and the appropriate one of MOLECULES or REACTIONS. The Menu will be displayed with the default Table. Modify the Table or, to start from scratch, point/select CLEAR (Figure 4/48).

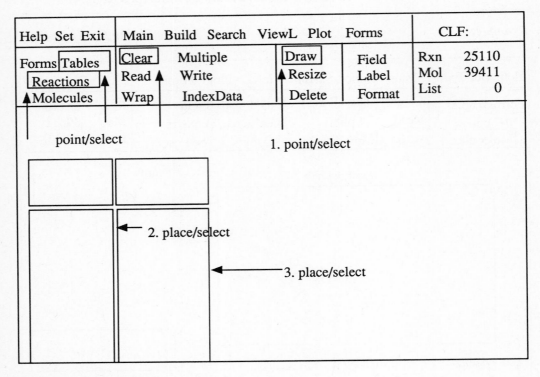

Figure 4/48. Table Menu

To draw a Table (Figure 4/48):
1. Point/select Draw.
2. Place/select the position of the right side of the column to be drawn.

REACCS will draw the column consisting of two boxes, the upper one for the label, the bottom one for the data of the specified datatypes.
To resize the columns:
1. Point/select RESIZE.
2. Place/select the column to be resized followed by place/select the new position of the right side of the column

ASSIGNING DATATYPE AND LABEL

To specify the datatype and label for each column:
1. point/select FIELD

CREATING FORMS AND TABLES

2. place/select the column
3. at the prompt: Field and label:
   type the entries

ASSIGNING LABEL

To assign label only for a column:
1. Point/select LABEL.
2. Place/select the column.
3. At the prompt: Column Label:
   type the label text

The default Molecule and Reaction Tables with the specified fields and labels are shown in Figures 4/117, 4/118, p. 712.

MULTIPLE DATA DISPLAY

Data in multiple datatypes are displayed in Tables in lines (e.g., Figure 4/119, p. 713). We can also display each line of the multiple data in separate columns (e.g., Figure 4/121, p. 715).

Multiple datatypes are assigned to columns by point/select:
MULTIPLE......displays all lines of data for a multiple datatype treename in Table
 System: Multiple datatype to repeat over:
 User: (enter the multiple datatype)
The datatype can be a single multiple datatype or a treename. The specified multiple will appear at the top of the Table.
For example, we have the following datatypes:
rxn:variation(1):reactant:grade
rxn:variation(1):catalyst:regno
rxn:variation(2):reactant:grade
rxn:variation(2):catalyst:regno
If we specify "variation" as the multiple for the Table, all lines of "variation" data will be displayed for the hit list (see Figure 4/119, p. 713).

Multiple can be used for only one datatype pathname per Table. If the pathname contains more than one multiple datatype, REACCS displays all lines of all multiple datatypes.
To display data over two or more multiple datatypes that are not along the same path, columns must be created for each of the lines, e.g., for multiple datatype "Variation" the datatype was specified:
Multiple datatype to repeat over: variation
for the Table created as shown in Figure 4/49.

REGNO	VARIATION	FIRST CAT	SECOND CAT
*regno	variation	catalyst(1)	catalyst(2)

Figure 4/49. Multiple datatypes in columns

MORE ON TABLE MENU OPTIONS

FORMAT.....changes the format of data displayed in Table

To display:	Use:
Real number	R
Range	R, '-', R
Integer	I
Date	DD/MM/YY any ordering
Fixed-length text	A
Text	'text'

INDEXDATA...lists datatypes defined in the database
WRAP........if on, all lines of text are wrapped around into multiple lines if too long for box. If off, the too long lines are truncated with @ at the line end

7 SEARCH TECHNIQUES

In REACCS, we can employ a variety of search methods to arrive at the best search results.
Searching in REACCS is a process which involves an object that is subjected to an action to produce a result. The object is a query (a question) which comprises certain conditions that must be met by the action (the search) to produce the result (the answer, the hits).
A search results in a List of registry numbers identifying the molecules or reactions that have been found with the query. The List of the registry numbers resulting from the search is the temporary list file named "Ln" by the system (see p. 585).
We can, therefore, say that the query is the List Descriptor since the query describes, identifies, and determines the entities that will be in the list Ln.
The query in REACCS may be:
1. a graphic query which is the structural graphs we create using a graphic input device
2. a numeric and/or text query that are the alphanumeric strings which we input via the keyboard
3. combination of both graphic and alphanumeric query

GRAPHIC QUERIES

When we formulate a graphic query to search for molecules or reactions in a database, the query structure is, in fact, the identifier of the registered molecules or reactions since the structure graph of the query must match (identify) the structure(s) in the database to produce a hit or hits. The query structure may be an exact molecule or an exact reaction, or it may be a substructure (fragment) of the structures of the molecules or reactions registered in the database.
REACCS uses the graphic query, we can imagine, as a template which is superimposed on the structures in the database. Those structures which are "identified" by this template are the candidates for the hit list "Ln." In Exact Structure Search the query must be identical with the structure as a whole in the database to produce a hit. In Substructure Search, it is necessary that the query be a fragment contained in its entirety in the structure(s) in the database to produce hits (see also the discussion on SSS, p. 665 and Part Two, p. 380).
In Similarity Search (see p. 673, 692), it is sufficient that a predefined percentage of keys in the query match those in the hits.

SEARCHING WITH GRAPHIC QUERIES

When we have a current molecule or reaction, the structures are ready to be used as the query for retrieval in Main Mode or to search in Search Mode. The displayed graphic structures are always the current entities which will be the subject of the search. We

simply point/select the appropriate option on the Menu and the action is performed without our further interaction with REACCS. If we wish to use the keyboard selection we can, of course, do so and type the option. For example, if a molecule is on display and we depress "k," REACCS will issue the prompt, e.g.:
Main:
User: find current
will perform the same retrieval as the graphic option FIND CURRENT.
However, all the conditions which must be met by the search to produce a hit are defined in the graphic query.
We will see in the following discussion of alphanumeric queries, that using keyboard selection and numeric or text queries gives us more flexibility in formulating the queries and in instructing REACCS how to proceed in the search. We will be able to use the REACCS "query language" to communicate with REACCS and to introduce specific conditions into the query which must be met to give a hit.

NUMERIC AND TEXT QUERIES

The registered molecules or reactions in the database can be also identified by other means than by their image (query structure). The data that are derived by REACCS from the structures or those we register in the datatypes can serve as their identifiers, thus, we can use these data to formulate queries.
The queries can be formulated using any descriptor which results in the hit list "Ln."
The descriptors may be the following:
1. list descriptor
2. data descriptor
3. data search expression
4. search expression

The descriptors are entered after we invoke an option and REACCS issues one of the following prompts:
Molecule ID:
Reaction ID:
List descriptor:
Datatype:
Data descriptor,'=',target value:
or after we switch to the keyboard mode by depressing "k" and receiving the prompt:
Search:
(It should be noted that these prompts may be issued by REACCS also in other than the true retrieve or search operations in Main and Search Mode, i.e. in Build, ViewList, and Plot Mode.)
The entries which the User inputs at these prompts can be composed of:
1. registry number of molecules or reactions
2. list file, listfile, current list, reference list
3. structure file (Molfiles, Rxnfiles, Query Files)
4. data descriptor

SEARCH TECHNIQUES

5. data-search expression
6. special data-search expression
7. search expression by query language

REGISTRY NUMBERS AS LIST DESCRIPTORS

The source of registry numbers can be:
1. single or multiple regnos entered via keyboard
2. list file, listfile, current list, or reference list

Registry Numbers

Registry number is the specific list descriptor which we can use to retrieve molecules or reactions from the database.
We have seen in describing the database structure (see p. 584) that each molecule and each reaction is assigned automatically by the system an internal registry number and that, in addition to this registry number, we can create a datatype for the external registry number assigned by the User to the molecules and reactions. Both of these registry numbers can be the specific list descriptor.
As an example, if we input at the prompt:
Molecule ID: 24
the default regno 24 will be the query for the search.
It should be noted that the "Molecule ID:" prompt is the specific prompt which is issued by REACCS in the Main Mode upon activating the FIND MOLECULE option. Since REACCS knows that we want to find a molecule, we can input the registry number without any further specification. Similarly, on selecting FIND REACTION we receive the specific prompt:
Reaction ID: 560
REACCS will retrieve the reaction of regno = 560.
However, in the operations in which REACCS does not know which of the entities we want to search, we have to specify whether the registry numbers are for molecules or reactions. For example, in the Search mode we can establish a hit list at the prompt:
Search:
We have to identify for REACCS the nature of the regnos by including "m" for molecules or "r" for reactions in the input, e.g.
User: m12-24,5,10 (m=molecule)
User: r12-24,5,10 (r=reaction)
The "m" or "r" specification is also required at the Molecule ID: or Reaction ID: prompt if the default registry number (see p. 699) has been changed.

List File, Listfile, Current List, Reference List

Since these lists contain registry numbers, we can use them as the entry after the prompts. For example:

Molecule ID: L4 (temporary list file containing molecule
   regnos)
Reaction ID: mylist (permanent listfile containing
   reaction regnos)
List descriptor: herlist (permanent listfile containing
   regnos corresponding to the action
   we are performing)

The option FIND in the Main Mode allows for retrieval of only a single molecule or reaction. Therefore, the Lists and Listfiles are of limited use in the Main Mode, unless they contain only one entry, because REACCS will use only the first entry in the list for the FIND operation if it contains multiple records. However, the lists are advantageous to use in the ViewList or Plot Mode to establish the Current List, or to search in the Search Mode, e.g.:

System: Search:
User: ref = mylist.lst
REACCS will create the Reference List (search domain) containing the regnos in mylist.lst
System: View:
User: list
System: List descriptor:
User: L5
REACCS will create Current List containing the regnos in L5.

Structure Files

The files containing structures will be the source of the graphics if we formulate a query composed of the names of the files, e.g.:
Search: sss=Q2 (temporary query file)
   sss=fstfile.mol (permanent molfile)
The structure in the Q2 file or in the fstfile is used by REACCS as the graphic query for the substructure search (SSS).
Note the format of inputing an option with the query via keyboard:
option = query

Data Descriptor

Data descriptor is required in many REACCS operations involving search and retrieval of data associated with the registered molecules or reactions.
The data descriptor is also used in data-search expressions (*vide infra*).
Data descriptor is the specification of the datatype treename (see p. 583) contaning the data we wish to retrieve. The datatype treename determines how specific the search will be.
The data descriptor may consists of:
1. the complete treename, e.g.:

rxn:variation(1):reactant(1):amount:weight
The query relates to the specific single datatype.
2. partial treename, e.g.:
   a. reactant(1):amount:weight
The query relates to all datatypes having the specified multiple datatype "reactant(1):amount:weight" in their treename, e.g.:
rxn:variation(1):reactant(1):amount:weight
rxn:variation(2):reactant(1):amount:weight
rxn:variation(3):reactant(1):amount:weight
but will not include:
rxn:variation(1):reactant(2):amount:weight
   b. weight
The query relates to all datatypes having the field "weight"
in the treename, e.g.:
rxn:variation:reactant:amount:weight
rxn:variation:product:amount:weight
rxn:catalyst:amount:weight
   c. catalyst
The query relates to all the datatypes having the parent "catalyst" in the treename, e.g.:
rxn:variation:catalyst:amount:weight
rxn:variation:catalyst:amount:volume
3. wild card treename
The wild card symbol "@" substitutes for any complete datatype in the treename or for any character in the datatype name for left-hand, right-hand truncation, or within the name, e.g.:
   a. temp@ or @perature or @perat@
The query relates to the datatype name "temperature."
   b. variation:@:weight
The query relates to all datatypes having the datatypes "variation" and "weight" and any datatype(s) between them.
   c. @
The query relates to all datatypes for the current molecule or reaction.

For example, if we point/select FIND DATA (see p. 706) in the Main Mode with the current molecule on display, we receive the prompt:
System: Datatype:
User: boil.pt
The query will retrieve the boiling point for the current molecule as follows:
MOL:BOIL.PT = 139.5 DEG C

Data-Search Expression

The data-search expression has the format shown in Figure 4/50.

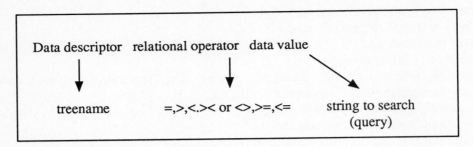

Figure 4/50. Data-search query format

In data-search expressions, the treename must include the field, i.e., the most specific datatype in the treename. The line numbers of multiple datatypes in data descriptors are not allowed.

Special Data-search Expression

We have learned that there are special data in REACCS databases which are derived by REACCS from the registered structures. They are:
1. formula
2. keys

In turn, the data can be made part of a query to search for molecules in the format:
formula=
keys=
e.g.:
System: Search:
User: formula=C9 H14 O2
    keys=14,18,48,50

SEARCHING WITH ALPHANUMERIC QUERIES

We have seen that retrieving/searching with graphic queries can be initiated by graphic selection of the options on the Main or Search Menu without further interaction with REACCS or by keyboard selection of the options after pressing "k". In the keyboard selection mode we can additionally use REACCS "query language." In Search Mode, REACCS is always ready to accept a query and performs the search operation:
User: k
System: Search:
User: sss=q2
REACCS performs the reaction substructure search using the structure query in the temporary query file Q2.
Search:
User: r1,2,5-7

SEARCH TECHNIQUES

System: **Preparing search query**
    Searching File
    5 reactions. Final. List is L1.
System: Search:
We can also initiate the search process any time in any Mode when it is appropriate to enter any of the descriptors (queries) discussed above. For example, in ViewList Mode:
User: k
System: View:
User: list
System: List descriptor:
User: r3,56,80
REACCS performs retrieval of reactions of registry numbers 3, 56, and 80 and the result becomes the Current List (see p. 652).

SEARCH TYPES

The examples of initiating the search process by keyboard input of the queries (*vide supra*) are indicative of the capability of REACCS to respond to direct entry of the search query and perform the search. REACCS enables the User, through the query language, to formulate more complicated search queries. Using the query language we can conduct the following types of searches:
1. simple search
2. conversion search
3. combination search

Simple Search

A simple search is the search with a query that does not contain any other conditions than the query itself, i.e., a graphic molecule or reaction, or an alphanumeric query (a list descriptor or data-search expression).
Examples of simple searches:
System: Search:
User: sss
  The current structure is subjected to the
  substructure search resulting in the hit list Ln.
System: Main:
User: find current
  The current structure (molecule or reaction) is subjected
  to the exact structure search resulting in the hit list Ln.
System: Search:
User: ref=list
System: List descriptor:
User: L2
  The registry numbers in the temporary list

L2 become the Reference List (search domain)
for the next search
System: Search:
User: author = @Jones@
The records with the author (J.,B.,A. etc.) Jones (Jr., Sr., etc.) will be retrieved.

Conversion Search

We have learned that REACCS databases can contain both molecules and reactions and that these entities are kept separate and have separate registry numbers and datatypes. These cannot be mixed in the operations we perform in REACCS. (Of course, regnos and datatypes from different databases also cannot be mixed.) However, there is a relationship between the two parts of the database. When we register a reaction, the molecules participating in the reaction are stored in the molecule part of the database and the reaction part of the database stores the reactions in form of a reaction scheme composed of the internal registry numbers of the molecules and their roles in the reaction (see p. 579). It would be advantageous to exploit the relationship between the two parts of the database for performing a search in both parts of the database with one query and in one search. REACCS offers such an advantage in the conversion search method.

The conversion searches convert a simple molecule search into a hit list containing reactions and a simple reaction search is converted into a hit list of molecules. The conversion is achieved by formulating the query containing one of the following statements:

a. as reactant
b. as product
c. as catalyst
d. as solvent
e. as agent (reactant, catalyst, or solvent)

It is to be noted that in REACCS the term "catalyst" means catalyst or reagent.

Examples of conversion searches:

Search: current as product
  The current query structure is a molecule. REACCS performs the search for the structure in the molecule part of the database and then searches in the reaction part of the database for all reactions in which the molecule is the product and creates a hit list (Ln) containing the regnos of the reactions.
  If the conversion phrase was not included, the "search: current" would retrieve the molecule matching the current structure but we would not have found out in what reactions the molecule had been produced.

Search: sss as reactant
  REACCS performs the substructure search in the molecule part of the database for the molecule(s) and then

SEARCH TECHNIQUES

performs the search in the reaction part of the database for reactions in which the molecule(s) is the reactant. A hit list (Ln) containing regnos of the reactions is created.
Without the conversion phrase, REACCS would perform only substructure search in the molecule part of the database for molecules.

Combination Search

A combination search is the search with queries composed of combined hit lists of simple and/or conversion searches, list descriptors, data descriptors, and/or data-search expressions. The combination is achieved with the Boolean Operators AND, OR, NOT (see p. 247). The lists used in the queries must be of the same entities, otherwise we have to include a conversion statement.
The following are examples of combination searches:
Search: L1 and L2
  REACCS creates a list (Ln) of the registry numbers that were common in L1 and L2
  e.g., L1(1,2) L2(1,2,3) Ln = 1,2
Search: L1 or L2
  REACCS creates a list (Ln) of the registry numbers that were both in L1 and L2
  e.g., L1(1,2) L2(1,2,3) Ln = 1,2,3
Search: L1 not L2
  REACCS creates a list (Ln) containing the registry numbers in L1 minus those that were common in L1 and L2
  e.g., L1(1,2,3) L2(1,2) Ln = 3
Search: L2 not L1
  REACCS creates a list (Ln) containing the registry numbers in L2 minus those that were common in L1 and L2
  e.g., L2(1,2) L1(1,2,3) Ln = 0
Search: m3,6,7-15 or fstlist.lst
  REACCS will search for molecules with regnos 3,6,and 7-15 and for the molecules in the permanent listfile fstlist.lst and creates the Ln which will contain regnos 3,6,7-15 in addition to the other in the fstlist.mol file.
Search: r56,78,90-95 and fstrxn.rxn
  REACCS will retrieve the reactions with regnos 56, 78, and 90-95 and the reaction in the Rxnfile fstrxn.rxn and will create a list Ln containing common regnos.

The following example of the combination search will give unique results:
If we are searching for product(s) that are actually formed in the reaction (i.e., the product

structural fragment is constructed in the reaction rather than the fragment being already present in the reactant), we use the following statement:
System: Search:
User: sss as product not sss as reactant
  REACCS performs substructure search with the current query
  structure which will be the product(s) of the reaction(s)
  (satisfying the "sss as product" requirement) but will exclude those reaction(s) in which the query structure was already contained in the reactant(s) of the reaction(s)
  (satisfying the "not sss as reactant" requirement).

We can combine the result of a single search or results of multiple searches, i.e., the hit lists (Ln), in a new search query, e.g.:
System: Search:
User: L1 not sss as product
  REACCS retrieves the regnos in the temporary list file
  L1 (note that they must be reactions since we are searching for reactions by the "sss as product" statement)
  and then searches for the current query structure which is
  to be the product of the reaction(s) and excludes from L1
  the reactions producing the query structure and creates the
  hit list, e.g., L2.

DESIGNATING TEMPORARY LIST FILES

We have seen that REACCS creates the hit lists (temporary list files) and automatically numbers them. If we choose so, we can force the system to assign the hit lists the numbers we specify.
If we perform a search in the Search Mode we can use the following format to assign our numbers to the hit lists:
System: Search:
User: L3=rss
System: Search:
User: L14=current as product

SEARCH IN MULTIPLE DATABASES

The database we select at the start of REACCS will be the database in which we perform all REACCS operations in the session. However, we can search in more than one database with one query in one search if we set the Setting (see p. 592):
GlobalSearch...turns on/off automatic searching of multiple
       databases. The default is off.
 If GlobalSearch is on, REACCS performs the search with
 the current query over the databases specified by the

SEARCH TECHNIQUES

Setting DBaseList. The last database in the list of
databases then becomes the current database.
DBaseList.......sets list of databases for GlobalSearch
  System: Enter database name to add to search list:
      Search database 1 [CLF:]:
  User: <cr> (to leave CLF:, the default database, on list)
      THEIL: (to replace CLF: with THEIL:)

Options to use: <cr>.....moves down the existing list
            clear....erases the entire list
            delete...erases the default [db] from list
            show.....displays entire existing list
If the DBaseList in the Setting Window is asterisked, it indicates that a list of databases has been set in the current UPR.

8   MOLECULE SEARCHING

Molecule searching in REACCS is based on similar principles as in the systems described in Part One, Two, and Three.
Those readers who studied Part Three will find it very easy to master molecule search techniques in REACCS. In fact, REACCS offers options that have the same names, and methods that are very similar to those in MACCS-II.
The discussion on stereochemistry in MACCS-II applies in REACCS as well.
We can search in the entire database or in a subset of the database.
REACCS permits searching in more than one database in a single search (see p. 646).
Although it is possible to search for molecules with data queries (see p. 697), graphic structure queries will be again the major method to search for molecules.
REACCS recognizes stereochemistry so that we are able to perform searches for:
1. enantiomers
2. diastereomers
3. racemates
4. geometric isomers

REACCS also recognizes tautomeric structures and these can be searched for similarly to MACCS-II searches for tautomers.
However, there is no Parent Search (see p. 511) in REACCS.

Molecule searching can be classified as follows:
1. Exact Structure Search
2. Substructure Search (SSS)
3. Similarity Search

EXACT STRUCTURE SEARCHING

Exact Structure search is performed if we wish to retrieve:
1. an exact structure with no stereochemical specification
   The search will give a hit if the structure in the database is an exact match with the query structure, i.e., no substitution by atoms and/or structural fragments is allowed in the retrieved structure. The query structure cannot contain any variables (see also p. 503 for discussion of Exact Structure Search)
   A query with no stereobonds and no CHIRAL label will retrieve a structure with:
   a. no stereocenters
   b. stereocenters of unknown absolute or relative configuration
   c. one asymmetric center of mixed configuration (racemate)

2. an exact structure with absolute or relative configuration
The search will result in a hit if the structure in the database matches exactly the query structure including the specified configuration.

A query with one stereobond at an asymmetric center and no CHIRAL label will retrieve a racemate.

A query with R configuration at one center and the CHIRAL label will retrieve the R isomer.

A query with S configuration at one center and the CHIRAL label will retrieve the S isomer.

A query with two or more stereobonds and no CHIRAL label will retrieve a structure with the relative configuration.

A query with two or more stereobonds and the CHIRAL label will retrieve a structure with the absolute configuration.

Note that if we wish to find out if there are any isomers of the query structure with indicated configuration, we have to perform the search in Search Mode (see p. 651).

The queries are prepared in Build Mode. We exit to Main Mode by activating the MAIN option on the Menu.

The structure is transferred to Main Mode and it is the current structure for exact search which is initiated by the FIND CURRENT option (Figure 4/51).

The search in Main Mode is always automatically performed over the entire database. The search domain is indicated in the upper right box by "List 0 M" which stands for the entire database (Figure 4/51)

DISPLAY OF SEARCH RESULTS IN MAIN MODE

The result, if any, of the exact search in Main Mode is displayed automatically on the screen as the graphic structure. The registry number of the retrieved structure is shown in the upper right corner of the Menu. The data associated with the molecule are displayed in the current form (see Figure 4/52).

One of the boxes in Figure 4/52 does not show any data (and label). If a datatype (and label) has been assigned to a box but there is no data registered for that datatype in the database (or if permission to read data has not been granted to the User), the box will be empty. If we wish to display the structure only, we set the Setting (see p. 592):

DataDisplay....turns on/off display of data (default is on)

As with the other Settings, we can use the keyboard:
User: k
System: Main:
User: set datadisplay (toggles the Setting)

Example: *Exact structure search*

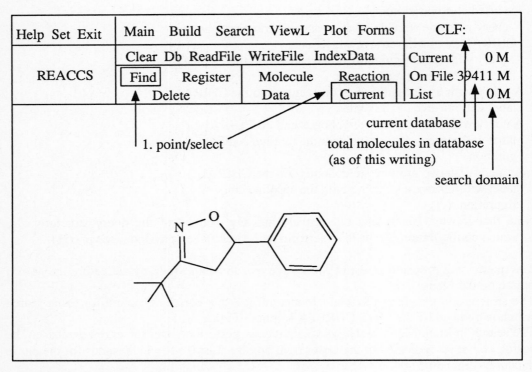

Figure 4/51. Exact structure search

Figure 4/53 shows exact search for the structure with absolute configuration.
Figure 4/54 shows exact search for structure with relative configuration.

SWITCHING DISPLAY FORMS

The Form whose filename is specified in the Settings will be used by REACCS for data display as the default. To select a Form file different from the default, set the Setting: FormFile......sets display formats for the current session
  System: Enter form file name:
  User: myform.frm

MOLECULE SEARCHING

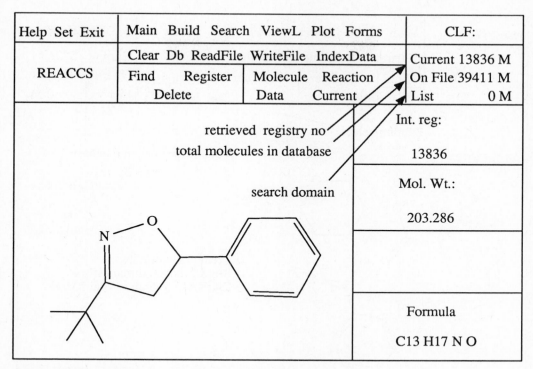

Figure 4/52. Retrived molecule and data

STEREOISOMER SEARCHING

We have discussed the search for exact structures with specified configuration (*vide supra*).
To perform a search for all stereoisomers of the query, we build the query structure with or without stereobonds in Build Mode and exit to Search Mode by activating the SEARCH option.
The structure is transferred to Search Mode and it is the current structure. The isomer search is activated by the keyboard option ISOMER (Figure 4/55).
The isomer search would retrieve the structures shown in Figure 4/56.

TAUTOMER SEARCHING

Tautomer search is performed in Search Mode. The tautomer search is, in fact, an Exact Structure Search. The query structure and the registered structure may differ in the position and type of bonds, stereochemical specifications, and the position of hydrogens, charges, and/or isotopes.
The search is initiated by the keyboard option TAUTOMER (Figure 4/57).

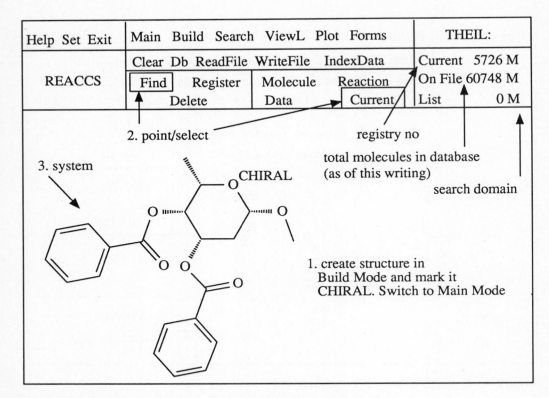

Figure 4/53. Exact structure search with absolute configuration

Tautomer search of the structure in Figure 4/57 would retrieve the structures shown in Figure 4/58.

DISPLAY OF SEARCH RESULTS IN SEARCH MODE

In Search Mode, the structure(s), if any are located in the database, are not displayed. The search result is reported by REACCS as shown in Figure 4/59. The hit list of the search in Figure 4/55 containing 2 retrieved records (Figure 4/56) is the current hit list L1 (for LIST 1).

To view the retrieved structures, we have to switch to ViewList Mode (Figure 4/60).

When we switch to ViewList Mode, the hit list (Ln) of the last search is transferred and becomes the Current List as indicated in the upper right box of the ViewList Menu. The structures in the List are ready to be displayed. To view the first item on the List, we point/select FIRST (Figure 4/61). The retrieved molecule is displayed with no data or with data in the current form depending whether the Setting DataDisplay is on or off (see p. 492, 649).

Note that it is always the last search hit list that is transferred to ViewList Mode (see also

MOLECULE SEARCHING

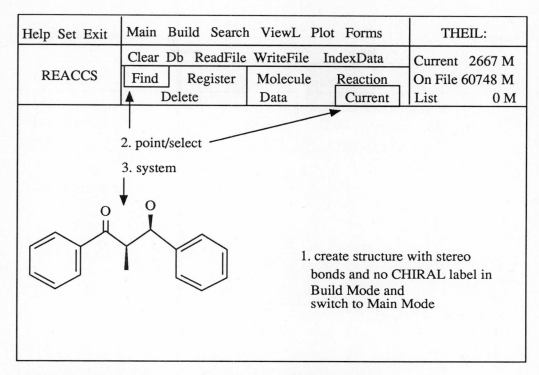

Figure 4/54. Searching exact structure with relative configuration

the discussion of LISTS, p. 658).

MORE ON DISPLAY OPTIONS IN VIEWLIST MODE

Figure 4/61 shows the option FIRST to view the first structure in the current hit list. The other graphic options are:
next......displays the next to the last displayed structure
         in the hit list
prev......displays the previous to the last displayed
         structure
item......displays an item of the hit list (e.g., 3 out of 5)
query.....displays the query structure used for the search
omit......removes the specified item(s) from the list and
         transfers it to a temporary omit list (see p. 585)
  To use the option, view the item, point/select OMIT. REACCS
  places the item into an omit list (Ln). To specify into
  which Ln the item is to be transferred, use the keyboard:
  User: k

Example: *Stereoisomer search*

Help Set Exit	Main Build Search	ViewL Plot Forms	THEIL:
REACCS	MolData Fmla RxnData SSS Similar RSS CurAsAgt CurAsPrd	Ref=List Db ReadList WriteList DeleteL IndexList	Current 0 M On File60748 M List 0 M

search domain (Reference List)

1. press k
2. system
3. type

Search: isomer

Figure 4/55. Searching for isomers

System: View:
User: omit L4
Omit lists can be displayed (listed) by point/select
INDEXLIST (*vide infra*). Omit list Ln can be used as
a List descriptor (see p. 639). Molecules and reactions
have separate omit lists. Omit lists from different databases cannot be mixed since they essentially refer to the
regnos of the database in which they have been created.

The following keyboard selection of options can be employed:
all.......displays all structures in the hit list one by one
        automatically with a time delay which can be set by
        the TIMER option
timer.....sets the length of time between displays by ALL
        (the default is 4 seconds)

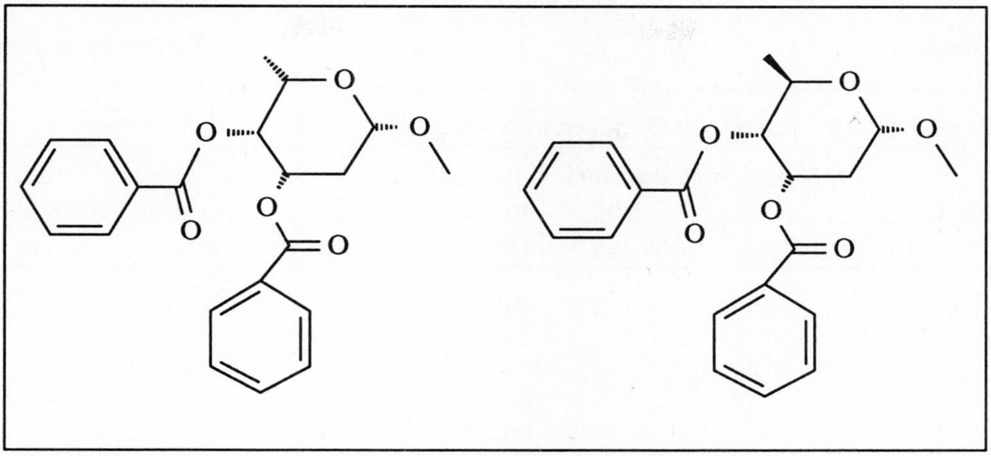

MOLECULE SEARCHING

Figure 4/56. Hits of query 4/55.

    press k
    System: View:
    User: timer
    System: wait time (sec):
    User: 15 (sets the time to 15 seconds between displays)
Note that we can set the display of simple-search hits so that we can see them already in the progress of the search. This might be useful if the search takes considerable time to completion and the number of hits is potentially large. If we see unfavorable hits, we can interrupt the search and reformulate the query.
The Setting (see p. 592):
  ViewHits......turns on/off display of simple-search results
        while a search is in progress.
    System: How should hits be displayed during search?
        Type a number of seconds displayed, FIRST for first
        hit only, or OFF for no display.
If the ViewHits Setting is "on," REACCS interrupts the automatic display upon completion of the search. We can then view the hit list as usual in ViewList Mode. If the ViewHits Setting is off, no hits are displayed during the search.

VIEWING HITS IN GLOBAL SEARCH

We have discussed the technique of searching in multiple databases with a query in one search (see p. 646).
REACCS reports the result of the search, e.g.:
8 rxns: Final. List is L3 (CLF:)
5 rxns: Final. List is L5 (JSM:)

Example: *Tautomer search*

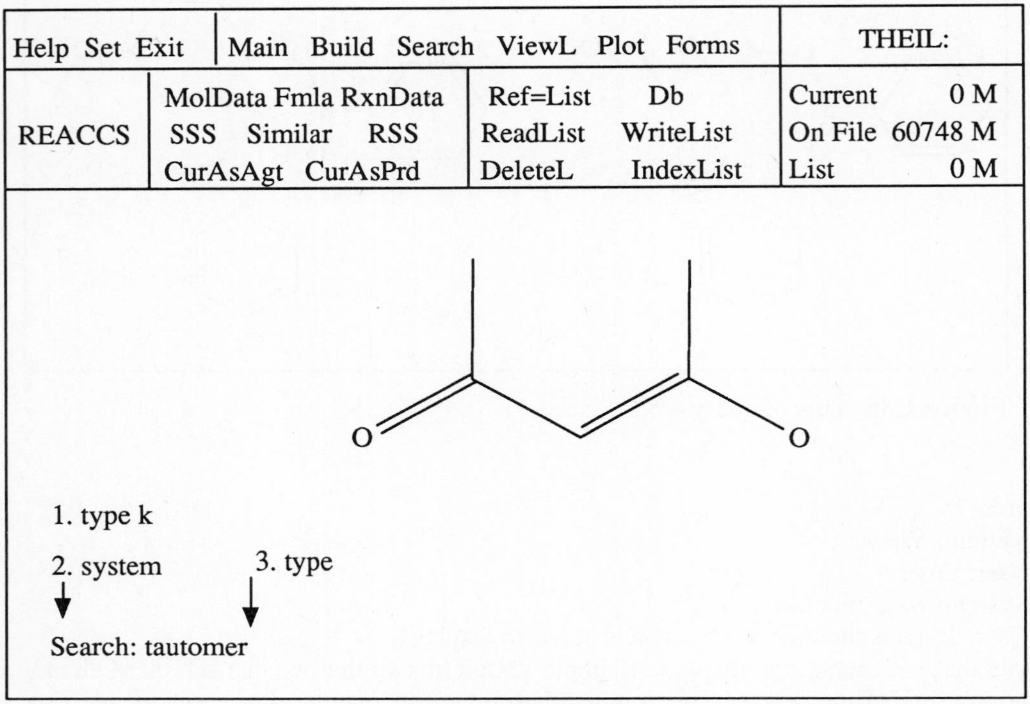

Help Set Exit	Main Build Search ViewL Plot Forms	THEIL:
REACCS	MolData Fmla RxnData SSS Similar RSS CurAsAgt CurAsPrd	Ref=List Db ReadList WriteList DeleteL IndexList	Current 0 M On File 60748 M List 0 M

1. type k
2. system
3. type

Search: tautomer

Figure 4/57. Tautomer search

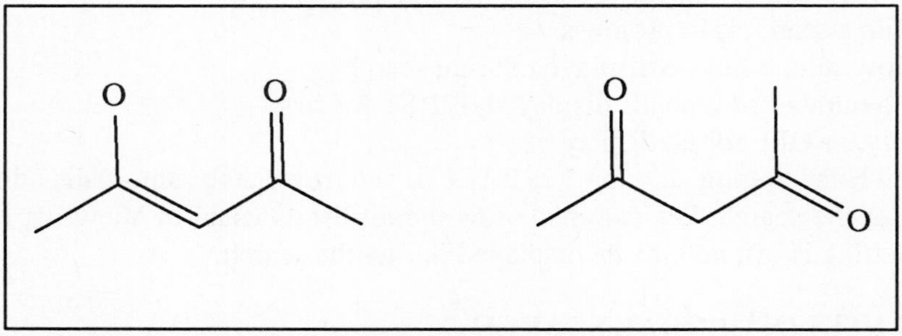

Figure 4/58. Tautomeric structures

To view the hits of this search, we switch to ViewList Mode and point/select LIST and at the prompt enter the Ln of the search, e.g.:
List descriptor: L3

MOLECULE SEARCHING

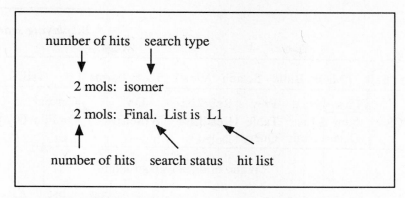

Figure 4/59. Search report

Example: *Switching to ViewList*

Help Set Exit	Main Build Search	ViewL Plot Forms	THEIL:
REACCS	MolData Fmla RxnData SSS Similar RSS CurAsAgt CurAsPrd	Ref=List Db ReadList WriteList DeleteL IndexList	Current 0 M On File 60748 M List 0 M

point/select

2 mols: isomer

2 mols: Final. List is L1

Figure 4/60. Entering ViewList Mode

Example: *Displaying search hits*

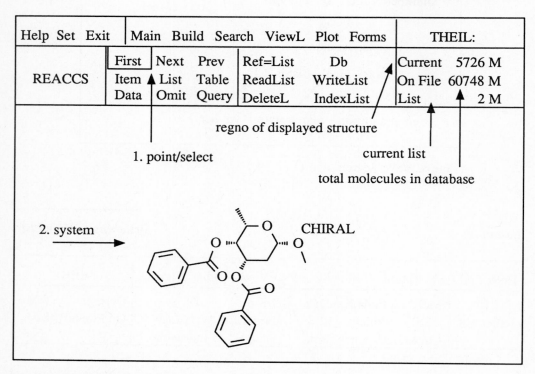

Figure 4/61. Viewing search hits

followed by any of the ViewList options.
REACCS will automatically get the hits from the database in which they have been retrieved.

WORKING WITH LISTS AND LISTFILES

There are two lists in REACCS which are monitored by the "List" indicator in the upper right corner of the Menus.
In the Main Menu and in the Search Menu the "List" refers to the Reference List (search domain), i.e., the number of records over which the search will be performed. If the indicator shows "List 0," the search domain (Reference List) is the entire database. In Main Mode, the search domain is always the entire database.
In Search Mode, the Reference List is by default the entire database. However, in Search Mode we can define a subset of the database to be the search domain and thus override the default. The Reference List (search domain) can be also defined in ViewList Mode.
In ViewList Menu, "List" indicator signifies current list (hits) of the most recent search.

MOLECULE SEARCHING

Defining the Reference List

A subset of the database can be arrived at by various means. For example, we can perform a formula search (see p. 701) over the entire database (List 0) to find molecules with a specified element count. The search will generate a hit list Ln. Or, we can perform a search with a broadly defined graphic structure query to generate the list Ln.

We then activate the REF=LIST option (Figure 4/62) to make the hit list the reference list (search domain) for the subsequent search. The search will be conducted only over the records in the hit list.

Example: *Setting search domain*

Help Set Exit	Main Build Search ViewL Plot Forms		THEIL:	
REACCS	MolData Fmla RxnData SSS Similar RSS CurAsAgt CurAsPrd	Ref=List ReadList DeleteL	Db WriteList IndexList	Current 0 M On File 60748 M List 0 M

1. point/select

2. system   3. type

List descriptor: L3

Figure 4/62. Establishing search domain

If the L3 temporary list contained 100 molecules, the search domain set up in Figure 4/62 would be 100 records and the count List 0 would change to List 100.

If we wanted to reset the search domain (the Reference List) again to the entire database, we would activate the DELETEL option and type "ref" (or L0) at the prompt (Figure 4/63). The List would then read 0. The same result is achieved by point/select REF=LIST to receive the prompt:

System: List descriptor

Example: *Defining reference list*

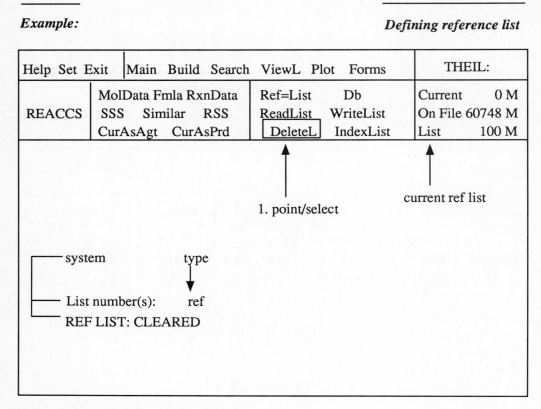

Figure 4/63. Defining reference list

User: m0 (or r0)
Similarly, we can define the Reference List in ViewList Mode by the same REF=LIST option. We receive the prompt:
System: List descriptor:
User: L3
We can also create the RefList in ViewList Mode by a search.
If we use the query language (see p. 637), we do not have to leave ViewList Mode to run the search. The search will be carried out in ViewList Mode and will result in the Reference List (Figure 4/64).
Note that if we are performing a Global Search (see p. 646) the Ref=List option is useful only if the List by which the reference is established is common for all the databases on the DBaselist (e.g., if we used regnos, they are not, usually, common to different databases). A Reference List can be, however, established for each database before we start the Global Search. During the search, with each database change the corresponding Reference List is used by REACCS.
Figure 4/65 shows examples of the status of "List" on moving from Search to ViewList

MOLECULE SEARCHING

Example: *Defining reference list*

Help Set Exit	Main Build Search ViewL Plot Forms	THEIL:
REACCS	First Next Prev — Ref=List — Db Item List Table — ReadList — WriteList Data Omit Query — DeleteL — IndexList	Current 0 M On File 60748 M List 0 M

1. point/select

```
         system            type
          │                 │
          ▼                 ▼
    ── List descriptor:  m12,14,25-34
    ── **PREPARING SEARCH QUERY
    ── **SEARCHING FILE
    ── [M]  11 hits                  ← established reference list
    ── Ref List [Mol]: 11 entries
```

Figure 4/64. Defining reference list

Mode.

Saving Hit List in Listfile

The temporary hit list Ln can be saved in a permanent Listfile by activating the WRITELIST option either in Search Mode or in ViewList Mode (Figure 4/66).
Note that in writing this list we had the hit list generated by the previous search. However, it is possible to write a list by direct input of the search query after the prompt:
List descriptor:
(see Search Techniques, p. 637).

Recalling Saved Listfiles

If we save a hit list in a Listfile, we can recall it in Search Mode or in ViewList Mode by the READLIST option. When we bring the hit list into the system from the Listfile, the list is assigned an Ln by REACCS to be able to work with the hit list (Figure 4/67) (compare

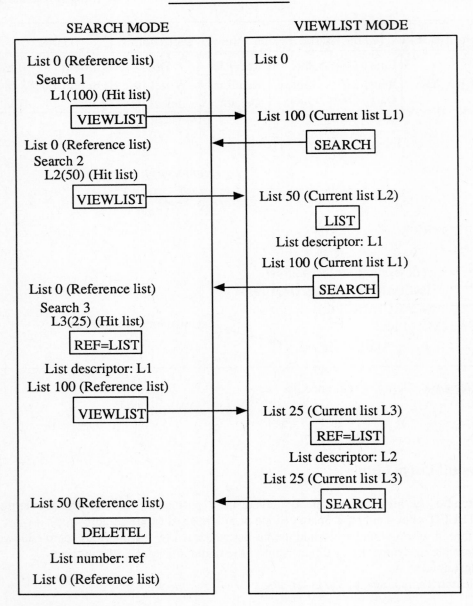

Figure 4/65. List status

SAVE and ACTIVATE in CAS REGISTRY, p. 240). The Ln is then used in subsequent operations, e.g., to view the hits in the list, we make it the Current List by activating the

MOLECULE SEARCHING

Example: *Saving hit list in listfile*

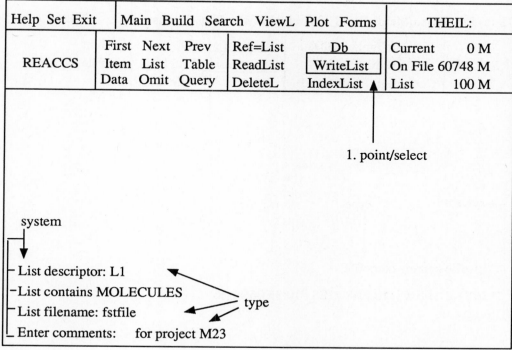

Figure 4/66. Saving list in listfile

LIST option (see p. 665).

Checking Hit Lists in Current Session

The Ln(s) that have been generated in the current session can be displayed by the INDEXLIST option both in Search Mode and in ViewList Mode. REACCS displays the lists, e.g.:

```
THEIL:
-----------------------------------------------
LIST 1       (RXNS)        12 ENTRIES   14:19   06/21/88
     RSS
LIST 3       (RXNS)         3 ENTRIES   14:21   06/21/88
     RSS
```

Example: *Recalling ListFile*

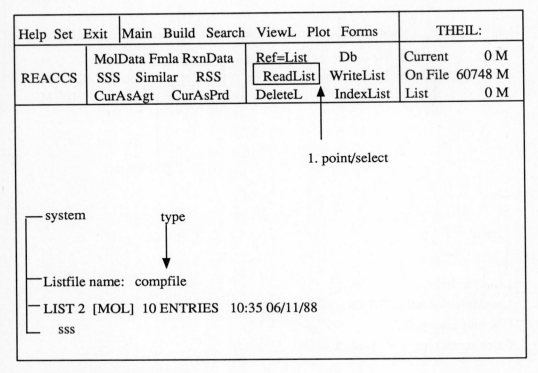

Figure 4/67. Reading a listfile

```
JSM:
----------------------------------------------------------
LIST 2          (RXNS)          5 ENTRIES    14:22  06/21/88
     RSS
```
Summary by database:
  2 lists for THEIL:
  1 list for JSM:

Deleting Lists

To delete temporary hit lists (i.e., the Ln designated lists), either in Search Mode or in ViewList Mode, we point/select DELETEL and respond to the prompt:
List number(s):
by the Ln number(s) we wish to delete. The Ln numbers can be input in the following formats:

MOLECULE SEARCHING

Ln
Ln,Ln,Ln....
Ln,Ln-Ln
Ln,Ln-Ln,Ln,Ln.....
Ln-Ln,Ln......
(We can also enter "ref" (or L0) to reset current Reference List to the entire database.)

Defining Current List

Similarly to defining a Reference List (see Figure 4/64), we can define the Current List in ViewList Mode. If we want to view a hit list from other than the last search, we point/select LIST and at the prompt:
List descriptor:
enter the Ln we wish to view.
The Current List can be also defined by a search in ViewList Mode. For example, the list can be created by entering registry numbers (Figure 4/68) or by other query language technique (see p. 637).

SUBSTRUCTURE SEARCHING (SSS)

Molecular Substructure Search (SSS) in REACCS is based on the same premises as we have discussed in Part Two (see p. 380).
The substructure search query is prepared in Build Mode if the substructure does not contain any variables and in Query Options Mode if we want to specify variables in the query structure. The SSS search is carried out in Search Mode (Figure 4/69).
We can check on the progress of the search by typing "?" at any point while REACCS is running the search. REACCS will report:

56 HITS SSS 35% COMPLETE ***(SEARCH CONTINUING)

If we for any reason want to stop the search, we type "!." If we wish to have the hits displayed during the search, we set the Setting ViewHits on (see p. 592).
When the final report is issued by REACCS, we switch to ViewList Mode to view the result of the search (Figure 4/70).

STEREOISOMERS IN SSS

We have discussed searching for exact structures with absolute or relative configuration (see p. 648).
In substructure searching we can search only for relative configuration. The query substructure is created in Build Mode/Query options as usual marking the stereocenters by the STEREO option (see p. 604). Note that we have to mark at least two stereocenters or both ends of the double bond in the query to retrieve structures with the same relative configuration. If we mark only one stereocenter, or if we do not mark the centers, any

Example: *Defining current list*

Help Set Exit	Main Build Search ViewL Plot Forms	CLF:
REACCS	First Next Prev Ref=List Db Item ⎣List⎦ Table ReadList WriteList Data ▲Omit Query DeleteL IndexList	Current 0 M On File 39411 M List 0 M

               1. point/select

- system
- Current List: 1   Item 0   type
- List descriptor: m78-130
- 53 mols: Final. List is L2

Figure 4/68. Defining hit list

configurational isomer will be retrieved. The bonds at the stereocenters have to be marked UP, DOWN, or EITHER before marking the stereocenters. The search is executed in Search Mode by the SSS option. It results in retrieval of structures containing the fragment matching the query and the configuration at the marked stereocenters of the query. The structures with unspecified configuration will not be retrieved. If we wish to search the database with a query with marked stereocenters and wish to retrieve structures with matching configuration as well as structures with no specified configuration, we set the Setting (see p. 592):

Stereosearch.....sets on/off search for stereochemical
        structures in SSS (and RSS, see p. 678)
   The default is "on", i.e. only structures with marked
   configuration will be searched for
   If the Setting is off, the subsequent searches will
   retrieve both stereo- and non-stereochemical structures.

Figure 4/71 shows Substructure Search for structures with relative configuration.
The result of the search in Figure 4/71 is 49 structures containing the fragment with the

Example: *Initiating molecule substructure search*

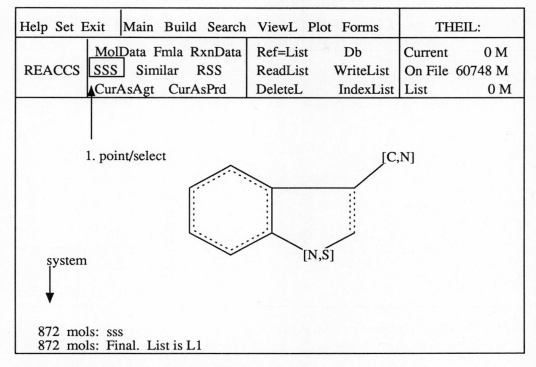

Figure 4/69. Searching with generic query

configuration matching the fragment of the query.
When we switch to ViewList Mode, the Current List will contain 49 items (List 49) (if no previous search was performed). Activating the FIRST option will show the retrieved record (Figure 4/72).
Figure 4/73 shows SSS with substructure query of relative configuration at the double bond.
One of the retrieved structures of the search in Figure 4/73 is shown in Figure 4/74.

SEARCHING MOLECULES IN DATABASE SUBSET

As we perform searches for molecules in REACCS, we find out that in some cases the initial query has been formulated in such a way that the search result is very broad giving either too many hits or retrieving structures which may satisfy some structural requirements but still have too many irrelevant fragments which we are not interested in. Or, the retrieved structures may have irrelevant associated data to satisfy our search goal.

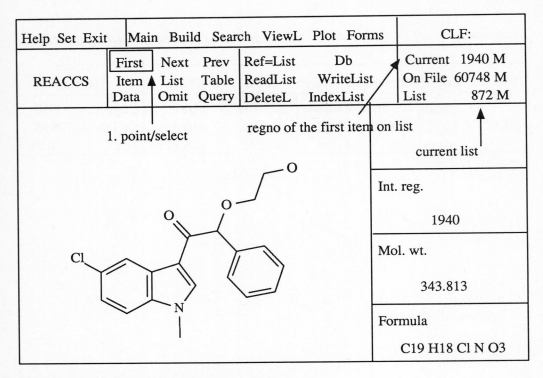

Figure 70. Displaying the first item in hit list

To eliminate these unwanted hits, we can employ techniques by which we narrow the result of the search.

We have learned how to define the Reference List which is the search domain (see p. 659). Therefore, we can make any hit list of a search the database subset over which the subsequent search can be carried out with a refined, more precise query that will eliminate the unwanted hits of the first search.

A database subset (compare "Subset Search in Reaction Searching," p. 675) can be defined by either a graphic query or an alphanumeric query. For example, we have a defined datatype in the database for the names of the registered structures as follows:
mol:name

If we first perform a search, e.g.:

System: search:

User: mol:name = @benzothiazin@

REACCS creates a hit list which contains all structures registered with the embedded name "benzothiazin." We then make the hit list the Reference List (search domain) for the subsequent search with a benzothiazine structure query to retrieve benzothiazines with the specified structural features. Or, we can use a formula search (see p. 701), reaction keyword (see p. 691), or reaction symbol (see p. 689) query to define a database subset.

MOLECULE SEARCHING

Example: *SSS for relative configuration*

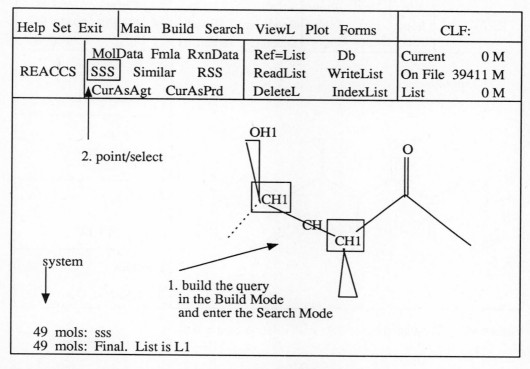

Figure 4/71. Substructure search for relative configuration

COMBINING LISTS

In the section "Search Techniques," we have shown how lists can be combined using the Boolean Operators, to merge, intersect, and/or subtract lists of registry numbers, i.e., the results of various searches. This technique can also be used to make our search more selective by adding and/or eliminating records from several searches in the same database. It also enables us to perform comparative searches over more than one database, assuming that there is correlation of the regnos. For example, we have hit list L1 from the current search in a personal database and we have a Listfile saved in a master/corporate database. We can compare and/or eliminate entries in the lists, e.g.:
System: List descriptor:
User: L1 and mylist.lst
This query will create a list which will have common regnos of L1 and mylist.lst file.

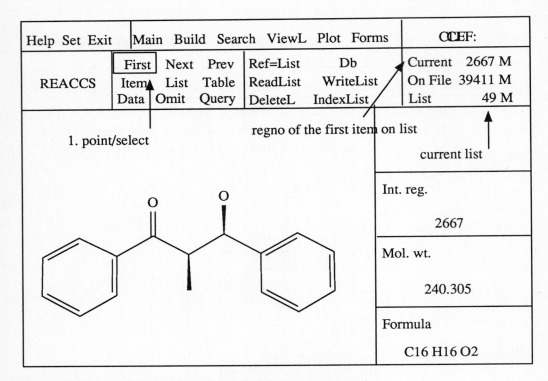

Figure 4/72. Displayed hit of SSS with relative configuration

COMBINATION SEARCHES

Selective searching can be also performed in one search by using the technique of combination searching.
The graphic structure query is not the only criteria we can specify for the search. As we have discussed in the section "Search Techniques" (see p. 637), we can employ the combination search to include additional conditions which the search has to satisfy to produce a result.
For example, we have a corporation database in which there are molecules synthesized by various departments in the organization. We have a datatype in the database:
MOL:DEPT:CHEMIST
We wish to perform a SSS but desire to retrieve only those molecules prepared in one of the departments by chemist XYZ.
We prepare the query structure and futher introduce into the query additional criteria:
System: search:
User: sss=current and dept:chemist = XYZ
REACCS performs the substructure search with the current structure and finds only those molecules synthesized by the chemist XYZ.

MOLECULE SEARCHING

Example: *SSS for relative configuration*

Help Set Exit	Main Build Search	ViewL Plot Forms	THEIL:
REACCS	MolData Fmla RxnData SSS Similar RSS CurAsAgt CurAsPrd	Ref=List Db ReadList WriteList DeleteL IndexList	Current 0 M On File 60748 M List 0 M

2. point/select

system

1. build the query
   in the Build Mode
   and enter the Search Mode

20 mols: sss
20 mols: Final. List is L3

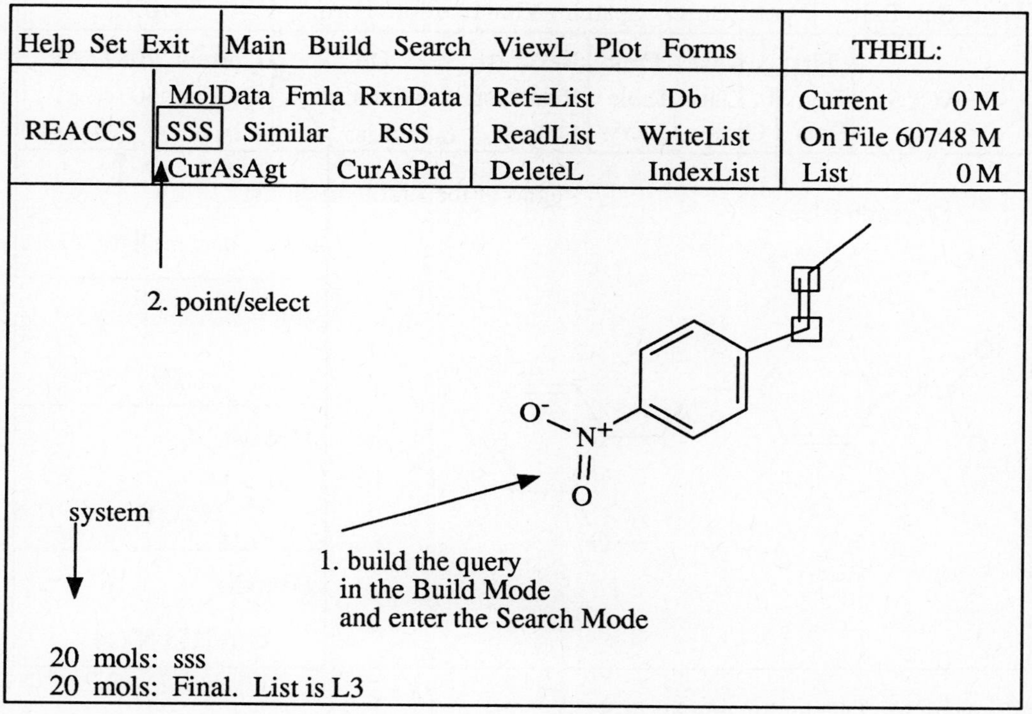

Figure 4/73. SSS for relative configuration

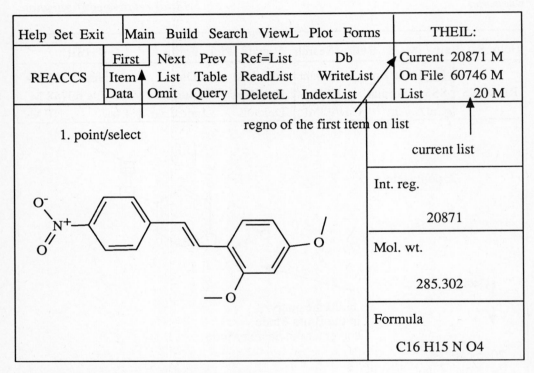

Figure 4/74. Displaying one hit of SSS

SIMILARITY SEARCH

We have discussed molecule searches with different degree of selectivity:
1. Exact Search in which the query structure must be an exact match with the entire structure of the hit
2. Substructure Search in which the query structure must be a fragment contained in the hits

We can perform yet another search with the least selectivity, i.e., the similarity search in which it is sufficient that a predefined percentage of keys in the query match those in the hits. The search technique will be discussed in detail in the section on "Reaction Searching."

9 REACTION SEARCHING

Reaction searching in REACCS can be carried out using a variety of methods. We have discussed the search methods which can be applied for molecule searching. Some of the techniques are applicable in reaction searching as well.
We now discuss the methods for reaction searching and the type of queries to retrieve reactions.
We can classify the search for reactions as follows:
1. Exact Reaction Search
2. Substructure Reaction Search (RSS)
   a. RSS without reaction center search
   b. RSS with reaction center search
3. Similarity Search

EXACT REACTION SEARCH

As in molecule searching, the Exact Reaction Search returns reactions in which the structures match exactly those in the query. Consequently, no variables are allowed in the query. The molecules in reactions will have the exact role in the reaction as we assign it in the query.
The Exact Reaction Search can be performed by:
1. Simple Search
2. Subset Search
3. Conversion Search
4. Combination Search
The difference between these searches is in the nature of the query.

Simple Search

Simple Search for an exact reaction is performed in Main Mode with a query which is the full reaction scheme containing all the exact structures involved in the reaction in their proper position in the scheme.
The reaction query is created as usual in Build Mode (see p. 608)(or we can recall a reaction from a RxnFile by READFILE). We then transfer into Main Mode where the query is now the current reaction. The search is initiated by the FIND CURRENT option (Figure 4/75). Note that this is the same option as for molecule Exact Search. REACCS knows by the query that we want to search for reactions.
Exact Reaction Search produces a hit which is displayed on the screen as the graphic reaction scheme (Figure 4/75) (note that for this and subsequent displays, the Setting DataDisplay (see p. 649) has been turned off).

Example: *Initiating exact reaction search*

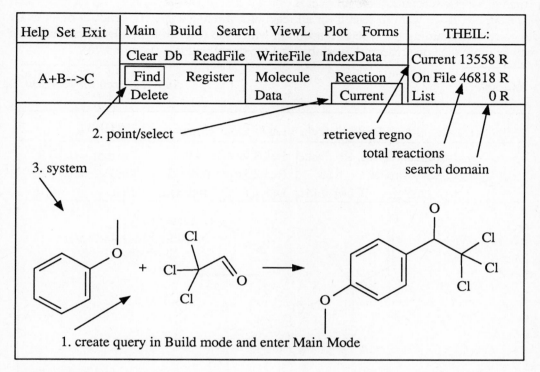

Figure 4/75. Searching exact reaction

Subset Search

Subset Search is a search with queries containing exact structures of reactant(s), product(s), or both. The query structures cannot contain any variables. The query can be formulated in the following variations:
1. A---->
2. ----->B
3. A---->B

Subset Search is performed in Search Mode and is initiated by the keyboard selection SUBSET.
The subset search will retrieve the reaction(s) which contains, at the minimum, the following structures:
1. for query A----->:
   the reactant A but may contain more than the single reactant A
2. for query ----->B
   the product B but may contain more than the single product B

3. for query A----->B
   the reactant A and the product B but may contain more
   than the single reactant A and product B

For example, if we want to search for all reactions utilizing a specific reactant of an exact structure (disregarding the type of reaction in which the reactant is involved and that there may be more than one reactant), we formulate the query option 1 (Figure 4/76).

Example: ***Subset search for reactant***

Help Set Exit	Main Build Search ViewL Plot Forms	THEIL:
A-->	MolData Fmla RxnData Ref=List Db SSS Similar RSS ReadList WriteList CurAsAgt CurAsPrd DeleteL IndexList	Current 0 R On File 46818 R List 0 R

```
                O
                ‖
                       Cl              1. create query
             ╲ ╱                          in Build Mode and enter
              C                           Search Mode
             ╱ ╲
                  Cl
           Cl

  2. type k         4. type
  3. system           ↓
      ↓
   Search:          subset
```

Figure 4/76. Subset searching

The Subset Search in Figure 4/76 would retrieve all reactions that employ the reactant specified in the reaction query, e.g., the reaction shown in Figure 4/77.

If we wanted to search for all reactions (disregarding the reaction type) producing a specific product of an exact structure, we would do the subset search with the reaction query -->B where B would be the exact structure of the product. For example, the product-producing reaction in Figure 4/77 would be retrieved with the following query (Figure 4/78).

If we wanted to search for reactions utilizing a specific reactant of an exact structure and producing a specific product of an exact structure (disregarding the reaction type), we would formulate a query with both the reactant and the product. The reactions would be

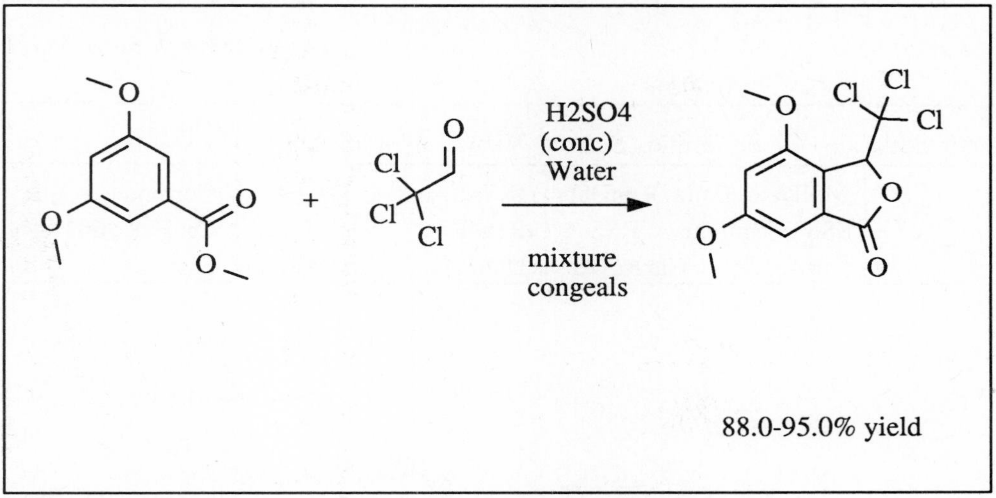

Figure 4/77. Hit of subset search 4/76.

retrieved in which at least the specified reactant would be a participant (there may be additional reactants) and at least the specified product would be produced (there may be other products produced as well). To retrieve the reaction in Figure 4/77 by this kind of subset search, we would formulate the query as shown in Figure 4/79 or the query shown in Figure 4/80.

Conversion Search

Conversion Search (see also Search Techniques, p. 637) allows for performing a search for reactions using a molecule query.
In REACCS, a search with a molecule query is, by implication, the molecule search. A search with a reaction query is, by implication, the reaction search. If we submit a molecule query to reaction search, REACCS issues an error message and does not perform the search.
The Conversion Search converts the molecule into a reaction query. The Conversion Search permits searches as in the Subset Search (*vide supra*), i.e., searches for reactants and/or products of exact structures disregarding the type of reaction in which they are involved.
Conversion Search is initiated by the graphic options CurAsAgt (Current As Agent) and CurAsPrd (Current As Product). The meaning of Agent in the CurAsAgt option is that of a reactant, a catalyst, or a solvent. Conversion Search can be also initiated by the keyboard selection, e.g.:
Search: current as reactant
Search: current as catalyst
Search: current as solvent

Example: *Subset search for product*

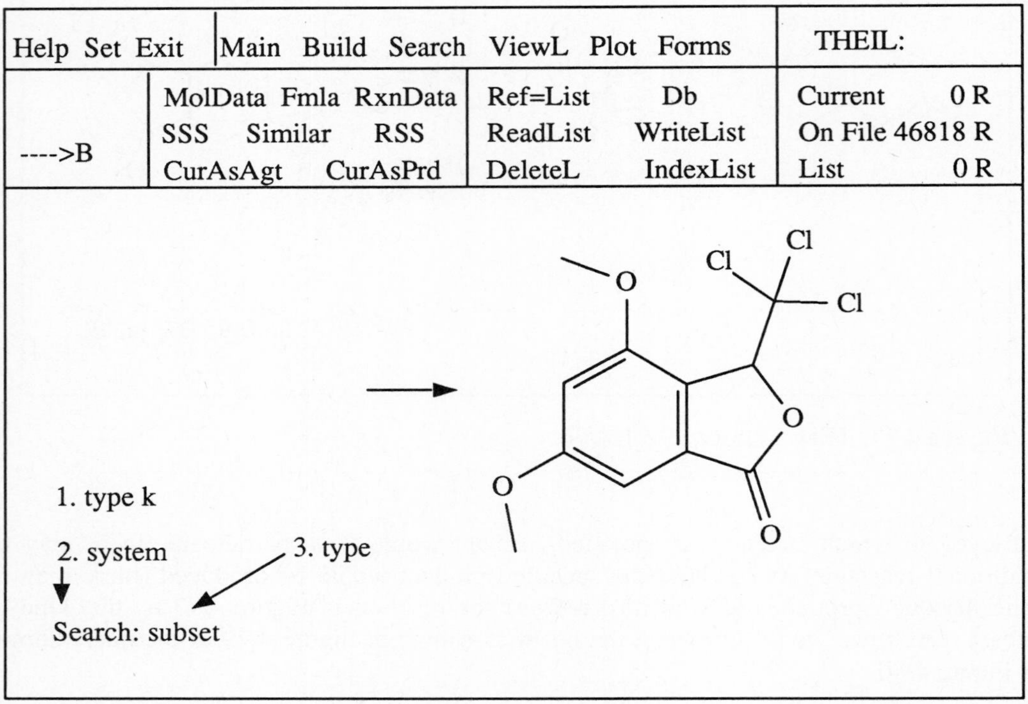

Help Set Exit	Main Build Search ViewL Plot Forms	THEIL:
---->B	MolData Fmla RxnData Ref=List Db SSS Similar RSS ReadList WriteList CurAsAgt CurAsPrd DeleteL IndexList	Current 0 R On File 46818 R List 0 R

1. type k
2. system
3. type

Search: subset

Figure 4/78. Subset search

Search: current as product
Search: current as agent
Thus, the reaction in Figure 4/77 can be searched in the conversion search as shown in Figures 4/81 and 4/82.
REACCS reports on the Conversion Search as shown in Figure 4/83.

STEREOISOMERS IN EXACT REACTION SEARCHING

We have learned that in molecule Exact Search we can search for absolute or relative configuration. Conversion Search enables us to similarly search for exact reactions involving molecules with specified absolute (Figure 4/84) or relative configuration.

SUBSTRUCTURE REACTION SEARCH

Substructure Reaction Search (RSS) will be probably the most used technique for reaction

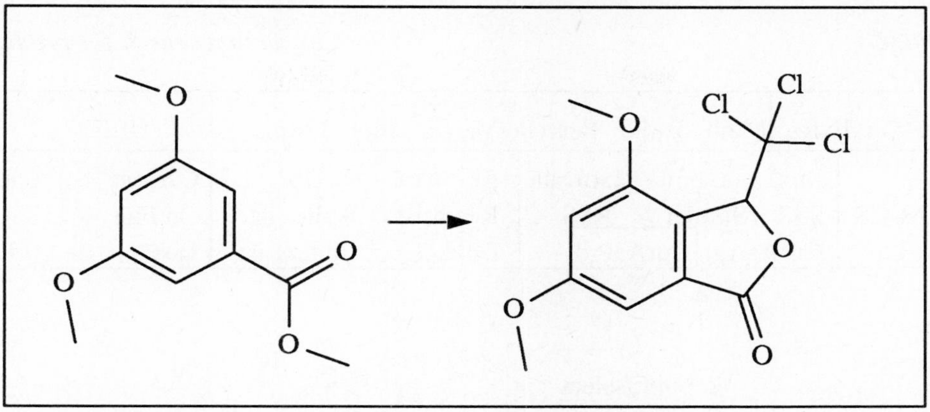

Figure 4/79. Query for subset search

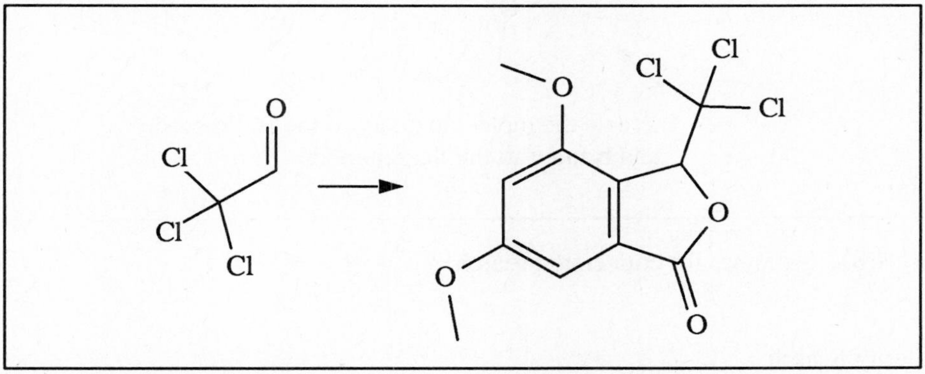

Figure 4/80. Query for subset search

searching. RSS is performed in Search Mode.

RSS retrieves reactions in which the structures of the molecules participating in the reaction, at a minimum, match the structures of the query. In addition, as was the case with the molecule Substructure Search (SSS) (see p. 665), the reactions will be retrieved in which the participating molecule(s) will contain the query fragment (see also the discussion on substructure searching p. 380).

The query for reaction substructure search can, of course, contain variables.

In RSS, we can employ the following type of search:
1. Simple Search
2. Conversion Search
3. Combination Search

Example: *Conversion search for product*

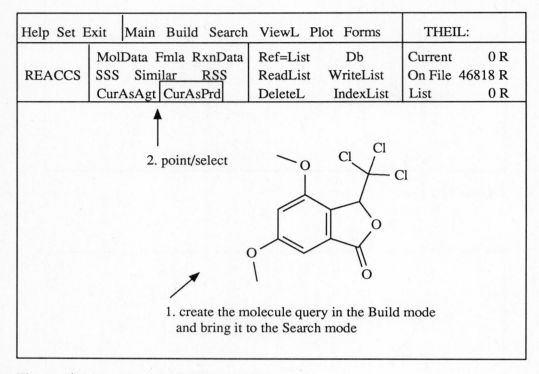

Figure 4/81. Structure for conversion search

4. Similarity Search

Simple Search

The query for a Simple Search can be composed of the substructures of:
1. reactant (A -->)
2. product (--> B)
3. reactant and product (A --> B)

If we were interested to find all reactions with molecules in which dihydroisoxazole is the fragment/reactant, the query in Figure 4/85 would retrieve, among others, the reaction in Figure 4/86 in which the dihydroisoxazole ring is opened to give beta-hydroxyketones. Note that any substitution is allowed in the RSS search unless we prevent it by explicit hydrogens or by the HYDROGEN option when building the query.
Should we be interested only in the formation of beta-hydroxyketones disregarding the starting material, we would formulate the query as shown in Figure 4/87.

Example: *Conversion search for reactant*

Help Set Exit	Main Build Search ViewL Plot Forms	THEIL:
REACCS	MolData Fmla RxnData Ref=List Db	Current 0 R
	SSS Similar RSS ReadList WriteList	On File 46818 R
	CurAsAgt CurAsPrd DeleteL IndexList	List 0 R

2. point/select

1. create the molecule query in the Build mode and bring it to the Search mode

Figure 4/82. Query for conversion search

Conversion Search

Conversion Search in reaction substructure searching is a shortcut to RSS search. We use a molecular substructure query instead of a reaction query (see also the discussion of Exact Search, *vide supra*).

The Conversion Search in Figure 4/88 is similar to the RSS search in Figure 4/85. It will retrieve all reactions in which the dihydroisoxazole structural fragment is in the structures of the reactants.

STEREOISOMERS IN SUBSTRUCTURE REACTION SEARCHING

In reaction substructure searching we can search only for relative configuration by performing simple or conversion search (Figure 4/89) with a molecule substructure query (see also the discussion on stereoisomers in Molecule Searching, p. 648).

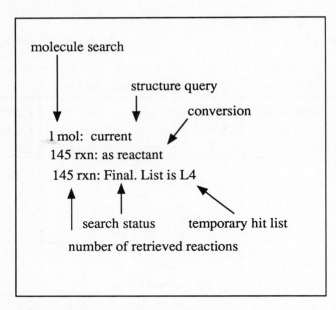

Figure 4/83. Report on conversion search

REACTION CENTER SEARCHING

In the examples we have discussed in RSS search (*vide supra*), we have not been concerned about the type of reaction we have been searching for. All we wanted to know was whether a substance represented by a structural fragment was involved in a reaction either as a reactant or as a product.

To make RSS more specific and to orient the search toward reaction type searching, we employ the Reaction Center and Atom/Atom Mapping search techniques.

In searching for reactions to find methods for converting specific functional groups into another specific functional groups or to convert a specific fragment in a structure participating in the reaction into another specific fragment, we employ the REACCS capability of searching for reaction centers.

Searches for reaction centers are executed in Search Mode by activating the option:
RSS (Reaction Substructure Search)
  performs the search with a query substructure(s) in which
  the bonds that change in the reaction are marked using
  the CENTER option in Build Mode (see p. 610), or the query
  is atom/atom mapped by AUTOMAPS (marking of bonds is
  automatic) or by MARKMAP (bonds are marked by CENTER
  in the HighlightRxn of Build Mode) (see p. 619)

For example, to search for selective reduction of aldehyde group in the presence of keto

REACTION SEARCHING

Example: *Stereoisomer searching in reactions*

Help Set Exit	Main Build Search	ViewL Plot Forms	CLF:
REACCS	MolData Fmla RxnData SSS Similar RSS CurAsAgt CurAsPrd	Ref=List Db ReadList WriteList DeleteL IndexList	Current 0 M On File 39411 M List 0 M

1. create query in Build Mode and exit to Search

CHIRAL

2. type k

3. system   4. type

Search: current as product

Figure 4/84. Searching for absolute configuration

group in steroids we build the query (marking the center (#) and notcenter (X) bonds) and enter Search Mode to activate the RSS option (Figure 4/90).

The query in Figure 4/90 would retrieve the reactions (one of them is shown in Figure 4/91) for reducing aldehyde group with the keto group intact.

To search generically for selective reduction of aldehydes in the presence of keto group (or for similar type of reactions), we can formulate the substructure query as shown in Figure 4/92. The query represents an aldehyde and a keto group in any environment, i.e., in any structure that is registered in the database while the query in Figure 4/90 limits the structures to the steroids and the locants indicated in the structures. Note that the query in Figure 4/92 is built by creating the two disconnected fragments at each side of the arrow as one structure and labeling the pair, in turn, as reactant and product.

The query in Figure 4/92 in RSS will retrieve the reaction in Figure 4/91 but also all other structures with aldehyde group selectively reduced, e.g., the reaction in Figure 4/93.

Example: *Reaction substructure search*

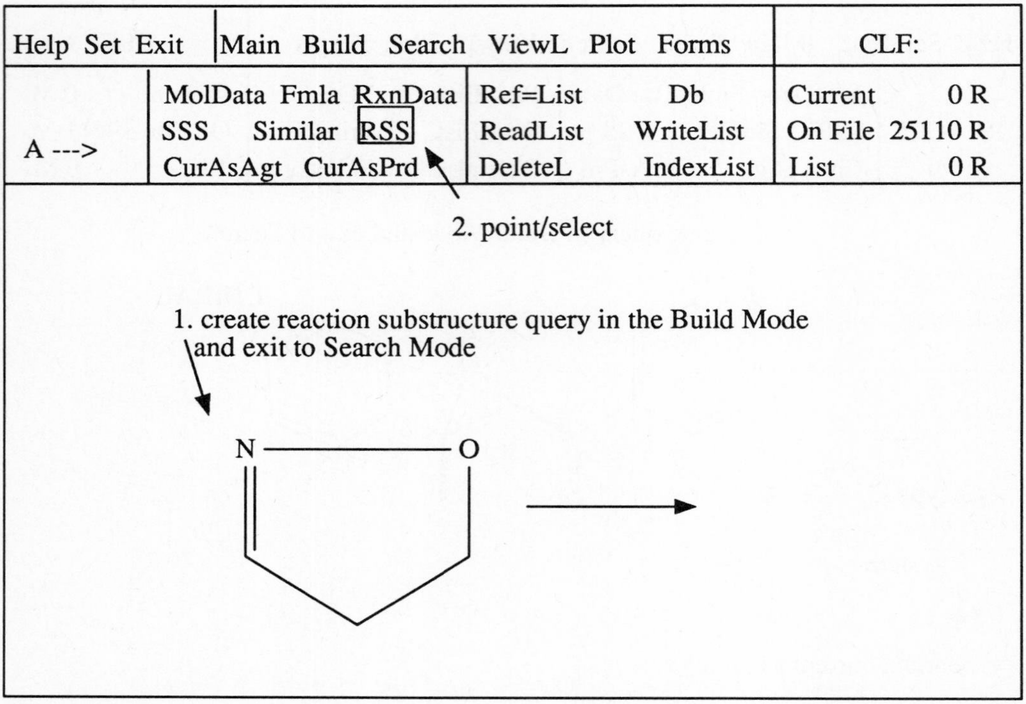

Help Set Exit	Main Build Search ViewL Plot Forms	CLF:
A --->	MolData Fmla RxnData Ref=List Db SSS Similar RSS ReadList WriteList CurAsAgt CurAsPrd DeleteL IndexList	Current 0 R On File 25110 R List 0 R

2. point/select

1. create reaction substructure query in the Build Mode and exit to Search Mode

Figure 4/85. Substructure reaction search

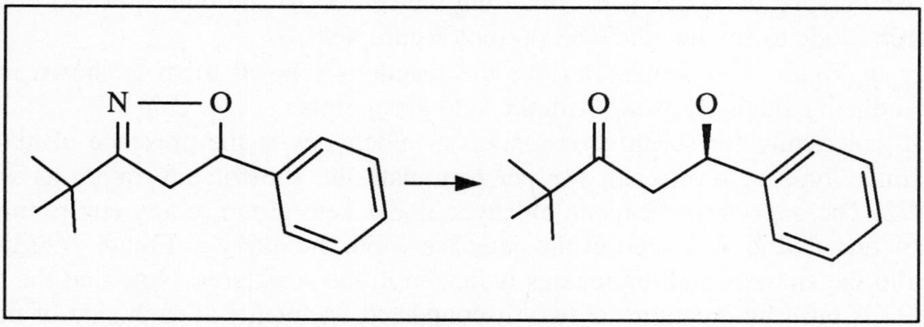

Figure 4/86. Retrieved reaction

REACTION SEARCHING

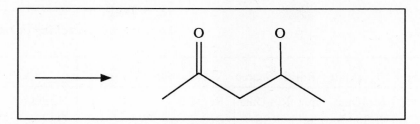

Figure 4/87. Reaction query structure

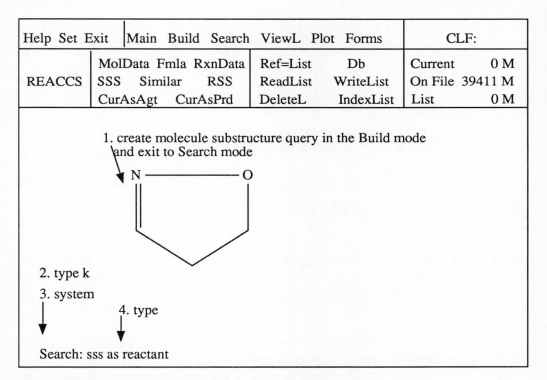

Figure 4/88. Conversion substructure search

ATOM/ATOM MAPPING IN RSS

Atom/atom mapping (see p. 616) greatly enhances selectivity and precision of RSS. For example, to search for regioselective reduction of dinitro coumpounds, we use

Example: *Stereoisomer searching in reactions*

Help Set Exit	Main Build Search	ViewL Plot Forms	CLF:
REACCS	MolData Fmla RxnData SSS Similar RSS CurAsAgt CurAsPrd	Ref=List Db ReadList WriteList DeleteL IndexList	Current 0 M On File 39411 M List 0 M

1. create query in Build Mode and exit to Search

2. type k

3. system   4. type

Search: sss as product

Figure 4/89. RSS for relative configuration

atom/atom mapping for the query in Figure 4/94. The mapped atoms .5., .6., and .7. in the reactant will correspond to the .5., .6., and .7. atoms in the product and the .8., .11., and .12. atoms in the reactant will be unchanged.

The query in Figure 4/94 in RSS would retrieve 7 reactions in the THEIL database one of which is shown in Figure 4/95.

Similarly, the Si-directed Nazarov cyclization can be searched for in RSS with the query consisting of the atom/atom mapped reacting fragments (Figure 4/96).

The query in Figure 4/96 would retrieve in the CLF database the reaction shown in Figure 4/97.

Note that if we atom/atom map a query, we do not have to mark the reaction center, although we can do so. The mapped atoms will be in the reaction center if the reaction involves those atoms and if we correctly map the atoms in the reactant and the product.

If we do not use atom/atom mapping and/or reaction center bond marking, we can still perform RSS search. However, we retrieve also reactions in which the structures will contain the query fragment(s) or functional groups no matter whether they have been

REACTION SEARCHING 687

Example: *Reacting center search - marked query*

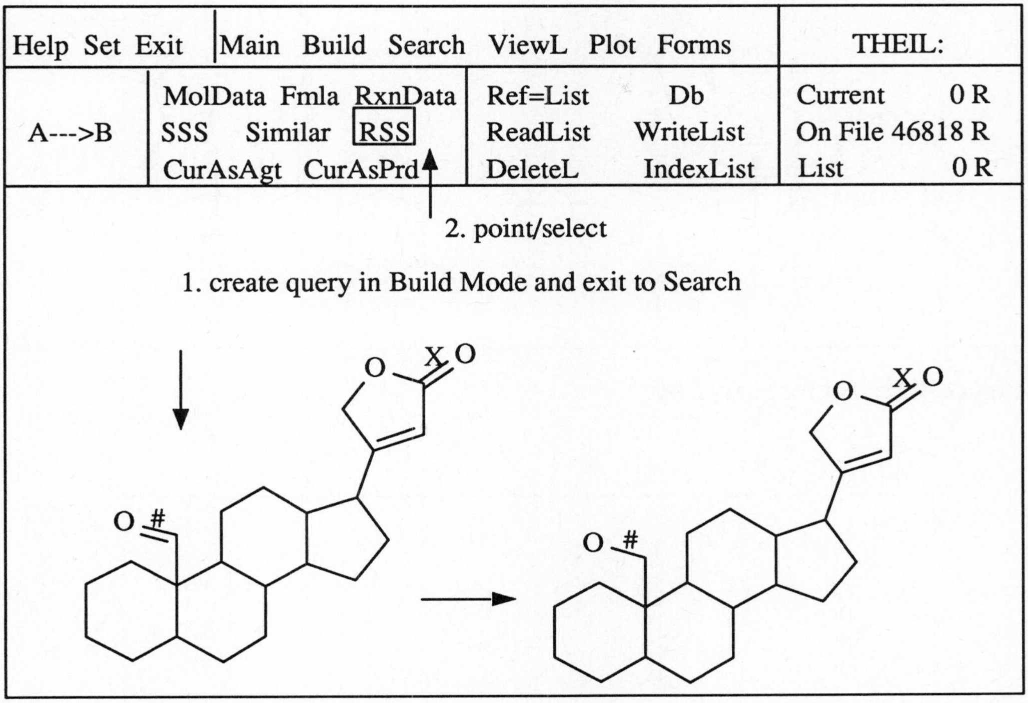

Help Set Exit	Main Build Search ViewL Plot Forms	THEIL:
A--->B	MolData Fmla RxnData Ref=List Db SSS Similar |RSS| ReadList WriteList CurAsAgt CurAsPrd DeleteL IndexList	Current 0 R On File 46818 R List 0 R

                           2. point/select

1. create query in Build Mode and exit to Search

Figure 4/90. Reaction center search

formed/changed in the reactions or have been already present in the reactants or products or remained unchanged.

Figure 4/97-A shows additional examples of atom/atom mapped queries for RSS searching of reactions involving rearrangement, ring expansion, and ring contraction. Figure 4/97-B shows the reactions retrieved with these queries in the CLF database.

COMBINATION SEARCHES

In the examples of RSS searching, we have shown searches with simple graphic reaction queries. In the Conversion Searches, we have shown examples of reaction searches using graphic molecule queries to search for reactions. The Conversion Search for reactants and/or products is, in effect, similar to RSS search using a reaction query in which we specify only the structure of the reactant or product with no reaction centers or atom/atom mappings.

In searching for reactions by the SSS Conversion Search we are frequently confronted by

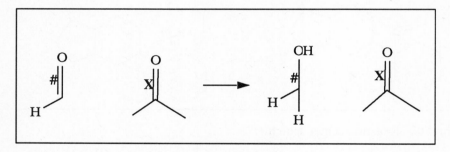

Figure 4/91. Hit for query 4/90.

Figure 4/92. Generic reaction center query

the need to selectively retrieve reactions in which the structural fragment of the reactant is changed in the reaction.

For this purpose, the Combination Search technique (see also p. 645) allows for refining our search queries. For example, the query in Figure 4/88 will retrieve reactions, in which dihydroisoxazoles were the reactants. However, in addition to the ring-opening reactions, there will be those in which the isoxazole ring will be unchanged and will appear in the reaction product structures. We can prevent it by formulating the combination search query for the substructure in Figure 4/88 as follows:

Search: sss as reactant not sss as product

In other searches, we wish to retrieve reactions in which a structural fragment (product) is formed rather than the fragment being already present in the reactant. For example, to search for all methods by which the benzoxazine system is constructed but to eliminate the reactions which use benzoxazines as the starting material we create the query and

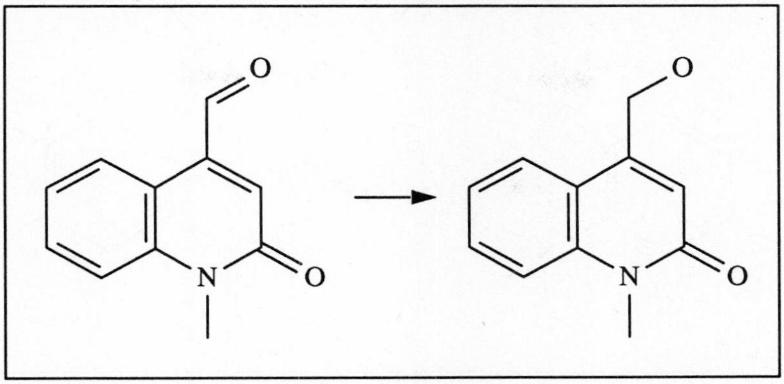

REACTION SEARCHING

Figure 4/93. Hit for query 4/92.

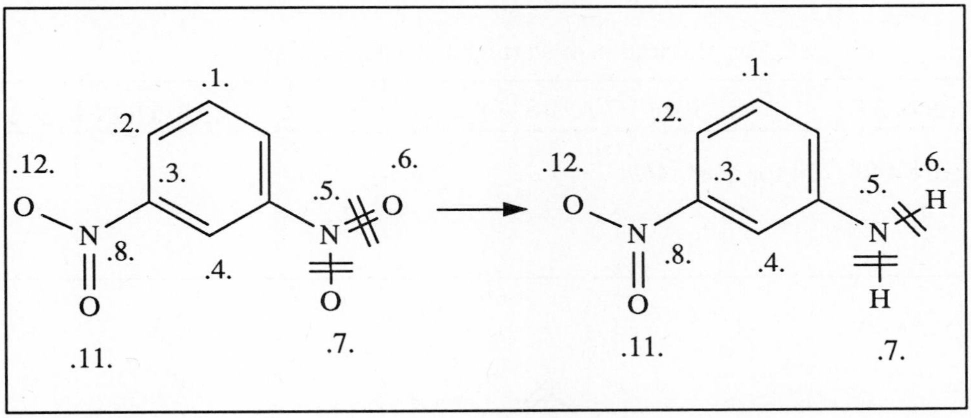

Figure 4/94. Atom/atom mapped query

formulate the search (Figure 4/98) by the query language.
Similarly, we could first perform the search:
Search: sss as product
The search will produce the hit list, e.g., L5. From this list we then eliminate the reactions using benzoxazines as reactants:
Search: L5 not sss as reactant
Looking for a comprehensive compilation of the chemical reactions of benzoxazines, we would formulate the query as follows:
sss as product or sss as reactant

We have seen that we can perform searches in a database subset by defining the Reference List using a variety of list descriptors, e.g., regnos, formulas, etc. We will show here

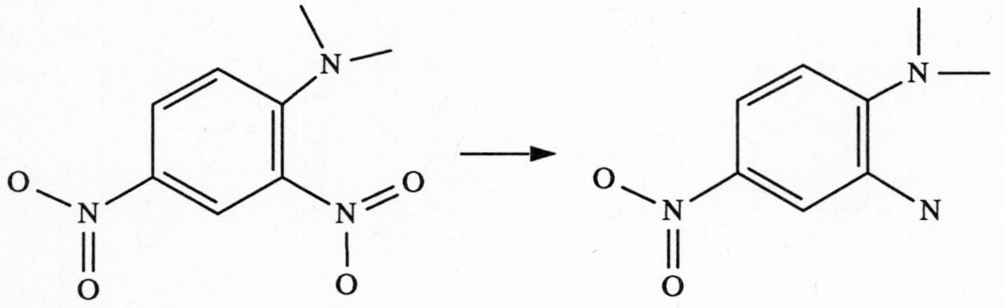

E E Ayling, J H Gorving, L E Hinkel, J Chem Soc p 755, 1942
L Sihlbom, Acta Chem Scand, 6 p 1528, 1952
Chem Abstr 1943, 11398j
Synth Meth, 1 (330)

For preparation of starting material see litref(3)

regno: 34	1024 1 7 A 1 STEP	VARIATION 1 OF 1

Figure 4/95. Hit for query 4/94

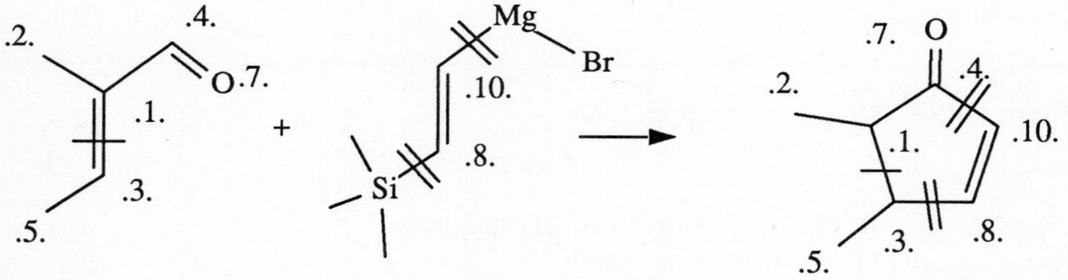

Figure 4/96. Atom/atom mapped reaction query

another useful method of how to define a database subset by searching for the reaction symbols which are used to classify the reactions in the printed version of Theilheimer. The datatype:

rxn:variation:rxnsym

has been defined for these data in the MDL THEIL database.

REACTION SEARCHING

[reaction scheme: cyclopentene carbaldehyde + vinyl silane with MgBr → bicyclic enone]
M Ramaiah, Synthesis (7), p. 529, 1984   S E Denmark, T K Jones, J Amer Chem Soc 104, p. 2643, 1982
regno: 12987   extregno-folder-path-step   VARIATION 1 OF 1   9103449   51A   3 STEP

Figure 4/97. Hit for query 4/96.

For example, to define a database subset comprising rearrangement reactions, we use the reaction symbol "REA" which stands for rearrangement and formulate the following query:
System: Search:
User: ref = rxnsym = @rea@
(The "@" wild card is used to substitute for the bonds formed and broken in the symbol, or we can specify the symbols for exact bonds formed and broken.)
The search domain (Reference List) will be created by REACCS containing all reactions involving rearrangements. This is the subset in which we now can search for rearrangement reactions giving a specific product or using a specific reactant by formulating a graphic query and performing the search, e.g.:
Search: sss as product
Search: sss as reactant

The MDL databases also contain a set of reaction keywords describing the registered reactions. The reactions which have been indexed with the same keywords have been collected in Listfiles. These Listfiles can be used to define a Reference List, e.g.:
System: Search:
User: ref = theil:dal.lst
The Reference List will contain all dealkylation reactions over which we can perform the more selective search using a graphic query.

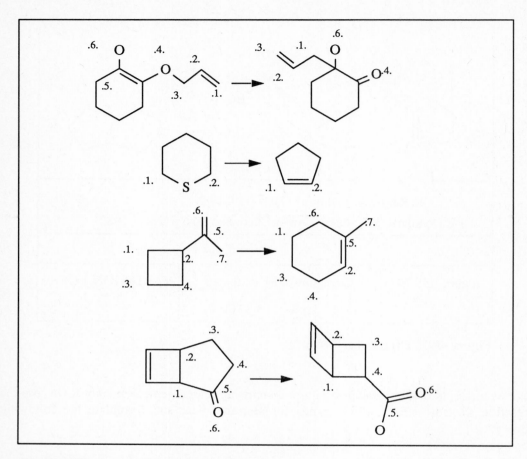

Figure 4/97-A. Atom/atom mapped queries

We can further select from the result of a search, e.g. L8, performed with a graphic query those reactions having specific data, e.g., reactions giving only high yields:
Search: L8 and yield > 80

SIMILARITY SEARCH

As we perform SSS or RSS, we find that in some cases even a broadly defined query in Substructure Search may give results that are too specific for our project.
Similarity searching serves as a technique to expand the criteria which are the basis for SSS and RSS.
In Substructure Search, the entire query fragment must be contained in the retrieved structures.
In Similarity Search, the query requirement is satisfied if only a predefined percentage of

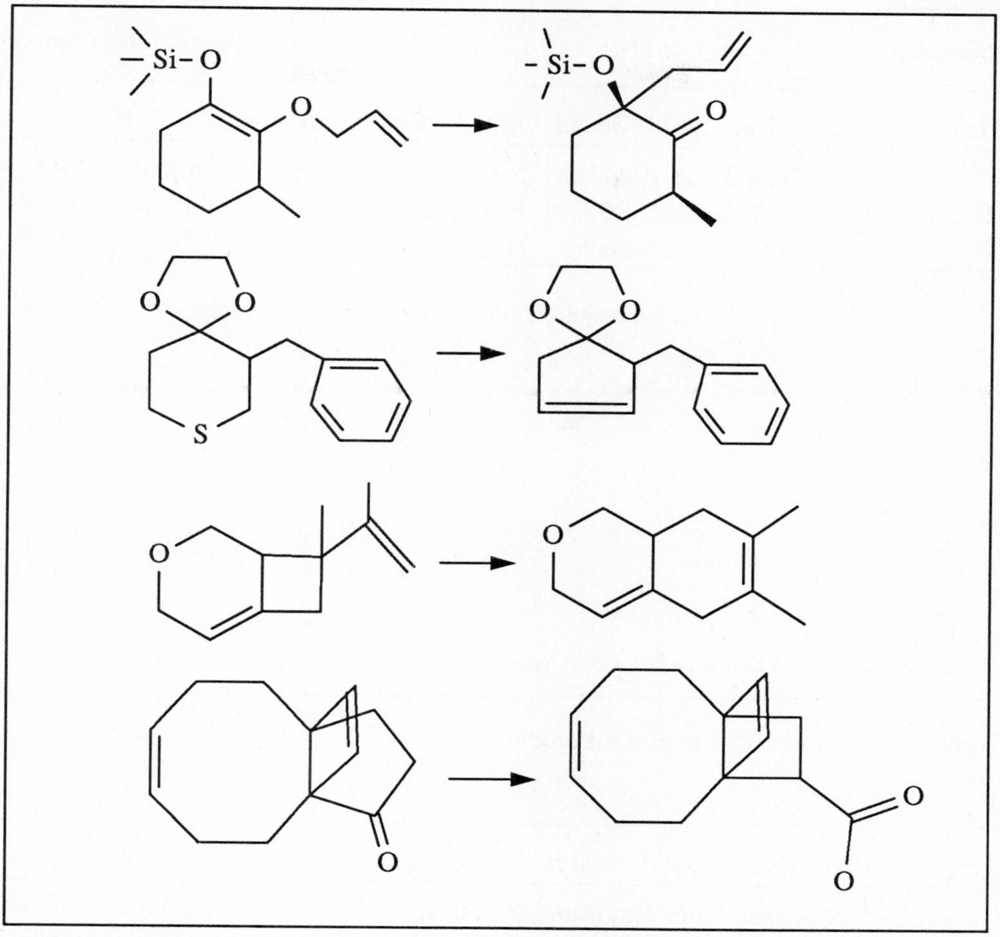

Figure 4/97-B. Retrieved reactions with queries 4/97-A.

keys in the query match those of the hit structures.
If we use the Similarity Search for molecule searches, the search compares the molecule query with the molecules in the database for structural similarity.
If we use the Similarity Search for reaction searches, the search compares the query with the reactions in the database for reaction center similarity and structural similarity.
Similarity Search uses the structural keys and reaction keys to arrive at the result by performing a statistical analysis of the database records (the reaction center and structural keys). For two entities to be similar, one entity must contain sufficient features in common relative to the total number of features in both entities.
Similarity Search can be conducted with a specified degree of similarity.

Example: *Combination search*

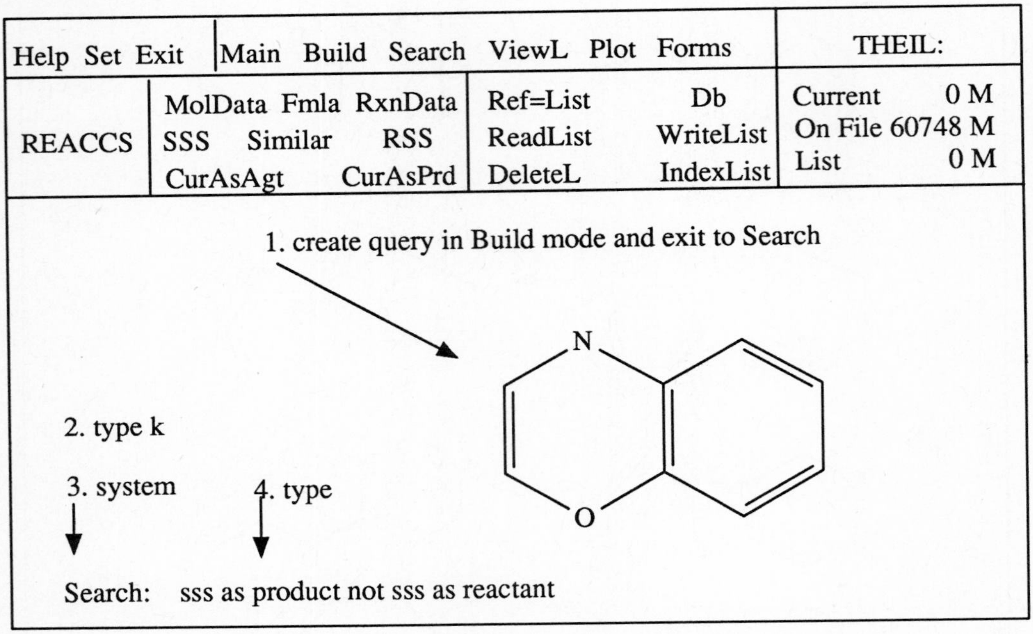

Help Set Exit	Main Build Search	ViewL Plot Forms	THEIL:
REACCS	MolData Fmla RxnData SSS Similar RSS CurAsAgt CurAsPrd	Ref=List Db ReadList WriteList DeleteL IndexList	Current 0 M On File 60748 M List 0 M

1. create query in Build mode and exit to Search

2. type k

3. system   4. type

Search: sss as product not sss as reactant

Figure 4/98. Conversion search for products

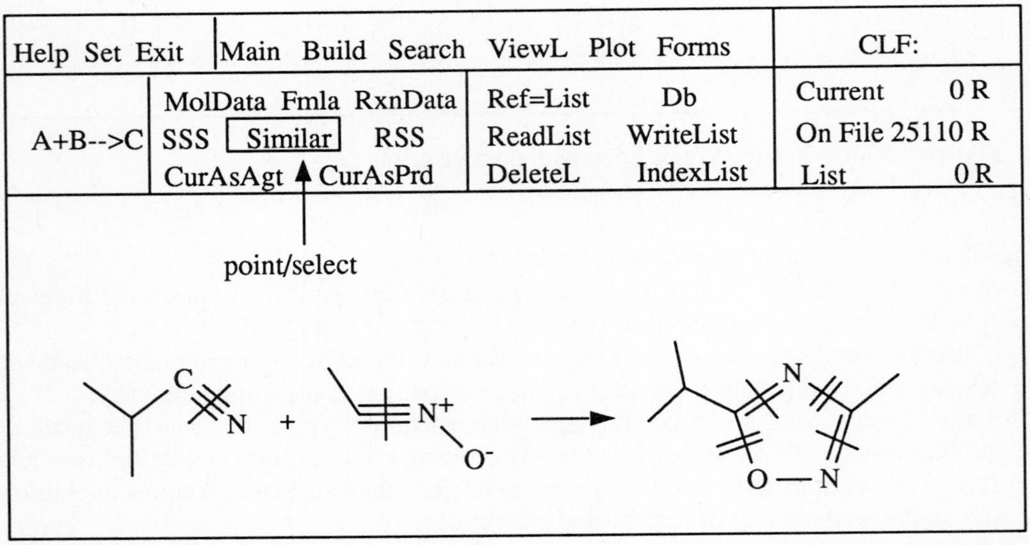

Help Set Exit	Main Build Search	ViewL Plot Forms	CLF:
A+B-->C	MolData Fmla RxnData SSS Similar RSS CurAsAgt CurAsPrd	Ref=List Db ReadList WriteList DeleteL IndexList	Current 0 R On File 25110 R List 0 R

point/select

Figure 4/99. Reaction similarity search

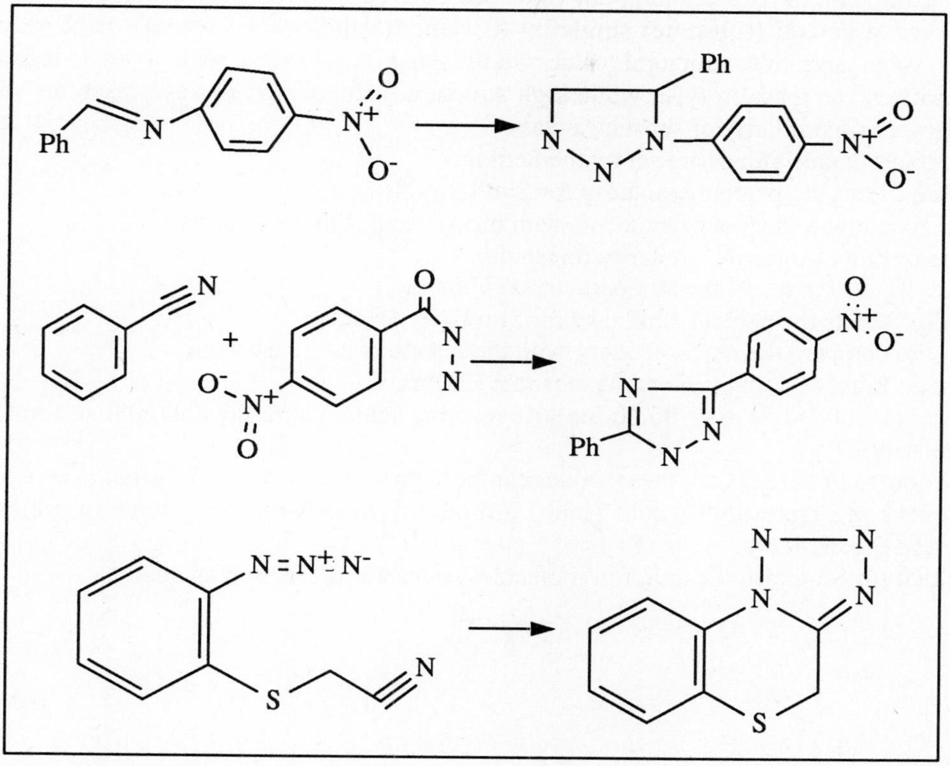

Figure 4/100. Retrieved reactions in similarity search

For molecule similarity searching, the degree value can be 1-100. The higher the value, the higher the similarity. The 0 value would invalidate the similarity; thus every molecule in the database would be a hit.

For reaction similarity searching, the value can be 1-100 for reaction center similarity and 1-100 for structural (substrate) similarity (0 value invalidates the search). High reaction center value and low structural value would retrieve reactions with a high degree of similarity of the reaction type, while high structural value would retrieve reactions with a high degree of similarity of substance class.

The default degree values are set by the Setting:

MolSim.....sets the percent similarity for Similarity Search
      to compare the query molecule with the molecules in the database
  System: Enter molecule similarity threshold:
  User: (1-100 (e.g., 50 for 50% structural similarity))

RxnSim......sets the percent similarity for Similarity Search
      to compare the reaction query with the reactions in the database
  System: Enter reacting center and substrate similarity:
  User: (1-100 1-100 (e.g., 80 20 for 80% reacting center similarity and 20% substrate similarity))

In the course of the session, these values can be set or reset by the SET option (see p. 594).

The query in Figure 4/99 would yield in Similarity Search reactions some of which are shown in Figure 4/100.

Note that for Similarity Search, full molecules or reactions are used as queries.

10 DATA SEARCHING

Data searching in REACCS can be categorized as follows:
1. searching data which have been generated by REACCS for the registered molecules and reactions.
   The data queries may consist of:
   a. registry numbers
   b. formulas
   c. keys
2. searching data which have been generated and registered by the User in defined datatypes for the molecules and reactions

Data searching can be performed in Main Mode and in Search Mode as is illustrated in Figure 4/101.
A data search of the category 1 results in the hit list which contains either exact molecules or reactions, or molecules or reactions in which a substructure fragment of the query is present. Therefore, in searching for data derived from the registered molecules or reactions, we can search for exact structures or substructures as in a search with a graphic query.

EXACT MOLECULE AND REACTION RETRIEVAL

Retrieval of a single exact molecule or reaction in the Main Mode is initiated by the FIND MOLECULE and FIND REACTION option. REACCS will prompt:
Molecule ID:
or
Reaction ID:
We can use the following descriptors at the prompts:
1. registry number
2. list file and listfile
3. exact formula
4. data query

Registry Number Query

Figure 4/102 shows the retrieval of a molecule using the internal registry number as the Molecule ID.
To retrieve a single reaction, we would input the reaction regno at the prompt:
Reaction ID:
which we receive after activating the FIND REACTION option.

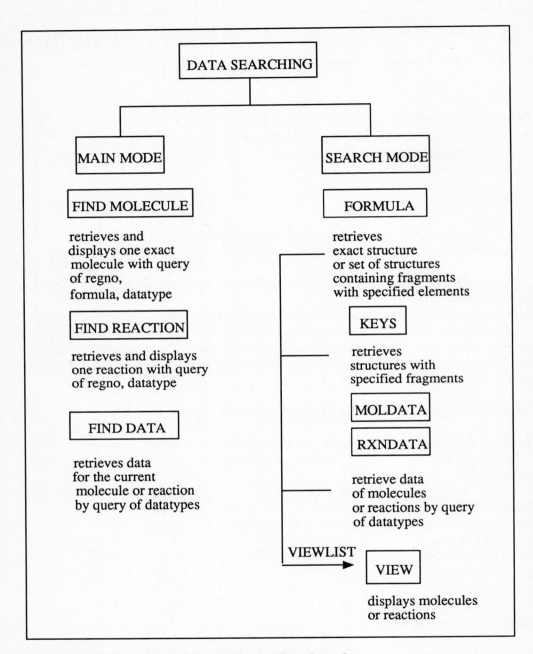

Figure 4/101. Data searching in Main and Search mode

DATA SEARCHING

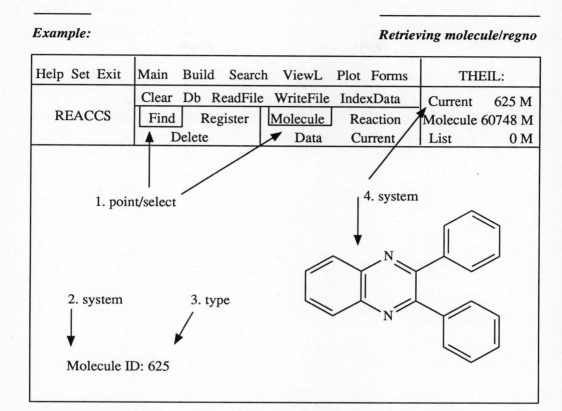

Figure 4/102. Retrieving molecule with regno query

In Search Mode, we can retrieve single registry number or a range of registry numbers (Figure 4/103) and thus establish the Current List or a Reference List.

The search in Figure 4/103 was specified for molecule intregnos 567 to 825. To specify reaction regnos, we would type "r567-825" to retrieve reactions in the range of intregno 567 to 825. The same search can be performed by selecting the Ref=List option and entering the regnos at the prompt:

List descriptor:

Default Registry Number

The default identifier for molecules and reactions is the internal registry number.

If we have defined external registry number for the database, we can use it as the Molecule ID and Reaction ID. However, we have to set the external registry number as the default by the Setting (see p. 592):

MolExtReg.........sets the default Molecule ID for the
    database

Example: *Searching registry numbers*

Help Set Exit	Main Build Search	ViewL Plot Forms	THEIL:
REACCS	MolData Fmla RxnData SSS Similar RSS CurAsAgt CurAsPrd	Ref=List Db ReadList WriteList DeleteL IndexList	Current 0 M On File 60748 M List 0 M

1. type k
2. system
3. type

Search: m567-825

Figure 4/103. Retrieving regnos in Search mode

For example, if we set MolExtReg to SK.NO (the external regno) and input at the prompt: Molecule ID: 789, REACCS will identify the molecule by the SK.NO = 789.

If the MolExtReg is set, REACCS recognizes the extregno SK.NO as the default Molecule ID if we input an integer at the prompt. If the MolExtReg is set and we wish to use the internal registry number as the Molecule ID, we have to precede the integer by the letter "m," e.g., m789 is the internal registry number 789 if the MolExtReg is set to another default identifier.

In fact, we can use any molecule datatype for setting the molecule identifier; however, the obvious choice is the external regno.

Similarly, to set the default Reaction ID for the reaction part of the database we use the Setting:

RxnExtReg.......sets the default Reaction ID for the database

If this Setting is set, use "r" to indicate internal regno for Reaction ID.

DATA SEARCHING

List File and Listfile query

The List File and Listfile queries must, of course, contain the appropriate registry numbers, i.e., of molecules or of reactions to be able to use them at the prompts:
Molecule ID: L2 (temporary list of molecule regnos)
Reaction ID: fstfile.lst (listfile with reaction regnos)
or at the prompt:
List descriptor:
after selecting Ref=List (we have to enter a descriptor with entities related to the operations we wish to perform).

Formula Query

Molecule ID: formula = C20 H14 N2
Note that in retrieving exact molecule, the formula must include hydrogens (*vide infra*) for formulas without hydrogens) and that if the formula represents more than one molecule, only the first one found is displayed.
In searching for a reaction with formula query of a molecule, we have to use the conversion statement (Figure 4/104).
The exact formula query can be also used in Search Mode
by activating the option FMLA (*vide infra*).

Data Query

The datatype we use in a data query must include the most specific part, i.e., the field.
To retrieve the molecule in Figure 4/102 we would formulate the query as follows:
Molecule ID: symbol = 2,3-diphenylquinoxaline
If we want to retrieve a reaction using the name of a molecule, we would use the conversion statement e.g.:
Reaction ID: symbol = 2,3-diphenylquinoxalin@ as product, or
symbol = 2,3-diphenylquinoxalin@ as reactant

SUBSTRUCTURE MOLECULE AND REACTION SEARCH

Formula Search

In Search Mode, we can perform a Substructure Search for molecules or reactions that have embedded in their structures a fragment which we specify as an element count (Figure 4/105). Note that in this kind of search, hydrogens are not included in the formula in contrast to the molecule Exact Search (*vide supra*) and that the specified elements, at minimum, must be present in the retrieved molecules. In addition, any other elements may be present unless we exclude them.
The search in Figure 4/105 would retrieve all molecules with 5 to 25 carbons, 2 nitrogens and 4 oxygens and any other additional elements. To exclude an atom from the molecule,

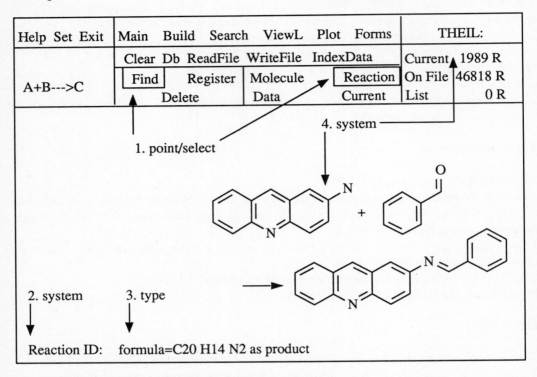

Figure 4/104. Retrieving reaction with formula query

we input the atom symbol with 0 count, e.g.:
C(5-25) Br0 N2 O4
To search for reactions, we include a conversion statement, e.g.:
C(5-25) Br0 N2 O4 as product

Keys Search

We have discussed in Part Three the keys that are assigned to structural fragments of the registered molecules (see p. 424). The same keys are operational in REACCS since the treatment of the structures on registration is the same in REACCS as in MACCS-II. There are 930 molecule keys and 300 reaction keys in REACCS. They can be partially (about 160 of them) displayed.
Type "indexkeys" at the Search: prompt in Search Mode to display the set of keys applicable to the REACCS database.
To search for fragments like those in the structure in Figure 4/106, we formulate the query

DATA SEARCHING

Example: *Substructure search with formula query*

Help Set Exit	Main Build Search	ViewL Plot Forms	THEIL:
REACCS	MolData Fmla RxnData SSS ↑Similar RSS CurAsAgt CurAsPrd	Ref=List Db ReadList WriteList DeleteL IndexList	Current 0 M On File 60748 M List 0 M

                 1. point/select

                        3. type

2. system   ↓
↓

Formula target: C(5-25) N2 O4

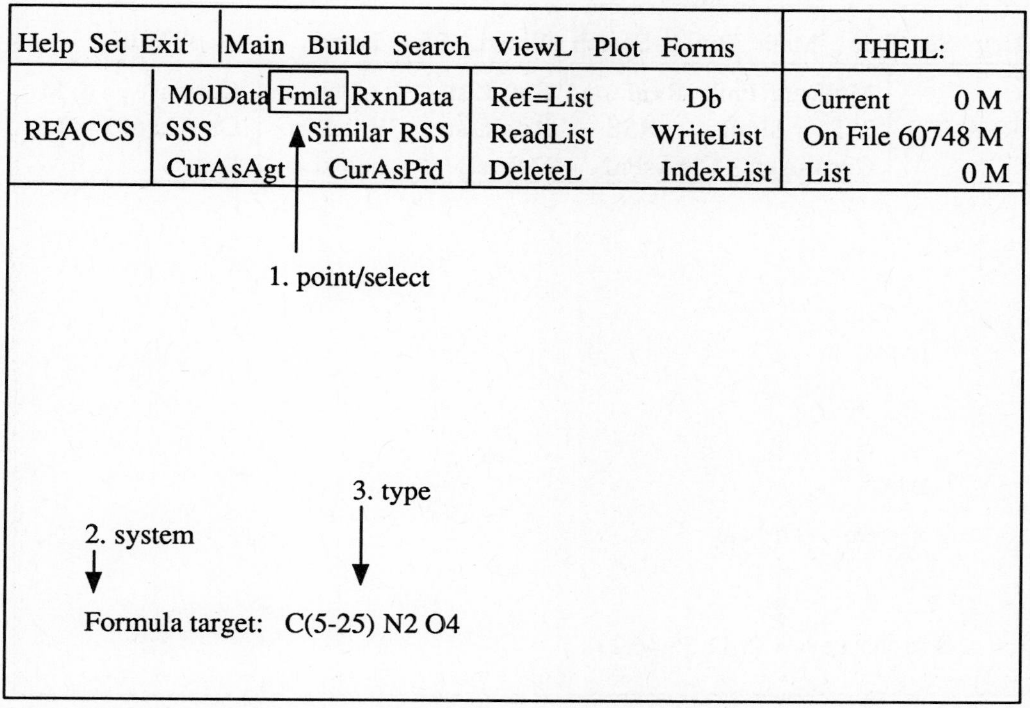

Figure 4/105. Substructure search with formula query

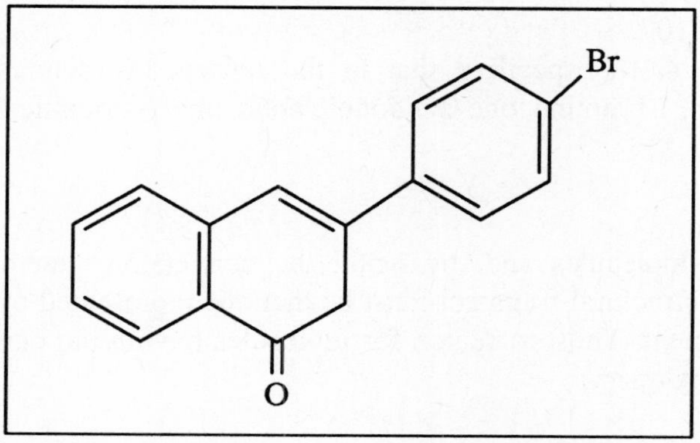

Figure 4/106

Example: *Substructure search with keys query*

Help Set Exit	Main Build Search	ViewL Plot Forms	THEIL:
REACCS	MolData Fmla RxnData SSS Similar RSS CurAsAgt CurAsPrd	Ref=List Db ReadList WriteList DeleteL IndexList	Current 0 M On File 60748 M List 0 M

1. type k

2. system   3. type

Search: keys = (5,11,35,46,29)

Figure 4/107. Searching with keys query

as shown in Figure 4/107.
The keys in Figure 4/107. specified that in the retrieved structures there must be a fragment of 1 oxygen, 1 bromine, one CO double bond, and 1 aromatic 6-membered ring.

Name Search

We can search for molecules and, by using the conversion statement technique, for reactions in which a structural fragment must be that one represented by a name or a name segment of the fragment. Thus, to search for molecules having the benzoxazine fragment we input the following query:
System: Search:
User: symbol = @benzoxazin@ (to search for molecules)
   symbol = @benzoxazin@ as product (to search for
      reactions)

DATA SEARCHING

Example: *Displaying datatype treenames*

Help Set Exit	Main Build Search ViewL Plot Forms	THEIL:
REACCS	Clear Db ReadFile WriteFile │IndexData│	Rxn 46818
	Find Register Molecule ↑ Reaction	Mol 60748
	Delete Data Current	List 0 M

1. point/select

─ system   3. type
                     ↓
2.
─ Datatype (or <cr> for all): grade

4. RXN:VARIATION:REACTANT:GRADE
   RXN:VARIATION:CATALYST:GRADE
   RXN:VARIATION:SOLVENT:GRADE
   RXN:VARIATION:PRODUCT:PRODAT:GRADE

Figure 4/108. Displaying datatype treenames

DISPLAYING DATATYPE TREENAMES

In the data searches we have discussed above and in those we will discuss below, we need to input the datatype treename(s) to formulate the query.
To find out which datatypes have been defined for the current database we have two options both in the Main Mode and the Search Mode:

INDEXDATA..........displays the complete treename of a specific field datatype
INDEXFULL..........displays complete datatype treenames, field datatype numbers, field datatype, and input/output format

If there is a current molecule, molecule datatype treenames will be displayed. If there is a current reaction, reaction datatype treenames will be displayed. If there is no current

Example: *Displaying datatype treenames*

Help Set Exit	Main Build Search	ViewL Plot Forms	THEIL:
REACCS	MolData Fmla RxnData SSS Similar RSS CurAsAgt CurAsPrd	Ref=List Db ReadList WriteList DeleteL IndexList	Rxn 46818 Mol 60748 List 0 M

```
    1. type k
 ┌── system                              type ──┐
 │                                               │
 ├── Search:    indexfull ──────────────────────┤
 │                                               │
 ├── Datatype ( or <cr> for all):  reactant ────┤
 │                                               │
 └── (15)  RXN:VARIATION:REACTANT:GRADE
           TYPE: Variable text
           INPUT PROMPT: Grade of reactant
```

Figure 4/109. Displaying datatype treenames and I/O format

entity, both molecule and reaction datatypes will be displayed (Figure 4/108, 4/109).

DATA SEARCHING

In the category 2 of data searching, we search for the data which have been registered in various datatypes (see p. 579 for discussion of datatypes and treenames).
Data searching gives different results depending on the Mode in which we perform the search:
1. FIND DATA option in Main Mode will retrieve data for
   the current (displayed) molecule or reaction

DATA SEARCHING

Example: *Data retrieval*

Help Set Exit	Main Build Search ViewL Plot Forms	THEIL:
REACCS	Clear Db ReadFile WriteFile IndexData	Current 2000 R
	Find Register Molecule Reaction	On File 46818 R
	Delete Data Current	List 0 R
colspan reaction scheme		

Reaction scheme: phthalic acid derivative + propylamine →(distill) N-propyl isoquinoline-1,3-dione

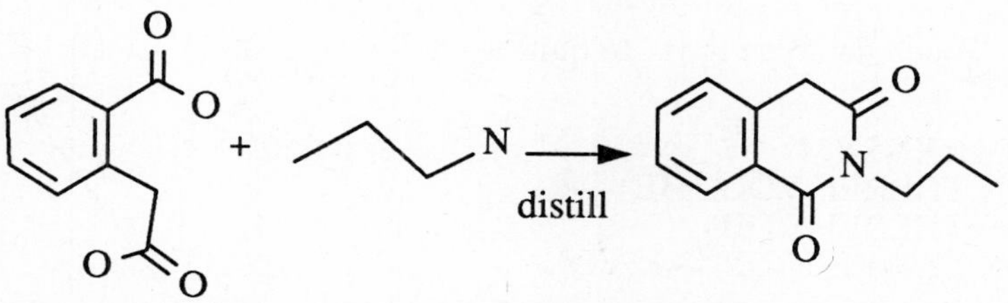

Buu-Hoi, Bull Soc Chim Fr, 12 p. 313, 1945
R B Moffet, J Org Chem, 14 p. 862, 1949

| regno 2000 | 2363 2 127A 1 step | VARIATION 1 OF 1 |

Figure 4/110. Retrieved reaction and data in Form

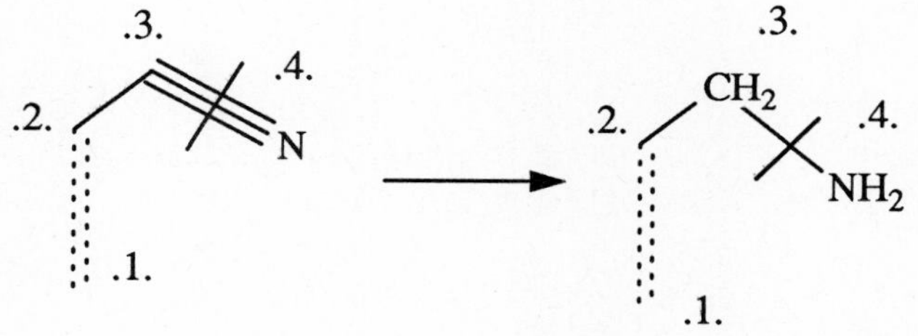

Figure 4/111. Reaction query

Example: *Reaction search with data query*

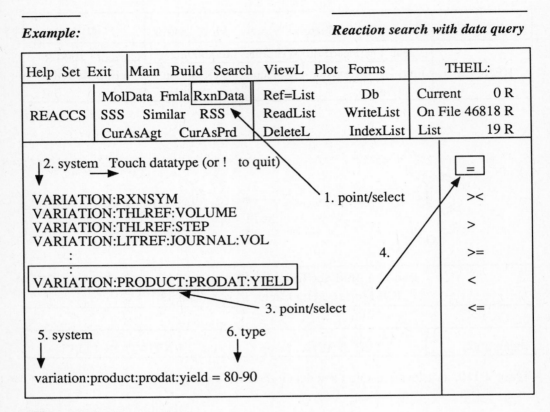

Figure 4/112. Search of data

DATA SEARCHING

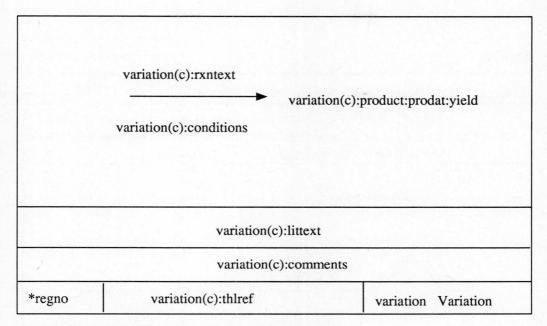

Figure 4/113. Default reaction form THL2.frm

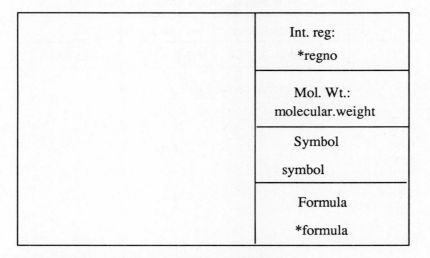

Figure 4/114. Default molecule form THL2.frm

Example: *Displaying data in Form*

Help Set Exit	Main Build Search ViewL Plot Forms	THEIL:
REACCS	First　Next　Prev　Ref=List　　　Db Item　List　Table　ReadList　　WriteList Data　Omit　Query　DeleteL　　IndexList	Current　　　0 R On File　46818 R List　　　　6 R

1. point/select

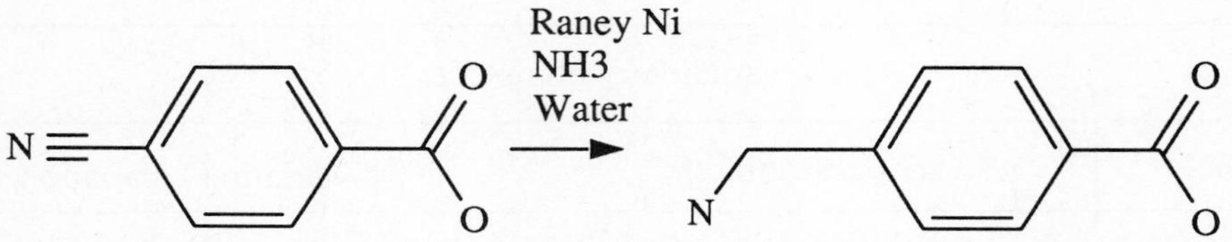

Raney Ni / NH3 / Water

A Albert, D Magrath, J Chem Soc p. 678, 1944
B S Biggs, W S Bishop, Org Synth 27 p. 18, 1947

For preparation of starting material, see liref (2)

| regno: 1591 | 2069　2　22A　1 STEP | VARIATION 1 OF 1 |

Figure 4/115. Displaying first record on the List

Example: *Displaying data for hit list*

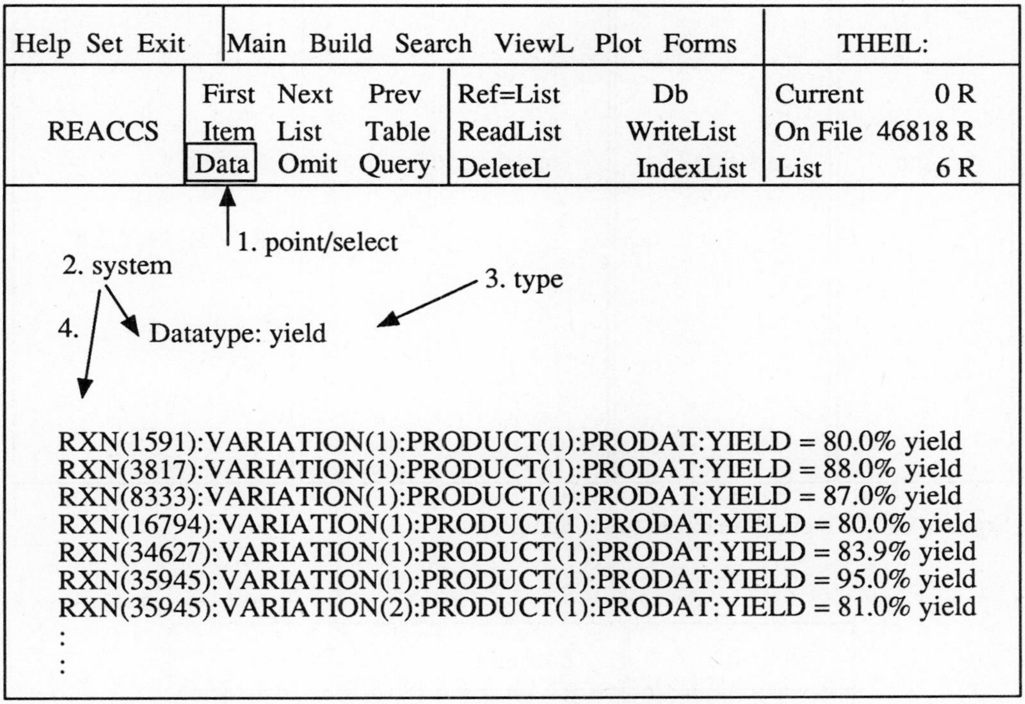

Help Set Exit	Main Build Search ViewL Plot Forms			THEIL:	
REACCS	First Next Prev	Ref=List	Db	Current	0 R
	Item List Table	ReadList	WriteList	On File	46818 R
	Data Omit Query	DeleteL	IndexList	List	6 R

1. point/select
2. system
3. type
4. Datatype: yield

```
RXN(1591):VARIATION(1):PRODUCT(1):PRODAT:YIELD = 80.0% yield
RXN(3817):VARIATION(1):PRODUCT(1):PRODAT:YIELD = 88.0% yield
RXN(8333):VARIATION(1):PRODUCT(1):PRODAT:YIELD = 87.0% yield
RXN(16794):VARIATION(1):PRODUCT(1):PRODAT:YIELD = 80.0% yield
RXN(34627):VARIATION(1):PRODUCT(1):PRODAT:YIELD = 83.9% yield
RXN(35945):VARIATION(1):PRODUCT(1):PRODAT:YIELD = 95.0% yield
RXN(35945):VARIATION(2):PRODUCT(1):PRODAT:YIELD = 81.0% yield
⋮
⋮
```

Figure 4/116. Displaying data for the current list

Regno	Variation	Catalyst	Yield
1	2	3	4
*regno	variation	catalyst:regno	yield R,'-',R,'%Yield'

Figure 4/117. Default reaction table Theil2.frm

RegNo	Formula	Symbol
*regno	*formula	symbol

Figure 4/118. Default molecule Table

DATA SEARCHING

Example: *Displaying data in Table*

Help Set Exit	Main Build Search ViewL Plot Forms	THEIL:
REACCS	First Next Prev Ref=List Db Item List Table ReadList WriteList Data Omit Query DeleteL IndexList	Current 0 R On File 46818 R List 6 R

2. system   1. point/select

```
Regno       Variation   Catalyst              Yield
[1591]      1           Raney Ni              80.0 % yield
[3817]      1           LAH                   88.0 % yield
[8333]      1           SnCl2                 87.0 % yield
[16794]     1           Pd(m)                 80.0 % yield
[34627]     1           NaBH4                 83.9 % yield
[34627]     1           BF3-Etherate          83.9 % yield
[35945]     1           B2H6                  95.0 % yield
[35945]     2           NaAl(i-Bu)2H2         81.0 % yield
[35945]     3           Raney Ni              91.0 % yield
  :
  :
```

Figure 4/119. Displaying reaction data in default Table

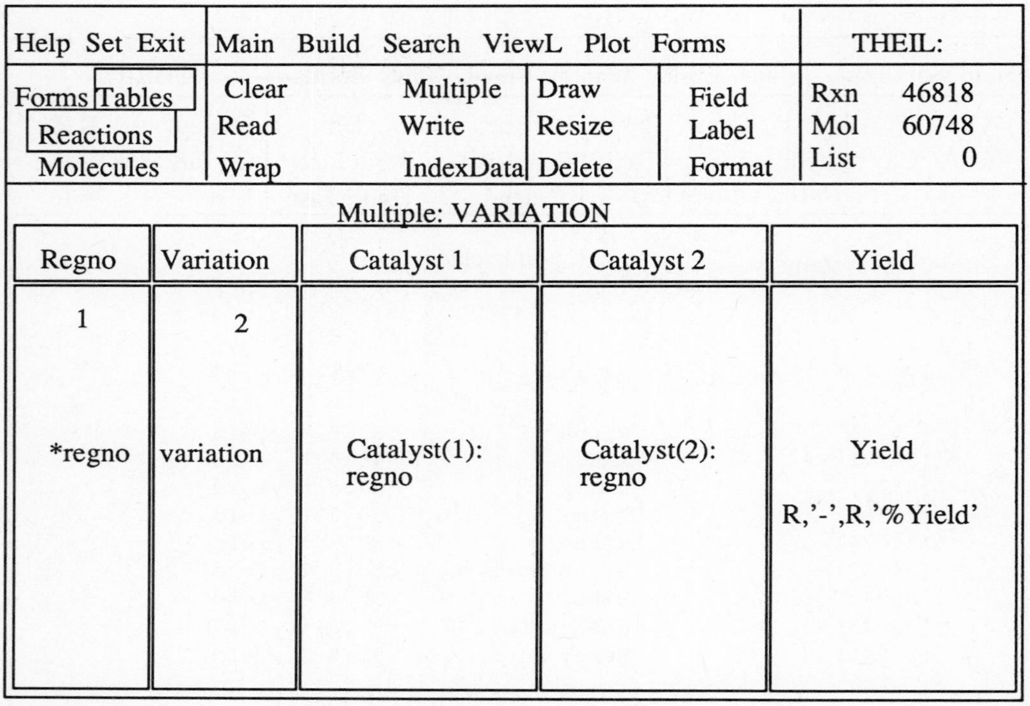

Help Set Exit	Main Build Search ViewL Plot Forms		THEIL:	
Forms Tables	Clear	Multiple	Draw	Field	Rxn 46818
Reactions	Read	Write	Resize	Label	Mol 60748
Molecules	Wrap	IndexData	Delete	Format	List 0

Multiple: VARIATION				
Regno	Variation	Catalyst 1	Catalyst 2	Yield
1	2			
*regno	variation	Catalyst(1): regno	Catalyst(2): regno	Yield R,'-',R,'%Yield'

Figure 4/120. Table for multiple lines of reaction data

DATA SEARCHING

Help Set Exit	Main Build Search ViewL Plot Forms			THEIL:	
REACCS	First Item Data	Next List Omit	Prev Table Query	Ref=List ReadList DeleteL	Db WriteList IndexList	Current On File List	0 R 46818 R 6 R

```
Regno      Variation  Catalyst 1   Catalyst 2    Yield

[1591]     1          Raney Ni                   80.0 % yield
[3817]     1          LAH                        88.0 % yield
[8333]     1          SnCl2                      87.0 % yield
[16794]    1          Pd(m)                      80.0 % yield
[34627]    1          NaBH4        BF3-Etherate  83.9 % yield
[35945]    1          B2H6                       95.0 % yield
[35945]    2          NaAl(i-Bu)2H2              81.0 % yield
[35945]    3          Raney Ni     Acetic anhyd  91.0 % yield
[35945]    4          LAH                        85.9 % yield
  :
  :
  :
```

Figure 4/121. Display of multiple reaction data in Table

When we retrieve a single molecule or reaction in Main Mode, the structures will be displayed in the Form including the data in the datatypes which have been assigned to the box(es) in the Form (e.g., Figure 4/97). The DataDisplay Setting must be on (see p. 649). If the DataDisplay is off, only the structures will be displayed. If we wish to retrieve for the current molecule or reaction the data in datatypes other than those assigned to the boxes, we have to set the Setting:

FindData......turns on/off default display of data in a Form
        File when FIND DATA is used to retrieve data
  When FindData is on, DataDisplay must be also on and
  FormFile must be set
  When FindData is off or FormFile is not set, data for
  the specified datatypes in the FIND DATA operation will
  be displayed in a line format on the screen.

For example, the reaction in Figure 4/110 was retrieved by FIND REACTION. The structures and data are displayed in the Form (DataDisplay is on, FindData is on).

To retrieve additional data for the reaction in Figure 4/110 in other datatypes than in those assigned to the boxes, we have to set FindData off followed by point/select FIND DATA (e.g. reactant grade):

System: Datatype:
User: grade
The screen will clear and the data will be displayed:

RXN(2000):VARIATION(1):REACTANT(2):GRADE = ANHYD SOLVENT

2. data searching in Search Mode will retrieve molecules
   or reactions that are associated with the data in
   specified datatypes.

For example, we perform RSS for reduction of nitriles (Figure 4/111) in Search Mode over the search domain established by:
1. System: Search:
2. User: ref = theil:red.lst
   (this query creates a List, i.e. search domain of all
   reductions classified by the keyword "red")

The graphic query search results in List 19R which we make the Reference List. In this search domain of 19 reactions we want to find those which give yield of 80-90% (Figure 4/112).

The search returned 6 reactions (List 6R) with which we enter the ViewList Mode.

DISPLAYING DATA IN VIEWLIST MODE

We already know that we can display molecules and reactions with the associated data in the default Form or Table. The datatypes that have been assigned in the boxes of the default Forms, are shown in Figure 4/113 and Figure 4/114.

Point/select FIRST displays the data of the first item in the current List in the default Form (Figure 4/115).
To display the data of all the records in the List, point/select DATA (Figure 4/116)
The default reaction Table is shown in Figure 4/117, the default molecule Table is shown in Figure 4/118.
Point/select TABLE will display the data in the current Table format for the reactions in the List (Figure 4/119).
To display the multiple data in Figure 4/119 in separate columns, we create columns for each line of the data and point/select MULTIPLE to specify the multiple datatype (Figure 4/120).
Figure 4/121 shows the display of multiple data in the format of Table in Figure 4/120.

11   PLOTTING STRUCTURES

The molecules or the reactions that we retrieve from the database can be plotted on the screen of our terminal or on a plotting device. We can also plot the structures we recall from a Molfile or a Rxnfile or the structures which we create in the Build Mode. The plotting is performed in Plot Mode (Figure 4/123) which we can enter from any of the Modes by point/select PLOT (Figure 4/122).

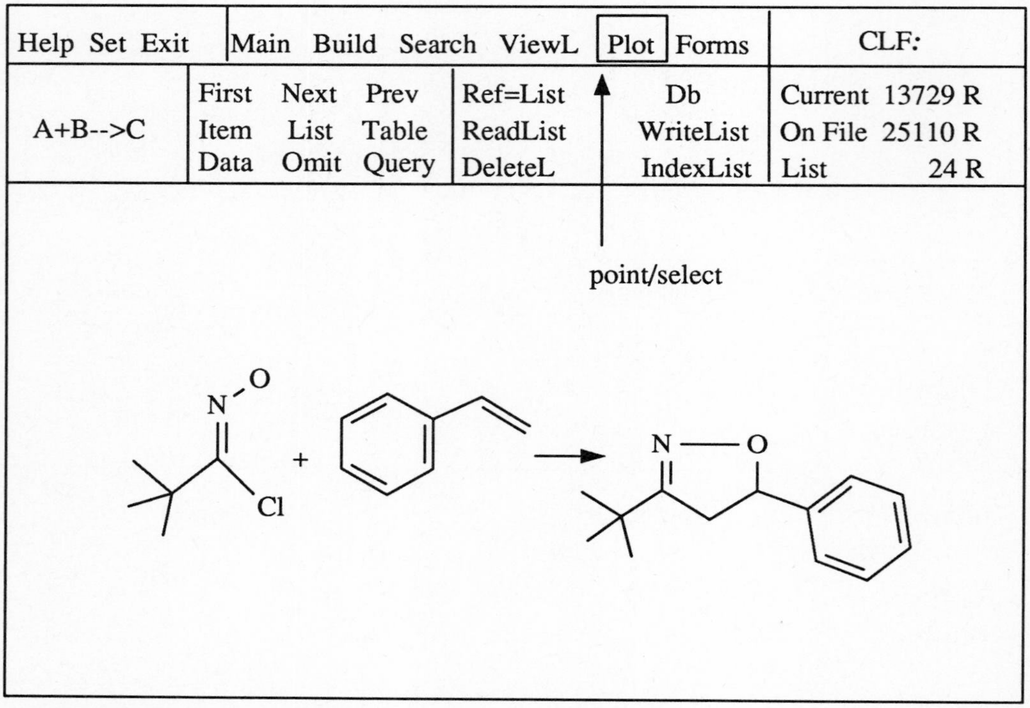

Figure 4/122. Entering Plot Mode

When we enter Plot Mode and select the option to specify which object we want to plot, REACCS will ask the following:

Comments:
Any text can be entered to be shown on the plotted page.

Example: *Plotting reaction and form*

Help Set Exit	Main Build Search ViewL Plot Forms	CLF:
A+B--->C	List <u>Current</u> Device Labels ReadFile 1 2 4 Numbers FindMol FindRxn NewPage Truncate	Current 13729 R On File 25110 R List 24 R

point/select

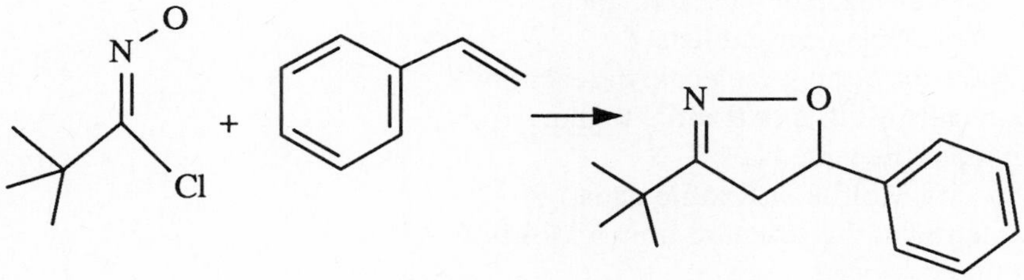

D P Curran, S A Scanga, C J Fenk, J Org Chem, 49(19), p.3474-, 1984

| regno:
13729 | extreg - folder - path - step
9105705   36A   2 step | variation 1 of 1 |

Figure 4/123. Plotting current reaction

Header on every page [N]?:
Type N (for No) - a header page consisting of:
User name, user address, date, time
will be printed for each group of plotted structures.
Type Y (for Yes) if the header is to appear on every page of the plotted structures.

The Settings (see p. 592) can be used to enhance the appearance of the plots:
address......sets address which will be printed on the plots
initials.....sets the initials to appear in the plots
name.........sets name to appear on the plots

The options on the Plot Menu have the following functions:

SELECTING STRUCTURES FOR PLOTTING

List........establishes the current list for plotting
   System: list descriptor
   User: (enter list descriptor)
        Ln - temporary list file
        fstfile - a Listfile
        m1-15,45 - a list of molecule regnos
        r1-15,45 - a list of reaction regnos
        ref - reference list
        L2 and fstfile - combination statement
        SSS as reactant - a search query
        ? to display current lists
Current.....plots the current molecule or reaction
ReadFile....recalls Molfile or Rxnfile to plot
   System: input filename
   User: (specify Molfile or Rxnfile name)
FindMol.....retrieves the structure from a Molfile
   System: molecule ID:
   User: (specify ID by the descriptor as in LIST)
FindRxn.....retrieves the reaction from a Rxnfile
   System: reaction ID
   User: (specify ID as in LIST)

SPECIFYING APPEARANCE OF PLOTTED STRUCTURES

Labels......displays and plots hydrogens in structures
Numbers.....displays and plots structure locants
Truncate....replaces common substituent fragments with
       text abbreviations

SELECTING PLOTTING DEVICE

Device......specifies the plotting device

 System: plotting device (L for list):
 User: L
 System:
 Remote:
   1. Versatec   2. Calcomp   3. Meta File
 Local:
   4. 4010/4014 Emulator   5. Tektronix 4662 6. HP 7221
   9. HP 747xA   40. Terminal screen
The device is selected by entering the number from the list.

FORMATTING PAGE FOR PLOTTING

1 2 4.......plot 1, 2, or 4 structures on the page
NewPage.....begin plotting of the next specified structures
    on a new page

The display of the plots on screen can be turned off while plotting by the Setting:
PlotDisplay.......turns on/off display on screen of items
    being plotted from a list

12   UPDATING DATABASES

Updating databases involves registration (entry) of new molecules, reactions, and data and/or changing/modifying the existing entries in the databases. The updating is performed in Main Mode.

REGISTERING MOLECULES

The register operation creates a new registry number under which the molecule will be registered.

To register a new molecule we create the structure as usual in the Build Mode. The structures for registration cannot contain variables (however, we can save a structure with variables in a Molfile). We can register molecules transferred from MACCS-II and/or ChemBase. We can register "no-structure" in which case there will be no molecules but the registry number will be used for registering data (we can register the molecule later). The molecule to be registered must be the current structure. To register "no-structure" the screen will be blank. The registration is initiated by the REGISTER CURRENT option (Figure 4/124).

If the molecule is already on file, REACCS issues the message:

OK to register duplicate [N]?:

or:

ALREADY ON FILE:

and gives the registry number for the molecule. The registration does not proceed.

If we register "no-structure," the message will be:

Wish to register NO-STRUCTURE [N]?:

Type Y(es) to accept the registry number under which we can register data (and a molecule later). Type N(o) or <cr> to abort.

The Current and On File counter reflect the completed registration (Figure 4/124).

The Setting MolInfo (see p. 594) determines how the "no-structure" will be displayed on retrieval of the record.

If we wish to register a molecule under one of the existing registry numbers, we establish the current molecule (it may be a new molecule or it may be a molecule retrieved from the database and modified) and activate the REGISTER MOLECULE option (Figure 4/125).

The structure in Figure 4/125 is the modified structure in Figure 4/124 and is re-registered under the same registry number we specify at the prompt (Figure 4/125). However, any existing regno can be input at the prompt. REACCS will ask whether the old structure under the spcified regno should be overwritten (Figure 4/125).

If we approve to overwrite (Y) and the molecule participates in a registered reaction(s), REACCS will list the reaction(s) and will prompt for approval to overwrite the molecule. If we approve to overwrite, all the reactions will have the new molecule as the reaction participant replacing the old molecule.

Example: *Registering molecules*

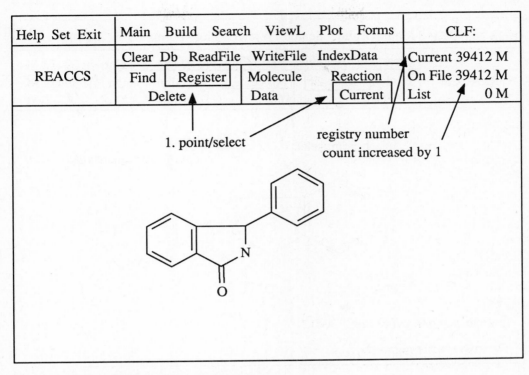

Figure 4/124. Registering current molecule

The data associated with the molecule/registry number are not altered in this operation. To alter data, use REGISTER DATA (*vide infra*).

REGISTERING REACTIONS

To register reactions, we establish a current reaction as usual and initiate the registration by the same option as for molecules, i.e., REGISTER CURRENT. To register "no-reaction," we leave the screen blank. REACCS checks for unregistered molecules participating in the reaction and issues the prompt, e.g.:

Molecule A and C will also be registered, continue [Y]?:

To proceed with registration, enter <cr>, or N to abort.
If the molecule(s) is already registered, REACCS gives the messsage as in the molecule registration (*vide supra*).
The option REGISTER REACTION functions similarly as we have described for the

Example: *Re-registering molecules*

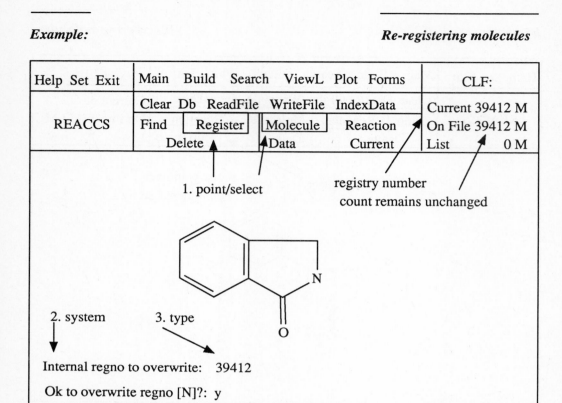

Figure 4/125. Re-registering molecule

REGISTER MOLECULE (*vide supra*).
Note that the data associated with the old reaction are not affected by this operation.

REGISTERING DATA

To register new data or modify the already registered data for a molecule or reaction, the molecule or reaction must be the current entity displayed on the screen (the Current counter must show the registry number). We establish the current entity as usual, i.e., by retrieving it from the database or if we have just registered a molecule or reaction, we can proceed with the data registration. If we wish to register data for "no-structure" or "no-reaction," we have to retrieve the registry number by FIND MOLECULE or FIND REACTION. The data registration is initiated by the REGISTER DATA option (Figure 4/126).
The prompt Datatype: (Figure 4/126) requires input of the datatype(s) into which we wish to register the data value(s) (see p. 579 for discussion of datatypes and treenames). To

UPDATING DATABASES

Example: *Registering data*

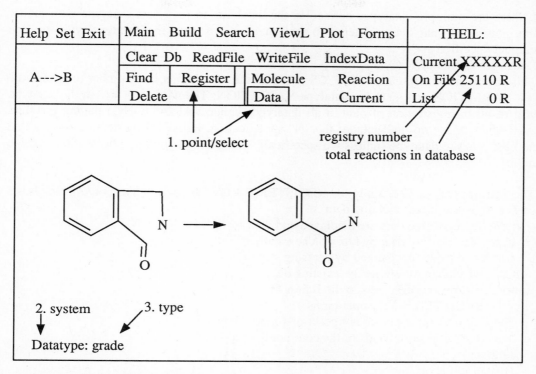

Figure 4/126. Registering data

verify which datatypes have been defined for the current database, use the INDEXDATA and INDEXFULL options (see p. 705).
REACCS will prompt for input of the data as in the following example:
RXN(XXXXX):VARIATION:REACTANT:GRADE
Grade of reactant: <cr>
(If we do not wish to register grade data, type <cr>.)
RXN(XXXXX):VARIATION:CATALYST:GRADE
Grade of catalyst: 99%
RXN(XXXXX):VARIATION:SOLVENT:GRADE
Grade of solvent: 98%
RXN(XXXXX):VARIATION:PRODUCT:PRODAT:GRADE
Grade of product: [65%]
(If there are already registered data values, they are displayed in square brackets. We can modify them by typing new value or leave them unchanged by pressing <cr>, e.g.):

RXN(XXXXX):VARIATION:PRODUCT:PRODAT:GRADE
Grade of product: [65%]: 100.0%

If there are multiple datatypes defined in the database, to register a datatype with higher line number to that one registered in the database, we input the datatype with the line number specified as "N":
Datatype: variation(N):@:amount
This specification will cause REACCS to assign the next higher line number (e.g., 2 if the last VARIATION was 1) to the datatype VARIATION. The wild card "@" in the datatype will result in registration of data in all datatypes under VARIATION(2) having the field AMOUNT. If we do not include the "N" specification, REACCS by default assigns line number 1 to the datatype (unless we specifically designate the line number in the datatype input).

The Settings (see p. 592) apply for REGISTER DATA: Overwrite....sets on/off the User's choice to overwrite the existing data
   If on, the registered data are displayed and we can:
     a. replace existing data by typing a new data in
     b. leave the data unchanged by pressing <cr>
   If off, we cannot overwrite the existing data
Editor........turns on/off access to the Editor in
        REGISTER DATA operations
   The Editor Commands are not described here. They are
   like most of the commands of the commonly used line
   editors.
   The default is off.
   If we do not use the Editor, we have to correct an
   error in the data entry by restarting the REGISTER DATA
   and input the correct data.

If the database contains datatypes with defined units, we can overrride the units or set new units by using one of the Settings (see p. 592).

DELETING MOLECULES AND REACTIONS

The delete operations enable us to remove molecules and associated data or reactions and associated data from the database.
DELETE CURRENT erases the record of the current molecule or reaction (displayed on the screen). REACCS will ask for approval to delete by displaying the registry number of the entity and a prompt:
Delete this molecule [N]?:
or
Delete this reaction [N]?:
The registry numbers of the deleted entities remain in the database and can be used to

register new molecules or reactions.
DELETE MOLECULE deletes the molecule and associated data and **all reactions** in which the molecule is a participant and associated data. The option can be used with a current molecule or reaction (on display) or, if there is no current entity, REACCS will prompt:
Molecule ID:
DELETE REACTION deletes the reaction and associated data. The reaction can be either the current one (on display) or REACCS prompts:
Reaction ID:
DELETE DATA deletes data for the current molecule or reaction. REACCS will prompt for the datatype descriptor (see p. 640) and then will ask for approval to delete the specified data.
For example:
Datatype: amount
RXN(20):VARIATION(1):REACTANT(1):AMOUNT = 312.0 gm 3.0 moles
Delete data [N]?: y
RXN(20):VARIATION(1):PRODUCT(1):AMOUNT = 421.0 gm
Delete data [N]?: y

TRANSFERRING STRUCTURES AND DATA

New entries can be registered in a database by transferring records from other REACCS databases or from MACCS-II or ChemBase.
The transfer operations basically involve writing files to store structures with or without data in the source database and reading the files in the destination database. To transfer files within REACCS, we simply use the REACCS options for writing and reading files (*vide infra*). To transfer files between REACCS and MACCS-II we use the options and operations of the systems on the host. To transfer files between ChemBase and REACCS, ChemTalk has to be installed on our micro and ChemHost (or other transfer protocol program) has to be installed on the host.
ChemTalk transfer operations are discussed in Part Five.

In transferring files from or to REACCS we encounter different qualities and quantities of data we will handle in the transfer. We can transfer a single molecule or reaction with or without the associated data. We can transfer multiple molecules or reactions with or without the associated data.

SINGLE MOLECULE TRANSFER

A single molecule stored in a Molfile (no data in datatypes can be stored in Molfiles) can be transferred into another REACCS database or into MACCS-II or ChemBase. To register the molecule from a Molfile into REACCS database, we use the READ FILE option to make the molecule current and then proceed with the registration as usual (*vide supra*).

SINGLE REACTION TRANSFER

To transfer a single reaction without the associated data between REACCS databases, we store the reaction in a Rxnfile and READ FILE to make the reaction current and register it into a REACCS database. To transfer a Rxnfile into ChemBase, we have to indicate that the Rxnfile we are writing is intended for ChemBase. When we respond to the REACCS prompt to input the filename, we include the qualifier "/pcrxn" to the filename:
System: Output filename:
User: myreact.rxn/pcrxn

REACCS can read the Rxnfiles originated in ChemBase.

MULTIPLE STRUCTURES/DATA TRANSFER

To transfer multiple molecules or reactions with or without associated data or single molecules or reactions with associated data, we use RDfiles.

RDFILES

Rdfiles are permanent files. They are used to transfer records between REACCS databases and between REACCS and ChemBase or MACCS.
The Rdfile is written in the Main Mode. The process is initiated by the keyboard selection of the BATCH option which places us in the Batch submode:
User: k
System: Main:
User: batch
System: Batch Input/Output of Molecules, Reactions, and Data
System: Batch:
At this prompt we can use the following Batch suboptions:
writeRD........writes Rdfile in REACCS using the specified
         list descriptor
readRD.........reads RDfile originated in REACCS or
         ChemBase and registers the data in the file
         into current REACCS database
quit...........completes the Batch process and returns to
         Main Mode
There are, similarly to the Rxnfile and PC-Rxnfile, two RDfiles we can write:
1. RDfile to use in transfers between REACCS databases (for molecules or reactions) or between REACCS and MACCS-II (for molecules only)
2. PC-RDfile to use to transfer data from REACCS to ChemBase (for molecules or reactions)

WRITING RDFILE

In writing RDfile, we have to instruct REACCS:
1. how to store the structures
2. which datatype(s) we wish to store
3. how to store catalysts and solvents

RDfile is written in the database from which we are transferring data (the source database). If we need to transfer to other than the current database:
User: k
System: Main
User: db
System: Enter name of new database:
User: (enter the desired database name)
System: Main:
User: batch
System: Batch:
User: writerd
System: Enter RDfile name (<cr> to exit):
User: myrdfile.rdf
System: List descriptor:
  We can use any list descriptor, e.g.:
  Ln (temporary list)
  mylist (a listfile)
  m34-156 (registry numbers for molecules)
  r58-245 (registry numbers for reactions)
  L0 or ref (reference list)
  a search query
System: Are you writing this RDfile for ChemBase [N]?
(Answer Y if writing PC-RDfile, *vide infra*.)
System: How should MOLECULES be stored [table], ireg, ereg?:
  (or: How should REACTIONS be stored [table], ireg, ereg?:)
     How should solvents and catalysts be stored: [table],
     ireg, ereg, or mi?:
table.....store structures as connection tables (default)
ireg......internal registry number is to be used to identify
      data records (use only for data transfer, not for
      structures, and only for databases with identical
      intregnos)
ereg......external registry number is to be used to identify
      data records (use only for data transfer, not
      for structures, and only for databases with
      identical extregnos). Note that the Settings
      MolExtReg or RxnExtReg must be set (p. 699)
mi........store catalysts and solvents as data rather than

structures/connection tables. Note that the MolInfo
   Setting (see p. 594) must be set.
System: Enter datatype name:
User: (enter datatype name defined in the source databa-
   se which we wish to store) and when finished entering
   datatypes, enter <cr>
   To specify all datatypes, type the wild card "@."
System: (writes the RDfile and displays the written records)
System: Batch:
User: quit
System: Main:

WRITING PC-RDFILE

In writing PC-RDfile, we have to instruct REACCS:
1. which datatypes we wish to store
PC-RDfile will not store some REACCS data that are not valid in ChemBase, i.e.:
*atom/atom mappings
*chiral labels
*inversion/retention specifications
*reaction center specifications
PC-RDfile will store catalysts and solvents only as data, not as structures. The type of data for catalysts and solvents will be determined by the MolInfo Setting (see p. 594). Multiple lines of REACCS data are stored in a single ChemBase data field. Data in the floating boxes of the REACCS form (see p. 628) will be stored as the *STEXT in ChemBase.

To write PC-RDfile, we initiate the Batch process as for the RDfile (*vide supra*). The REACCS prompt:
Are you writing this RDfile for ChemBase [N]?:
is answered Y(es).
System: Specify datatypes to be written to RDfile; end
   input with a blank line: (Your current form file
   will be used to place STEXT on the reaction.)
   Enter datatype name:
User: (enter REACCS datatype(s) to be stored)

DATATYPE TRANSLATION TABLE

In Part Two, we have discussed the necessity to define external field names for transfer of data from/to ChemBase if the data fields exchanging data are not identical. The ChemBase Form has the function of the translation table in REACCS which we are going to discuss here. If we are downloading data into ChemBase database, the ChemBase Form does the translation.
In uploading data from ChemBase to REACCS, we can use the ChemBase Form or we

can do the translation by writing a translation table in REACCS. Transferring data between REACCS databases with different datatypes requires a REACCS translation table.

Preparing REACCS Translation Table

The translation table is a list of entries establishing correspondence between the datatypes (data fields) in the source database and the destination database if there is conflict between the datatypes (data fields). The table is prepared in the Editor (or any other system for ASCII files) and is saved in a file. The file is then used in the process of reading the RDfile (*vide infra*).

To check for differences in the datatypes (data fields) of the source and destination databases, we can use the following two methods:

1. a. Write files of the datatypes (data fields) of the
source and destination databases:
In REACCS:
Main: indexdata @/out filename
  b. Print the files and check for differences.
2. Display the datatype (data field) in the destination.
REACCS database:
Main: indexdata <ChemBase field name>, or
    indexdata <REACCS treename>

The display will indicate whether the fields are the same or not. Example:
Main: indexdata workup
    (31) RXN:VARIATION:WORKUP

A single displayed treename indicates no conflict and the translation is not necessary.
Main: indexdata conditions
    (25) RXN:VARIATION:CONDITIONS:TEMP
    (26) RXN:VARIATION:CONDITIONS:ATMOSPHERE
    (27) RXN:VARIATION:CONDITIONS:TIME

Display of more than one treename indicates that translation is necessary.
Main: indexdata catalyst.name
    DATATYPE NOT FOUND
"Datatype not found" indicates that translation is necessary.

To achieve the translation for each of the "Datatype not found" or the multiple datatype display, we make entries in the translation table in the following format:

$source <datatype in the source database>
$destin <datatype in the destination database>

For example:
$source rxn:variation:conditions:temp
$destin rxn:variation:conditions:temperature

The line entries must begin with the "$" and can be up to 80 characters per line. To continue with the line, enter a <cr> after the 80th characters and type to continue the line. The entered datatype can be up to 240 characters for REACCS datatypes and 140 characters for ChemBase fields.
There can be up to 100 entries in one translation table.

Save the file with the translation table as usual.

READING RDFILES

In reading RDfiles and registering the records, we have to instruct REACCS:
1. how to register structures in the destination database
2. how to register data if there is a conflict between the data in the RDfile and the data in the destination database
3. to use translation table file for correspondence of datatypes (data fields) in RDfile and the destination database
4. how to register *STEXT data (if RDfile has been written in ChemBase)

We may encounter the following conflicts between the RDfile and the destination database:
1. The structure in RDfile is a duplicate of the structure in the database.
   Structure is not registered. If the associated data differ, they can be registered (*vide infra*).
2. The structure in RDfile is not a duplicate of the structure in the database.
   The structure is registered. The associated data can be registered (*vide infra*).
3. The data in the RDfile are to be registered but there are already data in the database.
   The data can be registered so that the data in the destination database is unchanged, changed, appended, replaced (*vide infra*).

RDfiles are read in the database which is to receive the data in the file (the destination database).
To read RDfiles in REACCS, we activate the BATCH process in the Main Mode:
User: k
System: Main:
  If the destination database is different from the current one:
  User: db
  System: enter name of new database:
  User: (select the destination database)

UPDATING DATABASES

System: Main:
User: batch
System: Batch Input/Output of Molecules, Reactions, and Data
System: Batch:
User: readrd
System: Enter RDfile name (<cr> to quit):
User: myrdfile.rdf
System: Enter file name of datatype translation table (<cr> for none):
User: mytable
System: How should structures, stored in RDfile as tables, be registered: [new], internal, or external? (If structure in RDfile are stored as regnos, press <cr>):

If in writing the RDfile we have elected to store the structures as regnos, the file contains only data and we cannot register structures, only data.
If in writing the RDfile we have elected to store structures as connection tables, we have the following choices how to register the structures:

new........(default) structures will be assigned new registry numbers on registration. If the structure in the RDfile has a duplicate in the database, the registration will not proceed.
internal...the structure in the RDfile will replace the registered structure of the same internal regno but only if the RDfile structure is not a duplicate.
external...the structure in the RDfile will replace the registered structure of the same external regno but only if the RDfile structure is not a duplicate (note that the MolExtReg or RxnExtReg Settings must be set (see p. 699).

REACCS prompts for instructions how to register data:
System: how should conflicting data be registered:
[add reline], add, ignore, replace, replace reline, or update?:

ignore......data in the RDfile are not registered, the data in the destination database remain unchanged
replace.....data in the RDfile replace the data in the destination database
replace reline......data in the RDfile replace the data in the destination database. REACCS reassigns line numbers in the treenames in the database
update......all data in the RDfile are registered. If a treename has the same line numbers as that one

in the destination database, the RDfile data overwrite the data in the identical treename.

add..........only data in datatypes with different line numbers from those in the destination database are registered.

add reline............if a treename in the RDfile has the same line numbers as the destination database datatype, REACCS appends the RDfile datatype and assigns new line numbers in the database treenames.
Single datatypes are not registered.
Add reline is the default choice since it always preserves the integrity of the destination database.

If we are transferring data from ChemBase, REACCS will prompt for instructions on how to handle the *STEXT data:

System: What datatype should be used to store ChemBase "STEXT" data? (press <cr> if the RDfile was written from REACCS database or if no "STEXT" data should be stored):
User: (enter a datatype in the destination database which will store the "STEXT" data)

If the data are transferred from REACCS database to another REACCS database, the prompt is answered by <cr>.

This concludes the instructions for reading the RDfile and REACCS proceeds with the registration and upon completing it returns the prompt:
System: Batch:
User: quit
System: Main:

PART FIVE

CHEMTALK

1 SYSTEM BACKGROUND

ChemTalk is one of the programs in CPSS (see p. 265) which we can install on IBM PCs or compatibles. ChemTalk is the communication manager which enables us to link a microcomputer with mini/mainframe computers (hosts) and communicate with them. The micro and the host are known as the online system.
ChemTalk has the following functions:
1. terminal emulation
2. data retrieval from MACCS-II databases
3. file transfer between the PC and the host

TERMINAL EMULATION

In Part One we have discussed the Emulation Softwares (see p. 14). ChemTalk can emulate the major terminals that are supported by most of the mini/mainframe systems:
1. DEC VT-100
2. DEC VT-640
3. Tektronix 4010
4. Tektronix 4105
5. DEC VT-220 (ChemTalk Version 1.3 only)

ChemTalk can be used for running text-based programs on mini/mainframes which support DEC VT-100 (or DEC VT-220 for ChemTalk Version 1.3).
ChemTalk can be used for running graphics-based programs, e.g., CAS Registry, MACCS-II, or REACCS which are described in this book (see Figure 2/1, p. 266), or other systems which support the terminals listed above.
ChemTalk allows for capturing text and graphics retrieved in the online systems, and saving them on disk. The saved data can be replayed and otherwise manipulated offline to include it in reports and manuscripts. We are able, for example, to capture the data we retrieve in CAS ONLINE, including chemical structures, and transfer them into ChemText, the word processing program of MDL (see p. 265). The captured text and chemical structures can be enhanced in ChemText and can be inserted in documents.
ChemTalk Plus (Version 1.3) enables the User to create graphic structure queries offline and upload them in CAS Registry (see p. 771) in similar operations as we have described for STN Express (see p. 184).

DATA RETRIEVAL FROM MACCS-II

Part Three describes the techniques for building graphic structure queries and searching and retrieving structures and data from MACCS-II databases. To perform these operations we have to be logged in and running MACCS-II system. We have to be familiar with

MACCS-II and its operations.

However, some Users may not need all the features of MACCS-II to search for structures and data. They have mastered ChemBase (described in Part Two) and maintain their databases in ChemBase and may only occasionally need to find exact structures and associated data in a corporate MACCS-II database and use them either in ChemBase or in creating manuscripts and reports in ChemText.

ChemTalk offers options whose selection results in retrieval of exact structures in the specified MACCS-II database. We can use queries of current structure, internal or external regno(s) to retrieve single or multiple records from MACCS-II databases. We can build structure queries in ChemBase and use them as the current structures for searching MACCS-II databases or we can save them in Molfiles and upload these queries for structure search in MACCS-II or REACCS (note, however, that ChemTalk cannot directly search REACCS databases). We also can retrieve the data associated with the structures.

ChemHost, the MDL program, has to be installed on the mini/mainframe computer for performing these structure and data retrieval operations in MACCS-II with ChemTalk.

FILE TRANSFER

ChemTalk can transfer any type of file from the PC to the host.

In running a system on mini/mainframe computer, ChemTalk can upload an ASCII file when the host system is ready for input (equivalent to keyboard input). This file transfer is not subject to error check.

However, the main function of ChemTalk is transfer of files containing structures and alphanumeric data between ChemBase and MACCS-II or REACCS. These file transfers include error check to ensure correct transfer of data. For this purpose, an error check file transfer protocol program has to be installed on the host, e.g., ChemHost, or Kermit.

Transfer of structural and alphanumeric data is accomplished by writing SDfiles and RDfiles from the source database, transferring the files, and reading the files in the destination database.

2 SYSTEM DESIGN

ChemTalk has been designed to allow for operations related to databases in ChemBase, MACCS-II, or REACCS. We use ChemTalk to work with molecules or reactions and associated data. The structures which are recognized by ChemBase (see p. 268) are recognized by ChemTalk. The concept of numeric and textual data, the database structure, and the nature of data fields discussed in ChemBase (see p. 269) will be useful in understanding ChemTalk. The output and input system and the I/O formats (Forms and Tables) (see p. 274, 346) are used in ChemTalk. The file manager Clipboard (see p. 287) is active in ChemTalk as well as the extensive Help System readily activated by the function key F1.

The design of Forms and Tables is important because in retrieval of structures and data from MACCS-II databases, only the data for which we have boxes in the Form or columns in the Table will be displayed on retrieval of a record from a MACCS-II database. In transfer of files from ChemBase, only those data for which we have boxes in the Form or columns in the Table will be written in RDfiles and SDfiles. In writing an SDfile or an RDfile in MACCS-II or REACCS, however, we can write all data fields or only the specified ones irrespective of the Form or Table. On registration of the records in the files in ChemBase, only those data will be registered for which we have boxes in the Form or columns in the Table and for which the data fields are defined in a ChemBase database.

CHEMTALK FILES

ChemTalk handles the CPSS files which we have discussed in Part Two, i.e., Molfiles (see p. 272), Form and Table files, (see p. 346), MACCS-II List files (see p. 430), Data files (see p. 273), SDfiles and RDfiles (see p. 273, 367). In addition, the type of the following files should be understood for ChemTalk functions within CPSS:

Data files with the .DAT file name extension are the files containing captured data from an online session using ChemTalk as the terminal emulator. The .DAT files can be converted by the MDL program FROM4010 into metafiles and ASCII files.

Metafiles contain graphics output (from CAS Registry, MACCS-II, REACCS or other systems) in a standard MDL metafile format which can be read into ChemText to include the graphics in the documents we create in ChemText.

Login files contain a set of commands that will be executed by ChemTalk at login (thus allowing for automated "login" for different hosts).

3   SYSTEM OPERATION

ChemTalk is operated similarly to ChemBase (see p. 278). Options are selected from the Menu using the mouse pointer. The function keys which are active in the particular mode are shown on the status line.

The cursor control keys move the cursor in one-character or in one-line steps.

Prompt boxes may be displayed by ChemTalk upon activating some of the options or keys. They are answered by Yes or No or by an input in the box.

ESC interrupts any selected option or operation in progress.

The ChemTalk Main Menu displays the default Form (we specify the default Form in the Settings window, see p. 742). We can switch from Form to Table by the VIEW option as in ChemBase. We can bring in other Forms or Tables, thus replacing the default ones by the USE option.

ChemTalk has two modes of operations:
1. Main Mode
2. Terminal Mode

MAIN MODE

Main Mode has a Menu with two menubars similar to the menus in ChemBase. They are shown in Figure 5/1 and Figure 5/2. They are interchangeable by clicking the menubar arrows.

Some of the commands on the Main Menu will work only when ChemHost, the MDL program, is installed on the mini/mainframe computer, i.e.:

FIND and LIST are active only with installed ChemHost on the host. ChemHost will be automatically initiated by ChemTalk when one of these options is invoked.

Sending or receiving files by SEND and RECEIVE options can be accomplished either with ChemHost or Kermit on the host.

ChemTalk comprises analogous features as ChemBase, i.e., Clipboard (see p. 287) and Settings (see p. 284).

ChemTalk Plus (Version 1.3) has two additional options in the Main Menu 2nd menubar: CAS and EDIT (see p. 779).

SETTINGS

The Settings window in ChemTalk which is invoked by point/ select SETTINGS is shown in Figure 5/3.

We have discussed the upper part of the window in ChemBase (see p. 284).

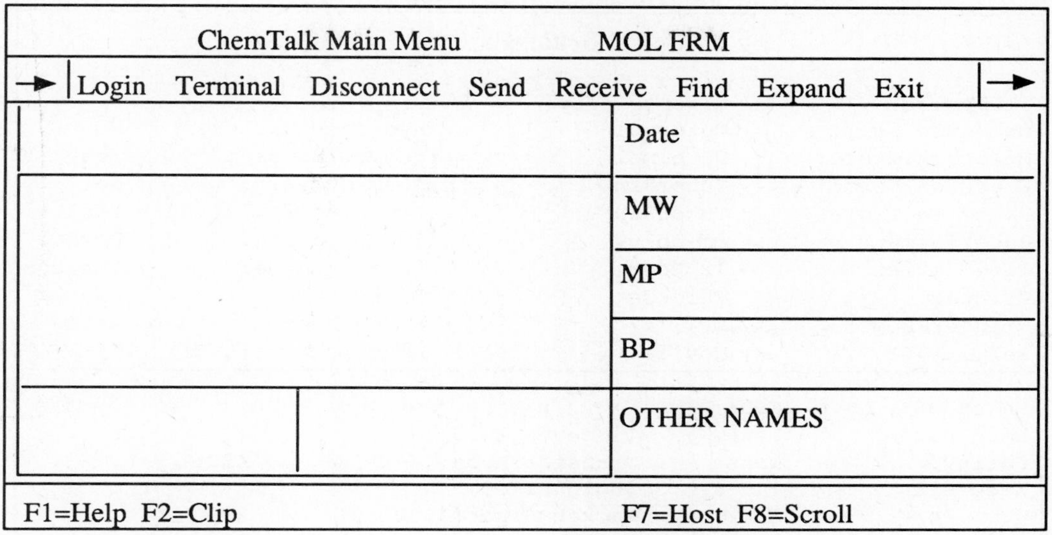

CHEMTALK

	ChemTalk Main Menu				MOL FRM			
→	Login	Terminal	Disconnect	Send	Receive	Find	Expand	Exit	→

	Date
	MW
	MP
	BP
	OTHER NAMES

F1=Help F2=Clip   F7=Host F8=Scroll

Figure 5/1. Main Menu (1st menubar)

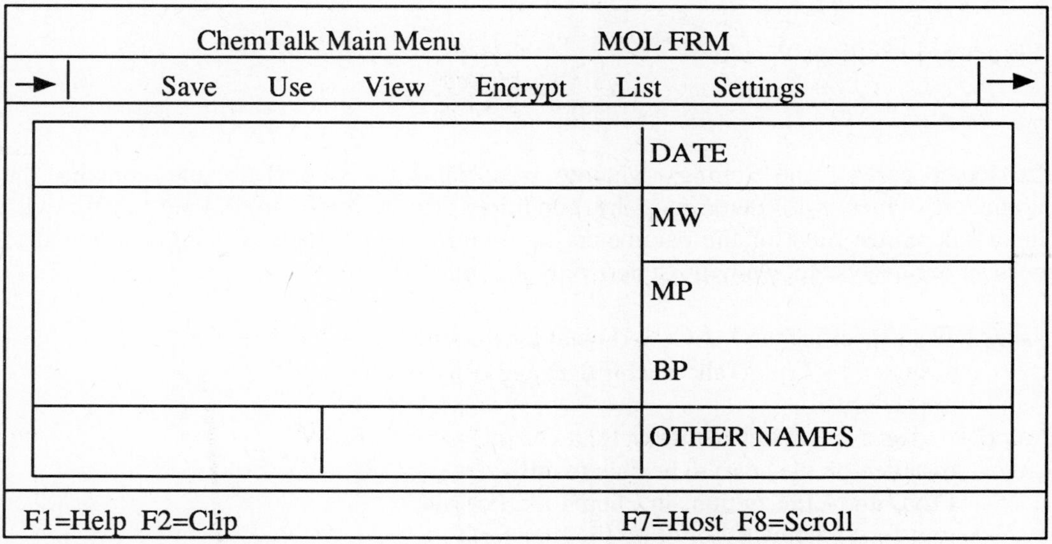

	ChemTalk Main Menu				MOL FRM		
→	Save	Use	View	Encrypt	List	Settings	→

	DATE
	MW
	MP
	BP
	OTHER NAMES

F1=Help F2=Clip   F7=Host F8=Scroll

Figure 5/2. Main Menu (2nd menubar)

```
┌─────────────────────────────────────────────────────────────────────┐
│          ChemTalk Main Menu              MOL FRM                    │
│                         Settings                                    │
├─────────────────────────────────────────────────────────────────────┤
│ Clipboard:           C:\cpss\clip                                   │
│ Molecule Form:       Tutormol      Hydrogen Lables:     No          │
│ Molecule Table:      Tutormol      Stereo Bonds         Yes         │
│ Reaction Form:       TutorRxn      Atom Numbers:        No          │
│ Reaction Table:      TutorRxn      Stext/Abbreviations: Yes         │
│ Date Format:         mm-dd-yy      Bond Spacing:        Normal      │
│ Time Format:         12 Hour       Scaling Mode:        Fit Box     │
│ Printer:             ThinkJet      Background:          Black       │
│ Printer Port:        LPT1:         Foreground:          White       │
│ Help Level:          Novice        Bond Length:[1-38]:  10          │
├─────────────────────────────────────────────────────────────────────┤
│ Maccs DB:      FCD:                                                 │
│ Baud Rate:     1200       Timeout [1-30]:           10              │
│ Parity:        None       Transfer Mode:            ASCII           │
│ Stop Bits:     1          Login File:                               │
│ Comm Port:     Com2       Packet Size [2000-50]     2000            │
│ Data Bits:     8          Emulator:                 ChemTalk        │
│ Keyboard Map:  PC         Lat Service                               │
│ Protocol:      ChemHost   Duplex:                   Full            │
├─────────────────────────────────────────────────────────────────────┤
│ F1=Help F2=Clip F3=Word F4=Line F5=Read F6=Save   F9=Undo F10=Done  │
└─────────────────────────────────────────────────────────────────────┘
```

Figure 5/3. Settings Window

The lower part of the Settings window is specific for ChemTalk and contains the parameters which will determine the conditions for running an online system with ChemTalk. Since most of the parameters are familiar to the readers using a terminal to acces an online system, we will list here only the following:

Maccs DB......specifies the MACCS-II database which will be
        accessed by ChemTalk when FIND and LIST options
        are activated
Protocol......must be set to ChemHost (and ChemHost must be
        installed on the host) to be able to utilize
        FIND and LIST options in ChemTalk. For file
        transfers, ChemHost or other file transfer
        protocol program, e.g., Kermit, can be specified
Login File....specifies the file containing ChemTalk login
        commands for the automated login
Emulator......specifies the graphics terminal we wish to

SYSTEM OPERATION

use and which is supported by the host. It can be:
VT-640, Tek 4010, Tek 4105, ChemTalk
VT-100 is always available in ChemTalk
as the text terminal

We have discussed editing of the Settings and/or saving the edited values for the current session or permanently (see p. 286).

ChemTalk Plus (Version 1.3) has a modified Settings window in which the Help Level Setting has been replaced by:
Atom Value Display Yes/No, and
Logout: manual/automatic Setting has been added to the window.

MORE ON MAIN MODE

Encrypt.......scrambles login files and protect them with
a password
Login.........starts automatic login based on the default
login file

TERMINAL MODE

By point/selecting TERMINAL on the Main Menu, or by pressing F7, we switch to the Terminal Mode. The terminal screen is shown in Figure 5/4. (In our example, a blinking cursor in the upper left corner indicates that we can issue the commands to run an online system.)

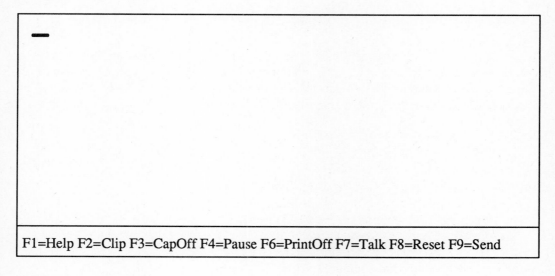

Figure 5/4. Terminal screen

The status line of the terminal screen shows the keys that can be activated in the Terminal Mode. We will discuss later those keys pertinent to a particular operation.

ChemTalk allows for keyboard mapping, thus establishing correspondence between the keys on the keyboard of the PC and the emulated terminal. The User can specify what sort of keyboard is being used: standard PC, PC-AT, or extended AT.

4 TERMINAL OPERATIONS

In order to be able to utilize ChemTalk as the terminal emulator for running MACCS-II, REACCS, or other systems (e.g. CAS ONLINE) on mini/mainframe hosts, we have to discuss first the terminal function of ChemTalk.
To run MACCS-II, REACCS, CAS ONLINE, or other programs that support ChemTalk terminal types, we have to connect with the host by login in the operating system of the host, initiating the program, and issuing commands to run the system.
To retrieve structures and data from MACCS-II database with ChemTalk, we still have to login in the operating system on the host, however, we do not have to initiate MACCS-II and select any MACCS-II options to achieve the retrieval. ChemTalk will do the work.
We already know that to perform retrieval with ChemTalk, ChemHost must be installed on the host.

SWITCHING TO TERMINAL MODE

The function key F7 is the gate/exit to ChemTalk terminal function. It can be pressed in both the Terminal Mode and the Main Mode, while the equivalent TERMINAL menubar option is available only in the Main Mode.
When we start ChemTalk we are in the Main Menu.
If we wish to start the login process to access an online system, we simply press the function key F7 (or select the TERMINAL option from the menubar) which results in display of the terminal screen (Figure 5/4). We are ready to input the first command to connect with the host.
If we are already running the online system and wish to use some of the ChemTalk functions, we again press F7. This will switch us to the Main Menu without interrupting the connection with the host. When we are finished with the Main Menu operations, we return to the online system by F7 or point/select TERMINAL and continue in running the online system. Even if we exit to one of the other CPSS programs, ChemBase or ChemText, the connection with the host is maintained.

TERMINAL SETTINGS

The Settings window (Figure 5/3) enables us to enter the parameters needed to efficiently run the online system. These default values will be loaded every time we start ChemTalk. If we wish, we can override these default values by entering new values for the current session followed by pressing F10. Upon exiting to DOS, they will be erased. Or, we can make them permanent by pressing F6 (save the file).
The terminal type which we wish to emulate in the current session can be specified either in the Settings window or by:
ALT-E......displays Terminal Emulator Window (Figure 5/5).

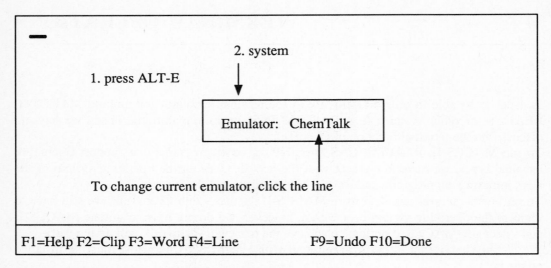

Figure 5/5. Terminal screen/emulator selection

Note that "ChemTalk terminal" is to be differentiated from ChemTalk program. The program is part of CPSS. ChemTalk terminal is an emulation mode that closely mimics Tek 4105 terminal but is specifically designed to optimally work with the MDL mini/mainframe systems, e.g. MACCS-II and REACCS. However, it works with CAS ONLINE as well.

ChemTalk Plus (Version 1.3) has an ALT-E window which shows the values that can be set for the VT-220 terminal.

Each of the emulated terminals has a Settings window which is displayed by:

ALT-S.......displays terminal settings window (Figure 5/6).
The function keys in Figure 5/6 have been discussed in Part Two (see p. 280).
Figure 5/6 shows the ChemTalk terminal values suitable for running MACCS-II or REACCS on VAX computers.
We shall point out here only the "Pause replay at:" value.
If we capture data from an online system, e.g., CAS ONLINE, in a disk file, we can specify the characters (e.g., PLEASE) or an escape sequence contained in the captured file to break the file content. When ChemTalk encounters this value on replay of the captured file, it will pause. Thus, we can display and print graphics and text from the captured file in segments.

ChemTalk Plus (Version 1.3) has an additional value in the ALT-S window:
Alpha plane: Auto/Manual/Up with GIN
These values allow for setting the screen to alpha mode (to display alphanumerics) and/or

TERMINAL OPERATIONS

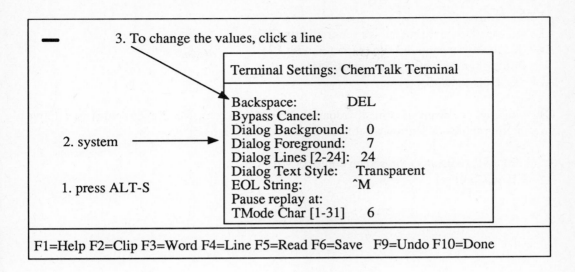

Figure 5/6. Terminal screen/terminal settings

graphics mode.
Auto is the default. The alpha mode goes down when the host issues a request for graphics.
Up with GIN - the alpha mode goes down with all graphics requests except for GIN request.
Manual - the screen has to be set manually with F10.

MORE ON TERMINAL FUNCTIONS

In the terminal mode, we can press function keys and ALT-character keys to invoke the following:

F3....turns on/off capture of host output in a file
F4....interrupts/resumes capturing data if F3 has been set on
F6....turns printer on/off (when on, screen text display in progress will be printed till F6 is again pressed)
F8....resets terminal screen into text mode. For Tek 4010 emulator, it is equivalent to the Page key.
F9....sends ASCII file when host is ready for input (equivalent to typing on keyboard)
F10...turns on/off the dialog box in MACCS-II or REACCS

ALT-B...sends a break to host
ALT-D...erases dialog box

ALT-G...erases all graphics
ALT-K...sets keyboard mappings for emulators/PCs
ALT-R...replays a captured file (captured with F3)
    (replay can be while online or offline)
ALT-X...exits to DOS

Occasionally, we want to refresh (repaint) the entire screen when in graphics to remove unwanted graphics or redraw what is on the screen:

*.......redraws, redisplays the screen content (MACCS-II, REACCS only)

5   MACCS-II DATA RETRIEVAL

In Part Three we have discussed in detail MACCS-II and the operations by which we retrieve structures and associated data from MACCS-II databases. ChemTalk enables us to perform certain simple searches without actually running MACCS-II and without having to know all the details of MACCS-II system. This has certain advantages. If we are running ChemBase or ChemText and need structures and/or data from a MACCS-II database, we do not have to exit from CPSS. We just exit from ChemBase (or ChemText) to ChemTalk to perform the retrieval and then return. If we wish to display the retrieved structures and data from a MACCS-II database in a Form or Table, we can use the Forms and Tables created in ChemBase. We do not have to create a Form or Table in MACCS-II and do not have to know and use MACCS-II at the command level to display data in a Form or Table and/or to write SDfiles. We can move the retrieved structures and data to ChemBase or ChemText through ChemTalk.

To retrieve structures and data from a MACCS-II database with ChemTalk, we have to set the correct communication parameteres in Settings and have to specify the MACCS-II database in Settings which we wish to use in retrieving structures and data (see p. 742).

We have to have a Form with boxes (or a Table with columns) for the structure and data we wish to display.

FORM DESIGN

ChemTalk, as the other CPSS systems, has the default sets of Forms and Tables. Checking the data fields (see Figure 5/12) in the MACCS-II Fine Chemicals Directory (FCD) database which we will use for our examples, we can see that we cannot display all of the data since the boxes do not correspond to the data fields in the database.

Figure 5/7 shows the data fields assigned to the boxes in the default Form. We know (see p. 422) that the asterisked special data are derived from or are attached to structures in MACCS-II databases. They are recognized by Chemtalk. These data will be displayed in the default Form. However, there are no boxes for CAS.Number, Supplier, and WLN, the other data fields in MACCS-II FCD database which we want to display. We have to create boxes and assign the data fields to the boxes as shown in Figure 5/8. (Note that these assignments are done in ChemBase or ChemText.) Since we intend to use the Form for registration into a ChemBase database the field names have to be the names defined in the ChemBase database. On registration, the Form will establish correspondence between the MACCS-II FCD and ChemBase fields. If we wish, we can assign labels for each box.

If we just want to retrieve structure and data and view them or print them in the Form or Table, we can specify only the external field names.

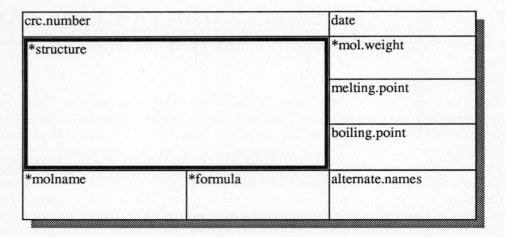

Figure 5/7. Default Form with assigned data fields

Field name: Internal regno External field name: *regno	Field name: Entry date External field name: date
Field name: *structure	Field name: CAS Regno External field name: CAS.Number
	Field name: Supplier
Field name: Name External field name: *molname	Field name: WLN

Figure 5/8. Form with boxes for FCD data fields

The MACCS-II special fields *structure, *formula, and *mol.weight are the same as in ChemBase. However, the MACCS-II *molname, *comments, and *regno are not recognized in ChemBase where they are regular data fields. Therefore, if we wish to display these in ChemTalk (and register them in ChemBase via SDfiles), we have to define field names for them in the Form (and in the ChemBase database). In ChemBase, ID field is the equivalent of extregno in MACCS-II.

ChemTalk uses the external field names to display data retrieved from a MACCS-II database by ChemTalk FIND or LIST if the ChemBase and MACCS-II fields do not match.

ChemBase uses the external field names in writing and reading SDfiles if there is no match of the fields. The field name is translated into the external field name which is written in the file. In registration, the external field names are translated into the field names and data are registered.

Similarly, we assign the field names and/or external field names in the columns of a Table.

EXACT STRUCTURE RETRIEVAL

Main Mode options in ChemTalk allow for retrieval of exact structure(s) and the associated data from MACCS-II databases. The FIND option in ChemTalk has similar function as the FIND option in MACCS-II (see p. 506), i.e., it retrieves exact structures.
We can use the following queries:
1. current structure
2. Molfile
3. internal or external regno(s)

To use a current structure as the query, we create the structure in the Molecule Editor in ChemBase (or ChemText) or retrieve a structure from a ChemBase database or extract it from a ChemText document, and exit to ChemTalk. The structure will be carried over and will be the current structure in ChemTalk.

The retrieval is initiated by dragging FIND(Current) (Figure 5/9).

ChemTalk will start ChemHost on the host mini/mainframe computer and will open the specified database and will ask for the password. If the exact structure is located in the database, ChemTalk displays the retrieved structure with the data for which we have boxes in the current Form (Figure 5/10).

SAVE(Form) will save the structure and data with the Form in a file which can be used in ChemText (or ChemBase).

Note that the box in Figure 5/10 which was assigned to the data field Supplier is too small to display all the text. We can expand any box in the Form by point/select EXPAND and selecting the box which we want to expand (Figure 5/10). The expanded box will be displayed (Figure 5/11).

To retrieve other data for which there are no boxes in the Form, we drag FIND(Field) (the current molecule must be on display) (Figure 5/12).

A list of the MACCS-II database data fields will be displayed (Figure 5/12) from which we select that one we want to retrieve. Data will be retrieved and displayed (Figure 5/13).

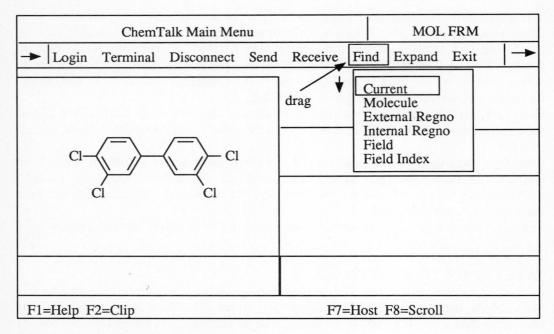

Figure 5/9. Exact structure retrieval

We also can retrieve exact structure by using a structure query which we have stored in a Molfile. Dragging FIND(Molecule) displays the Clipboard with .mol files. We select from the Clipboard the file containing the structure we wish to search. The structure becomes the current structure and the retrieval proceeds as with FIND(Current)(*vide supra*).
FIND(Current) or FIND(Molecule) retrieves a single exact structure.
We can retrieve multiple structures from MACCS-II databases with queries of internal or external registry numbers.
Dragging FIND(Internal Regno) will elicit a prompt for input of intregnos (Figure 5/14).
We can input a single intregno or a series of intregnos. The first item of the current (hit) list is automatically displayed in the Form (Figure 5/15.).
The items in the hit list (current list) can be viewed in the Form one by one by pressing the down arrow, or all of them in a Table.
If we are in Table view, all the entries will be filled in the Table without structures:

```
90  003264-82-2 D6O-NI-O ADJ D1 F1 B-& BD6O-NI-O ADJ D1 F1
91  015282-88-9 D6O-PB-O ADJ D1 F1 B-& BD6O-PB-O ADJ D1 F1
92  014024-61-4 D6O-PD-O ADJ D1 F1 B-& BD6O-PD-O ADJ D1 F1
93  017978-77-7 D6O-PR-O ADJ DXFFXFFXFFF FX1&1&1 B-&BD6O-PR
94  015492-48-5 D6O-PR-O ADJ DX1&1&1 FX1&1&1 B-&BD6O-PR-O A
95  00000000000 D6O-PR-O ADJ D1 F1 B-& BD6O-PR-O ADJ D1 F1 B
```

MACCS-II DATA RETRIEVAL

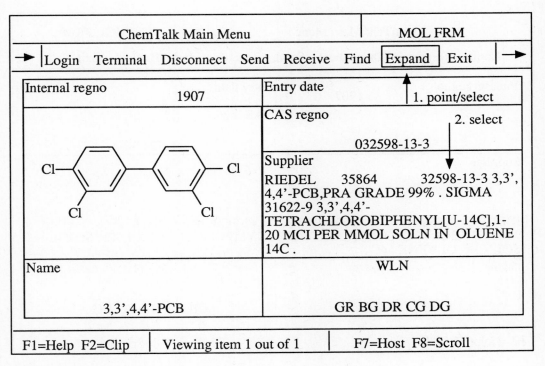

Figure 5/10. Exact structure retrieval

ChemTalk Main Menu	MOL FRM
→ Login Terminal Disconnect Send Receive Find Expand Exit	→

Internal regno	Entry date
1907	
	CAS regno
	032598-13-3

Supplier
RIEDEL   35864   32598-13-3 3,3',4,4'-PCB,PRA GRADE 99% . SIGMA 31622-9 3,3',4,4'-TETRACHLOROBIPHENYL[U-14C],1-20 MCI PER MMOL SOLN IN OLUENE 14C .

Name	WLN
3,3',4,4'-PCB	GR BG DR CG DG

| F1=Help F2=Clip | F7=Host F8=Scroll |

Figure 5/11. Expanded box

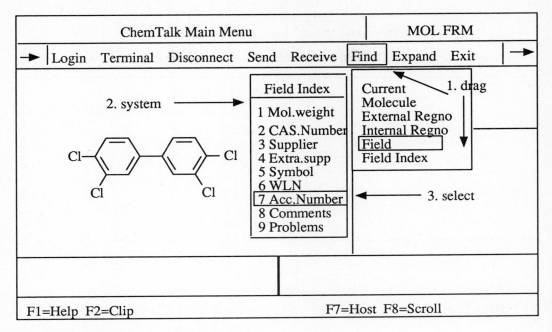

Figure 5/12. Additional data retrieval

Similarly, FIND(External Regno) retrieves a structure or a list of structures by the extregno(s).
Figure 5/16 illustrates the operations with ChemTalk FIND option.

CURRENT LIST IN CHEMTALK

When we perform retrieval in MACCS-II databases by ChemTalk FIND option, the result is a hit list. It is the same hit list as that one created if we run MACCS-II and use FIND (see p. 506), i.e., it contains the registry number(s) of the retrieved structure(s)/record(s). The list is the current list in ChemTalk. We can work with this list in ChemTalk by selecting the LIST option.
We have seen that in MACCS-II, we can save the hit list in a file by WRITE LIST (see p. 527). The Listfile will be saved in the current or specified directory of the mini/mainframe operating system.
In ChemTalk, we can save the current list resulting from ChemTalk FIND operation by:

LIST(Write).....saves current list (hit list) in a Listfile (see p. 430)(Figure 5/17)
Note that the Listfile will be saved **on the host.**
Thus, on the host, we can save MACCS-II hit lists (current lists) in Listfiles either from MACCS-II or from ChemTalk.
These Listfiles can be again read by both. In MACCS-II we select READ LIST (see p.

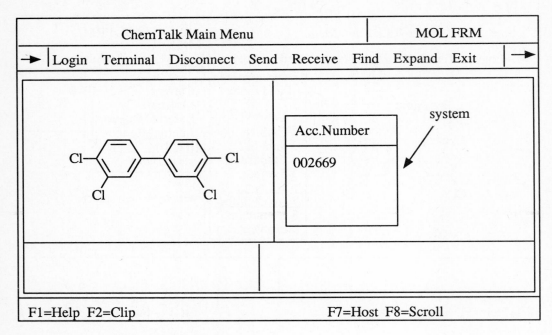

Figure 5/13. Additional data display

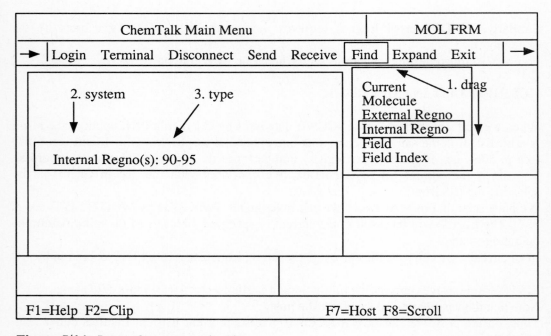

Figure 5/14. Internal regnos retrieval

Internal regno 90	Entry date
[structure of nickel(II) acetylacetonate]	CAS regno 003264-82-2 014024-81-8
	ALDRICH 28 Supplier 3264-82-2 NICKEL ACETYLACETONATE 98%/2, 4-PENTAN DIONE,NICKEL(II) DERIVATIVE . ALFA 22336 NICKEL(II) 2,4-PENTANEDIONAT ,ON POLYSTYRENE . ALFA 53129 3264-82-2 NICKEL(II) 2,4-PENT NEDIONATE . FISHER 1196336 3264-82-2 NICKEL(II)ACETYLACETONATE . FLUKA 72228 3264-82-2
Name NICKEL(II) ACETYLACETONATE	WLN D6O-NI-O ADJ D1 F1 B-& BD6O-NI-O ADJ D1 F1

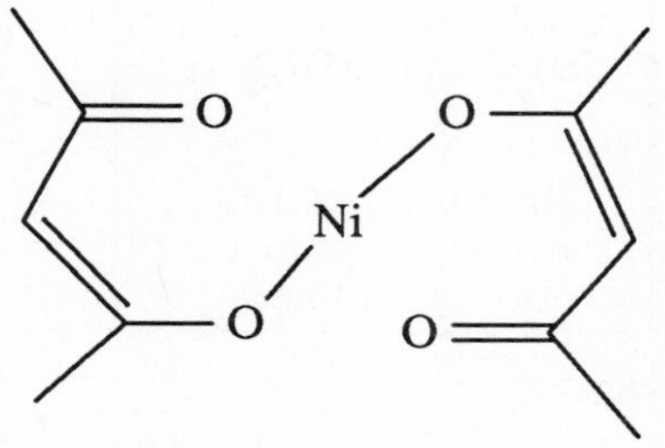

Figure 5/15. Retrieved first item in current list

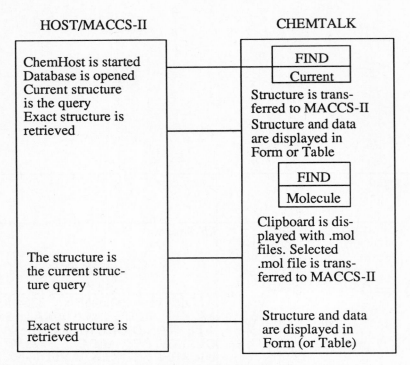

Figure 5/16. Concerted operations in ChemTalk FIND

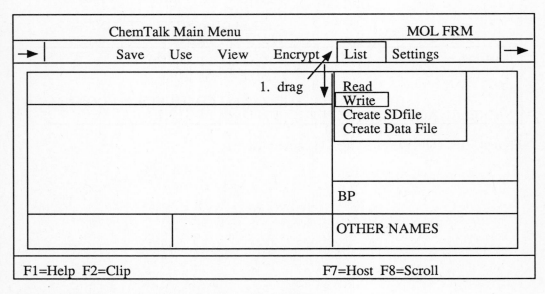

Figure 5/17. Saving ChemTalk current list in listfile

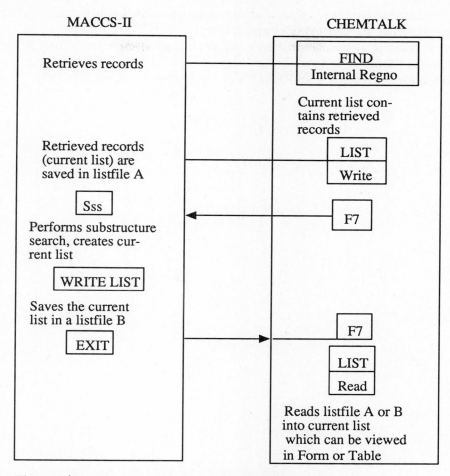

Figure 5/18. Working with lists in MACCS-II and Chemtalk

530).
In ChemTalk, we select:

LIST(Read)......reads a Listfile on the host and makes it current list in ChemTalk

The current list now can be displayed in ChemTalk either in Form or in Table.
The lists from MACCS-II databases are not compatible with the lists from ChemBase databases. Therefore, we cannot use MACCS-II lists directly to transfer structure and data from MACCS-II to ChemBase.
Figure 5/18 illustrates the operations with current (hit) lists in MACCS-II and ChemTalk.
We have seen that the FIND and LIST options enable us to retrieve structures/records from MACCS-II databases and to view the hits in ChemTalk in the current Form or Table

and/or write listfiles. However, if we want to register the retrieved structures and data in a ChemBase database, we have to do it through SDfiles.

Similarly, to transfer molecular structures or reactions and data from REACCS, we have to do it through RDfiles.

No direct retrieval of records from REACCS can be performed with ChemTalk.

6   TRANSFERRING FILES

Transfer of files containing data within CPSS does not require any special techniques. We simply write a file in one of the CPSS programs and read it in the other one.

When we exit from one CPSS program to the other, the current Form (including structures and data if there are any) is transferred into the other program where we can work with it.

With ChemTalk, we are able to transfer any file from the PC to the host. We have already seen that Molfiles can be transferred by the FIND operation (see p. 752). Similarly, we can transfer a Molfile by the SEND operation (see p. 766). If we transfer Molfiles containing generic structure queries from MACCS-II to ChemBase, ChemBase does not recognize:
1. Hydrogen Count (see p. 466)
2. S/A and D/A bonds (see p. 470)

Note that CPSS Rxnfiles are not compatible with REACCS Rxnfiles. However, REACCS can read CPSS Rxnfiles and can write Rxnfiles that can be used in CPSS (see p. 728).

Multiple molecules and data can be transferred from MACCS-II to ChemBase and from ChemBase to MACCS-II through SDfiles and from REACCS/ChemBase through RDfiles. Multiple reactions and data can be transferred from REACCS to ChemBase or from ChemBase to REACCS through RDfiles. If we transfer files from MACCS-II or REACCS to ChemBase through ChemTalk, the entries in the files can be just viewed in Form or Table or they can be registered in a ChemBase database.

Using ChemTalk, we can also perform data transfer from CAS ONLINE by capturing structures and data in a file in the course of the online sesssion and view them offline. We cannot directly register the structures and data in ChemBase. However, we can convert the capture file to a metafile (see p. 780) and, via the metafile, use the structures and data in ChemText.

Note that capturing data from CAS ONLINE is subject to the conditions of CAS ONLINE subscriber contract.

TRANSFER FROM MACCS-II TO CHEMBASE

Transfer of molecules and associated data from MACCS-II to ChemBase can be performed by three methods.
Method A
1. Retrieve structures and data from a MACCS-II database
   by using ChemTalk FIND (see p. 751). This will create a
   current list (hit list) in ChemTalk.
2. Store the current list directly in SDfile by the
   LIST(Create SDfile) option in ChemTalk. The SDfile is
   written in the current (or specified) directory on the
   host.

3. Transfer the SDfile from the host to CPSS Clipboard by selecting the RECEIVE option.
4. Exit from ChemTalk to ChemBase.
5. Register the records in the SDfile in a ChemBase database by selecting USE(SDfile) in ChemBase (see p. 367).

For example, the retrieval in Figure 5/14 creates ChemTalk current list which can be directly written in SDfile by switching to the second menubar and selecting LIST(Create SDfile) (Figure 5/19).

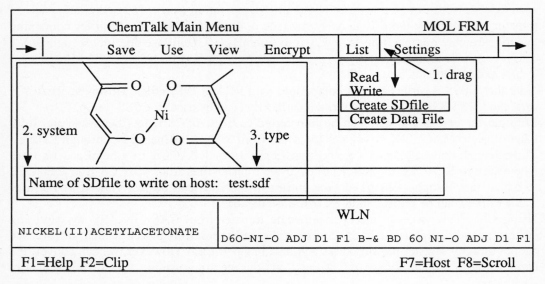

Figure 5/19. Writing SDfile

The SDfile has been written on the host and we are ready to transfer it to Clipboard (Figure 5/20.) (Note that in transfer, the file is actually copied so that the file remains on the host and its copy is in Clipboard after the transfer.)
ChemTalk displays a window showing the transfer in progress (Figure 5/21).
Figure 5/22 illustrates the operations for writing and transferring SDfiles.
Method B
1. Create current list in MACCS-II by running MACCS-II.
2. Save the current list in MACCS-II in a listfile.
3. Read the listfile in ChemTalk to make it the ChemTalk current list.
4. Write SDfile on the host by ChemTalk.
5. Transfer the SDfile to Clipboard.
6. Exit to ChemBase.

TRANSFERRING FILES

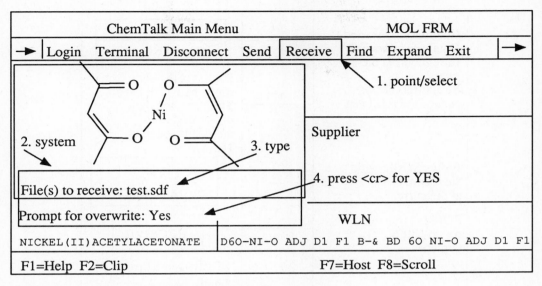

Figure 5/20. Transferring SDfile from host

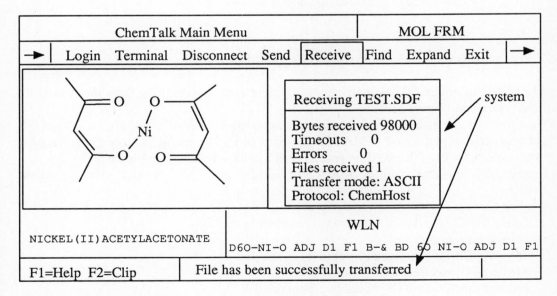

Figure 5/21. File transfer in progress and system messages

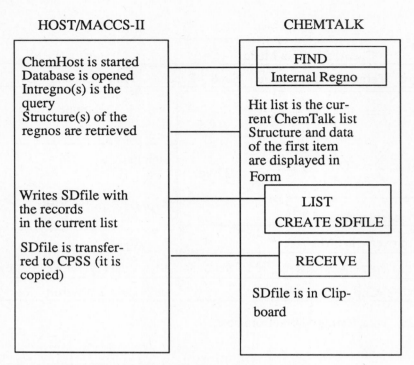

Figure 5/22. Concerted operations in SDfile transfer

7. Register structures and data in ChemBase by USE(SDfile).

The SDfile in Figure 5/19 was written using the current list resulting from the ChemTalk FIND operation (Figure 5/14).
However, we can also use listfiles that have been saved in MACCS-II after a retrieval or search operation performed in MACCS-II (see p. 661). The listfile is made the current list in ChemTalk by LIST(Read) (Figure 5/23) and SDfile is written (see Figure 5/19).
Figure 5/24 illustrates reading MACCS-II listfiles in ChemTalk and writing them in SDfile.
Method C
1. Run MACCS-II and perform retrieval or search operations to create current list.
2. Use MACCS-II at command level (not covered in this book) to write SDfile on the host.
3. Transfer the SDfile to Clipboard by ChemTalk RECEIVE.
4. Exit to ChemBase.
5. Register structures and data in ChemBase by USE(SDfile).

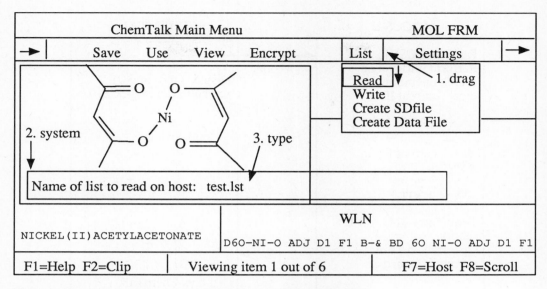

Figure 5/23. ChemTalk current list from MACCS-II listfile

SDFILE FORMAT

The SDfile is an ASCII file containing information about the structure and the data fields and the data itself.

In writing SDfiles in ChemTalk from a MACCS-II database, we can select some or all of the data fields which will be displayed (see Figure 5/12). It may be advantageous to write all data fields and selectively extract only those we wish to register in a ChemBase database by designing the appropriate Form or Table with only the data fields which we wish to get from the SDfile. Only the data fields for which we have a box or column in the Form or Table will be used from the SDfile.

In writing SDfiles from a ChemBase database, only those data fields for which we have boxes or columns will be written in the file and the field names will be translated to external field names (if there are any).

The *regno and *extreg special fields of a record are saved in an SDfile only if at least one data field in the record is present (in the SDfile, the data field is always preceded by the intregno, and if there is any, by the extregno). If there are no data fields, the regnos will not be written. However, the MACCS-II regular data field for external registry number is always written if we specify it as a regular data field in writing the file. Therefore, in transfers in which we are not sure that there are data fields for all records, the use of the regular extregno field ensures the regno transfer for all records.

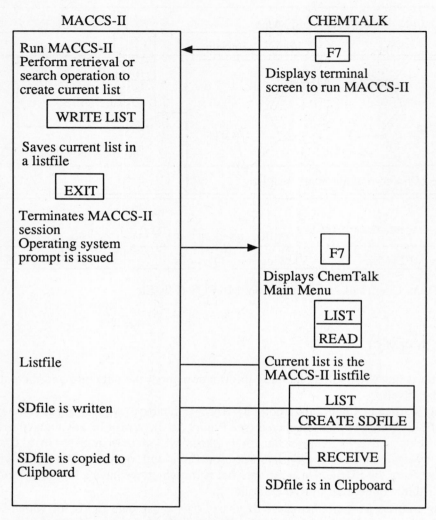

Figure 5/24. Using MACCS-II listfile to write SDfile

TRANSFER FROM CHEMBASE TO MACCS-II

An SDfile with records from a ChemBase database is written in ChemBase by the SAVE(SDfile) option (see p. 367). The file will be in the Clipboard as the .sdf file. To transfer the file to MACCS-II through ChemTalk, we select the SEND option (Figure 5/25). The file is copied so that the file remains in Clipboard and the copy is transferred. We point/drag the following options:

SEND(File)....displays Clipboard from which we can select any

TRANSFERRING FILES

        file we want to transfer to the host
SEND(Batch)...allows to specify any file(s) in any directory
        for transfer to the host (wild cards are
        allowed to transfer multiple files)

Note: SEND(Batch) transfers file(s) and **automatically overwrites** the file(s) with the same name on the host without warning.

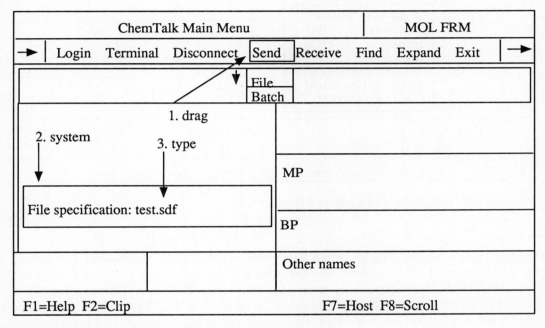

Figure 5/25. Transferring ChemBase SDfile to host

ChemTalk displays a window monitoring the transfer (Figure 5/26).
In transfers of files containing molecules from ChemBase to MACCS-II, the following apply:
1. Text descriptors attached to structures are
   not recognized in MACCS-II.
2. Nonstandard valence markings are not recognized in
   MACCS-II.
4. MACCS-II accepts the mm-dd-yy date format.

TRANSFER FROM REACCS TO CHEMBASE

ChemTalk cannot retrieve structures and data from REACCS or write listfiles or RDfiles. Transfer of data from REACCS is carried out via RDfiles writtent in REACCS.

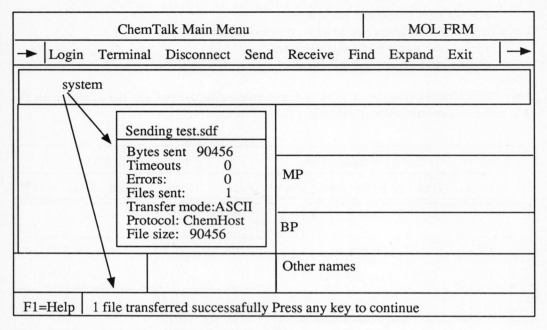

Figure 5/26. Transferring ChemBase SDfile to host

IN REACCS, we write RDfiles in the Batch mode (see p. 728). We have to specify that the file is intended for ChemBase. We can write RDfiles in REACCS with molecules and associated data or reactions and associated data. The RDfile is then transferred in ChemTalk as usual (see p. 766) to Clipboard. To ensure that the RDfile on transfer will be in Clipboard with the .rdf extension, we include .rdf in the filename. In ChemBase, we register (or just view) the records by USE(RDfile).

REACCS Rxnfiles are not compatible with ChemBase Rxnfiles. However, REACCS can read ChemBase Rxnfiles and can create Rxnfiles that can be used in ChemBase (see p. 728)

The following apply for transfer of REACCS PC-RDfiles containing reactions:

1. Atom/atom mappings, chiral flags, inversion/retention specification, and reaction center indicators are not recognized by ChemBase. They are not included in RDfiles created in REACCS.
2. Catalysts and solvents are not stored as structure connection tables, but as data. To register them in ChemBase, we have to define the fields in the ChemBase database. The MDL databases have the following datatypes for catalysts and solvents:
catalyst:regno
solvent:regno

TRANSFERRING FILES

3. REACCS datatypes which contain multiple lines of data will be in a single ChemBase field.
4. Data in the "floating boxes" (see p. 626, 628) will be in the *stext field.

Figure 5/27 illustrates the steps in transferring RDfiles from REACCS.

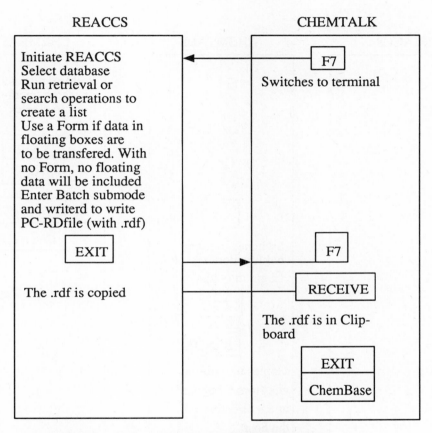

Figure 5/27. Concerted operations in RDfile transfer

Note that REACCS internal registry numbers are stored in the RDfile but cannot be registered in a ChemBase field. However, the external regnos can be registered in a ChemBase field.

TRANSFER FROM CHEMBASE TO REACCS

REACCS can read RDfiles created in ChemBase.
We write RDfiles in ChemBase for molecules and data or reactions and data by the

SAVE(RDfile). Note that only the data will be written for which we have boxes in the Form or Table. If needed, we establish correspondence by the external field names. The files are transferred to REACCS as usual (see p. 766). In REACCS, the files are read in Batch mode (see p. 728) to register the data.

REACCS does not recognize intermediates in reactions. Only reaction schemes with reactants and final products are accepted by REACCS.

Figure 5/28 illustrates transfer of ChemBase RDfiles to REACCS.

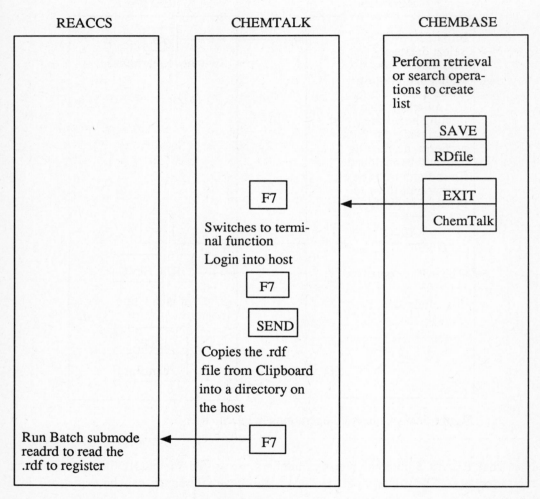

Figure 5/28. Concerted operations in transferring ChemBase RDfile

7   CHEMTALK AND CAS REGISTRY

TRANSFER FROM CAS ONLINE TO CPSS

We can run CAS ONLINE from the ChemTalk terminal screen both for the text (bibliographic) and graphics structure searches. Similarly, we can use the other terminals, VT-640 or Teks, emulated by ChemTalk.
Text and graphics from CAS ONLINE session can be transferred to CPSS and replayed in ChemTalk offline or they can be used in ChemText via ASCII files and metafiles. ChemBase Version 1.3 can utilize metafiles.
We can transfer from CAS ONLINE to CPSS the entire session or only the output after we issue the command DISPLAY (see p. 226).
When we perform the usual operations to retrieve records from CAS Registry File (see Part One for the operations) and are ready to display the retrieved records, prior to issuing the Display Command we turn on the capture function in ChemTalk by F3 (Figure 5/29).

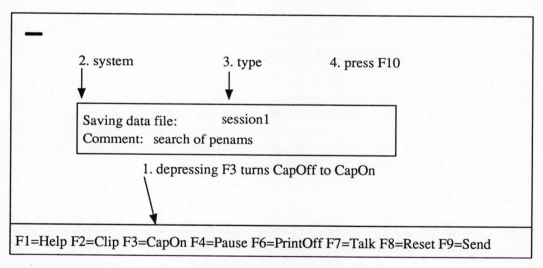

Figure 5/29. Capturing files from CAS ONLINE in ChemTalk

After we press F10 the file is opened and we are back in the Terminal Mode. From this point, the output from CAS ONLINE which is being displayed on the screen will be saved in the file. If we wish to interrupt the saving of the output we press F4(Pause). At the point of the output when we are again ready to capture the data, we press F4. We can perform any additional operations in CAS ONLINE after pressing F4 and start capturing the output in the same file by again pressing F4.

If we wish to save the output of the same CAS ONLINE session in different files, we press F3 for each separate capture and enter the name of the file at the prompt. The files are automatically saved when we exit CAS ONLINE and are in the Clipboard with the .DAT extension.

The captured files can be replayed in ChemTalk by ALT-R. (Note that the terminal emulator must be set to the same value as the one which was used for capturing the file.)

The .DAT file is selected from the displayed Clipboard. The output is shown on the screen in the same format as it was captured. We can use the "Pause Replay" terminal setting (see p. 746) to break the output into segments.

To utilize the captured file in ChemText, we have to use the MDL program FROM4010 (see p. 780) which converts the captured files into ASCII text and structure graphics metafiles that can be read by ChemText.

Thus, we can input directly the graphic chemical structures along with the text into ChemText documents. Part One contains numerous examples of the captured sessions and Display output of structures and data with ChemTalk followed by conversion into metafiles and ASCII files and using them in ChemText to prepare the manuscript for this book (see, e.g., p. 71, 232-238).

STRUCTURE QUERIES FOR CAS REGISTRY SEARCHES

The Users who have acquired ChemTalk Plus (Version 1.3) will be able to create structure queries offline and upload them in CAS Registry for search similarly to the operations we have described for STN Express.

ChemTalk Plus (Version 1.3) includes the Molecule Editor which we have described in Part Two (see p. 292). The operations by which we have created structures in ChemBase are valid for creating query structures for CAS Registry or Beilstein Files. However, CAS Registry does not allow for searching stereochemical structures with graphics queries. The stereobonds which we may indicate in structures created in the Molecule Editor are ignored in CAS Registry searches.

Creating Exact Structure Queries

We employ the usual techniques of the Molecule Editor to create structure skeletons; chains, rings, and fused ring systems. We can use the TEMPLATES.

We specify single atom values at nodes by dragging ATOM; bond values by dragging BOND. The BOND(Single, Double, Triple) are equivalent to SE, DE, and T in CAS Registry (see p. 90, 95).

We can specify CHARGE and MASS on atoms by selecting the CHARGE and ISOTOPE options. We specify HMASS (i.e., D and T) (p. 158-163) by ATOM(Other Special).

Models for building structures by recalling a Registry Number or a fragment from the CAS Registry Fragment File cannot be used.

We can insert in query structures the CAS Registry common substituents and functional groups (see p. 85) by employing the technique of atom alias (see p. 311).

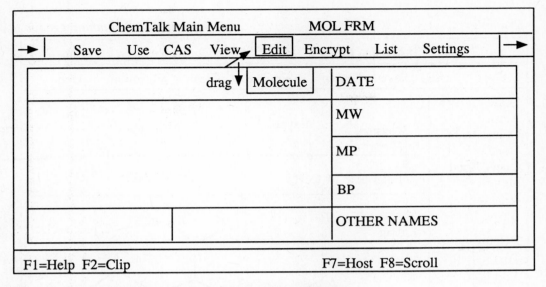

Figure 5/30. Main Menu (2nd menubar) of ChemTalk Plus

Figure 5/31 shows insertion of *m*-phenylene and phosphono groups.
Figure 5/32 shows the created query by the operations in Figure 5/31.
The Registry File allows for defining structure queries for EXACT Search with unspecified bond values. We use the Query Bond(Any) to indicate unspecified bonds in a query. We can also indicate that tautomerism (p. 97) may apply for a query by using the atom alias technique and a special indicator, i.e.,"CAS:"
To attach attributes to a query:
1. drag Text(Edit)
2. click outside of the structure
3. type CAS:t at the prompt

Creating Generic Structure Queries

We can employ the Query Atom(A, Q) to specify the CAS Registry File variables.
The atom alias technique specifies the CAS Registry File X and M variables (p. 119):
1. draw the structure query and assign to the desired node a halogen value
2. drag Text(Edit)
3. click the halogen node
4. type X at the prompt
The node having the halogen will be converted to generic X on searching with the query.
The Registry File variables CY, CB, and HY (p. 119) are similarly assigned:
1. draw a structure query
2. drag Text(Edit)
3. click the node intended for CY, CB, or HY

Example: *Inserting CAS common groups*

```
|                      Molecule Editor                              |
| Clear    Use    Save    Group    Template    Clean    Reaction    Done |
```

- Draw
- Atom
- Query Atom
- Bond
- Query Bond
- Move
- Rotate
- Resize
- Delete
- **Text**
- **Edit**
- **Move**
- **Delete**
- Isotope
- Valence
- Chiral

3. system → Enter text: M-C6H4   4. type   6. type → PO3H2

1. drag
2. place/select
5. place/select

F1=Help F2=Clip   F6=Rxn F7=Atom F8=Scrl F9=Switch F10=Rdrw

Figure 5/31. Specifying CAS Registry common substituents

4. type the variable symbol at the prompt

We can create sets of User variables by dragging Query Atom(List).

We can negate **one** node by using Query Atom(Not List).

We specify variable bonds by Query Bond(S/D, Any, Ring, Chain) and tautomeric bonds (p. 138, 142, 146).

We cannot specify the other CAS Registry bond value and bond type variables.

Defining Gk Groups

ChemTalk Plus allows for defining the Registry File Gk groups in generic structure queries (p. 119) which, however, are labeled R# (# = number).

In defining an R group, we have to have **at least two units** for the group definition.

Figures 5/33-5/36 show how to create in ChemTalk Plus the structure we have used in Part One, Figure 1/103 (p. 132).

CHEMTALK AND CAS REGISTRY

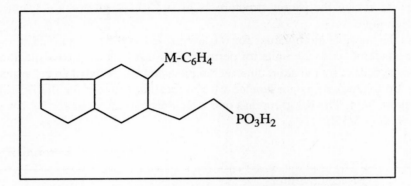

Figure 5/32. Specified CAS Registry common substituents

Example: *Specifying generic R groups*

Figure 5/33. Specifying generic R groups

Figure 5/33: we create the parent structure as usual. The R groups are specified by the atom alias technique.

Figure 5/34: the variable atom values for R1 are created in the Molecule Editor. Note that R1 has **two** attachments so the units for defining R1 must have also **two** attachments. The attachment is specified by the atom alias technique using the *# label (# corresponds to the R number). Both attachments are labeled *1 by repeating the step for the first attachment shown in Figure 5/34. The R2 group has **one** attachment so the units for R2 have also **one** attachment (Figure 5/35).

Example: *Defining R groups*

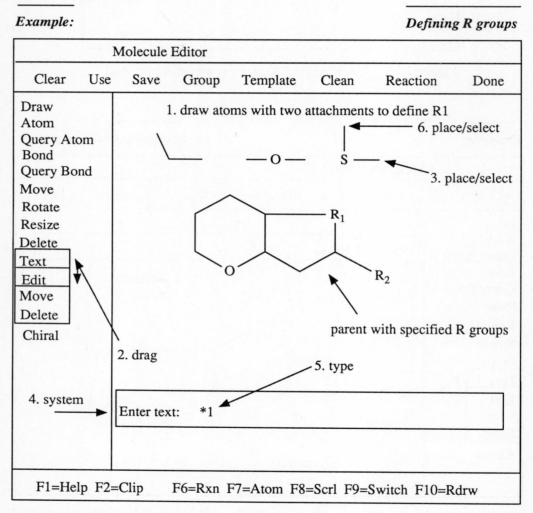

Figure 5/34. Defining R groups

Figure 5/36 shows the completed generic structure query with defined R1 and R2 groups.

CHEMTALK AND CAS REGISTRY

Example: *Defining R groups*

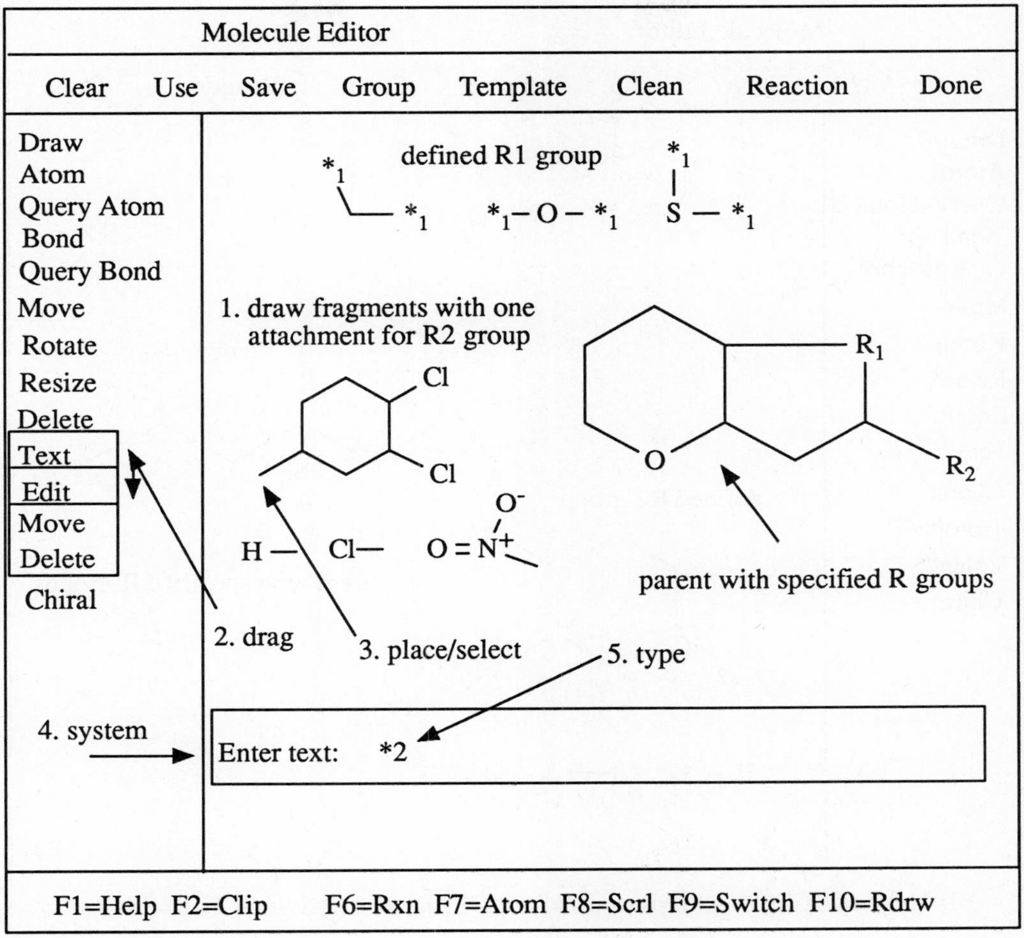

Figure 5/35. Defining R groups

UPLOADING STRUCTURE QUERY TO CAS REGISTRY

The structure query we have created can be saved in a Molfile and the Molfile can be recalled to upload the structure; or the current structure query can be directly used for upload and search.

Figure 5/37 shows the activated CAS option in the Main Menu 2nd menubar.

To connect from ChemTalk Plus to CAS Registry File we have to set up the parameters needed to call up either directly or through a telecommunication network. We select:

Example: *Defining R groups*

```
┌─────────────────────────────────────────────────────────────────────┐
│                         Molecule Editor                             │
├─────────────────────────────────────────────────────────────────────┤
│   Clear    Use    Save    Group    Template    Clean   Reaction   Done │
│                                                                     │
│  Draw                                                               │
│  Atom                       defined R1 group                        │
│  Query Atom                                                         │
│  Bond                                                               │
│  Query Bond                                                         │
│  Move                                                               │
│  Rotate                                                             │
│  Resize                                                             │
│  Delete                                                             │
│  Text                                                               │
│  Charge            defined R2 group                                 │
│  Isotope                                                            │
│  Valence                            parent with specified R groups  │
│  Chiral                                                             │
│                                                                     │
│                                                                     │
│                                                                     │
│                                                                     │
├─────────────────────────────────────────────────────────────────────┤
│   F1=Help F2=Clip    F6=Rxn F7=Atom F8=Scrl F9=Switch F10=Rdrw      │
└─────────────────────────────────────────────────────────────────────┘
```

Figure 5/36.   Defined R groups

CAS Settings........takes us through a question and answer process to set up the parameters:
        Enter Default Vendor (Direct Dial)/Telenet/Tymnet
        Enter Modem Baud Rate (300-19,200)
        Enter CAS Online Direct Dial phone number
        Enter Telenet phone number
        Enter Tymnet phone number
        CAS display scrolling mode (24 line Page/Continuous)
        Handling of illegal CAS Atom Labels (Ignore/Notify)

CHEMTALK AND CAS REGISTRY

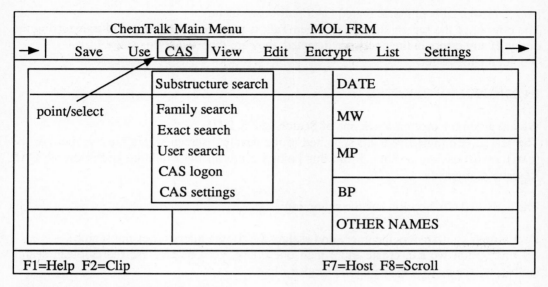

Figure 5/37. Activated CAS option

When we complete the question-answer session, we receive a message at the bottom of the screen:
CAS Settings Menu Complete.

We are ready to call up CAS ONLINE.
With a current structure query in the Form, we select:

CAS logon......automatically connects with STN according to the parameters in the CAS Settings (*vide supra*).

We can monitor the logon by the messages at the bottom of the screen.
After we answer the prompt for ID and Password and are connected to STN, we receive the message:

CONNECTED TO CAS ONLINE

The screen switches to the terminal screen and we receive the STN greetings. We are in the HOME File (see p. 20).
The last line then is the CAS prompt:
=>
At this prompt enter the desired file, e.g., file lreg (or reg or beilstein) (see p. 34).
When we are transferred to the file we selected and have the prompt:
=>
We press F7 to get back to the ChemTalk Plus screen and select the desired Type of

Search (see p. 217) in the activated CAS pull-down menu.
The bottom of the screen shows that ChemTalk is preparing the query and transferring the query. If any errors in the query are detected, they are reported at this stage.
We get the message:

Ln STRUCTURE CREATED

We are given the prompt for Scope of Search (see p. 207).
The search is initiated. We are switched to the terminal screen where we can monitor the search statistics (see p. 209). From this point we use all the available and desirable CAS Registry commands.

The activated CAS option has the suboption:

User search.......transfers the query and assigns Ln for the query (does not search).
By this option we can create more than one query, get Lns and use the Boolean search technique (see p. 247) to search the Registry File.

CREATING METAFILES

If we capture structures and text and numeric data (see p. 771), i.e., the result of our search, we can convert the graphics into metafiles and the text and numerics into an ASCII file by using the MDL program FROM4010 which is supplied on the Utilities diskette.
The captured files have the extension .DAT.
We load FROM4010 in the directory where it resides and follow the instructions issued by the program for converting the specified .DAT file. The metafiles and the ASCII files are stored in Clipboard or any other specified directory.
Note that selecting a name for metafiles does not require the .MET filename extension while ASCII filename requires the .ASC extension to be recognized by Clipboard.
Metafiles can be used in ChemText and in ChemBase Version 1.3.

INDEX

INDEX

ACTIVATE,CAS 240
ANSWER SET
   ACTIVATE,CAS 240
   CAS 210
   DELETE,CAS 241
   DISPLAY,CAS 241
   PRINT,CAS 242
      ORDER CANCEL,CAS 243
      ORDER DISPLAY,CAS 244
   RECALL,CAS 240
   SAVE,CAS 240
AROMATICITY
   CAS 95
   CHEMBASE 268
ATOM ALIAS
   CHEMBASE 311
ATOM/ATOM MAPPING
   AUTOMAP,REACCS 617,619,620
   BOND MARKING,REACCS 623
   IN RSS,REACCS 685
   MANUAL MAP,REACCS 617,621
   REACCS 616
ATOMS
   CHAIN/RING
      CAS 146
      MACCS 472,483
      UNSATURATION,MACCS 471
   DEFAULT VALUE,CHEMBASE 293
   NEGATION OF
      CAS 124,129,130
      CHEMBASE 315
      MACCS 465
      REACCS 604
   NUMBERS MACCS 447
   REPEATING
      CAS 132
      MACCS 475
   SPECIFY,CHEMBASE 293

VALENCE,CHEMBASE 309
VALUE ASSIGNMENT
   CAS 83
   CHEMBASE 293
   MACCS 449
     USE OF+OR-,453
   REACCS 600
VARIABLE
   CAS 118
   CHEMBASE 314
   MACCS 465
   REACCS 604
ATTACH MODE
   MACCS 458
ATTRIBUTES,CAS 146
   DELETE,CAS 174
BATCH SEARCH,CAS 215
BEILSTEIN,CAS 4
BONDS
   ALL IN RINGS,CAS 102,105
   ASSIGN IN RINGS,CAS 106
   CAS 29
   CHANGE,REACCS 623
   DEFAULT VALUE,CAS 29
   DEFINITION,CAS 96,145
   DELETE,CAS 170
   DISPLAY OF,CAS 144
   EXACT,CAS 100,103
   EXACT/NORMALIZED,CAS 139
      IN GENERIC STRUCTURES
      ASSIGNMENT,CAS 138,141
      DEFINITION,CAS 139
      DISPLAY,CAS 95,139
      SYMBOL FOR,CAS 144
   NORMALIZED,CAS 90,95,97,100
   SET BOND,CAS 101,102
   SPECIFICATION
      CAS 90,141

CHEMBASE 295
STN EXPRESS 190
SYMBOLS FOR,CAS 95
TYPE GENERIC,CAS 29,142
VALUE SPECIFICATION,CAS 100
VALUES AND TYPES,CAS 90,144
VARIABLE
   CAS 138,142
   CHEMBASE 315
   MACCS 470
   REACCS 609
BOOLEAN OPERATORS,CAS 247
BOX
   CHEMBASE 346
   CREATE,REACCS 627,632
   DATATYPES FOR,REACCS 629
   EXPAND,CHEMTALK 753
   IN TABLES,CHEMBASE 358
   OPTIONS,CHEMBASE 357
   STRUCTURE FIELD FOR,REACCS 629
BUILD MODE
   REACCS 588
BUILDING
   MOLECULES,REACCS 599
     EXACT 599
     QUICK WAY 606
   REACTIONS
     REACCS 608
      QUICK WAY 622
CA FILE 3
CAOLD FILE 3
CAS COMMANDS
   AT => 22
   AT : 23
CAS ONLINE 3
   FILES IN 3,5
CAS QUERIES
   CREATE,UPLOAD
     CHEMTALK 771
     STN EXPRESS 184
CASREACT 4,204
CHAINS
   BRANCHED,CAS 48,49
   CAS 44

DELETE,CAS 171
STRAIGHT,CAS 46,47
CHANGE
   BOND VALUES,CAS 175
   CHAINS,CAS 175
   NODE VALUES,CAS 175
   RINGS,CAS 175
   STRUCTURES,REACCS 601
CHARGE
   CAS 158
   CHEMBASE 309
   REACCS 600
CHIRAL
   CHEMBASE 308
   MACCS 418
   REACCS 652
CLEAN STRUCTURE
   CAS 181
   CHEMBASE 296
   MACCS 495
   REACCS 601
CLIPBOARD
   CHEMBASE 287
     CREATE 288
     SELECT FROM 289
COLUMN
   CHEMBASE 350
   CREATE,CHEMBASE 351
   REACCS 634
COMBINATION SEARCH
   REACCS 645,670,687
COMMAND SYNTAX
   CAS 24
COMMANDS
   CAS 21
COMPONENT STRUCTURE,CAS 27
CONFIGURATION FILE
   READ,CHEMBASE 287
   WRITE,CHEMBASE 286
CONNECT COMMAND
   CAS 152
   SELECTIVE FUSION,CAS 157
   SUBST. DEGREE,CAS 152
   SUBST. TYPE,CAS 155

INDEX

CONNECTION TABLE
    CAS 29-31
CONVERSION SEARCH
    REACCS 644,681,685,694
CORRECT/MODIFY REACTIONS
    CHEMBASE 338
CORRECT/MODIFY STRUCTURES
    CAS 168
    CHEMBASE 332
    MACCS 495
    REACCS 601
COUNTERS
    IN REGISTRY MODE,MACCS 506,507
    IN SEARCH MODE,MACCS 508,509
CPSS PROGRAMS 265
CPSS SYSTEMS INTERFACE 266
CROSSOVER
    CA, CAOLD 257
    CAS 10,11,257
    CASREACT 258
CURRENT LIST
    CHEMTALK 755
    DEFINE,REACCS 665
    SAVE,CHEMTALK 761
CURRENT-AWARENESS
    SDI,CAS 259
DATA
    DESCRIPTOR,REACCS 640
    DELETE RECORDS,MACCS 567
    QUERY FORMAT,REACCS 642
    REGISTER,REACCS 724
    RETRIEVAL FROM MACCS
        CHEMTALK 749
        REGNOS 751
    SEARCH EXPRESSION,REACCS 641
DATA FIELDS
    CAS 9
    CHEMBASE 272
    DATE,CHEMBASE 274
    DESCRIPTORS,MACCS 426
    EXTERNAL,CHEMBASE 359,368
    FIXED,MACCS 425,551
        DEFINE 551
    FIXED TEXT,CHEMBASE 272
    FLEXIBLE,MACCS 426,553
        DEFINE 553
    IN TABLES,CHEMBASE 357
    INDEX OF,MACCS 543
    LINK TO FORM,CHEMBASE 356
    NAMES,CHEMBASE 271
        CORRESPONDENCE FOR
        CHEMBASE 368
        REACCS 730
        CHEMBASE 359,368
    NUMBERS
        INTEGER,CHEMBASE 272
        RANGE,CHEMBASE 272
        REAL,CHEMBASE 272
    SET,CHEMBASE 360
    SPECIAL
        CHEMBASE 271
        MACCS 422
    TYPES,CHEMBASE 271
DATA FILE,CHEMBASE 273
DATA FORMAT
    FORMAT SYMBOL,MACCS 427
    MACCS 426
DATA RECORDS,CAS 9
DATA RETRIEVAL
    CHEMBASE 403
    MACCS/CHEMTALK 737,749
DATA SEARCH
    ANSWER DISPLAY,REACCS 716
    DATA FIELDS IN,MACCS 540
    DATA QUERY IN,REACCS 701
        CHEMBASE 405,407
    EXACT MOLS,REACCS 697
    EXACT RXNS,REACCS 697
    FORMULA,CHEMBASE 404
    FORMULA SEARCH
        CHEMBASE 406
        MACCS 540
        REACCS 701
    ID,CHEMBASE 403
    QUERIES,CHEMBASE 407
    RANGE TEXT,MACCS 546
    RANGE COLUMN,MACCS 545
    REACCS 697

    IN DATA MODE,MACCS 542
    IN SEARCH MODE,MACCS 542
    LIST FILE IN,REACCS 701
    LISTFILE IN,REACCS 701
    NAME SEARCH,MACCS 537
    NUMERIC DATA
        CHEMBASE 408
        MACCS 546,547
    REACCS 697
    REACTION KEYWORDS,CHEMBASE 411
    REACTION SYMBOL,CHEMBASE 410
    REGNOS,MACCS 535
    REVIEW OF
        MACCS 536
        REACCS 698
    STRUCTURAL KEYS SEARCH,MACCS 539
    SUBSTRUCTURE SEARCH,REACCS 701
        FORMULA 701
        KEYS 702
        NAME 704
    TECHNIQUES OF,MACCS 544
    TEXT DATA
        CHEMBASE 408
        MACCS 545
DATA TYPE
    MACCS 426
DATABASES
    BUILDING OF,CHEMBASE 361
    CHANGE OF,REACCS 598
    DEFINITION,CHEMBASE 340
    DELETE
        DATA,MACCS 567
        RECORDS,MACCS 562
    IN REACCS 577,597
    MODIFICATION OF,CHEMBASE 345
    MOLECULES IN,CHEMBASE 342
    NUMERICS/TEXT,CHEMBASE 344
    PASSWORD FOR,CHEMBASE 345
    REACCS 582
    REACTIONS IN,CHEMBASE 344
    REGISTER STRUCTURES
        MACCS 555
        REACCS 722
    RE-REGISTER RECORDS,MACCS 562

    REGISTER TEXT,MACCS 558
    SET UP
        CHEMBASE 340
        MACCS 550
    TEXT IN,CHEMBASE 344
    UPDATE
        MACCS 555
        REACCS 722
DATABASE STRUCTURE
    CAS 9
    CHEMBASE 270
    MACCS 422
    REACCS 584
DATABASE SUBSET
    REACCS 667,695
    SEARCH IN,REACCS 667
DATATYPES
    ASSIGN/BOX,REACCS 629
    ASSIGN/TABLES,REACCS 634
    ENTITY,REACCS 580
    FIELD,REACCS 581
    HIERARCHY,REACCS 580-582
    MOLECULES,REACCS 581
    MULTIPLE/TABLE,REACCS 714
    PARENT,REACCS 580
    REACTIONS,REACCS 581
    RECORDS,REACCS 581
    TRANSLATION,REACCS 730
    TREENAMES,REACCS 583
        DISPLAY,705
DATE,CHEMBASE 272
DATFILES
    FORMAT OF,MACCS 431
DELETE
    ANSWER SET,CAS 241
    ATOMS
        CAS 168
        CHEMBASE 332
        MACCS 495
        REACCS 601
    ATTRIBUTES,CAS 174
    BONDS
        CAS 170
        CHEMBASE 332

INDEX

    MACCS 495
    REACCS 600
  CHAINS,CAS 171
  DATA RECORDS,MACCS 567
  FRAGMENTS,CHEMBASE 332
  GROUPS,CHEMBASE 332
  LISTS,REACCS 658
  MOLECULES,REACCS 726
  NODES,CAS 170,171
  PRINT ORDER,CAS 243
  REACTIONS,REACCS 726
  RECORDS,CHEMBASE 365
  RINGS,CAS 172
  STRUCTURE
    CAS 173
    CHEMBASE 321,332,333
    MACCS 495,562
  STRUCTURE RECORDS,MACCS 562
  STRUCTURE ATTRIBUTES,CAS 174
DISCONNECT
  CAS 23
DISPLAY
  ANSWER FILENAME,CAS 241
  ATOM VALUES,CHEMBASE 294,295
  DATA FIELDS,MACCS 543
  DOCUMENT DATA,CAS 230
  FIXED DATA FIELDS,MACCS 508
  FLEXIBLE DATA FIELDS,MACCS 509
  FORMAT
    ABSTRACT,CAS 237
    ALL,CAS 237
    BIBLIOGRAPHY,CAS 235
    BROWSE,CAS 239
    INDEX TERMS,CAS 236
    REGNOS,CAS 232
    SAMPLE,CAS 234
    SCAN,CAS 231
    SEARCH ANSWERS,CAS 232
    SUB,CAS 232
    SYSTEM'S,CAS 228
    USER'S,CAS 229
  FORMATS FOR
    CHEMBASE 346
    REACCS 625

  HIT STRUCTURE,MACCS 506
  METAFILES,CHEMBASE 370
  MULTIPLE DATA,REACCS 632
  MULTIPLE HITS,MACCS 511
  R GROUPS,MACCS 494
  REACTIONS,REACCS 611
  SEARCH ANSWERS
    CAS 226
    MACCS 506
    REACCS 649,652
  SPECIFICATION OF,CAS 227
    ATTRIBUTES,CAS 165
    CON. TABLE,CAS 167
    STRUCT. IMAGE,CAS 31,165
    IMAGE+ATTR.,CAS 166
  STRUCTURE DATA,CAS 167
  SUBSTANCE DATA,CAS 228
DRAW
  CHAINS
    CAS 45
    CHEMBASE 292
    MACCS 447
    REACCS 601
  RINGS
    CAS 50
    CHEMBASE 294
    MACCS 448
EDIT
  BOX DATA FIELDS,REACCS 632
  REACTIONS
    CHEMBASE 321
    REACCS 608
  STRUCTURES,STN EXPRESS 192
EDIT COMMAND
  CHEMBASE 291
EMULATION SOFTWARE,CAS 14
EXACT SEARCH
  CAS 218
  CHEMBASE 374
  CHEMTALK 751
  MACCS 503
  REACCS 648
  TAUTOMERS IN,CHEMBASE 388

FAMILY SEARCH
   CAS 218
FILE COMMAND,CAS 34
FILES
   CAPTURE,CHEMTALK 739,771
   CHEMBASE 272
   CHEMTALK 739
   MACCS 428
   REACCS 584
   DATABASE FILES
      CAS 4
      REACCS 575
   DATA FILE
      CHEMBASE 273,378
      CHEMTALK 739
   DATFILES,MACCS 431
   FORM FILE
      CHEMBASE 273
      REACCS 587
   HOME FILE,CAS 20
   LIST FILE
      CHEMBASE 273
      REACCS 585
   LISTFILES
      MACCS 430
      REACCS 586
   LOGIN FILES,CHEMTALK 739
   METAFILES
      CHEMBASE 368
      CHEMTALK 739,780
   MOLFILES
      CHEMBASE 272
      MACCS 431
      REACCS 586
   OMIT LIST FILE,REACCS 585
   PERMANENT FILES,REACCS 585
   QUERY FILE,REACCS 585
   RDFILE
      CHEMBASE 273
      REACCS 587
   RXNFILE
      CHEMBASE 272
      REACCS 586
   SDFILE,CHEMBASE 273

   TABLE FILE,CHEMBASE 273
   TEMPLATE FILE,CHEMBASE 274
   TEMPORARY FILES,REACCS 584
      LIST FILES,REACCS 585
      OMIT FILES,REACCS 585
      QUERY FILES,REACCS 585
   TRANSFER OF
      CHEMBASE 368
      CHEMTALK 738,761
      MACCS 568
      REACCS 727
   USER FILES,MACCS 430
FIND
   CHEMTALK 751
   MACCS 506
FORM EDITOR
   CHEMBASE 278
   REACCS 626
FORMS
   CHEMBASE 346
   REACCS 625
   BOX IN
      CHEMBASE 346
      REACCS 627
   CREATE
      CHEMBASE 346
      CHEMTALK 749
      REACCS 626
   MOLECULES,REACCS 599
   REACTIONS,REACCS 608
   SAVE,CHEMBASE 354
   SWITCH OF,REACCS 650
FORMS MODE,REACCS 625
FRAGMENTS
   BONDING OF,CAS 45,50
   SEPARATE
      CREATION OF,CAS 44
FREE-HAND DRAW
   CAS 80
   STN EXPRESS 185
FULL SEARCH,CAS 210
FUNCTION KEYS
   CHEMBASE 280,283
   F7,CHEMBASE 304

INDEX

F9,CHEMBASE 298
FUNCTIONAL GROUPS
    CAS 84,89
FUSION
    CAS 47-50
    NORMAL,CAS 64,65
    ON ADJACENT RING NODES,CAS 70
    ON NONADJACENT RING NODES,CAS 72
    SPECIFIED,CAS 67
    TWO RINGS,CAS 50
    THREE RINGS,CAS 64
    >THREE RINGS,CAS 69
GENERIC STRUCTURES
    ATOMS IN,CAS 119
    BY STN EXPRESS 195
    CAS 118
    CHEMBASE 314
    CHEMTALK FOR CAS,773
    FRAGMENTS IN,CAS 124
    GK IN,CAS 119,122,130
    MACCS 465
        CREATE 465
        QUERY MENU 467,468,484
        ROOT 477
        R GROUPS IN 486,476
            ATTACHING TO ROOT 477
            DEFINITION OF 478,488
            MEMBERS OF 478,486
            OCCURENCE OF 481
        SYSTEM VARIABLES 465
        USER VARIABLES 465
    MODIFY GK,STN 199
    NODE VARIABLE,CAS 119
        ASSIGNMENT OF,CAS 120,121
    REACCS
        CREATE 604
    REPEATING GROUP,CAS 131
    RING SYSTEM VARIABLES,CAS 119
    SEARCH FOR
        CAS 224,245
        MACCS 519
        REACCS 665
    SYSTEM VARIABLES,CAS 119

VARIABLE ATOMS
    CAS 118
    CHEMBASE 314
    MACCS 465
    REACCS 604
VARIABLE BONDS
    CAS 138
    CHEMBASE 317,318
    MACCS 470
    REACCS 604
    VARIABLES IN,CAS 118
        DEFINITION OF,CAS 118
GEOMETRIC ISOMERS
    CHEMBASE 386
    MACCS 419
    REACCS 671
GIN,CAS 13
GK GROUPS
    ASSIGNMENT OF,CAS 118
    ATOMS IN,CAS 119
    ATTRIBUTES IN,CAS 163
    CAS 119,122
    DEFINITION OF
        CAS 118,119
        CHEMTALK 774
    DISPLAY OF,CAS 122
    FRAGMENTS IN,CAS 124
        ONE ATTACH.,CAS 124,132
        MULTIATTACH.,CAS 124,137
    GENERIC RINGS IN,CAS 124
    SHORTCUTS IN
        CAS 123
        STN EXPRESS 195
    VARIABLE COMMAND,CAS 122
    WITHIN GK GROUPS,CAS 130
GLOBAL OPTIONS,REACCS 589
GLOBAL SEARCH 655
GRAPH COMMAND,CAS 75
GRAPHIC QUERIES
    CAS SUBST.,CHEMTALK 772
    CREATE/CHEMTALK 776
    EXACT/CAS,CHEMTALK 772
    GENERIC,CHEMTALK 772

REACCS 637
UPLOAD TO CAS,CHEMTALK 777
GROUP
   ABBREVIATE,CHEMBASE 332
   MANIPULATE
      MACCS 498
      REACCS 602
GROUP MODE
   MACCS 497
HARDWARE
   CAS 12
   MACCS 416
   REACCS 597
HIGHLIGHT REACTION CENTER
   REACCS 616
HIT LISTS
   CHEMBASE 373
   DEFINE,REACCS 665
   DISPLAY,REACCS 663
   MACCS 507
HYDROGEN COUNT
   CAS 147
   MACCS 473
HYDROGENS
   CAS 29
   EXPLICIT,CHEMBASE 298
   F9 VALUES FOR,CHEMBASE 298
   IMPLICIT,CHEMBASE 298
   IN STRUCTURE,CAS 29
   MACCS 447
   REACCS 602
   STEREOSTRUCTURES,MACCS 456
INPUT/OUTPUT SYSTEM
   CHEMBASE 274
   DATABASE INPUT,CHEMBASE 276
   DATABASE OUTPUT,CHEMBASE 274
ISOTOPE
   CAS 160,161
   CHEMBASE 309
   MACCS 457
   REACCS 600
KEYBOARD COMMANDS
   CAS 39,43,49
   MACCS 441

   RINGS BY,CAS 49
KEYBOARD OPTIONS
   MACCS 440
   REACCS 591
KEYBOARD SWITCHES
   MACCS 438,444
   REACCS 590
KEYS
   MACCS 424
   SEARCH OF,MACCS 539
LINK DATA FIELDS
   IN FORMS,CHEMBASE 356
   IN TABLES,CHEMBASE 356
LIST DESCRIPTOR,REACCS 639
LIST FILES
   CHEMBASE 273
   REACCS 585,586
LISTFILES
   FORMAT OF,MACCS 530
   REACCS 639
   READ
      MACCS 530
      REACCS 661
   WORK WITH,REACCS 658
   WRITE
      MACCS 530
      REACCS 661
LISTS
   CHEMBASE 373
   COMBINE,REACCS 669
   CURRENT,REACCS 665
   DEFINE,MACCS 544
   DELETE,REACCS 664
   DESCRIPTOR,REACCS 639
   DISPLAY,MACCS 528
   IN SEARCH,MACCS 528
   LISTFILE FORMAT,MACCS 530
   MANIPULATION OF,MACCS 529
   READ,MACCS 530
   REFERENCE LIST,REACCS 639
   SAVE,REACCS 661
   STATUS,REACCS 662,663
   USE IN SEARCH
      CHEMBASE 391

INDEX

    MACCS 526
  WORK WITH
    CHEMTALK 755,759
    MACCS 526
    REACCS 658
  WRITE LISTFILE,MACCS 527
LOGIN
  CAS IN CHEMTALK 779
  STN 18
LOGOFF
  STN 19
MAIN MODE
  MACCS 432
  REACCS 588
MENU
  CAS 35
  CHEMBASE 278
  DESCRIPTION,CAS 36
  GRAPH COMMAND,CAS 75
  INITIATE,CAS 35
  NODE COMMAND,CAS 84
  OPERATION OF,CHEMBASE 278
  RING TEMPLATES,CAS 76
    SELECTION FROM
      CAS 36,38
      MACCS 437
    SWITCH TO KEYBOARD
      CAS 39
      MACCS 440
METAFILES
  CHEMBASE 368
  CHEMTALK 739,780
MODELS
  F7 KEY FOR,CHEMBASE 304
  FOR STRUCTURE BUILDING
    CAS 108,112,116
    CHEMBASE 300
    MACCS 451
    REACCS 600
  IN TEMPLATES
    CHEMBASE 301,303
    MACCS 452
  REACTIONS FROM,CHEMBASE 325
  RECALL OF,CAS 109,114

  SAVE,CAS 112,116
  SYSTEM'S
    CAS 109
    CHEMBASE 301
  TEMPLATES,CHEMBASE 304,307
  USER'S
    CAS 112
    CHEMBASE 305
MOLECULAR
  FORMULA,MACCS 425
  WEIGHT,MACCS 425
MOLECULES
  CAS 26
  CHEMBASE 268,372
  MACCS 417
  REACCS 578,599,606
MOLFILES
  CHEMBASE 272
  READ,MACCS 464
  WRITE,MACCS 463
MOVE
  ATOMS
    CAS 180
    MACCS 497
    REACCS 602
  FRAGMENTS,CAS 182
  STRUCTURES
    CAS 175
    CHEMBASE 335
    MACCS 495
    REACCS 602
NODE
  ASSIGN,CAS 83
    ALL 89
    MULTIPLE,CAS 83,89
    SINGLE,CAS 83
  ASSIGN BY MENU,CAS 87,89
  CAS 27
  DEFAULT VALUE,CAS 28
  DELETE,CAS 169
  GENERIC,SPECIFY,CAS 144
  NEGATION,CAS 124
  SPECIFIC,CAS 147
  VARIABLE,CAS 118

NORMALIZED BONDS
    ASSIGNMENT,CAS 101
    DELOCALIZED CHARGE,CAS 100
    IN RING SYSTEMS,CAS 95
    IN TAUTOMERS,CAS 97
NUMBERING STRUCTURE
    CHEMBASE 298
        MACCS 447
        REACCS 603
NUMERIC/TEXT DATA
    CHEMBASE 269
    MACCS 421
    REACCS 579
PAGE HEADING
    CAS 32
    SET,CAS 31
PLOT
    FORMAT PAGE
        MACCS 533
        REACCS 721
    FORMAT STRUCTURE,MACCS 534
    REACTIONS,REACCS 720
    SELECT DEVICE
        MACCS 532
        REACCS 720
    STRUCTURE
        MACCS 534
        REACCS 720
PLOT MODE
    MACCS 532
    REACCS 588,718
PRINT
    ORDER DELETE,CAS 243
    ORDER DISPLAY,CAS 244
    RETRIEVED DATA
        CAS 242
        CHEMBASE 375
        IN FORM 375
        IN TABLE 376
PROMPT BOX
    CHEMBASE 282
QUERY
    CREATE,CHEMTALK 772,773

RECALL,CAS 114
    CHEMBASE 383
    MACCS 516
    UPLOAD,CHEMTALK 777
RANGE SEARCH,CAS 212
R GROUPS
    DEFINE,CHEMTALK 774
    MACCS 478
    SPECIFY,CHEMTALK 774
RDFILES
    PC-RDFILE,REACCS 730
    READ,REACCS,732
    TRANSFER,REACCS,728
    WRITE,REACCS 729
REACTION CENTER
    BOND CHANGE,REACCS 620
    HIGHLIGHT,REACCS 612
    CENTER,REACCS 612
    MAKE/BREAK,REACCS 620
    NEGATION,REACCS 612
    NOT CENTER,REACCS 612
    SEARCH FOR,REACCS 682
    SPECIFY,REACCS 610
REACTION EDITOR
    CHEMBASE 278,319
REACTION FILES,CHEMBASE 272
REACTION KEYWORDS
    SEARCH FOR,CHEMBASE 411
REACTION SYMBOLS
    SEARCH FOR,CHEMBASE 410
REACTION TRANSFORM
    CHEMBASE 398
REACTIONS
    CHEMBASE 269
    CREATE
        CHEMBASE 319
        REACCS 608,622
    DISPLAY,REACCS 611
    EDIT
        CHEMBASE 321
        REACCS 608
    EXACT SEARCH
        CHEMBASE 394

INDEX

CONVERSION SEARCH IN,REACCS 677
REACCS 648,674
SIMPLE SEARCH IN,REACCS 674
STEREOISOMERS IN,REACCS 651,678
SUBSET SEARCH IN,REACCS 675
GRAPHIC DISPLAY
   CHEMBASE 322,323
   REACCS 579
REACCS 578
REARRANGE,CHEMBASE 338
REGISTER,REACCS 723
SAVE,CHEMBASE 325
SUBSTRUCTURE SEARCH
   CHEMBASE 396
   CONVERSION SEARCH,REACCS 680
   REACCS 678
   REL. CONFIGURATION IN,REACCS 681
   SIMPLE SEARCH,REACCS 680
REARRANGE
   REACTIONS,CHEMBASE 339
RECALL
   ANSWER SET,CAS 240
   DATA FILE,CHEMBASE 379
   FORM FILE
      CHEMBASE 353,378
      REACCS 633
   LIST FILE
      CHEMBASE 379
      REACCS 661
   MOLFILE
      CHEMBASE 378
      MACCS 510
   QUERY
      CAS 114
      CHEMBASE 383
      MACCS 516
      REACCS 600
   RDFILE,CHEMBASE 379
   REACTIONS,CHEMBASE 330
   SAVED ANSWERS,CHEMBASE 378
   SDFILE,CHEMBASE 379
   TABLE FILE
      CHEMBASE 355
      REACCS 633

REFERENCE LIST
   DEFINITION
      MACCS 544
      REACCS 659
REGISTER
   DATA,REACCS 724
   MOLECULES,REACCS 722
   REACTIONS,REACCS 723
   RE-REGISTER,MACCS 562
   STRUCTURES
      CHEMBASE 361
      MACCS 555
      REACCS 722
   TEXT DATA
      CHEMBASE 363
      MACCS 558
      REACCS 724
   DATA FILES,CHEMBASE 367
   MOLFILES,CHEMBASE 366
REGISTER MODE
   SEARCH IN,MACCS 505,506
REGISTRY FILE,CAS 3,7
REGISTRY NUMBERS
   CROSSOVER,CAS 10
   DEFAULT,REACCS 699
   EXTERNAL,MACCS 426
   ID IN CHEMBASE 271
   INTERNAL,MACCS 423
   LIST DESCRIPTOR,REACCS 639
   SEARCH FOR
      MACCS 535
      REACCS 639
   UPDATE ON,CAS 212
REPEATING
   CAS 135
   MACCS 474
REPEATING GROUP
   CAS 131
   STN EXPRESS 200
RESIZE STRUCTURE
   CHEMBASE 334
   MACCS 495
   REACCS 602

R GROUPS
    IN GENERIC STRUCTURES,MACCS 476
        ATTACHING TO ROOT 477
    DEFINITION OF,MACCS 478
    DISPLAY OF,MACCS 494
    MEMBERS OF,MACCS 486
    OCCURENCE OF,MACCS 481
    ROOT 478,486
RING SPECIFIC
    CAS 147
    MACCS 484
RINGS
    ATOM CONNECTIVITY IN,MACCS 472
    BRIDGED,CAS 73
    DELETE,CAS 172
    FUSED 4-8 NODES,CAS 47
    FUSED >8 NODES,CAS 48
    ISOLATED
        CAS 147
        MACCS 473,484
    SINGLE 3-8 NODES,CAS 46
    SINGLE >8 NODES,CAS 47
    SPIRO,CAS 75
    TAUTOMERISM IN,CAS 98
    TEMPLATES
        CAS 76
        CHEMBASE 303,304
        MACCS 454
        REACCS 600
SAVE
    ANSWER SET,CAS 240
    FORMS
        CHEMBASE 353
        REACCS 632
    LISTS IN LISTFILES,REACCS 661
    MOLFILES,MACCS 463
    QUERY TEMPORARILY,REACCS 605
        PERMANENTLY 605
    REACTIONS,CHEMBASE 325
    SEARCH ANSWERS,CHEMBASE 377
    STRUCTURES
        CAS 112
        STN EXPRESS 194

TABLES
    CHEMBASE 353
    REACCS 632
SCREENS
    CAS 8,255
    SEARCH WITH,CAS 254
    SPECIFY,CAS 256
SDFILE
    CHEMBASE 273
    FORMAT,CHEMTALK 765
    FROM LISTFILE,CHEMTALK 766
    TRANSFER,CHEMTALK 766
    WRITE,CHEMTALK 764
SEARCH
    ABSOLUTE CONFIGURATION
        CHEMBASE 374
        MACCS 511,514
        REACCS 651,665
    ATOM VALUES,CHEMBASE 295
    BATCH SEARCH,CAS 215
    BOOLEAN OPERATORS,CAS 247
    CAS 10,11,207
    COMBINATION,REACCS 645
    CONVERSION,REACCS 644,677
    DATA
        CHEMBASE 403
        MACCS 540,542
    DATABASE SUBSET
        REACCS 667
    DOMAIN
        CHEMBASE 372
        REACCS 659
    EXACT SEARCH
        CAS 218
        CHEMBASE 374
        MACCS 504
        REACCS 648
    FAMILY SEARCH,CAS 218
    FULL SEARCH,CAS 210
    GEOMETRIC ISOMERS
        CHEMBASE 386
        REACCS 671
        GRAPHICS,REACCS 637

INDEX

INTERRUPT/RESUME,MACCS 525
KEYS,MACCS 539
LIST DESCRIPTOR,REACCS 643
    MOLID 639
    RXNID 639
    REGNOS 639
LISTS,CHEMBASE 375
MOLECULES
    CAS 207
    CHEMBASE 372
    MACCS 503
    REACCS 648
MULTICOMPONENT QUERY,CAS 246,248
MULTIPLE DATABASES,REACCS 646
MULTIPLE QUERY,CAS 246
NAME,MACCS 537
ONE-COMPONENT QUERY,CAS 246
PARENT OF SALT,MACCS 511
PROJECTION,CAS 209
QUERY/TEXT,REACCS 638,642
    CHEMBASE 269
RANGE SEARCH,CAS 212
REACTION KEYWORDS,CHEMBASE 411
REACTION SYMBOLS,CHEMBASE 410
REACTION TRANSFORM,CHEMBASE 398
REACTIONS
    CHEMBASE 394
    REACCS 674
RELATIVE CONFIGURATION
    MACCS 525
    REACCS 604,651,653,669,686
R GROUPS QUERIES,MACCS 519
SAMPLE SEARCH,CAS 208
SCOPE OF,CAS 207
SIMILARITY SEARCH
    MACCS 522
    REACCS 673,692
SIMPLE SEARCH,REACCS 643,674
SINGLE QUERY,CAS 246
STATISTICS
    CAS 209
    CHEMBASE 372
STEPS IN, CAS 11

STEREOISOMERS
    CHEMBASE 296
    MACCS 511
    REACCS 651
    IN REACTIONS,REACCS 678
    IN SUBSTRUCT.,REACCS 681
STRATEGY,CAS 245
SUBSET,REACCS 675
SUBSIMILARITY,MACCS 524
SUBSTRUCTURE
    CAS 218
    CHEMBASE 380
    CONVERSION,REACCS 681
    MACCS 517,519
    REACCS 665,678
    REACTIONS
      CHEMBASE 394
      REACCS 674
    REL.CONFG,REACCS 686
    SIMPLE SRCH,REACCS 680
    REACCS 643
TAUTOMERS
    CHEMBASE 388
    MACCS 516
    REACCS 651
TECHNIQUES,REACCS 637
TYPE OF
    CAS 217
    REACCS 643
SEARCH MODE
    MACCS 505
    REACCS 588
SEARCH STRATEGY,CAS 245
SET FIELDS,CHEMBASE 360
SET PAGE,CAS 31
SETTINGS
    CAS 15
    CHEMBASE 284
    CHEMTALK 740,742
    MACCS 442
    REACCS 592
SHAPE STRUCTURES
    CAS 181

CHEMBASE 335
SHORTCUTS
   IN CAS 85
   IN STN EXPRESS 187
SIMILARITY SEARCH
   MACCS 522
   REACCS 692
SIMPLE SEARCH,REACCS 680
START
   CAS 17
   CHEMBASE 290
      EDIT COMMAND 291
      USE COMMAND 291
      VIEW COMMAND 291
   MACCS 442
   REACCS 596
STEREOBONDS
   CHEMBASE 296
   MACCS 418
   REACCS 600
STEREOCENTERS
   CHEMBASE 296
   MACCS 469
   REACCS 600
STEREOCHEMISTRY
   CAS 29
   CHEMBASE 269,296
   CHIRAL LABEL,CHEMBASE 269
   MACCS 418
   REACCS 578
   RULES FOR DRAWING
      CHEMBASE 297
      MACCS 421
STEREOISOMERS
   CHEMBASE 269
   CREATE,MACCS 455
   E-,Z-,MACCS 456
   HYDROGENS IN,MACCS 456
   R-,S-,MACCS 455
   RACEMATE,MACCS 455
   REL.CONFIGURATION,MACCS 469,477
   SEARCH FOR
      CHEMBASE 374
      MACCS 504,511,514,525

REACCS 648,651,665
UNDEFINED,MACCS 455
STN EXPRESS
   LIMITS IN,184
   OPERATION OF,184
STN INTERNATIONAL 4
   ACCESS 16
   FILES IN 5
STN OFFICES 17
STRUCTURAL KEYS
   INVERTED,MACCS 425
   MACCS 424
   SEQUENTIAL,MACCS 424
STRUCTURES
   BUILDING,CAS 33,37
   BY STN EXPRESS 184
   CAS 26
   CHEMBASE 268
   CREATE
      CAS 43
      CHEMBASE 292
      MACCS 444
      REACCS 599
   DELETE,CAS 173
   FIELD/BOX,REACCS 629
   FREE-HAND DRAW,CAS 80
   GRAPH,CAS 26
   MODELS
      CAS 108
      CHEMBASE 300
      MACCS 451
   NAME,MACCS 424
   NUMBERING
      CAS 45
      CHEMBASE 302
      MACCS 458
      REACCS 591
   REACTION QUERY,CAS 204
   RETRIEVAL FROM MACCS
      CHEMTALK 749
   SEMA NAME,MACCS 417
STRUCTURE ATTRIBUTES
   CAS 146
   CHARGE,CAS 158

INDEX

DELETION OF,CAS 174
DELOCALIZED CHARGE,CAS 159
GK,CAS 163
HCOUNT,CAS 147
HMASS,CAS 160
MASS,CAS 161
NODE SPECIFIC,CAS 147
RING SPECIFIC,CAS 147
VALENCE,CAS 162
SUBSET SEARCH
   REACCS 675
SUBSTITUENTS
   CHEMBASE 306
   COMMON,CAS 84
      SELECTION,CAS 89
   F7 USE FOR,CHEMBASE 307
   FUNCTIONAL GROUPS,CAS 84
      SELECTION,CAS 89
   TRUNCATE,REACCS 603
SUBSTITUTING
   CHAINS,CAS 48
   RINGS,CAS 50
SUBSTITUTION
   DEGREE OF
      CAS 152
      MACCS 467
   REACCS 604
SUBSTRUCTURE SEARCH
   CAS 218
   CHEMBASE 380
   MACCS 504
   MARKUSH STRUCTURES,CHEMBASE 381
   REACCS 669
   STEREOISOMERS IN
      CHEMBASE 386
      REACCS 665
   TAUTOMERS IN,CHEMBASE 388
SWITCH
   DISPLAY FORMS,REACCS 650
   GRAPHICS TO KEYBOARD,MACCS 440
   MENU TO KEYBOARD,CAS 41
   TO TERMINAL MODE,CHEMTALK 743
SWITCHES
   MACCS 438,444
   REACCS 590
SYSTEM BACKGROUND
   CAS 3
   CHEMBASE 265
   CHEMTALK 737
   MACCS 415
   REACCS 575
SYSTEM COMMANDS
   NOVICE VERSION,CAS 21
   EXPERT VERSION,CAS 21
SYSTEM DESIGN
   CAS 7
   CHEMBASE 268
   CHEMTALK 739
   MACCS 417
   REACCS 578
SYSTEM INTERFACE
   CAS 3
   CHEMBASE 266
   MACCS 414
   REACCS 576
SYSTEM OPERATION
   ATTACH,MACCS 433
   CHEMBASE 277
   CHEMTALK 740
   DATA MODE,MACCS 435
   DRAW MODE,MACCS 432
   MACCS 432
   PLOT MODE,MACCS 435
   QUERY MENU,MACCS 433
   REACCS 588
   REGISTER MODE,MACCS 434
   SEARCH MODE,MACCS 434
   STRUCTURE MENU,MACCS 433
SYSTEM PROMPTS
   CAS 21
   MACCS 440
TABLE
   ASSIGN DATATYPE,REACCS 634
   ASSIGN LABEL,REACCS 634,635
   CHEMBASE 349
      CREATE 351
      RECALL 353
      SAVE 353

    COLUMNS IN,CHEMBASE 350
    CREATE,REACCS 633
TAUTOMERS
    AROMATIC BONDS OVERLAP,CAS 99
    CAS 97
    CHEMBASE 388
    MACCS 420
    REACCS 651
    NORMALIZED BONDS IN,CAS 97
    REQUIREMENTS FOR,CAS 97
    RING SYSTEMS,CAS 98
TEMPLATES
    CHEMBASE 305
    FILES,CHEMBASE 274
    IN STN EXPRESS 187
    RINGS,CAS 76,79
    REACCS 600
    SAVE,CHEMBASE 305
    SELECTION,CAS 76
TERMINALS
    CAS 13
    EMULATION,CHEMTALK 737
    REACCS 597
    SELECT,CHEMTALK 745
    SETTINGS,CHEMTALK 745
TEXT
    DESCRIPTOR,CHEMBASE 311,331
    EDITING,CHEMBASE 283
    QUERIES,REACCS 642
TRANSFER
    DATA
        CHEMBASE 368
        MACCS 568
        REACCS 727
    FILES
        CAS->CPSS 771
        CHEMBASE->MACCS 368,766
        CHEMBASE->REACCS 368,767,769
        MACCS->CHEMBASE 761
        REACCS->CHEMBASE 767
    IN CHEMTALK 761
    IN DATA MODE,MACCS 569
    INVALID TRANSFERS,MACCS 570
    MULTIPLE RECORDS,REACCS 728
    RDFILE,REACCS 728
    SDFILE,CHEMTALK 766

    SINGLE MOL,REACCS 727
    SINGLE RXN,REACCS 728
    STRUCTURES,CHEMBASE 370
    VALID TRANSFERS,MACCS 569
TREENAMES
    DISPLAY,REACCS 705
    REACCS 583
UPDATE DATABASES
    CHEMBASE 363
    MACCS 555
    REACCS 722
    DELETE RECDS,CHEMBASE 365
UPLOAD QUERY
    CHEMTALK 752,780
    STN EXPRESS 201,204
USE COMMAND,CHEMBASE 291
USER PROFILE,REACCS 595
UTILITIES,STN EXPRESS 194
VALENCE
    CAS 162
    CHEMBASE 309
    REACCS 600
VARIABLE COMMAND
    CAS 122
VARIABLES
    CAS 118,123
    CHEMBASE 314,389
    MACCS 465
    REACCS 604
    STN EXPRESS 187
    SYSTEM DEFINED
        CAS 119
        CHEMBASE 315
        MACCS 465
    USER DEFINED
        CAS 119,142
        CHEMTALK 774
        MACCS 476
VIEW COMMAND,CHEMBASE
VIEWLIST MODE
    DISPLAY DATA
        REACCS 710-717
    DISPLAY OPTIONS,REACCS 653
    REACCS 588
VIEW SEARCH HITS
    REACCS 652,655